1993

The work of Nickolay Ivanovich Vavilov has formed the basis of much of the study of plant genetic resources that is carried out today. In recognition of his contribution to plant science, and to commemorate the centenary of his birth, a collection of all of Vavilov's works on the origin and geography of cultivated plant species was published in Russian in 1987. This English translation sees the publication of these seminal papers in their original form, but not original language, for the first time ever.

The structure of the book, with papers arranged in chronological order from 1920 to 1940, provides a unique opportunity to retrace both the development of Vavilov's theories on cultivated plants and his gradual creation of a definite terminology. The book will be of great interest to all those concerned with the development of cultivated plant species, not only in terms of the history of this discipline and its current status, but also its future direction.

ORIGIN AND GEOGRAPHY OF CULTIVATED PLANTS

Frontispiece:
Medal coined at the 100th anniversary of
the birth of N. I. Vavilov

THE ACADEMY OF SCIENCES OF THE U.S.S.R.
THE SECTION OF CHEMICAL, TECHNICAL
AND BIOLOGICAL SCIENCES

N. I. VAVILOV

Origin and Geography of Cultivated Plants

DEPARTMENTAL EDITOR
V. F. Dorofeyev
Academician of the V. I. Lenin All-Union Academy of Agricultural Sciences

Translated by
DORIS LÖVE

Published by the Press Syndicate of the University of Cambridge
The Pitt Building, Trumpington Street, Cambridge CB2 1RP
40 West 20th Street, New York, NY 10011-4211, USA
10 Stamford Road, Oakleigh, Victoria 3166, Australia

First published in Russian as
Proiskhozhdenie i geografiia kul'turnykh rasteniĭ
by Nauka, Leningrad Branch, © 1987 Nauka

First published in English by Cambridge University Press 1992
as *Origin and Geography of Cultivated Plants*

Printed in Great Britain at the University Press, Cambridge

A catalogue record for this book is available from the British Library

Library of Congress cataloguing in publication data
Vavilov, N. I. (Nikolaĭ Ivanovich), 1887–1943,
[Proiskhozhdenie i geografiia kul'turnykh rasteniĭ. English]
Origin and geography of cultivated plants / N.I. Vavilov :
departmental editor, V.F. Dorofeyev : translated by Doris Löve.
p. cm.
Includes bibliographical references and index.
ISBN 0-521-40427-4 (hardback)
1. Plants. Cultivated–Origin. 2. Plants, Cultivated–
Geographical distribution. I. Dorofeev, V. F. (Vladimir
Filimonovich) II. Title.
SB73.V3813 1992
630'.9–dc20 92-25039 CIP

ISBN 0 521 40427 4 hardback

SE

Contents

Abstract

FOR THE FIRST TIME all the major works by N. I. Vavilov, dealing with the problems of the geography of cultivated plants, the establishment of their centers and origins, are assembled in the present collection of papers. Arranged in chronological order, they furnish the reader with an opportunity to follow the development of this science. The methodological approach to an understanding of the material is specified and the tendency toward a definite terminology is outlined in each new paper.

This book is intended for use by geneticists, plant breeders and botanists.

Compiler:
A. A. Filatenko, Doctoral Candidate of Biological Sciences

I. A. Sakharov, *Doctor of the Biological Sciences*

S. R. Mikulinskiy, *Corresponding member of the Academy of Sciences of the U.S.S.R.*

A. A. Sozinov, *Academician of the Academy of Sciences and the V.I. Lenin All-Union Academy of the Agricultural Sciences*

V. I. Stukov, *Doctoral Candidate of Biological Sciences*

V. A. Trukhanov, *Doctoral Candidate of Biological Sciences*

Presentation of the English translation of N. I. Vavilov's Origin and Geography of Cultivated Plants

FOR WELL OVER a century, the Soviet Union and the United States have supported agricultural research programs aimed at crop improvement. Crop cultivars generated from these programs have contributed significantly to improving agricultural production in each country. In the Soviet Union, the N. I. Vavilov Institute of Plant Industry (VIR) has been the leading institution, whereas in the United States the U.S. Department of Agriculture's Agricultural Research Service (ARS) has taken the lead. Both organizations have joined in a cooperative effort on plant germplasm to further the interests of each country while benefitting the world as a whole.

In commemoration of the 100th anniversary of Nikolay Ivanovich Vavilov's birth, the Soviet Union has issued several volumes, documenting Academician Vavilov's work. The importance of Vavilov's scholarly work concerning systematics, evolution and geography of cultivated plants is recognized globally. Through a joint effort of VIR and ARS, *Origin and Geography of Cultivated Plants* has been translated and made available in the English language for the benefit of scientists and practitioners of the plant and agricultural sciences. We are pleased to be a part of this important effort to extend knowledge and science as a small tribute to Academician Vavilov.

V. A. Dragavtsev
Director, VIR

R. D. Plowman
Administrator, ARS

Translator's foreword

NICOLAY IVANOVICH VAVILOV was born into a Moscow family on November 23rd (13th, old style), 1887. He spent his childhood and youth in the capital city and attended a commercial middle school there. His student years were spent at the Moscow Agricultural Institute. After graduating, he remained at this institute in order to prepare for an academic rank under the tutelage of Prof. D.N. Pryanishnikov.

In 1913, he obtained a grant from the Moscow Agricultural Institute, which allowed him to study in England under W. Bateson for a fairly long period of time. One of his contacts there was none other than Francis Darwin, the son of the great Charles Darwin himself.

After returning to Russia, he worked at the Bureau of Applied Botany of the Department of Agriculture. In 1916, he organized his first expedition to Persia (now Iran) and also visited Turkestan and Pamir.

At the age of 30, in 1917, he was made professor at the Faculty of Agriculture of the University of Saratov and also taught genetics and plant breeding at the Voronezh Agricultural Institute. However, by 1921, Professor Vavilov was made head of the old botanical institute founded in St. Petersburg in 1894, which later became the Bureau of Applied Botany. The bureau was reorganized in 1924 into the All-Union Institute of Applied Botany and New Crops in order to implement Lenin's ideas about a renewal of agriculture in the U.S.S.R. It was the first link in a chain of institutes, which presently became associated with the V.I. Lenin All-Union Academy of Agricultural Sciences. In 1930, the institute was renamed the All-Union Institute of Plant Industry (VIR).

Under Vavilov's enthusiastic directorship, this institution underwent a remarkable development and grew to a colossal size. Reportedly, it had a staff of 20 000 and about 400 research laboratories and experimental stations dispersed all over the U.S.S.R. The so-called gene bank, established by this institute, where by 1940 over 160 000 samples of seeds and plants were kept, propagated and utilized for plant breeding, was a pioneering effort of its kind and a model for similar establishments elsewhere later on. The slogan of the institute was 'to bring the best plant resources of the world to the service of socialistic agriculture'.

For this purpose expeditions were sent out to around 180 areas all over the world as well as within the vast territory of the U.S.S.R. itself. Vavilov personally visited and made extensive research in no less than 52 different countries.

In 1920, he had published an important paper, 'The law of homological series in variation', which had assured him a firm position internationally within the scientific world. He became a frequent participant in scientific congresses and meetings within the U.S.S.R. as well as abroad, made valuable contacts and became a good friend of many foreign scientists. At the age of 36, Vavilov was, among others, made a corresponding member of the Soviet Academy of Sciences and five years later he was made a full member, the youngest ever to hold that rank. He was also chosen as a member of the British Society of Horticulturalists, the British Association of Biologists, the Academy of the Sciences at Halle, Germany, and the Czechoslovakian Academy of Agricultural Sciences. He was even made a Foreign Member of the British Royal Society, a rare honor since only 50 such are selected worldwide. He also received many honors, medals and prizes for his achievements within the U.S.S.R.

The publication of his book, *Centers of Origin of Cultivated Plants*, in 1926, was a major event in international scientific circles. In it he established the geographical centers of origin of the most important plant crops and their wild relatives, based on the differential taxonomic–geographical method established by himself and his colleagues at the Institute.

What is stated above indicates the enormous work load of the director of this large institute, but it should also be mentioned that his personal interests were very wide and that he, indeed, deserved the epithet 'renaissance man'. Although he frequently – even prophetically – complained that 'Life is so short and there is so much to do', he had time to become skilled not only in all aspects of botany and genetics, plant pathology, plant breeding and agronomy but also evolution and the history of the world as well as the sciences, archaeology, geography and even linguistics.

Vavilov was able to attract a number of first-class scientists to his institute and was a devoted and much-admired leader of all. Most of them remained loyal to him, even during the 1930s and 1940s when the work at the institute experienced a difficult time politically, since it was firmly based on Darwin's theories and Mendelian genetics. Vavilov was publicly accused of 'having failed in his duties of applying genetics to practical problems of crop improvement, of sending worthless expeditions to gather material for his collection instead of concentrating on local varieties . . . and of being unsympathetic to the theories of Michurin and Lysenko'. Michurin was a kind of Russian 'Luther Burbank' and Lysenko was a research worker under Vavilov who totally misinterpreted Vavilov's ideas but had won the ear of Stalin. Perhaps Vavilov at first did not realize how dangerous Lysenko was. After acting as an interpreter when the English scientist J.D. Harland interviewed Lysenko, Vavilov stated: 'Lysenko is an angry species; all progress in the world has been made by angry men so let him go on working . . . He does no harm and may some day do some good'.

However, in 1938 Vavilov was ousted from the Presidency of the Academy of the Agricultural Sciences of which he was the initiator and first president. He was replaced by Lysenko. At a meeting, defending his institute, Vavilov said in

1939: 'Let us be burned down, but we will not retract our scientific convictions'. Finally, during an expedition led by Vavilov to Chernovits, Ukraine, he was arrested for espionage and sabotage of Soviet agriculture and thrown into prison. He died of starvation in the prison at Saratov on January 26, 1943.

His loyal staff quietly continued his work (surviving members of his institute still meet and recall the glory days of the 'institute'), even during the siege of Leningrad. Some of them died of hunger rather than touch upon the rich stores of seeds in the gene bank. Among them was Vavilov's secretary, a Mrs. Brissenden from Los Angeles, who had left the U.S.A. in order to work with Vavilov.

By 1955, the political tide had changed and much thanks to the efforts of Vavilov's brother, at that time the president of the Academy of the Sciences, who in spite of much criticism stayed on as an Academician in order eventually to be able to help correct the wrong done, Vavilov was absolved from all 'crimes' and his name was cleared. The institute was again reorganized and now bears the name of the N.I. Vavilov Institute of Plant Industry. His institute was awarded the Order of Lenin in 1967 and in 1975 the Order of Friendship of the Peoples for its scientific achievement. The 100th anniversary of Vavilov's birth was celebrated in 1987 with a big conference attended by scientists from 40 foreign countries. The book translated here was published on this occasion, a medal was coined, a badge was made and a postage stamp was issued in his honor. His institute now flourishes and the work led by Vavilov during his short but dramatic life is continued and expanded on a global scale.

The present translation of Vavilov's papers dealing with the origin and geography of cultivated plants follows the original text very closely. The geographical names used by Vavilov have been retained, since they are easily found on older maps; unfortunately, Russian place names frequently change for political reasons. However, it should be noted that Soviet geographers make a sharp distinction between 'Tsentralnaya Asia' (Central Asia) and 'Srednaya Asia', which is also translated as 'Central Asia' in most English texts. I have chosen to use the term 'Inner Asia' for the latter territory, which covers the mountainous parts of the Turkmenia, Kazakhstan, Uzbekistan and Tadzhikistan, Iran, Afghanistan and Pakistan as well as India (mainly the mountains of Pamir, Hindukush, Karakorum and Himalaya). Central Asia is the territory east of Tien-Shan, bordering on Mongolia and China.

Russian measures of distance and weights have also been retained.

The Latin plant names used by Vavilov have not been changed, although in many cases they are outdated. Synonymy is, when it could be established, provided in the index of Latin plant names. However, common names of plants are routinely used in the U.S.S.R. Since they often vary from country to country (corn = grain in Europe, maize in America), and even within countries, an attempt has been made to add the Latin names (in brackets, e.g. 'garden rocket [*Eruca sativa*]') to them. I have done my best to identify the common names of exotic plants, spelled in Russian. I relied heavily on *A Checklist of Names for 3,000 Vascular Plants of Economic Importance* by E.E. Terrell, S.R. Hill, J.H. Wiersma and W.E. Rice (U.S.D.A. Agriculture Handbook 505, 1986) and, when that failed, on R. Mansfeld, *Vorläufiges Verzeichnis Landwirtschaftlich oder Gärtnerisch Kultivierter Pflanzenarten*, Academie Verlag, Berlin 1959. An index to the common names has been added.

All bracketed remarks and expressions are by the translator, all such in parentheses by Vavilov himself. Footnotes in the original text have been incorporated in the body of the text.

Acknowledgements

During the course of the translation I have enjoyed much encouragement and assistance especially from Dr. Douglas Johnson, USDA-ARS Forage and Range Research, Utah State University, Logan, Utah, for which I want to express my sincere gratitude. Valuable help with unusual Russian terms and expressions has also been given me by my friends in Leningrad and by Mr. Benjamin Kurolapnik, Northern Telecom Co., Santa Clara, California, to all of whom I am very grateful. Information and literature about the person of Nicolay Ivanovich Vavilov and his institute has graciously been provided to me by my Leningrad friends, Dr. Evgenia Dorogostaiskaya and Dr. M. Kirpichnikov, and especially by Dr. S.M. Alexanyan, head of the Foreign Relations of VIR, as well as by Dr. Patrick McGuire, Genetic Resources Conservation Program, Davis, California, to all of whom I am deeply indebted. The manuscript was read and typed by Ms. Gayle Swanson, Marketing Assistant, Samsung Electronics in San José, California. I owe many thanks to her and to that company for allowing the use of their computer equipment for this purpose. I am also very appreciative of the pleasant cooperation with the Cambridge University Press during the course of the work with the translation.

PUBLISHED SOURCES OF INFORMATION

1. Alexanyan, S.M. and Denisov, V.P. (1988). 'World scientists join in Soviet celebration of N.I. Vavilov centennial.' *Diversity*, **15**, 5–6.
2. Alexanyan, S.M. and Heints, G.G. (1988). 'Countries of the council for mutual economic aid collaborate in sphere of plant genetic resources.' *Diversity ?*, (**2–3**), 10–11.
3. Alexanyan, S.M. and Heints, G.G. (1989). 'Council for mutual economic aid commission establishes scientific technical council for plant genetic resources.' *Diversity V* (**2–3**), 9–10.
4. Anonymous, A. (1989). 'Vavilov, a centennary.' *Plant Genetic Resources News Letter*, **71**, 2–3.
5. Anonymous, B. (1990). 'An incomparable collection of genetic diversity.' *Geneflow* **1990**, 10–11.
6. Bakhareva, S.N. (1990). 'The USSR policy for exchanging genetic resources and the germ plasm collection procedures of the Vavilov institute.' *Diversity*, **6** (3–4), 10–12.
7. Gamkrelidze, T.V. and Ivanov, V.V. (1990). 'The early history of Indo-European languages.' *Scientific American*, **262** (3), 110–116.
8. Huxley, Julian (1949). *Soviet genetics and world science*. London: Chatto and Windus.
9. Heints, G.G. and Karyshev, O.V. (1987). *The N.I. Vavilov Institute of Plant Industry (Guide)*. Leningrad.

10. Krivchenko, V.I. (1988). 'The role of Vavilov in creating the national Soviet program for plant genetic resources.' *Diversity*, **16**, 5–7.
11. Vitkovskiy, V.L. and Kuznetsov, S.V. (1990). 'The N.I. Vavilov All-Union Research Institute of Plant Industry.' *Diversity*, **6** (1), 15–18.

Doris Löve *San José, California, March 1991*

N. I. VAVILOV

Preface

THE PRESENT ANNIVERSARY EDITION of the works of Academician N.I. Vavilov includes his papers about the problems dealing with the historical geography of cultivated plants. In the sense it is now understood, this science did not exist before N.I. Vavilov.

The theory about the origin of cultivated plants as worked out by Vavilov is a clear example of the creative power of a fundamentally basic concept. The papers by Vavilov, elucidating the origin and geography of cultivated plants, have served as a systematic collection of his work with plants and its presentation to the world in general. In 1940, the worldwide collection of the All-Union Institute of Agriculture (VIR), directed by Vavilov, consisted of 168 000 samples, which were studied, multiplied and distributed to plant-breeding institutions in the Soviet Union as material for introduction into cultivation. Plant breeders utilized the initial material, concentrated in the collection, and created hundreds of new varieties, based on its genetics. Many of these strains still exist today as masterpieces of Soviet plant breeding.

According to Vavilov, Darwin's ideas on evolution constituted the basis for searching for areas where the formation of types originally took place.

The first fundamental paper, 'Centers of origin of cultivated plants' (1926), was dedicated to A. De Candolle, the first person in the history of science to pose questions concerning independent centers of origin. However, he based this on a few data concerning only the distribution of various cultivated plants. Vavilov approached the solution to this problem from a completely new point of view.

The first expeditions sent to collect the plant resources of Iran and Inner Asia resulted in a rich material which could be mastered only by a research scientist with such a wide approach as that of Vavilov. He was a geographer, ecologist, florist and introducer of plants and knowledgeable both of cultivated plants and the history of the world's civilizations. At the same time, he was able to work out, brilliantly, the problems concerning agricultural production within all the zones of the Soviet Union.

Vavilov's wide-ranging research, based on the evolutionary concepts of Charles Darwin, led him to the establishment of entirely new data, radically changing the ideas concerning the most important cultivated plants. A proper theoretical and methodical approach was necessary for such studies.

The differential, phyto-geographical method became the guiding light towards an understanding of the origin of cultivated plants. According to its main principles, already taught by Vavilov in 1924, three centers could be distinguished, each common to whole groups of cultivated plants. In a public lecture on January 19, 1926, he convincingly proved the independence of these different centers.

However, the concept concerning the centers of origin of cultivated plants was, in a wider sense, set forth by Vavilov in his paper on 'Centers of origin of cultivated plants' (in 1926), for which the author was the first one to be awarded the Lenin Prize. The theory about the centers served as a major impulse for upgrading plant breeding in the Soviet Union during the Post World War I era. Valuable specimens of agricultural plants, included in the collection, served as basic material for the creation of frost-hardy, drought-tolerant and disease-resistant varieties. The phyto-geographical research conducted by Vavilov and the VIR under his direction led to a general development of plant breeding and to the advancement of agriculture in the Northlands and also in the dry, eastern areas of the Soviet Union, where a new agricultural area had become widely established by 1930.

Material from each new expedition, undertaken by Vavilov himself or by his colleagues at the VIR, and new results concerning the distribution of worldwide plant resources, were reflected in the papers by Academician Vavilov. He never fitted the data to the doctrines he elaborated but constantly introduced serious additions and made changes on the basis of new results and deep, often quite radical, thinking. In each of his new publications concerning the centers of origin of cultivated plants, a tendency toward concretization of methods and terminology can be observed, to which contemporary critics of his theory unfortunately did not pay due attention. It is a fact that during the analysis of the cultivated flora of any region, the use of a single criterion for determining the centers of origin, even that of a single crop, usually leads to an erroneous interpretation of Vavilov's concepts.

The present edition of the papers by N.I. Vavilov assembles all his major works, which deal with the problems of the origin of cultivated plants and the establishment of the centers where intensive processes of type-formation took place, i.e. those that resulted, to a major extent, in the cultivation of the plants mentioned. Arranged in chronological order, these publications offer the reader an opportunity to retrace the development of Vavilov's teachings where, on the basis of his studies of the origin of different plants, he is led to the establishment of cultivated floras that are specific to different areas.

He postponed the solution of many problems, among others those concerning the history of agriculture and the history of various crops, until the programs of the VIR expeditions had been completed, carefully collecting new botanical phytogeographical, ecological, genetical, phytopathological, archeological and linguistic data and many other results concerning cultivated plants.

The new material was immediately subjected to interpretation. We do not find any papers (except for two, dated to 1940) where there are no corrections or additions to the interpretations of the centers of origin of cultivated plants. Their number, distribution and even their names were never haphazardly given but were decided on the basis of logical conclusions concerning the generalization of the collections, made by each new expedition, and after checking them by means of other sciences.

As far as the distribution of the botanically investigated territories of many countries as well as the U.S.S.R. are concerned and by extension of the scientific work concerning the material collected, new centers of origin of cultivated plants were revealed not only as far as a specific cultivated flora and its accompanying weeds was concerned, but also with respect to the domesticated animals belonging to it as living testimonials of the creation and development of human civilizations. In 1932, Vavilov wrote (cf. p. 252): 'Many historical problems can be understood only because of the interaction between man, animals and plants', and later, in another paper from 1932: 'The centers of origin of the majority of the presently cultivated plants began in botanical areas where powerful type-forming processes are active. It is evident that primitive man walked around in these regions, which are rich in associations of plant species including large numbers of edible plants' (cf. p. 263).

The study of the laws of the geographical distribution of plant resources on Earth and the establishment of the enormous intraspecific diversity of the majority of crops allowed not only a determination of their localization but offered also an opportunity to judge the time of origin of the plants most important for cultivation. 'The history and origin of human civilizations and agriculture are, no doubt, much older than what any ancient documentation in the form of objects, inscriptions and bas-reliefs reveal to us. A more intimate knowledge of cultivated plants and their differentiation into geographical groups helps us attribute their origin to very remote epochs, where 5000 to 10 000 years represent but a short moment.' (cf. p. 12)

'The method of differential taxonomy offers an opportunity to trace the dispersal of some hundred cultivated plants and for unmasking their stages of evolution with respect both to the initial origin and their introduction into cultivation within different areas and their relation to wild species and varieties, but also with respect to the subsequent evolution of these plants when dispersed from the basic centers and undergoing changes under new conditions and further effects of natural and artificial selection.' (cf. p. 426)

The studies of the genesis of different cultivated plants led Vavilov to the establishment of new concepts, i.e. primary and more ancient crops in contrast to secondary ones, allowing him to characterize with great precision the centers where agriculture was born and the pathways along which it was dispersed.

The number of the centers listed in Vavilov's papers increased during a comparatively short period from three in 1924 to five during 1926–27, six in 1929–30, seven in 1931 and eight in 1934–35, but were again reduced to seven in 1940. Each publication appears to be the result of thorough consideration of new data.

Geographical regularities during the formation of cultivated plants are already revealed in the first paper (cf. p. 120): '. . . [an] interesting group of facts pointing in the general direction toward the type-forming processes peculiar to different plants in various regions'.

Examples can be cited, which illustrate the striking difference between the type-forming processes within different centers. These facts corroborate Vavilov's belief that agriculture arose independently in different countries. A general accumulation of dominant genes there seems to be common for such centers. 'By means of detailed studies of the racial composition of various Linnaean species, systems of characteristics can be identified which correspond to some extent to a system of genes and to geographical regions of concentration, of which investigators were unaware until recently.' (cf. p. 136).

Centers differ also with respect to the concentration of specific variation. Vavilov attached great importance to data indicating 'regions of major concentration of specific and generic variation'. During the arrangement of these according to the richness of cultivated floras, the Chinese center was put in first place and the Hindustani one in second (Vavilov, 1934) but more recent data (1940) led to the necessity for changing these places: 33% of all cultivated plants are concentrated in the southern Asiatic tropical center, which at his [Vavilov's] time nourished up to $\frac{1}{4}$th (now $\frac{1}{3}$rd) of the population of the world. In eastern Asia, the second most important center, 20% of the agricultural plants are grown. As far as the number of species introduced into cultivation is concerned, southwestern Asia follows, with 14%. However, Vavilov attached a particular importance to that center, since the composition of what is cultivated in the territory of the Soviet Union is a consequence of the influence from Asia in general but especially strongly of that from Asia Minor and Inner Asia. Vavilov determined the limits of that center.

In the papers published in 1934 and 1935, the division of southwestern Asia into two centers is suggested: an Middle Asiatic one and one covering Asia Minor. In 1935, the Middle Asiatic center was renamed the Inner Asiatic one. It belongs to one of the five major regions where cultivated plants originated in Asia and includes northwestern India, Afghanistan and the mountainous portions of Soviet Turkestan (Uzbekistan, Tadzhikistan and a part of Turkmenistan). This name does not agree with the centers of origin or with their subdivision in Vavilov's later papers. Its appearance is explained by the fact that, during that period and until recently, the exact spatial–geographical borders of Inner Asia had not been clearly outlined (Grach, 1984). As far as the investigation, especially of Central Asia proper, is concerned, Vavilov (1931) revealed the characteristics only of cultivated plants adopted mainly from Asia Minor and Inner Asia, and in part also from India and Central and Eastern China.

After rejecting the division of the southwestern Asiatic center, Vavilov (1938, 1940) discussed the composition of the complex of species formed by cultivated plants within the territory in question. He refers to the close relationship between Cis-Caucasus and Asia Minor: 'An enormous potential of species and even of genera is concentrated there, constituting genetically distinct units' (cf. p. 377). In addition to quantitative characteristics, Vavilov concentrated his full attention at that time on the specific composition of cultivated plants for each of which endemic genera, species and even forms occur. This direction of investigations was further developed in a book by E.N. Sinskaya: *Historical Geography of the Cultivated Flora* (1969). She extended considerably the comparative analysis of the centers of the cultivated flora, while refining the historical connections between the centers.

It is also appropriate to quote E.V. Vulf (1937) here: 'all the stages during the history of the Earth, including the time when vegetation first appeared upon it and, especially, the development of climatic conditions, are sharply reflected in the vegetation cover and modify both the quality and the quantity of the associations of developing species'. This applies equally well to the cultivated flora.

The ideas about the Mediterranean center have not undergone important changes since the time of its distinction. In 1926, Vavilov established that while the Mediterranean center had a tendency toward the formation of large-fruited, large-flowered and large-seeded forms, the southwestern Asiatic and the Afghanistani, Turkestani and Indian areas were characterized by a development of forms with small fruits and small flowers. Numerous examples confirm the differences in the processes of type-formation of many species of cultivated plants that belong to genera developing within the territories of both these centers. Gradually, cultivated plants (although of less economic importance) that had their primary center of origin in the Mediterranean area proper were discovered (e.g. garden rocket [*Eruca sativa*], lavender [*Lavandula vera*], artichokes [*Cynara scolymus*], carob trees [*Ceratonia siliqua*] and sulla [*Hedysarum coronarium*], etc.). The development of important genera (such as *Triticum* L., *Secale* L., *Hordeum* L., *Beta* L., *Brassica* L., *Daucus* L., *Lens* Adams., *Linum* L., *Mandragora* L., *Pisum* L., *Melilotus* Adams. and many others) to which belong a major proportion of the crops, forms the basis for agricultural production in countries around the Mediterranean; but most are intensely grown in the countries of Asia Minor and Inner as well as southwestern Asiatic centers.

After visiting Ethiopia, Vavilov became concerned with species which, with respect to the range of formation of species and varieties of cereals, were mainly derived during past epochs from Asia Minor and the Mediterranean center (e.g. wheat, barley, chick-peas [*Cicer arietinum*], lentils, flax, vetchlings and many others, composing the basic, edible resources of that area). Although limited in size in comparison with the other centers of agricultural territories, Ethiopia offered many plants cultivated by the natives (according to Vavilov, 1940, 4% of the total number of plants cultivated on Earth). African millet [*Eleusine coracana*], sorghum, teff [*Eragrostis tef*], giant beans, niger seed or ramtil [*Guizotia abyssinica*], chat [*Catha edulis*], coffee trees, sesame seed and many others belong to genera that have developed within paleotropical Africa. Ethiopia was the first country in which Vavilov became acquainted with the agriculture in this part of the African continent. Unfortunately, he was unable to study the countries to the south and west of Ethiopia. Archeological data indicating the significance of the ancient populations practicing agriculture in Equatorial Africa have appeared only during the last couple of years.

Phyto-geographical studies have revealed that, with respect to the areas of Africa south of the Mediterranean, there is a characteristic cultivated flora that is no less rich than those in other centers where agriculture arose. Many cultivated plants have undergone a secondary development there. Representatives of what are now known as endemic forms of cultivated rice, taro [*Colocasia* spp.], some species of bananas [*Musa, Ensete* ssp.], and so on, penetrated from southern Asia into Africa during the very remote past.

While developing Vavilov's ideas about the centers of origin, E.N. Sinskaya (1966) singled out the African region for the historic development of cultivated flora. Ancient Mediterranean elements (actually both Mediterranean and south-western Asiatic ones) predominated in the composition of the cultivated flora of Ethiopia but are not sharply delimited from those of other African areas. Elements from southern Asia occurred there as well. An American complex of plants is at present widely utilized within the agriculture of Ethiopia (Harlan, 1975). These facts convince us that it is now impossible to give Ethiopia a position as a basic, primary center of origin of agriculture and that this region should be considered as affected mainly by African and Mediterranean cultivated floras (Sinskaya, 1966, 1969).

During the last couple of decades, a major phyto-geographical investigation of western and central Africa has been executed for the first time by expeditions from VIR, since a comparative analysis of the cultivated floras in that area indicated that they were rather special. A rich variety of cultivated plants (217 species, belonging to 73 genera and 39 families) and the presence there of 339 endemic forms could be demonstrated. This convincingly confirmed the vision of E.V. Vul'f (1937), a very close friend of Vavilov's, that 'in the inexhaustible womb of the tropical flora treasures of plant products are hidden that are, in many cases, still unknown but valuable, and for the revelation of which only systematic, scientific investigations are necessary'.

Botanical investigations of the American continents, conducted by N.I. Vavilov, S.M. Bukasov, S.V. Yuzepchuk and other co-workers at the All-Union Institute of Plant Breeding, led to the establishment there of independent centers of origin of agricultural plants, based on species and even genera unknown in the Old World. The relative youth of the New World civilizations makes it possible to find cultivated species derived from associations of wild relatives. Vavilov distinguished two geographical centers of origin within the New World, where an amazing but localized type-formation of cultivated plants could be established: the Central American and the South American ones. In a later paper (1940), the latter was renamed the 'Andean center'.

Some words concerning the terminology used by Vavilov while solving the problems concerning the origin of cultivated plants: in all his papers, starting with the first and fundamental one – 'Centers of origin of cultivated plants' (1926) – and ending with 'Studies of the origin of cultivated plants after Darwin' (1940), Vavilov used the terms 'center', 'focus' and also 'area' of origin as synonyms. Their definition is important: the geographical centers are basic and independent foci where agricultural crops originated but are also geographic areas where cultivated plants are grown. He emphasized that 'the origin of cultivated plants and domesticated animals is primarily associated with time and space' (Vavilov, 1929). Passing from one of Vavilov's papers to another concerning the problem of the origin of cultivated plants, it is possible to conclude that the terms 'center' and 'focus' are mainly associated with large territories. In his last papers, he writes about 'areas of basic origin of cultivated plants' and about the conventional concept of 'center of origin' such as suggested by Darwin.

Summing up the data concerning the hundreds of cultivated plants from all over the world resulting from the systematic collection by the All-Union Institute of Plant Industry, Vavilov writes, in 1935: 'we can now speak with a considerably greater accuracy than dreamed of ten years ago about the eight ancient and basic centers of agriculture in the world, or, more accurately about the eight independent areas where plants were initially taken into cultivation'.

E.N. Sinskaya has continued the work of Nicolay Ivanovich [Vavilov] concerning the establishment of borders for the centers of cultivated plants and for the specification of the historical relationship between the centers. She has also made a comparative analysis of the cultivated floras within these centers.

The VIR continues also at present to send out expeditions, investigating the five continents of the world. An intensive study of species and their botanical, genetical, biochemical, physical–biological and historical status is conducted in order to provide material for the establishment of a new phytogenetic system, for intraspecific systematics and for more precise definition of distribution areas of species and genera. Thus, for instance, among the einkorn wheat of Asia Minor, a naked-grained diploid species similar to *T. monococcum* L. has been discovered for the first time, i.e. *T. sinskajae A.* Filt. and Kuzh. The spelt wheats [*T. sect. Speltae* Steudl.] found in Transcaucasia and Inner Asia but previously considered to be European cereals, emphasize the common character of the foci of origin of cultivated plants in Asia Minor, Inner and southwestern Asia and around the Mediterranean – an important fact for the interpretation of the development of Vavilov's theory about the centers of origin of cultivated plants. In an ancient agricultural oasis – Sin'tszyan, always isolated from any others – a new endemic species of wheat, *T. petropavlovskyi* Udacz., has developed. The examples can be continued for a number of other crops.

During his study of the geographical laws concerning the distribution of plants, Vavilov established that, in addition to general regularities typical of different crops, there are also various geographical variations, which Harland (1975) with good reason noted with respect to a number of African crops.

Vavilov writes that for crops such as wheats, divergence occurs at the species level. 'As far as other crops are concerned, there is with respect to such foci only a differentiation of genes, and in a third case, an adaptation of crops in this or that direction can occur.' These characteristics are extremely important for the elucidation of the phylogeny of various cultivated plants.

After a short time, considerable corrections were made by Vavilov and his co-workers concerning De Candolle's ideas about the genesis of important cultivated plants and his explanation of the origin of some hundred plants. However, such considerations became, for Vavilov, just material for preparing a path from the details to the whole, i.e. the establishment of basic, general centers of origin. In 1935, he had already described 'the basic, primary areas' – or, as we prefer to call them – centers of origin 'of the specific and varietal potential with a comparatively great accuracy' (cf. p. 324). Further on he states:

'As far as the introduction of new objects into the study is concerned, it becomes increasingly evident that there is a coincidence between areas for primary type-formation of species and even genera. In a number of cases, literally dozens of species can be referred to one particular area. The geographical studies led to the establishment of entirely independent floras of cultivated plants specific to different areas' (cf. p. 324).

At present, the botanical investigations concerning the centers of origin are continuing and the collections gathered are thoroughly studied. However, the new data have not provided any reasons for a revision of Vavilov's theories concerning the centers of origin of cultivated plants. E.N. Sinskaya (1966) writes that many amendments can be made but all amount only to a correction of details. 'The basic composition of cultivated plants, typical of this or that center, remains stable.'

As far as the historical character of Vavilov's works toward the establishment of the centers of agricultural crops is concerned, E.N. Sinskaya calls our attention to the prevalent use of the expressions 'historical–geographical area' or 'geographical areas of the historic development of a cultivated flora' which, as can be seen, appear regularly in Vavilov's papers, such as those presented in the present collection.

P.M. Zhukovskiy (1969) has introduced new terms: 'megacenters' in the ordinary sense of Vavilov's centers, and 'microcenters', i.e. foci where narrowly endemic species and even varieties were introduced into cultivation. It should be mentioned that for Zhukovskiy as well as for Vavilov, the idea of a discrete formation of varieties is well and thoroughly considered. However, the 'mega-centers' of Zhukovskiy are fairly wide: e.g. in the tropical belt almost the entire land area is enveloped by a single megacenter. Some megacenters have grown to the size of continents.

Historically, Zhukovskiy's distinction of new centers such as the North American, the Euro-Siberian or the Australian ones, appear unjustified. Vavilov knew very well the history of agriculture within the countries situated in these territories. He devoted himself for most of his life to the investigation of them, except for that of Australia. However, he was nevertheless interested in the rich, wild flora of that continent, from the associations of which eucalyptus trees, acacias and Australian pines [*Casuarina equisetifolia*] were taken into cultivation as recently as the nineteenth century. Agriculture there was like that of North America but also of parts of Europe based on crops originating from primary foci.

The introduction of a new concept by E.N. Sinskaya for the phyto-geography of cultivated plants also deserves attention, i.e. 'affected areas', to which the territories enumerated above also belong. Agriculture in North America developed on the basis of Mexican and Central American crops and, later on, on that of crops from the Old World. In central and northern Europe, on the Russian steppes and in Siberia, agriculture is based primarily on cultivated plants introduced from countries in Asia Minor and around the Mediterranean, etc. Agriculture in the 'affected areas' is never very old, although the period of development thereof is not really very short, something that can be judged from the large quantity of plants introduced into cultivation from the less rich, wild-growing flora of these territories.

So far, plant breeders have utilized archaeological data for confirming the centers of origin of cultivated plants. However, there are papers by archeologists and paleobotanists, who in their own way interpret the ideas about the existence of centers of domestication of animals and the introduction into cultivation of basic, edible or technical plants. B.V. Andrianov (1978) distinguishes the following major areas in the Old World: a 'western', an 'eastern' and an 'African' one, but also an area such as the 'New World'.

In the 'western area', Andrianov distinguished between a southwestern Asiatic and a Mediterranean one (not entirely distinctly subdivided), which, according to him, represent units of a historical-cultural region. This archeologist agrees, thus, in general, with the ancient Mediterranean area with respect to the development of cultivated plants such as those distinguished by Sinskaya, while indicating a complex of botanical associations in the flora of this territory on the basis of which agriculture was born there.

The 'eastern area' includes, according to Andrianov, two of the major geographical centers of Vavilov, the south-Asiatic tropical center and the east-Asiatic one, based on the similarity between cultivated plants there and the dispersal of agricultural processes.

However, botanists and plant breeders do not agree with this combination. The fundamental composition of the cultivated flora is typical for each of these regions. Eastern Asia is an enormous, independent area, where a cultivated flora developed that was associated not only with the very ancient civilization there, but also with the east-Chinese floristic region of the holarctic kingdom at the time when the south-Asiatic region, where a cultivated flora was also developed, was linked to the Indian, the Indo-Chinese and the Malayan floristic regions of the paleotropical kingdom of plants (Takhtadzhan, 1978).

Southern Asia, rich in endemic genera and species of cultivated plants, has provided worldwide agriculture with rice, tea, citrus fruits, bananas, cotton, linseed and many leguminous plants and spices. Eastern Asia appears to be the native land of many species, originating within this region and having primary or more-or-less secondary centers of development in that area. Many cultivated species are linked to genera for which the native land is found outside eastern Asia or has not yet been established. These have found a main area of development here (e.g. millet, shama millet [*Echinochloa colona*], broom corn [*Sorghum bicolor*], soy beans, Chinese and Japanese radishes, ginseng and many others). From this center, plums, cherries and almonds dispersed over Eurasia and showed a tendency to form secondary species and varieties there.

It has been demonstrated by archaeologists and paleobotanists, as well as by botanical investigations, that in prehistoric time there was already an exchange of primary, cultivated plants between southern and eastern Asiatic areas.

It is difficult to determine accurately the northern limit of the southern Asiatic region, but it can be stated on the basis of botanical data that a union of the two regions discussed is not appropriate. As far as the agricultural methods used are concerned, Andrianov has distinguished zones in India where a hoeing-and-irrigation agriculture, flooded during high-water, was practiced in connection with the Indian city civilizations and where there is a hoeing-and-irrigation agriculture based on water drilled from wells, but also where there is a non-irrigated hoeing agriculture in Central India and the Ganges valley. However, in eastern Asia there is a stick-and-hoe, slash-and-burn, non-irrigated agriculture as well as a stick-and-hoe agriculture of rice using irrigation but also doing without it.

The southern areas of China, Indo-China, the Moluccas and Malaysia should, according to Vavilov, be referred to the southern Asiatic region, although they differ slightly from what is found in India proper. Stick-and-hoe agriculture was also characteristic of these areas. In addition, many crops (rice, cotton, citrus fruits, tea, eggplants, cucumbers, bananas, etc.) were dispersed from southern and eastern Asia, possibly together with the agricultural practices associated with them.

The four areas distinguished by B.V. Andrianov return us, in fact, (although with some corrections) to those established by A. De Candolle, but pay no attention to the qualitative composition of the cultivated floras of these centers or to their definite affinity: 'The actual localization of the basic cultivated plant material proved to be more distinct and delimited than presented at the time of De Candolle (1882) for whom it was a matter of entire continents. Direct studies of the plant resources by means of the differential method has allowed us to determine exactly the main areas within the continents where cultivated plants originated' (cf. p. 354–5).

It is necessary to keep in mind that Vavilov constantly worked under the specific concept of centers of origin of cultivated plants. In Vavilov's later works, the term 'center' becomes increasingly associated with more wide-ranging geographical territories. He referred emphatically to the division of the globe into floristic regions and subregions such as accepted by conventional phyto-geography.

Based on the results of his work, it became evident that it was necessary to extend Vavilov's investigations into the fields of geography and history of cultivated plants. Specific interpretations of the solution of problems concerning the origin of cultivated plants have been found for a number of questions of general biological importance (e.g. questions about species, type-formation, resistance to diseases, etc.) that had been developed by Vavilov in other papers.

Because of the accumulation of new data, the great importance of the theory about the centers of origin of cultivated plants has become evident. All the work done at the All-Union Institute of Plant Breeding is built up on the basis thereof.

It is impossible not to mention the exceptional influence and importance that the theory about the centers of origin of cultivated plants has had on plant breeding in the U.S.S.R. Vavilov developed a new theory based on the appearance of regularity in the geographical distribution of plant resources in our world, on the accumulation of data confirming the law of homologous series within the hereditary variability and on the development of the representation of species as complicated systems. This theory became fundamental for plant breeding in the Soviet Union and its management. Worldwide collections, representing the richest varietal and specific multiformity of various cultivated plants were, from the very beginning of their creation, distributed among plant breeders. New crops have been introduced that became valuable new strains, occupying large areas of our country. At present, the worldwide gene-bank of cultivated plants in the U.S.S.R. consists of more than 300 000 samples.

Under the present conditions, VIR sends out expeditions for replenishment of this collection of plants, according to a plan providing for special searches for original material and data for future programs toward the development of selective plant-breeding work. Special attention is paid to the collection of early-ripening hybrids of maize, varieties of sunflowers resistant to diseases, high-quality and drought-tolerant as well as disease-resistant productive varieties of wheat, rye, barley, oats and hybrids of rape, annual and perennial fodder grasses, productive varieties and hybrids of potatoes, vegetables, melons, fruits, berries and grapes possessing the highest edible and technical qualities.

At present, the following kinds of material are being accumulated by means of expeditions and assembled in the five gene-banks: 1. wild species and weeds related to cultivated plants; 2. local varieties that are still grown in some countries within elevated mountain areas and that derive from primary centers of origin; 3. strains, presently being bred, and hybrids combining complexes of agriculturally valuable character and biological traits of basic interest for plant breeding; 4. genetical sources and donors of agriculturally valuable characteristics, selected from the varietal diversity existing and obtained experimentally by geneticists and plant breeders in the U.S.S.R. as well as in foreign countries; and 5. rare botanical varietal mutants and genetic lines of various categories in the form of products of experimental biology.

Samples from the worldwide gene-bank, which have been studied with respect to the most important and agriculturally valuable characteristics and biological traits by departments and laboratories of the VIR but also at its experimental stations, have for a long time served as original material for plant-breeding establishments. This shows the positive effect of the development of our native plant breeding.

Utilizing the ideas basically elaborated by Vavilov, VIR carries out wide-ranging and complex studies of the worldwide material in the gene-banks by means of modern methods used for biological and agricultural sciences. These are delegated toward three basic tasks: evaluation of the samples as original material for creating new varieties of a high quality, elaboration of a theoretical basis for plant breeding and classification of cultivated plants.

Special attention is paid at VIR to studies that deal with problems of original material, i.e. the genetic basis for modern plant breeding. For this purpose, agriculturally valuable characteristics such as yield, early-ripening, resistance to diseases and pests, high quality production, drought- and frost-tolerance, straw-strength, etc. are selected at plant breeding stations, in the departments of plant resources and at theoretical laboratories. Specific gene-banks are created for each of these characteristics and the material is passed on to all plant breeding centers in the country.

During the last couple of years of work with the VIR gene-bank, the research associated with the verification of the varieties distinguished by donor characteristics has been strengthened by plant breeding and genetic analyses. New methods allow us to find donors of agriculturally valuable traits and, thus, to affect more actively the development of plant-breeding work.

As a result of the complex research at the worldwide gene-bank concerning cultivated plants and the use of methods belonging to botany, genetics, physiology, biochemistry, molecular biology and phytopathology, VIR not only selects varieties valuable for plant breeding and hybrids and strains with complex characteristics that are agriculturally valuable, but also works out new methods for the evaluation of original material and even manages active research within the areas of theories concerning plant breeding and introduction of plant material, toward which the far-reaching scientific work of the founder of this institute, Academician Nicolay Ivanovich Vavilov, aimed.

The content of Vavilov's papers corresponds to the texts as originally published, except for some minor corrections, which are indicated by the editors. The geographical, botanical and other epithets as well as units of length and volume have been retained as used at the time when the papers were written.

V.F. Dorofeyev, Academician of the V.I. Lenin All-Union Academy of the Agricultural Sciences, and
A.A. Filatenko, Doctoral Candidate of Biological Sciences.

BIBLIOGRAPHY

1. Andrianov, B.V. (1978). *Zemledenie nashikh predkov* [*The agriculture of our ancestors*]. Moscow.
2. De Candolle (1883) (publication date, Oct. 1882). *L'origine des plantes cultivées* [*The origin of cultivated plants*]. Paris.
3. Grach, A.D. (1984). Tsentralnaya Asii i obschcheye i osobennoye v sochetanii sotsial'nikh geograficheskich faktorov [Central Asia, generalities and specifics of social and geographical factors] In: *Rol' geograficheskogo factora v istorii dokapitalisticheskikh obshschestv* [*The role of geographical factors in the history of post-capitalistic management*]. Leningrad.
4. Harlan, J. (1975). Geographic pattern of variation in some cultivated plants. *J. Hered.* **66** (4).
5. Sinskaya, E.N. (1966). Uchenia N.I. Vavilova ob istoriko-geografiche skikh ochagax razvitiiya kul'turnoy flory [The theory of N.I. Vavilov concerning historical-geographical centers of origin of cultivated floras]. In *Voprosi geografii kul'turnykh rasteniy i N.I. Vavilov*, [*The problems of the geography of cultivated plants and N.I. Vavilov*]. Moscow, Leningrad.
6. Sinskaya, E.N. (1969). *Istoricheskaya geografiya kul'turnoy flory (na zare zemledelya)* [*Historical geography of cultivated plants (at the dawn of agriculture)*]. Leningrad.
7. Takhtadzhan, A.L. (1978). *Floristicheskie oblasti Zemli*, [Floristic areas of the world]. Leningrad.
8. Vul'f, E.V. [Wulff, E.V.] (1937). Opyt deleniya zemnogo shara na rastitel'nye oblasti na osnove kolichestvennogo raspredeleniya vidov [An attempt to a division of the world into vegetational areas on the basis of quantitative distribution of the

species]. *Trudy po prikl. botan., genet. i selek.*, [*Papers on applied botany, genetics and plant breeding*], Ser. 1, (no. 2).

9. Zhukovskiy, P.M. (1969). Mirovoy genofond rasteniy dlya selektsii (megatsentry i mikrotsentry), [World-wide genebank of plants for plant breeding (mega-centers and micro-centers)]. In *Voprosy geografii kul'turnykh rasteniy i N.I. Vavilov*, [*The problems of the geography of cultivated plants and N.I. Vavilov*]. Moscow, Leningrad.

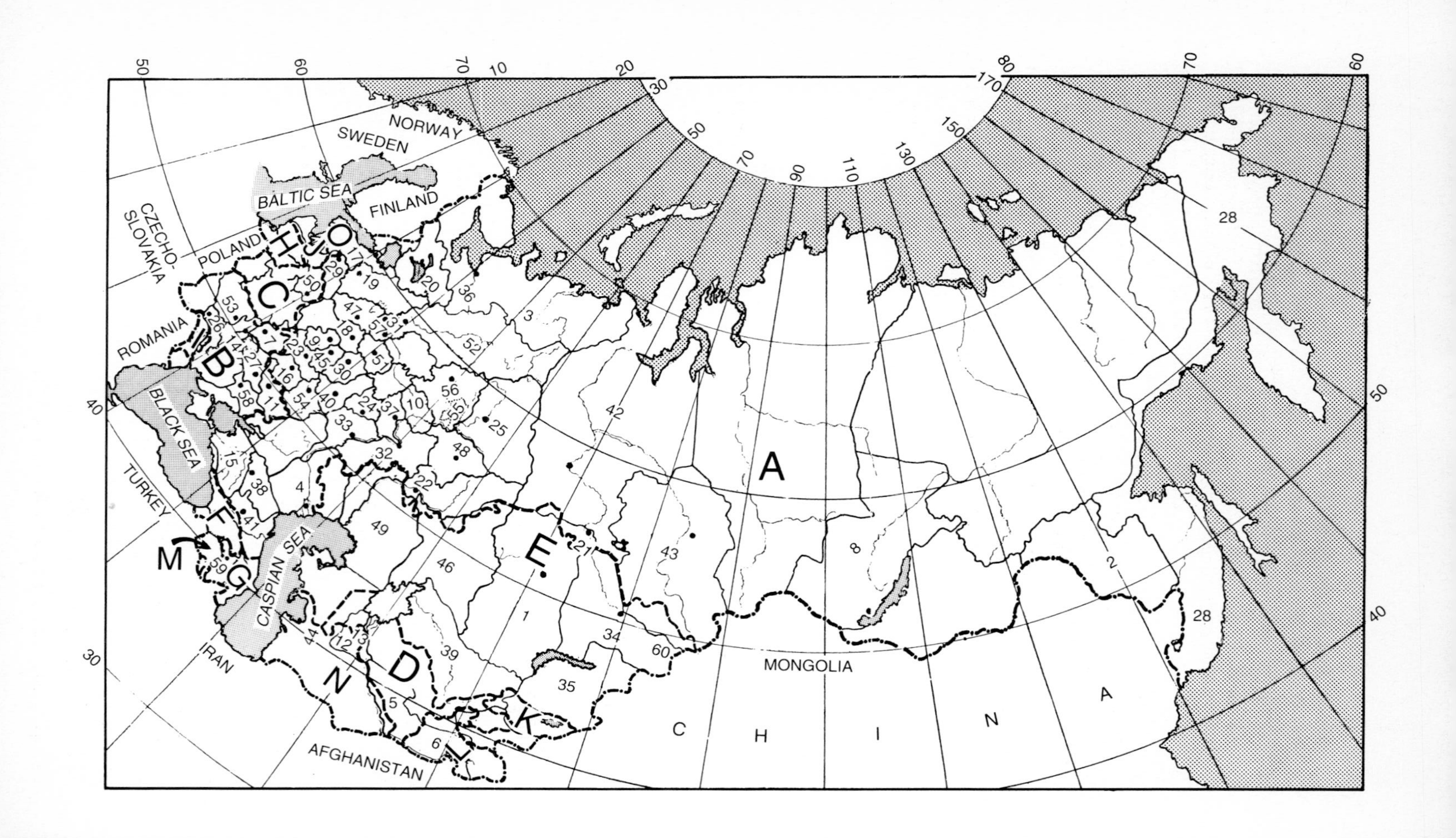

NORWAY
SWEDEN
FINLAND
BALTIC SEA
POLAND
CZECHO-
SLOVAKIA
ROMANIA
BLACK SEA
TURKEY
CASPIAN SEA
IRAN
AFGHANISTAN
MONGOLIA
C H I N A
A
B
C
D
E
F
G
H
J
K
L
M
N
O

1 *Akmolinsk* District (now: *Tselinograd* Region in Kazakhstan)
2 Amur District
3 Arkhangelsk Governate
4 Astrakhan Governate
5 Bukhara
6 Mountainous Bukhara
7 Chernigov Governate
8 Irkutsk Governate
9 Kaluga Governate
10 Kazan Governate
11 Kharkov Governate
12 Khiva
13 Khorezm (borders not clearly defined)
14 Kiev Governate
15 Kuban District
16 Kursk Governate
17 *Leningrad* Governate (now: *St Petersberg*)
18 Moscow Governate
19 Novgorod Governate
20 Olonets Governate
21 Omsk Governate
22 Orenburg Governate
23 Oryol Governate
24 Penza Governate
25 Perm Governate
26 Podolsk Governate
27 Poltava Governate
28 Primorye District
29 Pskov Governate
30 Ryazan Governate
31 Rybinsk Governate (correct: Rybinsk district of Yaroslavl Governate, borders not clearly defined).
32 *Samara* Governate (renamed *Kuibyshev*, now again *Samara*)
33 Saratov Governate
34 Semipalatinsk Governate (correct: Semipalatinsk District)
35 Semirechye District (now *Alma Ata* & *Taldy Kurgan* Regions in Kazakhstan)
36 Severnaya Dvina Governate (borders not clearly defined)
37 *Simbirsk* Governate (renamed *Ulyanovsk*)
38 Stavropol Governate
39 Syr-Darya District
40 Tambov Governate
41 Terek District (central city: Vladikavkaz, renamed Ordzhonikidze, now again Vladikavkaz; part of Russian Federation)
42 Tobolsk Governate
43 Tomsk Governate
44 Trans-Caspian District (now: part of Kazakhstan and Turkmenia)
45 Tula Governate
46 Turgai District (now: Aktyubinsk Region in Kazakhstan)
47 *Tver* Governate (renamed *Kalinin*, now again *Tver*)
48 Ufa Governate
49 Ural District
50 Vitebsk Governate
51 Vladimir Governate
52 Vologda Governate
53 Volyn Governate
54 Voronezh Governate
55 Votskaya Autonomous Territory (now: Udmurtiya)
56 *Vyatka* Governate (renamed *Kirov*, now again *Vyatka*)
57 Yaroslavl Governate
58 *Yekaterinoslav* Governate (renamed *Dnepropetrovsk*)
59 *Yelizavetpol* Governate (renamed *Kirovabad*, now *Gyandzha*)
60 Zaysan Territory (now: East Kazakhstan Region)
A. Russian Federation
B. Ukraine
C. Byelorussia
D. Uzbekistan
E. Kazakhstan
F. Georgia
G. Azerbaijan
H. Lithuania
I. Moldavia
J. Latvia
K. Kirgizia
L. Tadzhikistan
M. Armenia
N. Turkmenia
O. Estonia

– Borders of the Soviet Union
– Borders of Governates and old Districts
– Borders of modern Republics
– Border of Votskaya Autonomous Territory
● – Central cities of Governates and Districts

On the Eastern centers of origin of cultivated plants

A GREAT INTEREST in the study of Asia Minor, the Middle East and China exists not only due to the natural desire for penetrating into the life of the ancient peoples, whose fates are inscribed in the early history of Man, not only because sooner or later the East will lead the same life as the West, but especially because the key to the understanding of many important contemporary natural-historical and cultural-historical phenomena is found just in the East.

The present plant and animal life in Europe, Siberia and Turkestan is intimately associated with the plants and animals of the East. To understand the evolution and the origin of the plants and the animals in our own country, it is impossible to avoid associating them with the plant and animal life in the mountain areas of southwestern and southeastern Asia.

This is especially obvious with respect to our important cultivated plants and domesticated animals. The ancestors of our domestic animals are the wild horses, the wild camels and the wild yaks discovered by Przheval'skiy in central Asia. In the mountain areas of Asia, travelers are struck by the amazing variety among the species of domesticated cattle. Nature has collected a living museum there of all kinds of existing breeds. This is also, to a great extent, manifest in the case of cultivated plants.

The objective of the present short article is first and foremost to sum up our knowledge about the relationship between the European cultivated plants and those of the East.

For a long time, historians, archaeologists, philologists, agronomists and botanists have been drawn to the problems concerning the origin of our cultivated plants.

The questions concerning the origin of a given cultivated plant – how it became cultivated, where its original native land was, where the sources of the development of varieties were found, and where the clues to the wealth of forms could be discovered – are not only of general importance for explaining the historical destiny of peoples but also of actual and practical importance for the present agronomical work toward the exploration of varieties for plant breeding.

During the decades following the classical works by De Candolle and Hehn,

First published in *Noviy Vostok* [*The New East*], no. 6, pp. 291–305 some interesting reviews have been published, e.g. Buschan: Yorgeschichtliche Botanik (1895); Laufer: Sino-Iranica (1919); Sturtevant: Notes on edible plants (1919); Carrier: The beginnings of agriculture in America (1923); A. Schulz: Die Geschichte der kultivierten Getreide (1913), and Gibault: Histoire des legumes (1912), and so on.

Until recently, the problem concerning the site of the origin of plants was decided mainly on the basis of archeological and linguistic work and, in part, on the basis of historical data and the localities of the appropriate, cultivated plants in their wild state. The antiquity of plants was determined mainly by means of archaeological discoveries. Dispersal of the plants from one country to another and from one people to another was decided on the basis of comparisons between the names of the plants in various languages.

Investigations during the last couple of years by the Department of Applied Botany and Plant Breeding of the State Institute for Experimental Agronomy have allowed the introduction of new, more exact methods for determining the centers where the development of different kinds of cultivated plants took place, i.e. methods for identifying the centers of origin according to the *concentration of the variation of types*. A detailed review of the investigations concerning cultivated plants is at present being prepared for publication by this author.

In order to establish the centers of origin of different genera and families of animals and plants, botanists and zoologists usually determine the geographical distribution of the species and the maximum concentration of specific diversity within different areas. Such methods, although more differentiated, have also been applied by us in respect to cultivated plants.

While basing our work on the description of the centers where the development of the variations of different *species of cultivated plants* took place and in the light of the present representatives of the Linnaean species in the form of collective units, we have had to concentrate on the variety of species, races and modifications, i.e. on the small subdivisions, constituting the Linnaean species. If, for instance, it is necessary to establish the center of origin and development of varieties of cultivated rye, the verdict reached allows us to find the geographical center *of the maximum type-formation of rye*.

Investigations during the last couple of years have shown that, in spite of an important industrialization of cultivated crops, in spite of repeated migrations and colonization by peoples, and in spite of the antiquity of agriculture, it is still possible to establish the areas of endemic varieties and species and the areas of the original, primary development of forms. It has been demonstrated that, in general, the answer to the problem concerning the center of origin can be arrived at by the same methods as the botanists and zoologists apply for wild species, but by more profound and differentiated ones including an inventory of such forms and varieties for which the botanists and zoologists have had little interest.

The geographical concentration within various regions can sometimes be secondary, and in order to determine exactly the *actual center, where the development of forms took place*, it is necessary to supplement the data on the varieties of the species studied with data on the geography and distribution of the varieties of adjacent, genetically closely related species and to compare the areas of contiguous wild and cultivated species. Thus, for instance, in order to determine exactly the center for the development of forms of soft wheat (*Triticum vulgare*) it is necessary to simultaneously clear up the center of variation of the closely related species of shot and club wheats (*T. compactum*, *T. sphaerococcum*).

The method of differentiating detailed phyto-geographical analyses of plants furnishes results immeasurably more exact than the methods used by archeologists and linguists who do not always distinguish between the species of plants, which are already botanically differentiated. Philologists, archeologists and historians speak of 'wheat', 'oats' and 'barley'. The present state of botanical knowledge demands that the cultivated wheat species be distinguished into 13 species, and oats into six species, all quite different and not able to cross with each other, and geographically originating from different centers, and so on.

It is natural that this kind of investigation requires an enormous number of specimens of all kinds of cultivated species from different countries and areas for a strictly systematic geographical study. The species must be studied not only in herbaria but also by cultivating them in order to be able to evaluate all the morphological and physiological as well as hereditary differences. Due to the inadequacy of the investigations of the many interesting areas of ancient agriculture in Asia and Africa, our information is so far fragmentary and often insufficient, and many countries are in need of special studies. However, since this method of actually establishing the centers of origin and development of forms on the basis of botanical data furnishes an absolutely exact answer to where the origin of a given plant is located, it can be utilized for practical breeding work and for searching out practical and valuable modifications from the primary source of type-formation of the plants in question.

De Candolle placed the native land of wheat in Mesopotamia. Investigations, applying the differentiating botanical methods used by me and my co-workers in the course of the last couple of years and based on a large amount of material collected by the Department of Applied Botany during special expeditions to Persia, Bokhara, Turkestan and Mongolia, as well as being provided with, thanks to the U.S. Department of Agriculture, an extensive collection of material from the rich collections in Washington assembled from various countries in Asia and Africa, have resulted in some unexpected results: it could be concluded that cultivated wheat has *two* basic centers. For one group of wheat species, in particular the hard ones such as our brands 'Byeloturka' and 'Kubanka', the original center of variation is found in *northern Africa*. For the most important group, i.e. that of soft wheat – the kinds that appear in the form of the common European wheats – a center of variation could be identified in *southwestern Asia*. It turned out that the more we approached southwestern Asia and the areas of Bokhara and Afghanistan, the greater was the variation in the species. The different modifications of the soft wheat increased as well. At the same time as the granaries of Russia, especially in the southeast and in the Samaran and Saratovan provinces, display six or seven varieties of soft wheat, the number thereof in Persia, Bokhara and Afghanistan amounts to 60. The maximum variability, including all the varieties so far known, the majority of which are not cultivated in Europe, is found in the area between northern India and western Persia. There, in particular, a rich variation is found in the mountain areas of Bokhara bordering on Afghanistan as well as in Afghanistan itself. The Persian and the Mesopotamian areas, distinguished by a large number of species, all yield in this respect to Afghanistan and the mountain areas of Bokhara, where an exceptional wealth is found not only of soft wheat types but also of varieties of shot wheat (*T. compactum*), genetically close to the former. These are at present lacking in Persia. In addition, varieties of wheat are found there which are presently unknown in other countries (e.g. some with extremely simplified leaves – *eligulatum*). All characteristics of the cultivated kinds of soft wheat in Siberia, European Russia and western Europe are met with in southwestern Asia, although in different combinations. In Africa, soft wheat has been grown only recently by European colonizers. The New World has cultivated wheat as well as the majority of other agricultural plants (rye, barley, oats, millet, flax and peas) only since being occupied by the Europeans during the nineteenth century.

In the mountain areas of southwestern Asia there were discovered not only endemic varieties unknown to Europeans but also many typically European forms with easily-shed seeds. In Shugnan and Roshan, in the mountains of Bokhara, it was possible to observe typically East-Siberian small-grained kinds of wheat, indicating that the native land of the Siberian varieties of wheat is located in the mountain areas of southwestern Asia. Expeditions during 1922 for studying plants cultivated in Mongolia and led by a specialist from the Department of Applied Botany, i.e. V.E. Pisarev, confirmed definitely that all the wheats of eastern Siberia originate from the mountains to the south and were introduced into Siberia via Mongolia.

A great variation within the soft wheat types characterizes Georgia, Armenia and all of Transcaucasia. This is similar in composition to that in Persia. The center of development of one third of the types genetically close to the soft wheats, but belonging to the club wheat (*T. sphaerococcum*) was, as could be expected, found also in areas contiguous with Afghanistan and Bokhara, i.e. in northern India.

In the illustration showing the geographical distribution of the wheat varieties (see Fig. 1, p. 37) there are various symbols corresponding to different groups of variation, embracing a large number of species. The areas of the closely related varieties of soft wheat are indicated by different outlines. The illustration shows that all the groups center around a region in southwestern Asia where the maximum variation of the species and the corresponding center of origin of the soft wheat are found (for more details, cf. Vavilov, 1922).

Thanks to the geographical studies of wheats, a new aspect has opened up on Persian wheat (*T. persicum* Vav.) and on the great variety of its forms recently observed (in 1922) in Georgia and Armenia by the director of the Tiflis Botanical Garden, P.M. Zhukovskiy. Because of its physiological and morphological traits, this species forms a link between the African and the Asiatic groups of the wheat species.

These investigations are of great practical importance since they establish where it is necessary to search for initial material for breeding, and since they definitely determine that the European and Asiatic groups of cultivated wheats originated from southwestern Asia. During very ancient times, Egypt and countries in northern Africa had developed their own groups of cultivated wheat, sharply isolated from Asiatic wheats.

Investigations using the method of differential evaluation for the varieties and races of *cultivated barley* in the Old World have established that – just as in the case of the wheats – we can define *two basic centers* for its development of forms. On the one hand, it is *southeastern Asia*, including China, Japan and Tibet; on the other hand, an independent center of barley cultivation turned out to exist in the *northeastern part of Africa*, in the mountain areas of Abyssinia.

The asiatic Sino-Japanese center gave rise first to all the naked-grained, awnless and semi-awned barleys. Among these we encounter a multitude of peculiar forms, widely distributed in northern India and, in part, as early-ripening forms, reaching as far as the Altai and western Himalayas. The Abyssinian center gave rise initially to the European hulled barleys. The special expedition led by Dr. Harlan to Abyssinia has just returned. It was sent out by the U.S. Department of Agriculture to study cereal grasses and confirmed that this interesting country has indeed, in recent times, enriched our information about cultivated plants in an important way.

It has been definitely established that the native land of *millet* – both the common one (*Panicum miliaceum*) and the foxtail (*P. italicum*) – is found in eastern Asia, in China and in the countries contiguous with it. All the variable forms of millet, known in European Russia and in western Europe, originally came from eastern Asia, where, in addition, an enormous number of endemic, original kinds of millet, unknown in Europe, are found. In Mongolia, a millet exists that is characterized by its easily shed seeds, reminding us of wild grasses. Side by side with early-ripening mountain species, China, Manchuria and the Far East cultivate a large number of very late-ripening ones, unknown in Europe, e.g. groups of strongly and loosely spreading kinds with peculiar seeds. In some parts of Afghanistan and Bokhara, they cultivate curious (from a practical point of view) kinds of millet with thin membranes and easily shed hulls – analogous to the naked-seed barley. It is impossible to avoid mentioning that southeastern Asia (China) appears to be the native land also of large-grained, naked-seeded oats.

In the case of flax [*Linum*], the existence of *two independent centers* could also be established: one in *southwestern Asia,* the other in *northern Africa*. To the first one belong the common European forms of long-staple flax [*L. usitatissimum*], and a variety, distinguished by small seeds, small capsules and small flowers. To the second group belong large-seeded, large-capsuled and large-flowered races, typical of Egypt, Tunisia, Algeria, Morocco and the European coast of the Mediterranean but almost unknown among the groups grown in Russia.

Investigations have revealed that the species with such a distribution and such numbers of independent centers were also characteristic of flax in the remote past. The African flax species are characterized by large capsules and seeds (up to

6mm long) and are distinguished from the Asiatic types with small seeds (ca. 3 mm long). Archeological discoveries in Egypt from the time of the different dynasties of pharaohs revealed only large-seeded and large-capsuled types.

As far as it has been possible to demonstrate, all the varieties of flax in Asia are concentrated in Bokhara, India and Afghanistan where, in addition to the common types of flax, we also meet with white-seeded ones with wrinkled petals and extraordinarily late-ripening, low-growing and productive flax plants at the same or an even higher elevation, all common mainly in the high mountains.

Studies of Asiatic and European flax species revealed interesting facts about the regular distribution of the types in the Old World and explained the concentration toward the north of flax, cultivated mainly for its fibers, as opposed to the southern types, cultivated for their oily seeds. It turned out that the distribution from the centers of variation in the direction from the south toward the north occurred by virtue of natural selection with respect to the vegetative period. The length of the stem is, as a rule, closely associated with the vegetative period: the shorter the growing period, the earlier the species ripens and the longer the stem of the plant becomes and vice versa. Our northern fiber flax turned out to be the earliest-ripening type; in contrast, the southern linseeds, cultivated for oil and seeds, are all late-ripening and do not thrive in the north. In the mountain areas of Asia, where the maximum variation of both linseed and fiber flax is found simultaneously, the fiber flax tended to become more alpine, while the concentration of late-flowering native large-seeded kinds producing oil increased in a southerly direction. The cultivation of fibers moved naturally northward into central Russia and western Europe. On the other hand, in India, a concentration of linseed crops appeared, cultivated exclusively for seeds. Just as in the case of the cereal grasses, the mountain areas of southwestern Asia are the center of variation of flax species.

During the studies concerning the origin of *rye* we encountered a group of curious phenomena which subsequently turned out to be of general importance.

The rye cultivated in Europe, European Russia and Siberia is botanically entirely homogenous and represents, in essence, a single botanical variety (var. *vulgare)*, marked in Fig. 6 by a 'V' (see p. 80).

The study of specimens from different areas showed that all the variation and wealth of rye forms were concentrated into those areas where rye was either of secondary importance or had never been important as a cultivated plant.

The main botanical variation of rye was found in *Persia*, *Afghanistan*, *Turkestan*, *Bokhara*, and in *Georgia* and *Armenia* (Transcaucasia), where rye is not cultivated but is known mainly as a weed contaminating wheat and barley and, especially, the crops of winter wheat and winter barley.

Among the *weedy* types of rye it was possible to establish an exceptional variety of types that had no representatives in Europe. In those areas, red-spiked rye, rye with long and coarse awns and rye with velvety hulls are very frequent. In the mountains of Bokhara are found types of rye with reduced leaves and without any ligulae. Recently, a black-spiked rye has been found in Georgia.

In short, in southwestern Asia, as well as in Transcaucasia, the investigator comes in contact with an exceptional variety of rye, which reminds him of some types of soft wheat, and, in addition, the rye he finds there is mainly in the form of weeds among wheat and barley crops.

We must, without question, acknowledge that the main center for the variation and development of forms of *rye crops* occurs in southwestern Asia, i.e. within the same area where the development of some types of wheat took place.

The areas in which soft wheat and rye both developed coincide in striking detail. Thus, for instance, non-ligulate rye is found within the same area where non-ligulate wheat occurs, especially in northern Afghanistan. Rye with non-shedding, enclosed grains that is distributed in Turkestan and Persia is located also within the same areas where non-shedding soft wheat is found.

Local studies in Turkestan, Persia, Afghanistan and Bokhara reveal that rye is mainly a weedy plant. Rye is not cultivated in India, China, Syria or Palestine. It was not known in Turkestan until it was introduced by the Russians.

Moreover, the very epithets of rye used by different peoples in Turkestan, Persia, Afghanistan, Arabia and Turkey, i.e. 'gandum-der', 'dzhoy-der' and 'chou-der', mean literally 'plants preying upon or contaminating wheat or barley'. The names of rye in the languages of the inhabitants of these countries, thus, bear witness to the interference of this plant with the wheat and barley crops in a remote past. The wide distribution of these epithets all over the East, from India to Arabia, shows that the local inhabitants from time immemorial knew rye as a weed that was hard to get rid of.

Grains of wheat are hard to distinguish from those of rye even by the best of specialists. Therefore, complaints about rye, heard by travelers in Bokhara, Persia and Turkestan, can be understood.

However, rye does not always occur exclusively as a weed in southwestern Asia. As one ascends into the alpine areas of Turkestan, rye can be found which has become a commonly cultivated plant.

Rye is a crop plant in habitats high up in the mountains of Shugnan and Roshan, in Bokhara and in southern Fergana. Its epithets are already very different: the peoples in the valleys still use the names 'gandum-der' and 'chou-der', i.e. 'weeds', while the Tadzhiks of Pamir call the rye 'kal'p' or 'loshak'.

Due to its very nature and its difference from wheat and barley, rye appears to be a biologically more frost-resistant, rougher and hardier plant. Therefore, it is natural that, in the course of cultivation of the former, the rye began, under more difficult conditions, to entirely replace the more tender wheat and barley plants. Rye tolerates lower temperatures and poorer soils than does wheat. It can be grown on low-yield, clayey or sandy soils where wheat either does not thrive or fares poorly.

Natural processes such as those that have taken place in the mountains have also occurred in the past during the dispersal toward the north. During the dispersal of crops of wheat from the center toward the south and north, east and west, the rye contaminating the wheat gradually displaced it and developed into crops of pure rye.

Involuntarily, Man made a weed into a crop during the northward dispersal of his crops. On poor soils and under severe conditions in the north, rye had, in the words of the Taeers, become a 'better gift from the Gods' and the people began, by necessity, to sow rye instead of wheat. Similarly, at present, the farmers in northern and central Russia, also by necessity, plant rye rather than winter wheat.

Considering the above facts, the history of the origin of cultivated rye is extremely simple and easy to understand. The ancient crops of wheat and barley brought with them the rye in the form of a weed during the dispersal of the crops into more severe conditions with colder winters and with poorer soils; rye began to out-compete the more tender wheat and barley crops. In Mongolia, the contamination of the spring wheat with weedy rye still occurs, and there is a progressive increase in rye in crops the farther north one goes (V.E. Pisarev).

It is interesting that, at the boundary of competition between winter wheat and winter rye, farmers have for a long time, and do even now, purposefully sown a mixture of rye and wheat, called 'surzhy' or 'surzhik' [the 'hardy one']. While not relying on obtaining a good yield of wheat every year, the farmer knowingly sows a mixture of rye and wheat in the hope that, should the wheat crop fail due to a harsh winter, at least the rye will survive and give a reasonable yield. Such 'surzhy' can still be seen used in the Terian region, in the Balkans and in Normandy (Vavilov, 1917).

It may be possible to trace, in detail, all the processes of the introduction of rye into cultivation from southwestern Asia and from Transcaucasia. From the Caucasian mountain chain, rye spreads southward as a weed but northward as a cultivated plant.

In order to understand the origin of cultivated rye, it is thus necessary to associate it with the cultivation of wheat. Cultivation of rye was a result of the cultivation of wheat and, in part, of barley. It is therefore self-evident that the distribution areas of the varieties of soft wheat and rye coincide.

In practice, it turns out that the greatest interest for plant breeders must be concentrated to the fields of weedy rye [*Secale cereale*] in southwestern Asia. For the selection of new types, the weeds among the crops of wheat and winter barley in Persia, Afghanistan, Georgia and Armenia hold the greatest hope. From the weedy types of rye, special drought resistant races can be selected that deserve to be introduced into cultivation in southeastern European Russia and in the Ukraine (Vavilov, 1922).

On the whole, in accordance with the origin of rye from weeds among the wheat and barley crops, the fact is that, as far as data obtained from linguists and archeologists are concerned, cultivation of rye arose much later than that of wheat or barley, since it has long been known that the cultivation of rye originated from weeds among the former crops. We have barely enough data for an exact identification of the local origin of cultivated rye, but, evidently, this process occurred simultaneously and independently in different localities.

An analogous picture develops when investigating the cultivation of *oats*.

While traveling in Persia in 1916, we happened to come upon crops of emmer [macaroni wheat] with weedy oats. The emmer itself (a particular species of ancient wheat, *Triticum dicoccum*, with grains closely surrounded by its hulls) was met with in fairly small quantities, mainly in western Persia among the

Armenian settlements. As a rule, the crops of emmer were badly contaminated with such types of oats as are now cultivated. Emmer without oats was, on the whole, not seen in Persia. Neither Persians nor Afghans knew how to grow oats; in Turkestan the cultivation of oats was introduced by Russian settlers.

Studies of oats collected from the emmer fields revealed that they belong in part to an original, new kind of oat cultivated in Europe, i.e. *Avena sativa*. This weedy type is distinguished by the shape of its panicle, the structure of its hulls and represents, in general, its own kind of cultivated oat, easily crossing with the common types of cultivated oats. It has now been distinguished as a special variety.

This fact attracted our attention and forced us to investigate the weedy plants in emmer fields in other areas. As a result, we could establish that wherever emmer was grown, it was, as a rule, accompanied by the weedy oats; cultivated oat, thus, is represented by a specialized weed among emmer. Studies of the main Russian center of emmer cultivation in the Pri-Kama region (in the Kazakh, Siberian, Permian and Vyatsk provinces) revealed the presence of a large number of new forms of oats among emmer crops, which, until then, were not known in cultivation. It turned out that forms of oats were found that represented their own extreme morphological evolution as far as the attachment of grains in the spikelets is concerned, reminiscent of this structure in emmer itself. The different flowers adhere tightly and the grain itself can only with difficulty be dislodged from the glumes. When threshed, the spikelets remain whole.

Studies of the emmer in Georgia and Armenia resulted in discoveries of a number of different kinds of oats, unknown among the ordinary oat crops in Europe. In other words, it was evident that the center for selection of oats must be retraced to new kinds of oat varieties among crops of emmer, where the majority of the forms not found in pure cultures could be discovered.

Oats occur as bad weeds among emmer, lowering the quality of the meal and the flour. A very frequent phenomenon, particularly in the north, is the repression of emmer due to wild oats. In literature from the eighteenth and nineteenth centuries, we find many indications of displacement of emmer by wild oats. Thus, for example, an observer over a period of 40 years writes: 'Who among us does not know that oats overpower emmer so that from an imperceptible number of grains of oats, by chance found among the grains of emmer, after a few or up to some ten harvests thereof, the field seems to be sown with a crop of oats into which, by chance, some emmer is mixed' (Velikov, 1810). The ancient authors knew oats only as a weed among the crops of other cereals. Hehn explains the German epithet of oats, 'Hafer', as 'Bockskraut', i.e. a 'devilish weed'.

Thus, the interesting fact is that crops of emmer which, as far as can be established, have their center of origin in northern Africa, areas of Abyssinia, Asia Minor and Transcaucasia, when dispersed toward the north brought with them oats, in the form of an inescapable attribute, which during the northward transfer of the crops displaced the emmer.

Oats appear to be hardier plants, more resistant to cold and less demanding on the soil. As definitely demonstrated through archeological discoveries, emmer represents a more ancient crop, which is presently to a major extent extinct. In all the world only a few centers remain for the cultivation of emmer in mountain areas (e.g. in Abyssinia, the Pyrenees, the Balkans and in the Caucasus) as well as in the Pri-Kama region among the Chuvaks, the Mordovinians and the Tatars. The cultivation of emmer occurs in small fields in isolated areas and among backward people, such as the Basques, the Chuvaks, the Mordovinians, the Ossetians, the Armenians and the tribes populating Abyssinia, who have preserved the ancient crops up to the present.

When dispersed over the Old World in the very remote past, emmer brought with it a collection of weedy plants; during the transfer of the crops toward the north, these weeds turned into separate crops. Thus, the cultivation of oats arose in a number of areas. In Transcaucasia and in other areas, wheat and barley are often contaminated by oats and the emanation of oats from the weedy crops of those cereals is very likely. For instance, in the province of Bratsk, many local crops remind one strikingly of speltoid forms of wheat. All information clearly indicates that oats are a more recent type of crop. The earliest information about cultivated rye and oats dates from the beginning of the Christian era. The comparative youth of oats as a crop becomes easy to understand, once we realize that this plant was brought along with other ancient crops (of emmer, wheat and barley), when they were dispersed from the south toward the north. There are many indisputable data in favor of the fact that the cultivation of oats arose from weeds: the ancestral origin of cultivated oats appears, consquently, to be the weedy oats, which until relatively recently contaminated emmer, wheat and barley.

Thus, in order to understand the evolution of the two very important northern crops, oats and rye, it is necessary to associate it with the development of cultivated emmer, wheat and barley.

We still don't have enough facts to confirm that all cultivated oats had this kind of an origin. Oats have a polyphyletic origin and, perhaps, different kinds of it (e.g. the African forms) became cultivated along other routes, which demand further research.

A comparison of facts reveals that such a kind of origin of other cultivated plants from *weeds*, contaminating old, basically southern crops, has happened fairly widely. No doubt, whole groups of the oil-producing cruciferous plants arose from weeds, contaminating other basic, but older crops. Garden Rocket (*Eruca sativa)*, false flax (*Camelina sativa*), corn spurrey [*Spergula arvensis*] and sea rocket [*Cakile maritima*] all originated from weeds contaminating the basic crops of flax. Excellent research within this field has been done by N. V. Tsinger (1909).

Perhaps the vetches [*Vicia*], vetchlings [*Lathyrus*] and ordinary peas [*Pisum*] have also originated from some wild leguminose plants.

Tatarian buckwheat [*Fagopyrum tataricum*] is a serious weed among the crops of common buckwheat [*F. esculentum*] although it is at the same time a cultivated plant in some localities in the Altai.

In Crimea, einkorn [*Triticum monococcum*] contaminates the fields of emmer just as do oats. It is also found in southern France. At the same time einkorn is also known as a pure crop in mountain areas.

In summing up, it is generally possible to separate all the cultivated plants into two groups. One is a group of basic, ancient crops known to man exclusively as pure crops in a cultivated state; this is the group of *primary crops* to which belong wheat, emmer, barley, rice, soy beans and flax. To the *secondary* crops, no less extensive, belong all plants evolved from weeds that infested the primary, basic crops, especially rye, oats, false flax, spurrey, garden rocket, tatar buckwheat, Narbonne vetch [*Vicia narbonensis*], coriander and many cruciferous plants.

The natural route for the development of this second group of weeds into independent crops was the dispersal of crops from the valleys into elevated mountain habitats or the dispersal from the south toward the north.

During the transfer of crops toward the north, a differentiation took place, of course, selecting out the frost-resistant, tougher and hardier plants as well as those that ripened earlier. The process was assisted by the needs of the peoples. Agriculture has taken this into account as a fact.

In general, such basic major civilizations as were associated with major water basins or major rivers were believed to set the scene for the development of cultivated plants. Lev Menchikov expressed this idea very nicely in his book, *Civilizations and the Great Rivers*.

The main civilizations of the Old World were distributed mainly along the basins of the Nile, Tigris, Euphrates, Ganges, Indus, Yangtzekiang and Hwang-Ho and, of course, the idea therefore arose that the cultivation of plants in the Old World spread outward from these great river basins.

The most recent research in the Orient, southwestern Asia and northern Africa indicates that all the high-quality varieties of field and vegetable crops could equally well have originated from the mountainous areas of southwestern and southeastern Asia and from the mountainous areas of northern Africa. Since all the high-quality varieties occur in the Asiatic (including Caucasian) and African mountain areas, it is evident that it is basically wrong to assume that the crops developed in the valleys.

The mountain areas offer, of course, optimum conditions for bringing out high-quality varieties and for the differentiation of varieties and forms. All the variation in the kinds of our main agricultural plants is concentrated in the Asiatic and African mountain regions. Europe and Siberia do not have such definitely endemic forms. The fact that, to the collectors and the herbarium curators, the mountain areas appear to be not only the center of high-quality variation thanks to the variation of the conditions in those areas, but also the historical center of origin is supported by the different composition of the crops and the species in the various mountain areas of Asia and Africa.

When considering the process of agricultural evolution we must inevitably acknowledge that the period during which major civilizations united the multi-tribal societies of people was preceded by a natural period of solitary life in secluded areas. To this end, the mountain areas may have served as ideal refuges. Restriction to the large rivers, the taming of the Nile, Tigris, Euphrates and other major waterways demanded an ironhanded organization to build dams and regulate the seasonal flooding. It also needed organized mass action of which the primitive agriculturalists (farmers) in northern Africa and southwestern Asia could not even dream. It is therefore more likely that agriculture was focused on the various forms found in mountainous areas. For instance, in Shugnan and Roshan, the conditions for irrigation are very simple and convenient.

The mountain areas are not only the center of variation in respect to the cultivated species of plants but also that of the variation of human tribes. The diversity of ethnographical societies still favors a differentiation of races.

Most of all, preliminary research on the distribution of the species in Turkestan and the Caucasus reveals that the dispersal within the mountain habitats parallels that of the longitudinal and latitudinal distribution.

Siberian and Archangelian types of wheat and barley can be found in the elevated mountain areas of the Caucasus, Turkestan and Bokhara.

In isolated areas of Cis-Pamir, Siberian types of delicate, soft-awned wheat with easily shedding grains are still kept. The peas of Pamir remind one of the Siberian and north-European types of 'pelyoshki' [pea-pods?]. The mountain areas display all the extent of variability that is capable of populating the large territory stretching from Mesopotamia to Siberia and reaching up to the limit of agriculture in the northern parts of European Russia.

The elucidation of the centers of type-formation and the origin of cultivated plants permit us to touch objectively upon some cultural-historical problems and to establish the basic centers of human civilization. There are arguments about whether the autonomous Egyptian civilization was adopted from Mesopotamia or vice versa; the problem concerning the independence of the Chinese and Indian civilizations can be solved by objective investigations of the kinds of plants cultivated there. The plants and their varieties are not so easily carried from one area to another; in spite of the many thousands of years of tribal wanderings, it is not difficult to establish the basic centers of type-formation in the case of the majority of cultivated plants. The presence in China, in southwestern Asia and northern Africa of major, endemic, genetically sharply isolated groups of species and varieties of cultivated plants can also settle arguments about the autonomy of those civilizations from a cultural-historical point of view.

The history of the origin of human civilizations and agriculture is, of course, much older than the documentation in the form of relics, inscriptions and bas-reliefs can tell us about the past. The more recent acquaintance with cultivated plants and their differentiation into geographical units compels us to attribute the very origin of these to distant epochs where periods of 5–10 000 years mean only a short time.

It is possible that we shall succeed in establishing the main centers of groups of cultivated plants by following detailed studies of geographical species and their composition within the different continents, and after investigation of the still sparsely explored areas of Africa, Asia and South America. It is already possible to distinguish the north African center from the southwestern Asiatic one in respect to many plants. And, within the borders of Asia, we can recognize both southwestern and southeastern centers.

The faint familiarity with the Orient and the exceptional interest in the practical and scientific attitudes to the studies of cultivated plants for the purpose of exploring mountain areas for types to be introduced into cultivation compel us to turn our attention in the near future to the organization of special expeditions for studying the cultivated plants of Afghanistan, Asia Minor, Persia, Bokhara, Khiva, Tibet and China. Areas of Abyssinia are also exceptionally interesting since they are the only really mountainous area of Africa, and since these may be a center of origin for a number of our cultivated plants.

Many riddles concerning the origin of our main cultivated plants and domesticated animals have been untangled in the East.

We do not doubt that the results of our investigations will furnish rich material for finding new kinds of cultivated plants. Every single packet of grain brought from an uncontaminated area, every handful of seeds and every bundle of ripe spikelets is of important scientific value.

REFERENCES

Tsinger, N.V. (1909). O zasoryayushchikh posevy l'na vidakh *Camelina i Spergula* i ikh proiskhozhdenii [On the origin of Camelina and Spergula, infesting crops of flax]. *Tr. Botan. Muzeya Akademii Nayk. Sib.* [*Papers from the Botanical Museum of the Siberian Academy of the Sciences*], No. 6.

Vavilov, N.I. (1917). O prokhozhdenie kul'turnoy rhzi [The origin of cultivated rye]. *Tr. Byuro po prikl. botan.* [*Papers from the Bureau of Applied Botany*], **10**, 7–10.

Vavilov, N.I. (1922). Polevie kul'tury Yugo-Vostoka [Field crops of the south-east]. *Tr. po prikl. botan. i selek.* [*Papers on applied botany and plant breeding*], No. 23.

Velikov, V. (1810). *Pererozhdenie rasteniy i vozrozhdenie ikh cherez obrazovanie novikh raznovidnostey* [The degeneration of plants and their regeneration through the generation of new varieties]. Moscow.

On the origin of cultivated plants

THROUGH THE PRESENT report I wish to share with you some results of the work presently being conducted in the Department of Applied Botany and Plant Breeding at the State Institute of Experimental Agronomy (the GIOA).

The problem concerning the origin of cultivated plants is one of the subjects we are investigating. In order to elucidate the genesis of plants and for practical breeding purposes, we need to know where each crop originated and where to find the geographical centers, whether a given crop is associated with wild species, and when the innumerable types arose that eventually produced cultivated plants.

Forty years have passed since the first publication of De Candolle's book on the origin of the cultivated plants appeared. It was an investigation that was the first to embrace extensively the problem of the origin of cultivated plants.

Since then, the data available to De Candolle have been supplemented by much historical and botanical information, but, on the whole, the solution to the problem of the origin has been only slightly advanced. Later investigations did not point out any new methods nor any new ways for solving this generally complicated problem.

Investigation of the variability of cultivated plants, establishment of a system describing the specific varieties and their morphological and physiological characteristics and, if possible, attempts to make a complete botanical description of the varieties of important cultivated plants have prompted us to pose questions regarding the centers of type-formation and the problem of the origin of cultivated plants. During the search for missing characters and shapes of different species we succeeded as far as was possible during the present difficult work conditions[1] to expand the phyto-geographical analysis of the varieties of cultivated plants all over the world.

The search for the centers of type-formation in the Old World led unexpectedly to the establishment of primary centers of variation for many of the plants as well as to the discovery of an enormous variety of unknown forms, which have not been previously described. Objective solutions were also found to

Lecture read to the 1st convention of agronomists, graduated from the Leningrad Agricultural Institute and the University associated with it; Leningrad – Detskoye Selo, Jan. 19, 1926. First published in: *Novoye v Agronomii* [*News in Agronomy*], pp. 76–85, 1926.

[1] The post WWI era. – D.L.

historical problems concerning the origin by using exact methods that turned out to be more specific than it had, perhaps, been assumed earlier, and to be of more practical application for the purpose of plant breeding.

At first, we restricted ourselves to using the same methods for determining the centers of origin and of type-formation of cultivated plants.

De Candolle had considered it his main objective to clear up the problems concerning the native lands of cultivated plants and the localities where they were first introduced into cultivation.

For this purpose, botanical, archeological and paleontological, as well as historical and linguistic, methods were available to De Candolle. However, he gave the last two methods only a secondary role. The majority of the plants was first cultivated during historical times or during very remote historical epochs. Neither historical nor linguistic methods can be properly applied. Paleontological data are usually entirely unavailable for cultivated plants and archeological data are very poor and fragmentary.

Detailed studies of the species from a physiological and morphological point of view, and an analysis of the systems of species, their varieties and races as well as their regularities led us to the problems concerning the origin of cultivated plants. During our search for missing links in the systems we unintentionally approached an explanation for the distribution of races and varieties and a solution for the problems concerning the centers of type-formation.

During the last couple of years, the problem of the geographical origin of cultivated plants has been roughly settled. To solve it, physiologists, historians and archeologists utilized historical data, archeological discoveries and linguistic uses and comparisons.

Our most recent studies of the distribution of cultivated plants and their varieties convinced us, however, how inaccurate and sketchy many of the statements were that De Candolle and many later authors made. Detailed studies of the species of cultivated plants within Asian, European and African countries, and of the considerably larger amount of material available to us now than to De Candolle at his time, have revealed the need to break down the problems concerning the origin of cultivated plants into areas such as those dealing with the origin of Linnaean species and with genetical groups of varieties and races. In order to clear up the problem of the centers of origin of wheat, we had to break it down, during the progress of the work, into questions concerning the origin and the centers of type-formation of hard wheat, soft wheat and so on. In order to understand the origin of oats, we needed to study separately the development of the African species and the Asiatic and European groups which, to a great extent, proved to be independent. Later investigations established that very often the centers where the forms of various isolated species belonging to this or that genus (e.g. wheat, barley, flax, etc.) developed, could be found on different continents. One species of cotton originally came from America, another from Asia. One species of wheat is typical of Asia, another of Africa. Inevitably, the investigations led us to a polyphyletic origin of many of the major cultivated crops.

By means of phyto-geographical methods used for determining the centers of origin of various groups of species and genera, their distribution within different areas became apparent, while, as far as possible, eliminating the conditions imposed on the species due to interference by humans. Because of the concentration of species within different areas and the localities of their greatest variation, the botanists were able to establish the centers where the forms of a given group of plants developed. In order to establish the centers of type-formation of cultivated plants, it was necessary to apply a differentiating method and take into consideration not only the species as such, but also its different varieties and races.

In spite of the internationalization of many crops and the major roles played by migration and colonization of people in the dispersal of species, we can nevertheless still encounter species endemic to different areas: North Africa, Abyssinia, Algeria, India, Mesopotamia, Afghanistan, Bokhara and China contain the majority of the varieties known so far. In mountain areas and among almost inaccessible civilizations, as well as in isolated countries of Asia, Africa and the Caucasus, an enormous number of original species, not known to Europeans, have been preserved to date. Thanks to detailed studies of the distribution of varieties, paying particular attention to the investigation of ancient crops and the ancient centers of agriculture so far hardly touched upon by the investigators, it may be possible to determine the centers of a surprising amount of variation, including forms rarely seen by Europeans and, in addition, a number of new forms and characteristics that have not yet reached Europe.

To do this, it is necessary to assemble collections of a large number of specimens from various localities and countries as well as to make detailed systematic studies of all their physiological and morphological peculiarities.

During the last couple of years we have succeeded in spreading some light on the composition of cultivated plants in Bokhara, Pamir, Persia and Turkestan, as well as a portion of Afghanistan (Badakhshan), Mongolia and southern Russia, to where special expeditions have been sent out.

Because of the difficulties we have had during the past couple of years, the organization of the expeditions could not be as extensively set up as desired, and a great deal remains unfinished for various reasons.

Along with the studies of the cultivated plants themselves, we must take an interest in and give significance to an investigation of the closely related wild species and their variation and racial composition when we try to elucidate the genesis of cultivated plants. Fortunately, wild species of many plants such as oats, wheat, rye, barley, millet, hemp, etc. are still preserved.

The studies revealed many new facts and allowed us to approach the problem concerning the origin of cultivated plants from a new angle.

Thus, we could definitely establish that one large group of wheat species, in particular the soft wheats (*Triticum vulgare*) and the club and shot wheats (i.e. *T. compactum and T. sphaerococcum*), originally came from southwestern Asia. Their center of variation appears to be in northern India, Belutchistan, Afghanistan, Bokhara and Persia. In the mountains in those areas, forms and characters are hidden which, in part, reach as far as Europe but, in part, are still unknown to Europeans.

It turned out, however, in the case of hard wheats and adjacent Linnaean species, that the center of variation was in northern Africa, i.e. in Algeria and Abyssinia. These areas still appear to be the main foci for all the variation of hard wheat.

In spite of the large area covered by the cultivation of hard wheat in the southeastern and southern parts of Russia, the Russian strains proved to be represented only by an insignificant number of forms in comparison to those found in Africa. During the search for valuable types of hard wheat to cultivate, it became evident that Russian plant breeders must first and foremost turn to northern Africa.

The case of oats was similar, although it showed an even more complicated picture for the development of its forms. Following detailed studies by methods using cross-breeding and immunity-tests, it was established cytologically (by A.G. Nikolayev) that there was an extraordinarily sharp differentiation of oats into many species which were unable to cross with each other. The different species have different centers of variation. For some, e.g. *Avena sativa*, *A. orientalis*, *A. nuda*, *A. fatua* and *A. ludoviciana*, the center where the forms developed could be placed in Asia and Europe, but as far as others, e.g. *A. byzantina* and *A. algeriensis*, are concerned, the center appeared to be northern Africa.

The group of oats species comprising *Avena strigosa* var. *brevis* and *A. nudibrevis* is concentrated mainly in northern Europe, as far as their variation is concerned.

As far as barley is concerned, it turned out that the main center of all the variation, comprising almost all the characteristics by which the forms known in Europe and Africa are distinguished, is located in Abyssinia. The mountain habitats of Abyssinia still preserve everything necessary for plant breeders to create new cultivated strains, including a multitude of endemic varieties and races not yet touched upon by European plant breeders. Eastern Asia (including China and Japan) turned out to have a secondary, more or less isolated center, especially of naked-grained barleys.

The cultivation of flax and fava beans apparently developed independently on two continents. Africa and Asia each have large, independent and original groups of flax and beans.

Using differentiating methods when studying a large number of specimens collected from primary sources and detailed studies of vegetative, physiological and morphological traits, it could be definitely established where the sources of the variation were found and, consequently, also where the sources of type-formation exist.

We have tried to study plant after plant in that manner. The most difficult thing to do is to obtain material from the most interesting but, at the same time, least accessible areas. The mountain areas of Asia (e.g. Afghanistan, Belutchistan, Tibet, China and Asia Minor) as well as of Abyssinia and Sudan hide the majority of the forms not explored so far. Some investigations of a number of Asiatic countries and of odd specimens we have obtained – in particular with assistance from the U.S. Department of Agriculture – made this fact particularly obvious to us. Special expeditions to study this wealth of plant resources are necessary since these areas have not yet been fully explored.

At present we are putting together maps – to some extent already completed – of the distribution of the varieties and characteristics of the plants most important for field cultivation. An enormous amount of material, thousands of specimens from different countries, is needed for this type of map. They can be really accurate only when based on a large amount of material, which makes it possible to clear up problems concerning centers of type-formation.

Establishment of cycles of variability, characterizing different Linnaean species, allows us to predict the discovery of a particular variety and to search for it in an intelligent manner. When sending out an expedition, its eventual success may be ascertained beforehand when looking for various forms with different characteristics.

We already know which characteristics are inadequately represented within the systems of barley and wheat and we do not doubt that, sooner or later, we shall find them. That kind of study is also of great practical importance as we have seen, for instance, in respect to some disease-resistant species from Africa.

The study of the association between cultivated plants and the wild species closely related to them revealed an interesting fact. *Many presently cultivated plants originated from weeds infesting the crops of more ancient cultivated species*. Thus, it was first and foremost established that cultivated rye originated from kinds of weedy rye that infested winter barley and winter wheat in southwestern Asia and in Transcaucasia.

Before the Russians arrived, neither Persia nor Turkestan were familiar with cultivated rye, nor was it grown in India or China. Nevertheless, rye in all its forms, which are astonishing to Europeans, is concentrated especially within these countries and in the form of weeds among wheat and barley crops. The very epithets of rye in the Persian language, adopted also in Turkey, India and Turkestan, are 'chou-dar', 'dzhou-dar' or 'gandum-dar', which literally mean 'the plant tormenting the barley and wheat' and, consequently, at the same time 'a weed'. This weedy rye includes an enormous number of forms, unknown both in Siberia and in Europe, e.g. red-spiked, black-spiked, velvety and non-ligulate as well as, in particular, drought-tolerant kinds of rye. Forms with variable vegetative characteristics, reminding one of those of barley (e.g. with large auricles), are met with especially among the weedy rye.

In the high mountain areas of Pamir and Badakhshan and in some mountains in Asia Minor, weedy rye begins, because of its greater frost-resistance, to naturally displace wheat and turn into a pure crop of rye. The same phenomenon, although on a larger scale, takes place also in a northerly direction. When wheat crops dispersed northward from southwestern Asia, i.e. from the basic center of soft wheat into Siberia and European Russia, weedy rye began to replace the wheat. Thanks to its inherent winterhardiness in nature, which still characterizes wild Persian rye, it supplanted wheat, due to a great extent to man's actions. Just as happened in the elevated mountain areas, rye turned into pure crops. At the limits for cultivation of winter wheat and winter rye, some farmers still plant a mixture of rye and wheat, a so-called 'mixed crop', which promises a more reliable harvest than wheat alone when over-wintering under severe conditions.

This process has been successfully retraced in detail. As it happened, it was recently established that, as far as origin is concerned, cultivated rye could not be separated from wheat. Wheat outcompeted rye in crops in southwest Asia and Transcaucasia. However, among its weeds there, plant breeders have for a long time looked particularly for valuable forms of rye. Our experiments in Saratov have shown, for instance, that there are especially drought-tolerant races among such types of rye.

The coincidence between the geographic centers of origin of soft wheat and rye extends into details. Thus, the African and Bokharan mountain centers, characterized by a large group of non-ligulate types of wheat, include, simultaneously, a large group of non-ligulate types of rye, at present not known in Europe.

An even more interesting fact became obvious when we studied emmer, i.e. *Triticum dicoccum* [also called macaroni wheat, D.L.]. This crop from the remote past has been preserved in 'islands' of cultivation in Russia, the Caucasus, the Pyrenées, the Balkans and in Abyssinia. In Russia, it is still grown in considerable quantities in the Pri-Kama region, mainly among backward people such as the Chuvashs and the Mordovinians. In Persia, it is often found among Armenian settlers and in the Pyrenées it is grown by the Basques. During our studies of this type of crop, we happened across the interesting fact that oats occurs as a constant companion to emmer in the form of a weed.

Investigations of weedy oats unexpectedly revealed a wide variety of forms. A large number of new forms of oats were discovered together with the strains that did not differ from those presently being cultivated. The new forms differed in many characteristics that, until then, were not known in cultivation. The maximum variation in oats was found especially among emmer crops. There, very peculiar races with characteristic morphological traits were found, e.g. with grains very tightly enclosed by the hulls and in that respect analogous to emmer itself, which does not shed its grains, or with flowers that remain paired even after threshing (twin-grained oats).

During the investigations of emmer we encountered, in general, the curious fact that, in order to discover interesting types of oats, it was necessary to study the weeds in among the emmer, especially within those areas where crops of oats were not grown.

It turned out that, in particular – just as in the case of rye – oats appeared to be a hardier plant in the more northerly countries: it displaces emmer crops in both the northern and the central provinces of Russia. It is still possible to find statements by old investigators indicating that oats can out-compete emmer. In the literature from the mid-eighteenth century we found an observation stating: 'Who does not know that oats can displace emmer?' It becomes, then, even more apparent that emmer, as an ancient crop, brought oats along when dispersing from the south toward the north. As a weed, oats already infested emmer in ancient times and turned into a pure crop, supplanting the emmer, subsequently developing into cultivated oats. In view of the obviously polyphyletic origin of cultivated oats, this problem calls for more detailed studies. However, it is, in any case, without question a fact that in his search for interesting forms, the plant breeder must turn his attention to weedy oats amongst the emmer, without whose presence it is difficult to understand the origin of cultivated oats.

Thus, the process of infestation of one crop with weeds of another potential crop and the displacement of the originally cultivated crop by such weedy specimens, now cultivated, can be observed also in the case of lentils (*Lens*). Weedy vetches (*Vicia*) frequently outcompete the originally cultivated lentils.

The same phenomenon occurs in respect to the Tatarian buckwheat [*Fagopyrum tataricum*], which infest the crops of common buckwheat [*F. esculentum*]. In mountain areas, the Tatarian buckwheat displaces the common buckwheat and can become a special crop of its own.

Crops of false flax [*Camelina*] and garden rocket [*Eruca sativa*] arose in the remote past from weedy plants that infested flax crops.

It has become evident that all crops can be distinguished schematically into *primary*, basic and very old crops to which belong wheat, barley, rice, lentils, flax, etc., and *secondary* and younger kinds of crops, which arose from the weedy plants infesting the basic crops and, during their dispersal to the north or into high mountain areas alongside the primary crops, by pure chance, eventually displaced the basic species and developed into crops of their own. Many cultivated plants have developed in this manner. Therefore, the study of weedy plants is of great interest to plant breeders for discovering strains that are valuable for cultivation under severe conditions, and also for clearing up the genesis of cultivated plants.

This kind of investigation has attracted our attention, especially during the last couple of years. It demands a large number of specimens, specialized studies of different kinds of crops and the organization of expeditions.

On the basis of parallelism in variability we have succeeded in retracing, in detail, the process of how hemp (*Cannabis*) and millet (*Panicum miliaceum*) became cultivated. We were lucky enough to find existing wild hemp and wild millet (in Mongolia), characterized – as in the case of most wild plants – by their ability to self-propagate and by their peculiar morphological traits with respect to their seeds, which favor shedding upon maturing (e.g. types with elaiosomes). In some cases, the problem of the origin of cultivated plants, such as that of hemp, has become absolutely clear. We succeeded in finding a whole series of forms among wild hemps from which, by breeding, cultivated forms could be selected: e.g., non-shedding and large-grained types. Detailed studies revealed that such hemp plants were naturally associated with Man, multiplying mainly in fertilized sites and accompanying nomadic camps, eventually being introduced into cultivation by the involuntary actions of humans.

In other cases, such as with respect to oats or rye, the entire process of the origin of their cultivation occurred due to voluntary actions by humans.

Thus, gradually, step by step, during the search for forms that could be expected on the basis of general conformity with respect to variation, the investigator can find and establish the centers of type-formation of cultivated plants and clear up along which main routes they became cultivated.

In spite of the many thousands of years that have elapsed since the beginning of civilization, we can perhaps still, in many cases, retrace the stages of introduction into cultivation.

Phyto-geographical methods used in investigating the problem of the origin of cultivated plants and their centers of type-formation appear to be important for elucidating the general evolution of cultivated plants and for practical considerations for the purpose of establishing areas whence material for plant breeding work can be obtained.

An exact determination of geographical centers of origin can also be established from the known extent of similar centers of human civilizations.

The cultivation of crops and of plants paralleled the general civilization of Man. The question of the independence of different civilizations, such as, for instance, the Mesopotamian and the Egyptian ones, reveals extremely well and beautifully the presence of both in these and other areas of genetically independent groups of cultivated plants. The autonomy of different phyto-geographical centers of cultivated plants, established on the basis of specific associations and detailed studies of species, can, to a great extent, solve the problems concerning the autonomy of the different foci of human civilizations as well.

Centers of origin of cultivated plants

Devoted to the memory of
ALFONSE DE CANDOLLE
Author of
'Phytogéographie rationale' (1855)
'Phytogéographie' (1880), and
'L'origine des plantes cultivées'(1882)

Introduction

WHEN STUDYING THE specific composition of cultivated plants it became necessary to establish a definitely natural system relating to the development of forms within the limits of Linnaean species and genera (Vavilov, 1920, 1922). By means of a thorough study, the variation of the inherited shapes and forms turned out to be subordinate to a well-composed system. As a result of collective work, we succeeded in constructing systems of varieties and races, and general schemes of the inherited variability, for entire families of the most important cultivated plants.

Table 1, which illustrates the variation within the family Papilionaceae, can be used as an example of such a system. It illustrates the diversity of forms and differentiating characteristics according to which taxonomists concerned with cultivated plants can build up their own systems, not to mention the innumerable combinations into which the different alternative characteristics can be combined. In order not to make the table too unwieldy, not every characteristic has been included but only those most obvious and distinct, which are easy to recognize. (A more detailed scheme for the variability of grasses was furnished in my paper 'Toward an understanding of the soft wheats'; Vavilov, 1923.) The traits proposed for the description of the morphological and physiological characteristics, presently existing, i.e., the forms of different Linnaean species, are, correspondingly, indicated by the symbol '+' in the table.

Similar systems have been constructed for the grasses (Vavilov, 1924), Cucurbitaceae (Vavilov, 1924), Solanaceae (Bukasov, 1925), Cruciferae (Sinskaya, 1924), Linaceae and others. They are equally applicable to self-pollinating plants as to cross-pollinating ones. Studies of the system of variability revealed a striking resemblance between a number of morphological and physiological characteristics of closely related genera and species, which represent different

First published in *Tr. po prikl. bot. i selek.* [*Papers on applied botany and plant breeding*], vol. 16, no. 2, 1926

Table 1. *General scheme of the variability within the species belonging to Papilionaceae*

Variable hereditary characteristics	*Pisum sativum*	*Vicia sativa*	*Vicia faba*	*Ervum lens*	*Lathyrus sativus*	*Cicer arietinum*	*Phaseolus vulgaris*	*Soya hispida*	*Medicago sativa*	*Trifolium pratense*
Flower characteristics										
Color										
White	+	+	+	+	+	+	+	+	+	+
Pink	+	+	−	−	+	+	+	+	+	+
Red	+	−	+	−	+	+	−	+	+	+
Violet-blue	+	+	−	+	+	+	+	−	+	+
Yellow	−	+	+	−	+	+	−	−	+	+
Variegated (flag differs from wings)	+	+	+	−	+	+	+	+	−	−
Flag and wings spotted or striped	−	+	+	−	+	−	+	−	+	−
Dimensions										
Large	+	+	+	+	+	+	+	+	+	+
Small	+	+	+	+	+	+	+	+	+	+
Pod characteristics										
Wall structure										
With a papery layer	+	−	−	−	−	−	+	+	−	−
Without a papery layer	+	+	+	+	+	+	−	−	−	−
Shape										
Linear	+	+	+	−	−	−	+	+	−	−
Rhombic	−	−	−	+	+	+	−	−	−	+
Sickle-shaped	+	−	+	−	+	−	+	+	+	−
Swordlike	+	−	−	−	−	−	+	+	−	−
Moniliform	+	−	−	−	−	−	+	+	−	−
Hairiness										
Hairy	−	+	+	−	−	+	+	+	+	−
Glabrous	+	+	+	+	+	−	+	+	+	+
Color of unripe pods										
Yellowish	+	−	−	−	−	−	+	+	−	−
Greenish	+	+	+	+	+	+	+	+	+	+
Violet-brown (with anthocyanin)	+	+	−	−	−	−	+	+	+	+
Color of ripe pods										
Yellowish-green	+	+	+	+	+	+	+	+	+	+
Black (dark brown)	+	+	+	−	−	−	+	+	−	+
Spotted (or striped)	−	−	−	−	+	−	+	+	−	−
Dimensions										
Large	+	+	+	+	+	+	+	+	+	+
Small	+	+	+	+	+	+	+	+	+	+

Table 1. (*cont.*)

Variable hereditary characteristics	*Pisum sativum*	*Vicia sativa*	*Vicia faba*	*Ervum lens*	*Lathyrus sativus*	*Cicer arietinum*	*Phaseolus vulgaris*	*Soya hispida*	*Medicago sativa*	*Trifolium pratense*
Surface										
(a) Smooth	+	+	+	+	+	+	+	+	−	+
Uneven	+	+	+	−	−	−	+	−	−	−
(b) Convex	+	+	+	+	+	+	+	+	+	−
Flat	+	+	+	+	+	−	+	+	−	+
Seed characteristics										
Shape										
Orbicular	+	+	+	+	+	+	+	+	+	+
Ovate (or egg-shaped)	+	+	+	−	−	−	+	+	+	+
Cylindrical	−	−	+	−	−	−	+	−	−	−
Flat (disk-shaped)	+	+	+	+	−	−	+	+	+	+
Angular	+	+	+	−	+	+	+	−	−	+
Kidney-shaped	−	−	+	−	−	−	+	−	+	+
Surface										
Smooth	+	+	+	+	+	+	+	+	+	+
Wrinkled	+	+	+	−	−	+	+	−	−	−
Color										
White	+	+	−	+	+	+	+	−	−	+
Yellow	+	+	+	+	+	+	+	+	+	+
Green	+	+	+	+	+	−	+	+	+	+
Grey	+	+	+	+	+	−	+	+	−	+
Pink	+	+	−	+	−	+	+	−	+	+
Red	+	+	+	+	−	+	+	+	−	+
Chestnut (brown)	+	+	+	+	+	+	+	+	+	+
Black	+	+	+	+	+	+	+	+	−	−
Dimensions										
Large	+	+	+	+	+	+	+	+	+	+
Small	+	+	+	+	+	+	+	+	+	+
Pattern										
Marbled	+	+	−	+	+	+	+	+	+	−
Dotted	+	+	+	+	+	+	+	+	−	−
Striped	+	+	+	+	+	+	+	+	−	+
Color of cotyledons										
Green (dull)	+	+	−	+	+	−	+	+	−	+
Yellow	+	+	+	+	+	+	+	+	+	+
Red (orange)	+	+	−	+	+	−	+	−	−	−
Color of hilum										
White	+	+	+	+	+	+	+	+	−	−
Brown	+	+	+	+	+	−	+	+	+	+
Black	+	−	+	+	−	−	+	+	−	+

Table 1. (*cont.*)

Variable hereditary characteristics	*Pisum sativum*	*Vicia sativa*	*Vicia faba*	*Ervum lens*	*Lathyrus sativus*	*Cicer arietinum*	*Phaseolus vulgaris*	*Soya hispida*	*Medicago sativa*	*Trifolium pratense*
Vegetative characteristics										
Leaf structure										
With tendrils	+	+	–	+	+	–	–	–	–	+
Without tendrils	+	+	+	+	–	+	+	+	+	+
Leaf shape										
Linear	–	+	–	+	+	+	–	+	+	+
Cuneate (rhomboid)	+	–	–	–	–	–	+	+	+	+
Ovate	+	+	+	+	–	+	+	+	+	+
Dimension of leaves										
(a) Long	+	+	+	+	+	+	+	+	+	+
Short	+	+	+	+	+	+	+	+	+	+
(b) Wide	+	+	+	+	+	+	+	+	+	+
Narrow	+	+	+	+	+	+	+	+	+	+
Leaf edges										
Entire	+	+	+	+	+	–	+	+	+	+
Toothed	+	+	+	–	–	+	–	–	+	+
Hairiness of leaves										
Hairy	–	+	–	+	–	+	+	+	+	+
Glabrous	+	+	+	–	+	–	–	+	+	+
Color of stipules										
Green	+	+	+	+	+	+	+	+	+	+
With anthocyanin	+	+	+	–	+	+	+	+	–	+
Leaf color										
Yellowish	+	–	–	–	–	–	–	+	+	+
Green	+	+	+	+	+	+	+	+	+	+
Waxy cover on plant										
With wax	+	–	–	–	+	–	–	–	–	–
Without wax	+	+	+	+	–	+	+	+	+	+
Stem structure										
Straight	+	–	+	+	–	+	+	+	+	+
Coiling	+	+	–	–	+	–	+	+	–	–
Growth shape										
Upright	+	+	+	+	+	+	+	+	+	+
Creeping	+	+	–	+	–	+	+	+	+	+
Height of plant										
Tall	+	+	+	+	+	+	+	+	+	+
Intermediate	+	+	+	+	+	+	+	+	+	+
Low	+	+	+	+	+	–	+	+	–	+

Table 1. (*cont.*)

Variable hereditary characteristics	*Pisum sativum*	*Vicia sativa*	*Vicia faba*	*Ervum lens*	*Lathyrus sativus*	*Cicer arietinum*	*Phaseolus vulgaris*	*Soya hispida*	*Medicago sativa*	*Trifolium pratense*
Hairiness of stem										
Hairy	−	+	−	−	+	+	+	+	+	+
Glabrous	+	+	+	+	+	−	−	+	+	+
Shape of stem										
Cylindrical	−	−	−	−	−	+	−	+	+	+
Square	+	+	+	+	+	+	+	−	−	−
Fasciated	−	−	−	−	−	−	+	−	−	+
Color of shoots										
Green	+	+	+	+	+	+	+	+	+	+
With anthocyanin	+	+	−	+	+	+	+	+	+	+
Color of stem										
Green	+	+	+	+	+	+	+	+	+	+
Violet (with anthocyanin)	+	+	−	+	+	+	+	+	+	+
Biological characteristics										
Vegetative period										
Short (early-ripening)	+	+	+	+	+	+	+	+	+	+
Long (late-ripening)	+	+	+	+	+	+	+	+	+	+
Albinism	+	−	+	−	−	+	+	−	+	+

hereditary forms. The system of forms described resulted in Linnaean species and not in randomly composed races. The same applies not only to phenotypic but also to genotypic variation as long as it is subjected to study according to the method of hybridization.

The compilation of such systems for the purpose of building up strains based on a botanical classification revealed a large number of 'missing links', so far not known to taxonomists or not yet discovered but the existence of which could be predicted as a result of the laws of variability in the case of Linnaean species. During the search for the 'missing links' in the system, we succeeded in clearing up the geographical distribution of races and characteristics, the problem of the geographical centers for the development of forms and the process concerning the origin of the geographical centers of origin of cultivated plants. Consequently, questions arose: where to find the primary centers of variation, where to look for 'missing' forms, where to locate the maximum specific variability of a given Linnaean species and from where and how all the varieties of the forms of cultivated plants could be obtained.

The problem concerning the origin of cultivated plants has interested many scientists beginning with De Candolle (1855) and Darwin. Botanists, historians, archeologists, physiologists and agronomists have been preoccupied with it as well. The origin of cultivated plants has given rise to an extensive literature among which, up till now, the classical synoptic work, 'L'origine des plantes cultivées' by De Candolle (1883; *note*, the first abbreviated edition appeared in 1882) is outstanding, although the comparatively scanty information available to De Candolle concerning the different groups of plants, has since been substantially augmented by new historical and archeological data and a number of botanical and genetical observations (Schweinfurth, 1872–73; Woenig, 1886; Buschan, 1895; Joret, 1897–1904; Solms-Laubach, 1899; Höck, 1900; Hoops, 1905; Bailey, 1906; Helweg, 1908; Reinhardt, 1911; Hehn, 1911; Gibault, 1912; Trabut, 1913; Schulz, 1913; Thellung, 1918; Laufer, 1919; Sturtevant, 1919; Zade, 1921; Schiemann, 1922; Schweinfurth, 1922; Carrier, 1923; Safford, 1925; and Cook, 1925). It should also be mentioned that among this extensive literature a wealth of data on the geographical centers of some plants can be found.

While concertedly approaching the problem concerning the origin of the cultivated plants for the purpose of utilizing the data on the geographical centers of origin in order to locate 'missing links', we soon became convinced of the imperfection of the majority of the data and of the discrepancies between the statements in the literature concerning the geographical centers of origin as well as of the definite inaccuracy of much data with respect to the important cultivated plants, not excluding those concerning the cereals.

Most recently, critical studies have compelled us to renounce the majority of the statements made and to make an attempt to solve the old problem about the native lands of cultivated plants from a fresh point of view and to combine with it the question about the origin by using, as far as possible, new and more accurate and objective methods for this purpose. When extending knowledge of the composition and differentiation of the species, the very problem of the origin of cultivated plants becomes much more complicated than it appeared at the time of De Candolle. At the same time it became absolutely clear that both the understanding of the dynamics of the evolution of cultivated plants and the genesis of the Linnaean species, as well as, to a great extent, the practical work of the plant breeders, depend on the correct solution of this problem. We have become convinced that a direct and exact establishment of the geographical centers of origin will open up extensive opportunities for a practical utilization thereof when the sources of a wealth of valuable strains are discovered.

The present paper is intended as a short review of our investigations concerning the origin of cultivated plants in the Old World and an account of the methods used for this research.

CHAPTER 1 METHODS USED FOR DETERMINING CENTERS OF ORIGIN OF CULTIVATED PLANTS

To determine the native lands of cultivated plants, De Candolle (1883) relied, in principle, on the location of specific cultivated plants while distinguishing between the state of running wild (becoming naturalized) and the primitive wild state, and separating the original discovery in the wild state from a later naturalization of a species in an area new to it. This was, in essence, the botanical method for determining the native land of the plants, such as understood by De Candolle. Archeological, historical and linguistic methods were used by him only as secondary methods, applied without any botanical basis.

In view of the present knowledge of the composition of the Linnaean species and their differentiation, such botanical methods evidently need to be revised, supplemented and changed.

Criticism of the methods used for determining the native lands of cultivated plants

Neither De Candolle nor authors unfamiliar with botany distinguished strictly between the different Linnaean species *sensu strictu* and the different genetic groups of varieties into which the majority of cultivated plants have at present been broken up. The problem of the origin has conventionally been solved in relation to the plant in the wide sense, i.e. all the species and genetic groups composing it such as, e.g., all the cultivated kinds of barley, flax, oats and so on. Under such conditions, when the cultivated plants, e.g., wheat or oats, are represented by several Linnaean species, sharply differentiated and physiologically isolated from each other and unable to hybridize, it is evident that uniting them cannot but lead to definitely erroneous conclusions. Primarily, this circumstance makes many of the native lands established for some cultivated plants highly inaccurate. It was especially easy to make mistakes when utilizing historical, linguistic and archeological methods where an evaluation of the specific differences is absolutely impossible. However, De Candolle was the first to distinguish the origin of some species in the New World from that of the same species in the Old World while remarking, e.g., that individual species of cotton and grapes were typical of America, while others were typical of Asia. Sufficiently accurate distinction of geographic and genetic groups within the limits of the Old World was apparently not feasible 40 years ago. Later investigations did not solve this problem either.

The method used by De Candolle and other authors for determining the native lands, where habitats of a given cultivated plant in its wild state could be found, are far from always to be considered reliable; this is primarily so because many cultivated plants are not known except in a cultivated state and, secondly, because what is called 'wild ancestors' are only limited groups of forms with a not very large range of varieties, which are rarely genetically isolated and not able to explain all the rich variation that, as a rule, is shown by the cultivated plants themselves. More often, such plants appear to be wild genera or related forms but are not ancestors in the true sense.

The wild wheat, *Triticum dicoccoides* with 28 chromosomes, found by Aaronsohn in Syria, corresponds to some extent in its variation to the cultivated species of wheat with the same number of chromosomes. However, the existence of this wild emmer does not explain the occurence of the large, genetically as well as physiologically isolated group of soft wheats with 42 chromosomes. In essence, the relationship of *T. dicoccoides* to all the species with 28 chromosomes is not even entirely clear or even that to the cultivated emmer (*T. dicoccum*) with which Körnicke united it into one Linnaean species, because when crossed with cultivated emmer or hard wheat it results in first and later generations in a sharply expressed inclination toward sterility.

Wild barley (*Hordeum spontaneum*) is very close to one of the smaller groups of cultivated barley, i.e. the one with hulled, distichous and awned varieties, but can by no means be included among all the varietally differentiated species of cultivated, naked-grained and hulled barleys such as the awnless, furcate or two-rowed *deficientes* or the four-rowed or six-rowed barleys. The entire group of cultivated, naked-grained barley is isolated morphologically and physiologically from the wild distichous and hulled barley (*H. spontaneum*) by many characteristics. The presence of a naked-grained group in eastern Asia does not explain the existence in southwestern Asia of a large amount of wild, two-rowed and hulled barley. On the other hand, it is necessary to keep in mind the fact that wild barley has not been taken into cultivation to the same extent as wild wheat or oats since it does not lose the inherited brittleness of its spike and since it remains distinct from the cultivated forms. This makes its role as a direct ancestor more than doubtful.

An attempt by Schindler to cultivate mountain rye (*Secale montanum*), which botanists consider an ancestor of cultivated rye (*S. cereale*), did not succeed in making it into an annual from its perennial origin and did not turn its brittle spikes into those of the cultivated rye, however much the experimenter hoped it would. Just as unsuccessful were – and will be – the many attempts to turn *Avena fatua* into *A. sativa* or *Triticum dicoccoides* into *T. dicoccum*, and so on. Wild 'ancestors' or forms with adapted or induced self-propagation represent interesting groups of related forms corresponding to the cultivated plants and sometimes linking them (e.g. in the case of barley) to a member of cultivated forms within the limits of Linnaean species and tying them together with the wild species. However, the fact that they could be turned into cultivated forms, such as was believed by De Candolle and all the later authors, could not be proven. Therefore, the discovery of such questionable ancestors, which have been found during the last couple of years in the case of wheat, millet and hemp, does not yet solve the problem of the origin of cultivated plants at their centers of geographical variation.

Most of all – as shown by the examples – the investigation has demonstrated that the distribution areas of all these wild 'ancestral' forms are usually either fairly limited and confined to restricted and isolated areas, or, on the other hand, fairly wide, when attempts are made to use them for localizing the primary centers of cultivated plants. The full range of varieties of wild wheat, *Triticum dicoccoides*, has been preserved in Syria and Palestine, but only a few of its varieties have been found in Persia and Transcaucasia. The center of variation of the different species of cultivated wheat is far away from Syria and Palestine. On the other hand, wild barley, *Hordeum spontaneum*, is closely related to the cultivated kind, occupies a wide area stretching all the way from northern Africa to Asia Minor and is particularly common in the area of southwestern

Asia, the region of the specific center of development of strains of cultivated barley. As will be shown later, this center is confined to a fairly small geographic area. Where there is an infestation of wild barley in the foothill areas of Turkestan, in northern Afghanistan and in the mountains of Bokhara, we found a striking scarcity of varieties of cultivated barley. Wild melons (*Cucumis trigonus*), ancestral to the cultivated ones, are widely distributed from the Aral Sea to Fergana in areas where cultivated melons show an astonishing variation. The composition of cultivated melons in this area is rather a demonstration of remoteness from the source for their variation. The wild watermelon, *Citrullus colocynthis*, close to the cultivated *C. vulgaris*, embraces within its distribution area not only Africa – the geographical center of origin of watermelons – but reaches into Asia and is widely distributed in the desert areas stretching from Hindukush to Afghanistan, Persia and Belutchistan. However, we tried in vain to locate any native land for watermelons in Asia.

A number of wild species, corresponding to cultivated ones, have enormous distribution areas, extending all over Eurasia, such as, e.g., in the case of red clover (*Trifolium pratense*). Of course, such distribution areas do not furnish any real support for localizing the process of their introduction into cultivation.

Quite often the present distribution areas of the wild 'ancestors' do not coincide with the real centers where the strains developed, as will be seen, e.g., in the following discussion concerning cereals and other plants in the Old World.

Methods suggested for establishing centers of origin of cultivated plants

Detailed investigations during the last couple of years of the botanical composition of races and varieties of cultivated plants have allowed us to approach the problem concerning their origin from a fresh point of view, while enhancing and extending the methods used by De Candolle. After some decades of research done by the experimental agronomists at the Institute of Applied Botany and at other Russian experimental stations, the botanical and the qualitative structure of many plants within the European and Asiatic parts of the U.S.S.R. including Caucasia and Turkestan has to a great extent been cleared up. Expeditions have investigated agricultural areas of Persia, Bokhara, Korea, Mongolia, Afghanistan, Asia Minor and Mexico. English experimental stations in India, in particular the work done there by A. and G. Howard, Leach and Watts, have established the composition of many cultivated plants both in India itself and in Belutchistan, as well as in adjacent parts of Persia. The composition of plants in northern Africa has to a major extent been elucidated by American and French scientists. Published data from various countries in Europe, Africa and Asia permit – although far from completely – a more thorough approach to the questions regarding the centers where the strains of cultivated plants developed.

Studies of the cultivated vegetation of the Old World revealed one extraordinarily important fact, i.e. that in spite of the internationalization of cultivated crops and in spite of all the human migrations and colonizations as well as the very antiquity of agriculture, it is still possible for students of taxonomy and geography to establish regions of definitely endemic varieties and races, to discover regions where the maximum primary variation of strains occurred and to establish a number of regularities concerning the distribution of inherited characteristics. Comparisons between data on the geography of varieties have revealed that the cultivated races, varieties and species are characterized by distinct geographical distribution areas and that, for a solution of the problem concerning centers of typification, the same methods can generally be used as those that botanists and zoologists apply for species of wild plants and animals.

As a basis for determining the centers of origin of this or that group or genus of animals and plants, taxonomists concerned with geography use the distribution of the species and the maximum concentration thereof within various areas. Naturally, just such a method for establishing the centers by means of regional concentration of varieties and by the use of extensive differentiation has already been applied to cultivated plants as well.

As a basis for determining the centers where the forms of different Linnaean species develop and in the light of the present representatives of the species as well as of polymorphic systems, we consider it necessary not only to establish the distribution area of an entire species, irrespective of the composition of its elements as developed up till now, but most of all to establish the distribution areas of all the parts composing it and the geographical concentration of all the inherited characteristics of the forms of this species. In other words, as a basis for the determination of the center, where the development of the forms of a given species took place, we must provide differentiating methods for establishing the racial composition of the species in question and the geographical distribution of the racial variation within countries and regions.

Just as a botanist or a zoologist establishes the exact specific composition of a genus and the distribution of its different species over the entire world when determining the geographic center of type-formation of a particular genus, so it is necessary, with respect to cultivated plants and their various Linnaean species, to apply differentiating methods for the racial and varietal composition, the racial variability of the species and the geographical distribution of all the Linnaean species, the so-called linneons.

Detailed botanical knowledge of the specific composition, of the complex of morphological and physiological characteristics and of the distribution within a region is necessary for the solution of those types of problems for every plant investigated.

Such a differentiating geographical method requires, of course, an enormous amount of botanical material from various countries, and studies of the different races by means of crops grown under identical conditions in order to ascertain inherited differences. In general, this always appears tedious and time-consuming but it furnishes distinct ideas about the actual centers where those strains were formed, which can then be used by plant breeders for practical work.

Acquaintance with the distribution area of the species *sensu latu* gives little information about the center where the forms developed. The distribution area of wild wheat, *Triticum dicoccoides*, reaches – as recently demonstrated – from

Palestine to western Persia and Transcaucasia as well as Asia Minor (according to P. M. Zhukovkskiy) but the entire diversity of its races and varieties is, as far as is known, typical only of Syria and Palestine. Out of the immense diversity of the Syrian forms in Georgia and Persia only a few varieties are found. The area covered by cultivated rye is very large, the common types of it are widely cultivated from the Himalayas all the way to the Polar Circle; the area where all the racial variation of rye is concentrated is, as we will see below, considerably more limited.

It is natural that differentiation into varieties and races must precede the strict differentiation of a given cultivated plant into Linnaean species. Thus, by applying various methods such as cytology, hybridization, determination of resistance to parasites and the usual ones of taxonomy and morphology, the investigators have learned to delimit specific groups exactly, to isolate the actual Linnaean species from the artificial, vernacular-named species. The distribution areas of the genetically sharply isolated species of cultivated plants are often very different and not rarely typical of different continents.

The differentiating, taxonomical–geographical method suggested for the determination of centers of development of forms can be applied for various plants and consists of:

(a) differentiating a given plant into Linnaean species and genetic units by means of various methods such as morphological–taxonomical ones, parasitological ones, etc.;
(b) determining, if possible, the distribution areas of these species in the remote past when communications were more difficult than at present;
(c) determining in detail the composition of the varieties and races (more exactly, the inherited variation of the characteristics) of each species and the general system of inherited variability; and
(d) establishing the distribution of the inherited variation of the forms of the species in question and determining the geographical centers of accumulation of varieties. The region of maximum variation, usually including a number of endemic forms and characteristics as well, can usually also be considered as the center of type-formation.

Some examples follow below and illustrate the application of the differential method for determining the center where the forms developed.

For a more exact establishment of the centers of origin and the development of forms, a determination of that geographical center is necessary where there is a concentration of genetically closely related species. Just as with respect to wild plants, as a rule, the more closely the species are related, the more they are characterized by having adjacent or often overlapping distribution areas. It can also be understood that investigations of such species furnish valuable information for the determination of the centers of origin. Thus, as will be shown below, e.g., in the case of soft wheat (*Triticum vulgare*), the center of variation coincides with the geographical center of variation of shot wheat (*T. compactum*) and club wheat (*T. sphaerococcum*), i.e. species quite close to soft wheat, easily crossing with it and forming a single genetical unit together with it.

Finally, using necessary corrections and any supplementary information available when determining the areas of origin, establishment of the areas of variation of closely related wild varieties or species of certain plants can be effected, while applying also the differential method of racial studies to them.

Usually this means components of such Linnaean species by which the cultivated plants are represented, but with additional characteristics that enable them to remain in the wild condition (e.g. elaiosomes at the base of the seed, brittle spike, grain splitting open when ripening, etc.). Botanists usually attached exceptional importance to the studies of such forms, viewing them as ancestors that still exist, and believing that once the wild form of the cultivated plant was found, the question of its origin could at the same time be answered.

While setting this not yet proven hypothesis aside for later discussion, there is no doubt that the wild forms, ancestral to the presently cultivated species, deserve the same research efforts as given the cultivated species.

On the basis of this kind of analytical botanical research, data from archeology, history and linguistics have acquired a great value. When related to a definite, botanical point of view, they can often, and in an important manner, supplement and enhance information about the past of cultivated plants.

The center of origin of a given group of cultivated plants is often characterized by many special parasites, widespread within that area and typical of that particular group of cultivated plants. That is, the centers of variation of the parasites, typical of a particular group of cultivated plants, coincides with that of the hosts. The greatest variation among the smuts of rye is found in southwestern Asia, the center of variation for this crop. Similarly, out of the 10 species of smut living on species of sorghum, the majority (*Tolysporium filiferum*, *T. volkensii*, *Sorosporium ehrenbergii* and *S. simii*) is found in Africa only, i.e. the native land of the sorghum (Reed and Melchers, 1925).

Primary and secondary characteristics of centers of diversity

As far as particular groups of plants are concerned, the centers of botanical diversity cannot always be considered as the primary foci of type-formation. There may be occasions when the present racial variation is the result of the similarity of different species and their hybridization with each other. In the case of cross-pollinating plants, where under natural conditions mainly dominant characteristics are present, perhaps an opportunity for isolation and exposure to external variation can arise far from the center of origin, or due to artificial breeding, e.g. inbreeding of recessive forms.

The maximum variation of some garden plants can often be found in horticultural nurseries. As a rule, this is apparently connected with the practice of hybridization. For example, in the case of the fruit fly, *Drosophila*, an exceptional variation is known to occur as a result of mutations within artificial surroundings. On the basis of recent cytological investigations of *D. melanogaster*, Jeffrey arrived at the conclusion that the exceptional inclination of this group toward mutations was linked to natural hybrids between this and other species of the same genus. His research revealed the same picture of abnormalities during mitosis and meiosis as found for *Oenothera lamarckiana* (cf. American Naturalist, 1925).

It is necessary to be prepared for the existence of such genera and to separate critically the secondary development of forms from the primary one. However, actual investigations of a significant number of cultivated plants demonstrate quite definitely that the process of evolution occurred both in time and space. We have demonstrated by the examples above that the routes of geographical dispersal of forms, the foci of the genes and the basic centers of origin can still be established. Data from the geography of the specific variation of the cultivated plants do, on the whole, not oppose the basic condition for Willis' (1922) theory on the role of age and area for the evolution of plants.

Information on the center of variation of a species investigated and augmented by data on other ancestral species and wild forms and the simultaneous utilization of data from archeology, history and linguistics allow us to approach the establishment of a definite center of origin while keeping in mind the historical period most accessible to us. It may be possible to verify that many cultivated plants differentiated and formed different Linnaean species within the course of a certain period of time. It is possible also that other Linnaean species, that once originated from a single common geographic center, are separated today on different continents, although at present they are available for research only from the recent past. While interpreting the present distribution of a cultivated plant, it may be possible to start a reconstruction of its history in a more remote past. The interest in specific distribution and the foci of the genes in present time as well as during a historical period close to us can be of practical value.

Below we shall examine a number of actual examples of how to determine the centers of type-formation of the most important crops in the Old World by means of the differential–geographical method and show how we arrived at the facts established for a number of species.

CHAPTER 2 GEOGRAPHICAL CENTERS FOR THE DEVELOPMENT OF FORMS OF CULTIVATED PLANTS IN THE OLD WORLD

Geographical centers of type-formation of wheat

We shall start by establishing the geographical centers for the type-formation of the most important cultivated plants in the Old World, i.e. the wheats.

De Candolle believed that all the cultivated wheats had their origin in Asia. While distinguishing between hard, soft, English and emmer wheats, he did nevertheless not pay special attention to their differentiation into separate species. He was familiar with the experimental hybridization of wheats by Vilmorin but did not see any proof of an adequately expressed isolation of the individual species thereof. In his opinion, the absence of specific epithets for English and hard wheats in the Antique indicated that they were all *Triticum vulgare*. The hard wheats were, according to his judgment, established during the present era in Spain and Africa and underwent there a corresponding modification into *T. durum* and *T. turgidum*. De Candolle considered it necessary to single out only einkorn, *T. monococcum*, for which he pointed to Serbia,

Greece and Asia Minor where it is found in a wild state. Its definite physiological isolation, expressed by the difficulty of crossing einkorn with soft wheat, made it, in De Candolle's opinion, necessary to single it out.

In his comprehensive monograph, Percival (1921) was not concerned with the problem of the geographical centers of origin although the book contained much valuable information pointing toward a solution of the geographical problem. This question has also been pondered by other taxonomists and investigators (A. Schulz *et al.*) concerned with the problem of the origin of cereals.

Research during the last couple of decades has cleared up the differentiation of cultivated and wild wheats into individual Linnaean species.

By means of systematic investigations of the varieties of wheat, after many years of interspecific crosses and results obtained from cytology, parasitology and studies concerning the resistance of the species to diseases as well as serology, we have finally succeeded in distinguishing all the present kinds of wheat into the following genetic groups of species:

I With $n=21$ chromosomes	II With $n=14$ chromosomes	III With $n=7$ chromosomes
1 *T. vulgare* Vill.	5 *T. durum* Desf.	13 *T. monococcum* L.
2 *T. compactum* Host	6 *T. turgidum* L.	
3 *T. sphaerococcum* Perc.	7 *T. polonicum* L.	
4 *T. spelta* L.	8 *T. dicoccum* Schrank	
	9 *T. pyramidale* Perc.	
	10 *T. orientale* Perc.	
	11 *T. persicum* Vav.	
	12 *T. dicoccoides* Koern	

If there are doubts about the equivalence of the species or about the fact as to whether a given species is a subspecies or a species in the Linnaean sense, then they can be left within the limits of the group in which they are listed. On the whole, all three groups appear absolutely and without question physiologically and genetically isolated from each other and the division of wheats into three groups is not an arbitrary classification but an indisputable fact, proven simultaneously by various methods (Tschermak, 1914; Zade, 1914; Vavilov, 1914, 1919, 1925; Percival, 1921; Sax, 1921; Nikolayeva, 1923).

Group I (the *vulgare*) is characterized by $n=21$ chromosomes and all the species within its limits are easy to cross with each other and give rise to absolutely fertile offspring. On the whole, this group appears susceptible to infection by fungal diseases: smut, rust and mildew.

Group II (the *durum*) is characterized by $n=14$ chromosomes; all the species are easy to cross with each other and produce fertile offspring but differ considerably in their susceptibility to fungal diseases. They can only with difficulty be hybridized with species of groups I or III and display a more or less sharply distinct sterility in the first as well as later generations of the hybrids. *T. dicoccoides* shows some isolation resulting in a considerable degree of sterility when crossed with cultivated emmer, hard wheats or *T. persicum*. According to the characteristics of the hybrids between *T. durum*, *T. dicoccum* and the other species of group II, they remind one of hybrids between distant species which, in spite of a considerable sterility, are able to produce a multitude of new types.

Group III (*monococcum*) has a karyotype with $n=7$ chromosomes. It is distinguished by a strong resistance to infectious diseases and by sterile hybrids when crossed with species of the other two groups.

Let us now turn to the determination of the geographical centers of origin of these physiologically isolated groups.

Centers of soft, shot and club wheats

During the course of the last nine years I and my colleagues (E.I. Barylina, G.M. Popova, A.A. Orlov and others) have worked out details of the distribution area from a botanical–geographical point of view in the case of the most complicated and intimidating of the species, i.e. *T. vulgare* (Vavilov, 1923). This species has lately been split by us into 67 botanical varieties, differing in the color of the spike, the grains and the awns, in the shape and size of the awns, in the pubescence of the glumes as well as the presence or absence of ligulae on the leaves. In addition, the varieties of soft wheat were studied from the point of view of racial composition. Not less than 166 characteristics were distinguished for the spike, the grains and the vegetative traits according to which different races could be separated and, thus, a botanical analysis was made – as far as possible in minute detail – of this polymorphic species.

Data from the special expeditions sent out from the Department of Applied Botany to Turkestan, Bokhara, Khorezm, Mongolia, Persia, Afghanistan and Asia Minor as well as literature data and the worldwide collections of the Dept. of Applied Botany were used as a basis for this research.

While not dwelling on details, in part already published (Vavilov, 1923), a slightly improved picture of the present geographical distribution of the varieties of soft wheats was obtained and is geographically represented by the resulting Fig. 1.

The taxonomic–geographical analysis demonstrated definitely that the varietal composition as well as the racial variation of soft wheats were concentrated in southwestern Asia. Studies of the geographical varieties and the racial characteristics demonstrated that the specific diversity increases in the direction toward Turkestan, India and Afghanistan. Proportionally, the farther away from central Europe and Siberia toward southwestern Asia, the more the number of botanical varieties and new, original races increases. All the diversity of the morphological and physiological characteristics of the soft wheats, existing at present, were found in the mountain areas of southwestern Asia. All the traits of the races, cultivated anywhere and typical of any variety of *T. vulgare*, are met with, although in different combinations, within this region where, together with the common European and Siberian forms, a large number of endemic varieties can be seen such as original, new non-ligulate forms with reduced leaves, or a number of new varieties with short awns or with colored edges to the hulls, never found except in this area.

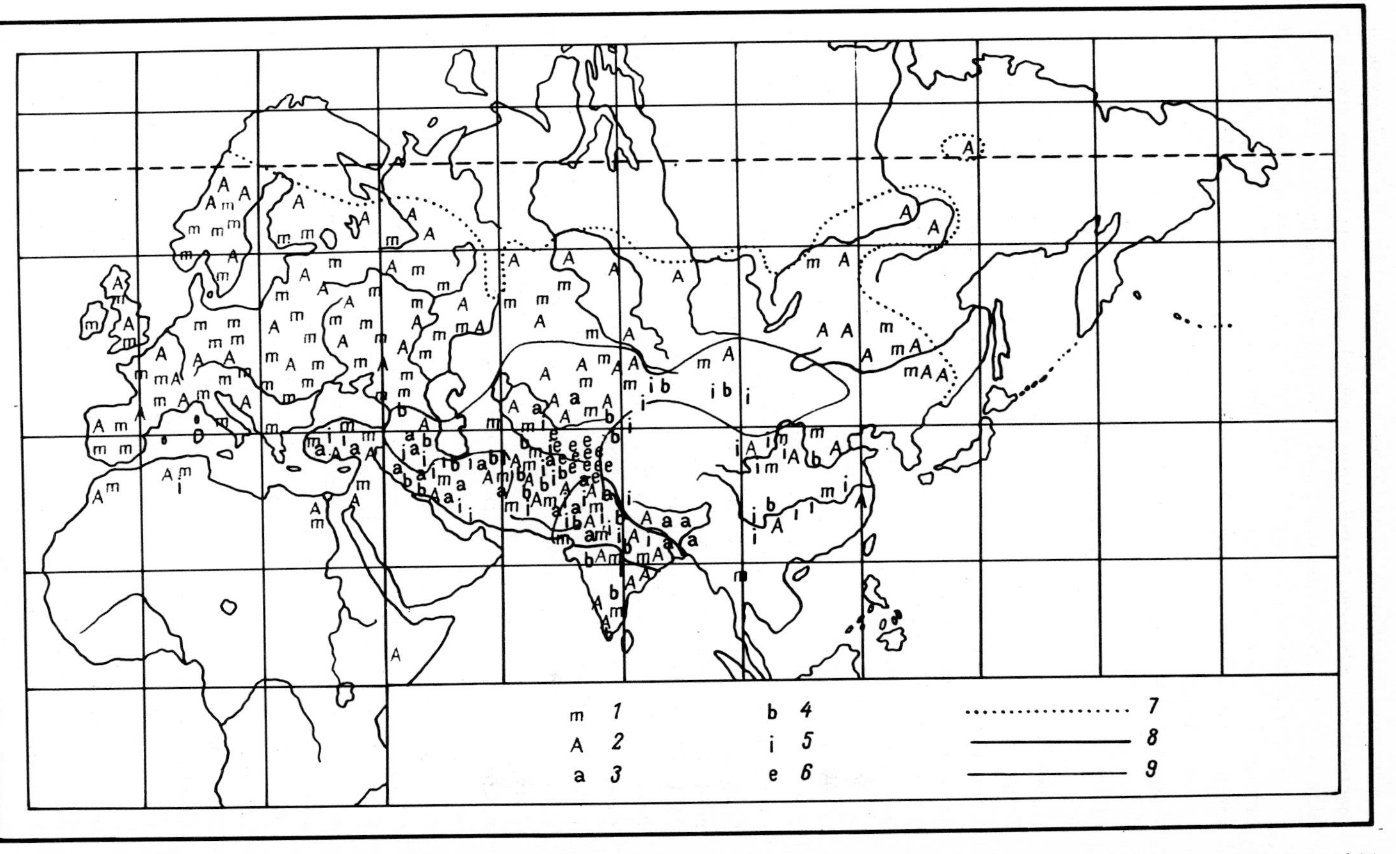

Fig. 1. Center of origin of the soft wheats. Geographical distribution of the varieties of soft wheat, including *Triticum sphaerococcum* Perciv. and *T. compactum* Host. 1. *Triticum muticum* Al.; 2. *T. vulgare aristatum* Al. (var. *europeae*); 3. *T. vulgare aristatum* (vars. *asiaticae, endemicae*); 4. *T. breviaristatum* Vav.; 5. *T. inflatum* Flaksb.; 6. *T. eligulatum* Vav.; 7. the northern limit of wheat growing; 8. the distribution area of *T. sphaerococcum* Perciv.; 9. ditto of *T. compactum* Host.

After studying the wheats of Afghanistan in 1924, it could be demonstrated that there was an especially rich variation of soft wheats in the area adjacent to the eastern part of Hindukush in the direction toward northwestern India. As far as we were able to find out, the non-ligulate forms of soft wheat are typical exclusively of Badakhshan, Chitral and the areas adjacent to them in Bokhara and, possibly also in Kashmir. They were not found in Turkestan, Khorezm, Persia, Mongolia or India, not to mention Europe or Siberia.

The number of types and the quality of the different racial characteristics decrease definitely from Afghanistan toward Mongolia (data from the expedition led by V.E. Pisarev) and central India (according to A. & G. Howard) as well as China (according to Percival).

All the European parts of the U.S.S.R., including northern Caucasus and Siberia, share only a small number of varieties of soft wheat. The number of varieties established for *T. vulgare* in Persia and Afghanistan amounts to 50, yet all of Europe has only 15–20. As could be expected because of the geographical and historical connections with Armenia, Georgia and Azerbaidzhan as well as southwestern Asia, the varietal composition of the wheats approaches that of Persia and southern Turkestan only in Transcaucasia.

As demonstrated by recent research by French authors (Boeuf, Miège, Duchêsne), soft wheat was not known in Africa until the colonization by Europeans. The introduction of different types into the oases of Sahara has not changed the picture of the world-wide geography of soft wheat. The New World adopted the cultivation of wheat from the Europeans and the botanical composition of cultivated wheats in the Americas is, in spite of the wide distribution of this crop in the U.S.A., Canada and Argentina, very poor in comparison with that in southwestern Asia (Clark *et al.*, 1922).

Thus, the method of differential studies of races and varieties very definitely allows us to approach the establishment of actual centers of variation and type-formation of the soft wheats. When making their way from Europe into southwestern Asia, the investigators, so to say, approach the center where the soft wheats arose and an advance from northwestern India toward Punjab and Upper Indus is accompanied by discoveries of more and more new types.

The study of the geographical distribution of the varieties of club wheat, *T. compactum*, which is closely related to soft wheat, fully corroborated our conclusions concerning the geographical center of soft wheat. The distribution area of club wheat is comparatively small. Club wheat is characterized by an important variation in the Khibinskiy oasis, and it is occasionally cultivated in the mountains of Bokhara, Transcaucasia and southern Turkestan, and occurs even in China. In Persia it is found just as a single variety in the Urmia area (A.A. Grossheim) but, in spite of special investigations, club wheat has not been found anywhere else in that country.

A detailed study by myself in Afghanistan in 1924 corroborated fully the theoretically predicted discovery of crops of club wheat in Hindukush. Crops of *T. compactum* were widely distributed there and – what is very important – many new, at that time unknown varieties of multi-awned and inflated club and shot wheats were also found there. The largest number of varieties of *T. compactum* occurred mainly in an area adjacent to northwestern India.

While bearing in mind that also a third species, closely related to the soft wheats, i.e. *T. sphaerococcum*, recently described by Percival, is found in its entire diversity only in northern India, it is evident that there is a complete overlapping of the areas covered by the varieties belonging to these three species. This confirms even better the veracity of applying the differential–geographical method for solving problems of origin.

Plant breeders must search for new types of soft wheat in the mountain areas of southwestern Asia. In the mountains of southwestern Afghanistan and the foothills of southwestern Himalayas, the process of type-formation of soft wheats and the species genetically related to them has been revealed. Northwestern India and the areas adjacent to it in Afghanistan, Chitral and Kashmir, display still the greatest diversity of the varietal and racial characteristics, many of which appear to be endemic. This area has a wealth of varieties and races. As demonstrated by *in situ* studies, all the enormous variety of spring and winter types of soft, club and shot wheats are represented here, together with the majority of the new races, which can be useful for practical breeding, if not directly then at least for the purpose of hybridization.

T. spelta is so far known only as a relict form, preserved as isolated crops grown in the mountains of Schwaben, Switzerland and, very rarely, in Urmia. It has been found by archeologists in a fossil condition in Egypt and in Asia Minor (Buschan, 1895). To establish exactly the center of its variation seems impossible at present. First of all, it may be assumed that this species, genetically related both to soft, shot and club wheats, should also have its center of variation and type-formation in the areas of southwestern Asia. However, in spite of painstaking efforts, this species has not been found in Khorezm, nor in Turkestan, Afghanistan, Persia or Asia Minor, or even in Mongolia. Additional research in this respect is desirable both in Asia Minor and northern India. Perhaps this species will turn out to be of hybrid origin?

The center of hard wheats

Let us now turn to the study of the center of hard wheats. Summing up the results of research concerning the native land and the antiquity of the hard wheats, Buschan (1895) stated that this problem could not be solved for lack of data. Botanists at the Department of Applied Botany have studied this problem in detail under the guidance of my long-time assistant, A.A. Orlov (1923).

While carrying out similar work using differential–botanical analyses of hard wheats on the basis of a large number of specimens (ca. 1500) we were able to establish that the main source of the racial variation of hard wheats appeared to be northern Africa and, in general, the coast of the Mediterranean. There, a multitude of original and endemic types of this species, not known in Europe or Asia, are met with, and at the same time all the racial and varietal characteristics, typical of the entire species of *T. durum* including its European and Asiatic races, are concentrated here. All of Asia and also the area of Mesopotamia (according to Whimshurst) are very poor with respect to the variation of hard wheats. India, China, Turkestan, Bokhara and Siberia are characterized by a sharply reduced variation in this species. Hard wheat is absent in the major part of southwestern Asia. In spite of painstaking efforts, we did not find a single spike of hard wheat among the innumerable fields of wheat we investigated in Afghanistan and Khorezm in 1924 and 1925. In spite of the striking variation in conditions for cultivation, varying from subtropical ones to the limits of agriculture in the mountains, Afghanistan with its primitive crops of wheat appeared to have been bypassed by this wheat when it dispersed. As we could see above, Afghanistan has, at the same time, a concentration of varieties of soft, club and shot wheats.

On the other hand, the basic focus of the types and the central area of all the polymorphism of *T. durum* seem to be found along the coasts of the Mediterranean, especially in Abyssinia, Algeria, Egypt and Greece. Such types as varieties of hard wheat with purplish-violet grains, pubescent leaves and sheaths or long-awned forms with elongated cilia on the auricles or at the base of the leaf blade, or with shoots distinguished by a brownish-violet color due to anthocyanin are known only from northern Africa.

A peculiar, non-ligulate hard wheat has, so far, been found only on the island of Cyprus (by Flaksberger, 1926). At the same time as endemic non-ligulate soft wheat was found by us in southwestern Asia in a secluded area of Badakhshan, endemic, non-ligulate hard wheats were discovered in the Mediterranean region, where we expected to find them.

According to investigations by A.A. Orlov, all the varieties of hard wheats can be determined according to 78 characteristics, including those of the grains, the awns and the spikes as well as vegetative ones (e.g. non-ligulate types) and reproductive ones.

All the races, characterized by the presence of a combination of the 78 characteristics, have been found only in the countries along the Mediterranean coast as well as in Abyssinia, whereas in Asia the diversity of the racial characteristics of hard wheats can be determined using only 51 traits.

Only the African region has a valid representation of the full range of polymorphism within *T. durum*. Only there the significance of the hereditary variability occurs to an extreme degree and the most extreme variants of the racial characteristics can be found there alone. Thus, e.g. there, groups of races can be met with which have awns more than 20 cm long and, at the same time, there are very short-awned races as well as 'semi-awned' types with awns less than 10 cm long. Exactly the same can be said about every characteristic typical of the races of hard wheat in Africa. The presence in northern Africa, and the islands close to it, of all the variations and all the forms of hard wheat, indicates, thus, that just this area is the main geographical center of the type-formation of the species of *T. durum*. It is necessary to study especially the countries along the Mediterranean coast and the banks of the Nile River all the way to Sudan in order to establish, in detail, the center of variation of the hard wheats.

The species close to *T. durum* are, unfortunately, still not as well studied in detail due to the lack of an adequate number of specimens, especially of *T. dicoccum*, *T. dicoccoides*, *T. turgidum* and *T. polonicum*, which are all typical of the same area. Large amounts and a great variety of the wild wheat, *T. dicoccoides* with $n=14$ chromosomes, have been found in Syria and Palestine. Individual strains only of this species have reached as far as western Persia and Transcaucasia in the form of one or two varieties.

Cultivated emmer increases in diversity toward N. Africa. In Abyssinia, black-spiked, white-spiked and red-spiked types and forms with fringed, dark-pigmented hulls, not known anywhere else, have been found there. Emmers, easy to thresh, also exist in that area (according to Harlan). An enormous amount of spike-types of emmer have also been discovered in the tombs of the Egyptian pharaohs.

All the varieties of the so-called poulard wheat, *T. turgidum*, are typical of the Mediterranean coasts.

The Polish wheat, *T. polonicum*, is characteristic of Egypt and the countries contiguous with it and all the strains of *T. pyramidale* are typical of northern Africa. *T. orientale* is also a Mediterranean species.

In general, this group of species which are difficult to delimit, easily hybridize with each other, are characterized by their resistance to infectious diseases, have the same number of chromosomes ($2n=28$), and have distribution areas along the coast of the Mediterranean (see Fig. 2).

Some years back, I distinguished a new species of hard wheat under the name of *T. persicum*. It has recently been discovered in a great number of varieties in Georgia, Armenia, Daghestan and Persia (Vavilov, 1919; Atabekova, 1925; Dekapredelevich, 1925). P.M. Zhukovskiy (1923) described white-spiked, black-spiked and red-spiked varieties as well as a multitude of races, distinguished by vegetative and other generative and biological characteristics. This species represents a morphological as well as a transitional group of forms between hard and soft wheats. Due to its external aspect, it was previously unhesitatingly referred to the soft wheats by the taxonomists but, according to the hereditary, physiological characteristics and the number of chromosomes, this species is undoubtedly closer to the hard wheats. *T. persicum* hybridizes easily with hard wheats and produces completely fertile offspring. The geographical distribution area and the center of variation of Persian wheat (i.e. Armenia, Georgia and Asia Minor) occupy an area intermediate between the Afro-Mediterranean area of the hard wheats and southwestern Asia, the center of the soft wheats (Vavilov and Yakushkina, 1925). In addition, *T. persicum* has been found between Khorosan and Seistan in Persia. This species of wheat, occupying a position genetically intermediate between groups I and II of the wheat species, possibly originated from hybridization of this species and some other species belonging to either of these two groups, such as could be expected from the geographical concentration of its variation along the borders between the areas of groups I and II.

The center of einkorn

Finally, group III of the wheats, einkorn (*T. monococcum*), known mostly in its wild form, is found in large numbers in a wild state in Asia Minor, Syria, Kurdistan, Transcaucasia, Crimea, Greece, Mesopotamia and Palestine (Flaks-

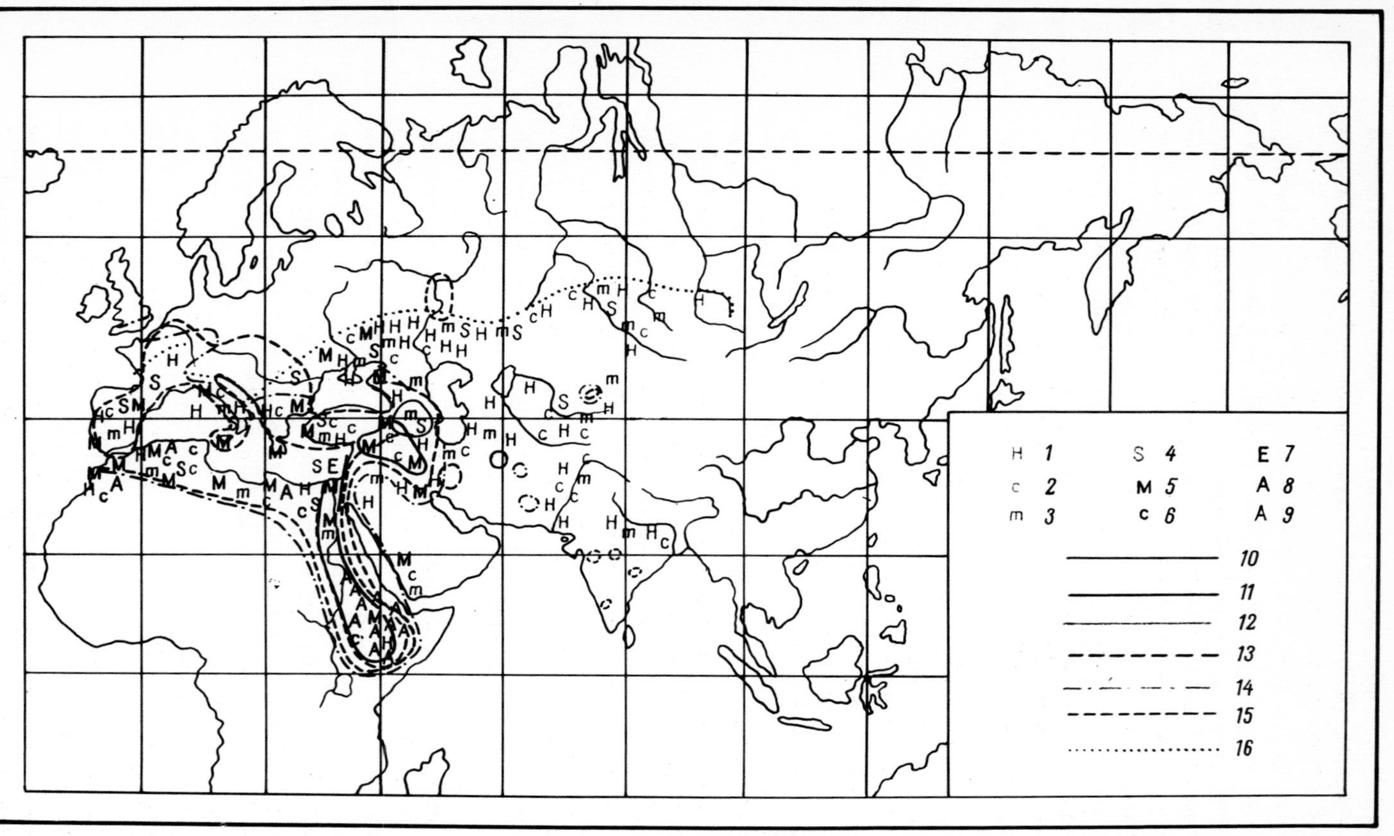

Fig. 2. Center of origin of the hard wheats. Geographical distribution of *Triticum durum* Desf. and the groups of species with $n = 14$ chromosomes. 1–3: *Communes*: *hordeiforme* Host (1), *coerulescens* Bayle (2), *melanopus* Al. (3); 4. *subcommunes*: e.g. *leucorum* Al., *leucomelan* Al., *affine* Koern.; 5. *mediterranea*: e.g. *africanum* Koern., *aegypticum* Koern., *italicum* Al., *apulicum* Koern.; 6. *circumflexae*; 7. *eligulatae* var. *aglossicon* Flaksb.; 8. *africanae*: *duro-compactum* incl. *T. pyramidale* Perciv.; 9. *abyssinicae caryopsididibus violaceis*; 10. *T. dicoccoides* Koern.; 11. *T. pyramidale* Perciv.; 12. *T. persicum* Vav.; 13. *T. polonicum* L.; 14. *T. turgidum* L.; 15. *T. dicoccum* Schrank; 16. northern limit of the distribution of the hard wheats.

berger, 1925). In addition, it occurs in Spain, in the mountain areas of France, in southern Germany (Würtemberg, Tübingen, Baden and Brandenburg), in Switzerland and northern Caucasus (in the Terskaya district), Morocco and Algeria (Miège, 1924). During prehistoric time, einkorn was distributed in the mountain areas of Asia Minor, the Balkans and the Alps as well as the Carpathians. Einkorn was known to Aristoteles as forage for pigs. Galen, Theophrastes and Dioscorides also mentioned it (Buschan, 1895).

In spite of special searches for it, einkorn has not been found either in a wild or in a cultivated form in southwestern or eastern Asia. The distribution area and the center of einkorn are without doubt placed outside the center of type-formation of the soft wheats and isolated from the area of the greatest variation of the hard wheats as well. Most likely, the region of Asia Minor and the areas adjacent to it appear to be the center of einkorn variation. At first, the greatest variation of einkorn was found in Crimea, where it had been subjected to a special investigation (by P. Larianov and E.I. Barylina). Subsequent research concerning the distribution area of einkorn has, however, shifted the center of variation to Asia Minor, where this species appears to be a common weed in the crops of the cereal grasses (P.M. Zhukovskiy).

Thus, the differential, taxonomic–geographical method clearly outlines three independent groups of wheat and furnishes a comparatively accurate picture of the concentration of the centers of type-formation of the different groups thereof. The distribution of the centers of variation of wheat species bears witness to the isolation of two groups of Linnaean species over a span of many thousands of years. There can be no doubt that the ancient inhabitants of northern Africa and southwestern Asia based and developed their agriculture on different kinds of wheat. The present absence of hard wheat in Afghanistan, and in many areas of India, China and Khorezma is indicative of this. The ancient agricultural civilizations of Mesopotamia and along the Indus were developed on the basis of both soft and hard wheats, while those of Egypt, Crete and around the Mediterranean were, as they are now, based on the group of hard wheats.

The ecological type of hard wheat is different from that of the soft ones. The strains of hard wheat are (as a unit) plants that require moisture, particularly during the early stage. In a mature state they tolerate drought well, in this respect not yielding to, but definitely surpassing the soft wheats. This difference is apparently linked to the native area of hard wheats. The maximum precipitation along the coasts of the Mediterranean occurs usually during the autumn and spring, the times of seeding and development of the early stages, whereas in Europe and Asia, the precipitation falls mainly during the winter months (Scharfetter). The vegetative rhythm of the species does, as demonstrated by Scharfetter, usually correspond to the climatic rhythm of the region where the plants originated.

The facts stated above indicated to the plant breeder where he should search for material suitable for plant breeding both for hard and for soft wheats. At the same time, these facts reveal that to unite hard and soft wheats on the basis of their origin, which is still done, while not taking into consideration the geographical peculiarities of the species, leads simply to confusion in respect to the history and origin of wheats.

The distribution area of emmer

The taxonomic–geographical investigation of emmer (*T. dicoccum*), the bread wheat of the ancient agricultural peoples, turned up a very interesting group of phenomena for the elucidation of the antiquity of wheat cultivation.

As is already known from archeological discoveries, emmer was cultivated in the past in considerable quantities in Egypt (Schulz, 1913). Grains of it have also been found from the Stone Age in Germany (at Worms) and from the Iron Age in Italy (at Aquilaia) and in Switzerland (at Auvernier) according to Buschan (1895).

The racial composition of emmer and the history of its cultivation have, during the past couple of years, been thoroughly studied at the Department of Applied Botany (Stoletova, 1925). The investigations have revealed that out of the ancient crops of emmer in the Old World only isolated sites remain. They are almost exclusively associated with backward or isolated populations. As indicated by statisticians in the U.S.S.R. and in other countries, this plant is becoming extinct and is being replaced by other crops.

Thus, in Russia, the area under emmer was reduced by almost 50% between 1888 and 1916 in spite of the fact that the areas of other field crops were increased by almost the same figure during the same period. A similar decrease took place in Serbia. Only the following isolated areas remain where emmer is grown in the Old World: in respect to the area covered, the Pri-Kama region takes first place, i.e. the areas along the rivers Volga and Kama, including the Tatar and the Bashkir republics, the autonomous territory of the Chuvashi (the upper Kazán, Simbirsk and the Ufa Governates) and a portion of the Votskaya Autonomous Territory as well as the Province of Perm. In 1916 up to 170 000 hectares of emmer were grown there. Important areas of emmer are found along Lower Volga (more than 20 000 ha) and in the Cis-Uralian region (more than 15 000 ha). Sites with emmer occur also in northern Caucasus and Transcaucasia but especially in Armenia, Georgia, and Azerbaidzhan (in Caucasus up to 5000 ha). Insignificant localities under emmer remain in Crimea (see Stoletova, 1925, for more details). In western Europe, emmer occurs mainly in the mountains of Switzerland, Germany (Baden, Schwaben, Würtemberg), Spain, France and Italy, and in the Balkans in the mountain areas of Serbia, Bulgaria, Chernogora and Croatia. In the Pyrenées it is still grown by the Basques. In addition to Asia Minor, emmer is found in small areas of Persia, according to our experience almost exclusively among Armenian settlers of the area. In India, emmer is occasionally grown in Bombay, Madras and Mysore as well as in the central provinces (A. & G. Howard, 1909). In Africa, it is grown mainly in the mountain areas of Abyssinia (Kostlan, 1913).

The absolutely certain remainders of the ancient cultivation of emmer are linked, in the Old World, to the ethnographical picture. In the U.S.S.R., emmer is cultivated by the Tatars, the Bashkirs, the Chuvashi and the Votyaki, in the

Caucasus by the Yaphetids [Indo-European peoples]: Armenians, Gortsians, Georgians, Pshabi, Tushini, Rachintsi, Khevsuri and Ossetini; in Germany it is grown by backward Schwabians and in Spain by the Basques. In Abyssinia, emmer is grown by tribal Masai and Amharans and along the Tigris and Euphrates by the Kurds, i.e. by people associated with remote antiquity in respect to their customs and civilizations. V.S. Sboyev (1856), the explorer of the Chuvashi, called them 'repositories of Antiquity'.

Studies of specimens, collected by us from all the areas where emmer is cultivated, revealed that in Europe and Asia only two botanical varieties are almost exclusively grown, i.e. *T. dicoccum* var. *farrum* and var. *rufum*. In addition, a black-spiked winter-emmer, *T. dicoccum* var. *atratum*, is occasionally grown in France. In other words, at first glance the varietal composition grown in Europe and Asia is very poor and monotonous. Only Abyssinia constitutes in this respect an exception, as demonstrated by recent research done by Harlan (1923). This revealed a major variation in the forms of emmer in that country.

Detailed, multifaceted research at the Institute of Applied Botany on the forms of emmer from various areas has shown that, although emmer is grown in different countries and belongs either to a variety of *farrum* or *rufum*, the racial composition of these is sharply differentiated. After detailed studies by means of crops grown it could be shown after some years that the different races within the limits of these varieties differed in some dozens of well-expressed morphological and physiological, hereditary characteristics. We shall present the characteristics of the major traits of the geographical races belonging to var. *farrum*.

Table 2 demonstrates, with a description of the geographical races, what profound differences can be found behind the 'exterior' of a single variety. At the same time as both the botanical varieties and the conventional taxonomic units differ in one or two characteristics (e.g. white, red, or black spikes), the races within the limits of a single variety can differ by dozens of traits. Simultaneously, the table as set up, illustrates facts concerning the sharp divergence of the races depending on 10 or more morphological and physiological traits. The presence of such a divergence of characteristics can only mean that the species of *T. dicoccum* was once represented by a major, polymorphic species as in the case of the other species of wheats and that, apparently, it was widely distributed in the Old World. The present existence of differentiating racial characteristics linked to different areas is evidence for the past diversity of emmer. The various geographical races remaining and the isolated areas, where major crops of the polymorphic species are grown, indicate the profound antiquity of emmer and of all the crops of wheat. Some races, remaining in isolated areas of emmer cultivation and differing in many characteristics, create an opportunity for a reconstruction of the entire system of alternating characteristics, making up the species of *T. dicoccum*.

Although it may be possible to establish the original center of variation of some other species of wheat, such as *T. vulgare*, *T. compactum* or *T. durum*, in the case of emmer the initial composition of the variation can be restored only by summing up the complex of all the characteristics of the different races found in the Old World.

Table 2. *Races characteristic of the variation within* Triticum dicoccum *var.* farrum

Characteristics	Races found in					
	Abyssinia and India	Persia	Transcaucasia	The Pri-Kama region	Germany	Spain
Color of shoots	Violetish, with anthocyanin	Violetish, with anthocyanin	Violetish or green	Violetish	Green	Green
Growth shape	Upright	Upright	Upright	Upright	Spreading	Spreading
Color of stem	Violetish	Violetish	Violetish	Violetish	Green	Green
Height of stem	Low	Medium	Tall	Tall	Tall	Tall
Culm diameter	Medium	Slender	Slender or medium	Slender	Slender	Slender
Culm filled or not filled with pith	Filled	Filled	Not filled	Not filled	Not filled	Not filled
Hairiness						
of nodes	Glabrous	Very hairy	Very hairy	Hairy or glabrous	Glabrous	Glabrous
of leaves	Smooth but with small prickles along veins	Very hairy, almost woolly	Very hairy	Very hairy	Very hairy	Very hairy
of secondary axis	Hairy along the segments	Hairy along the segments	Glabrous or hairy segments	Hairy at the base of the spike	Hairy along the segments on a significant portion of the axis and at the base of the spike	Similar to the German race
of sheaths	Glabrous	Hairy	Hairy	Hairy or glabrous	Glabrous	—

Consistency of leaves	Rigid, leathery	Soft, velvety	Soft, velvety	Soft, velvety	Soft, velvety	Soft, velvety
Color of leaves	Yellowish-green	Bluish	Bluish	Bluish	Bluish	Bluish
Color and hairiness of nodes	Colored, hairy	Colored, hairy	Colored, hairy	Colored, hairy	Not colored, glabrous	—
Type of awns	Coarse	Coarse	Average	Tender	Tender or coarse	Coarse
Diameter (in μm) of stomata	Large, 65.1	Average, 53.1	—	Average, 56.2	Small, 49.2	—
Life type	Spring-type	Spring-type	Spring-type	Spring-type	Spring-type	Winter-type
Vegetative period	Very short	Medium long	Medium or long	Medium	Long	—
Size and shape of teeth on glumes	Short, pointed	Short, pointed	Short or long, hooked	Short, blunt or conspicuous	Elongated (forms exist with hooked or with straight teeth)	Elongated, straight
Resistance to						
Erisyphe graminis	Strong	Weak	Strong	Medium	Strong	Strong
Puccinia glumarum	—	Weak	Strong	Weak	Strong	Strong
P. triticina	Strong	Weak	Strong	Weak	Strong	Strong
P. graminis	Strong	Weak	Strong	Weak	Strong	Strong

The crops of emmer associated with ethnic differentiation appear to be interesting examples of the geographical distribution of species about to become extinct and represent instructive examples of the expediency of the differential phyto-geographical method for studying cultivated plants.

Thus, the solution of the problem of the geographical origin of cultivated wheat can be drawn up. We shall select the two basic, independent centers of type-formation with respect to the two major groups of Linnaean species of wheat with $n = 14$ and 21 chromosomes, respectively, and belonging to two continents. The theory of a polyphyletic origin of wheat according to Solms-Laubach (1899) borders on the acknowledgement of a miracle and seems less than believable. It was confirmed only by a method that Solms-Laubach himself had outlined but of which he could not take full advantage due to the lack of data at that time, i.e. a comparison of the botanical composition between the species in northern Africa and those in Asia. Instead of confirming the assumption of Solms-Laubach about the unity of the wheat species of Abyssinia and those in eastern Asia, where he was inclined to place the native land of wheat, a comparison of the species, varieties and races of the wheats on the two continents confirms the opposite fact, i.e. that there is a sharp difference between the Asiatic and the African groups of wheat.

De Candolle cursorily launched a hypothesis that the soft wheats, in his opinion originally from Mesopotamia, changed into hard wheats. This reasoning is at present impossible to substantiate; all objective and irreproachable data force us to speak of independent type-formation within either center. We may later succeed in linking together both centers, to understand the origin in general of all the species of wheat and to tie together the genesis of all the species of wheat into one genus, *Eutriticum*, but this is a matter for future investigations. It is a problem of general importance, which affects all the questions concerning the origin of Linnaean species. There is no doubt that this problem can also be solved by painstaking studies of the entire composition of the varieties and races of all the wheats in the Old World by means of detailed geographical investigations, especially of the mountain areas in Africa and Asia and by means of experimental syntheses of the species after hybridization.

In any case, plant breeders may be satisfied with a temporary solution to the problem regarding the centers of type-formation of wheats since the detailed geographical data allow us to build practical plant breeding work from them.

In order to understand the origin of the Linnaean species of wheat it is necessary to include not only the species of wheat but also those of the genus *Aegilops* [the goat grasses] within the scope of the investigations. Some species of this genus hybridize easily with wheats, although this results mainly in sterile offspring.

The distribution areas of many species, belonging to *Aegilops*, overlap those of the wheats. As demonstrated by research made by G.M. Popova, species of *Aegilops*, although polymorphic, remind one in this respect of the species of wheat. Most of all, their differentiation into groups is similar to the division of the wheat species. Species such as *Ae. cylindrica, Ae. squarrosa* and *Ae. crassa* correspond in their response to fungal parasites and to 'hollow straw' disease to that of the soft wheats; in their resistance to these diseases as well as in the character of the pith, *Ae. triunciales* and *Ae. ovata* remind one of the hard wheats (Popova, 1923). Apparently, the former easily hybridize with the soft wheats as well. By selecting similar species of the genus *Aegilops* and resorting to complicated hybridization between them and wheat, an investigator can perhaps come close to a solution of the problem concerning the speciation of these closely related genera. Under the conditions in Turkestan, extensive hybridization between wheat and *Aegilops* is a common phenomenon. However, these hybrids are usually sterile and cannot give rise to fertile forms. The role of species belonging to *Aegilops* in the origin of the cultivated species of wheat is at present highly problematic and requires very complicated research.

The geographical centers of type-formation of barley

De Candolle looked for the native land of barley in the area where the distichous, hulled barley, *Hordeum spontaneum*, grows in its wild state.

At present we are able to add considerably to the information available to De Candolle about the geographical area of wild barley; it extends across northern Africa all the way from Morocco to Abyssinia and on into Asia Minor, Turkestan, Bokhara, Persia and northern Afghanistan as well as Transcaucasia (Vavilov, 1918).

Our studies of wild barley, *H. distichum* var. *spontaneum*, in southwestern Asia (Persia, Bokhara, Turkestan and Afghanistan) have revealed a number of new varieties and races. In addition to the usual yellow-spiked form, we know now of a black-spiked variety of this wild barley, i.e. *H. distichum* var. *transcaspicum*. In northern Afghanistan, there is a new variety with brown glumes, a genetically dominant characteristic. In the Transcaucasian area (Turkmenistan) around Anau, our expedition (under D.D. Bukinich) found a race of distichous barley with a comparatively rigid rachis. In addition to the ordinary races of winter barley, we also found a spring barley.

Nevertheless, studies comparing wild and cultivated barleys in the Old World have demonstrated that, in general, the former is represented by a very limited group of forms, composed of a series of distichous, hulled barleys, but in no way corresponding to the rich diversity and polymorphism found among cultivated barleys. The large number of cultivated varieties and races described during the last couple of years is characterized by a multitude of varietal and racial traits, both recessive and dominant when crossed, but not typical of the wild barleys so far described. For instance, such characteristics as compact spikes, multistichous spikes, underdeveloped lateral spikelets (*deficientes*), the formation of three-lobed appendages instead of awns (*trifurcatum*), short awns, no awns, smooth awns, naked grains, wide leaves (as in the case of many races of var. *coeleste*) and broad glumes have been noted.

Plant breeders have looked in vain for forms of wild barley suitable for practical purposes within the major portion of its distribution area where it occurs as a weed on loess soils in northern Afghanistan, Transcaucasia and Bokhara. While ecologically clearly related to the cultivated races of distichous, hulled barley, wild barley (*H. spontaneum*) displays only a fraction of the hereditary variation typical of *H. sativum (s. lat.)* in which plant breeders are interested.

A group of barley, sharply isolated from the wild one in southwestern Asia, is represented by the naked-grained barleys of southeastern Asia and the elevated mountain areas of Central Asia. These have wide leaves and thin-walled stems of an anatomically peculiar structure and coarse, rough awns (Vavilov, 1921). They are also characterized by a comparatively satisfactory resistance to mildew (*Erysiphe graminis*). The wild barley, i.e. *H. spontaneum*, is usually not common in the mountain areas of Afghanistan, Bokhara and Turkestan above 1800–2000 m.s.m. Naked-grained barley, on the other hand, is typical at the limits of cultivation. As a rule, it is grown in southwestern Asia above 2500 m. In the alpine areas of Tibet, Ladakh and Hindukush, it occurs up to 4000 m elevation.

Perhaps future studies of the genetics will help solve the problem concerning the wild ancestors of the cultivated barleys. At present it is obvious that the existence in nature of a large amount of distichous, hulled barley, *H. spontaneum*, in southwestern Asia does not solve the problem concerning the origin of cultivated barleys or where the source of the wealth of strains of cultivated barleys can be found.

Investigations at the Department of Applied Botany during the last couple of years by means of special expeditions to Mongolia, Afghanistan, Persia, Turkestan, Bokhara and Asia Minor, and studies of north-African material sent to us from the Department of Agriculture in the U.S.A. (through Dr. Harlan) as well as a comparison of materials from various countries obtained from experimental stations in India, Burma, China, Algeria, Tunisia, Egypt and other countries, have helped us approach the problem of the geographical centers of varietal diversity of cultivated barleys. Our access to this material of strains revealed, thanks to detailed taxonomic–geographical studies that there are *two* basic centers of variation as far as cultivated barleys are concerned.

One center of type-formation appears to be related to northeastern Africa, especially the mountain areas of Abyssinia, where there is a particular wealth of forms of hulled barley. A whole series of peculiar varietal and racial characteristics are found exclusively among the barleys cultivated in Abyssinia. Varieties with broad glumes, distichous barley with entirely under-developed lateral spikelets, forms of barley with pubescent paleas (the pubescence is visible even in the field) or with brightly anthocyanin colored stems are known only from Abyssinia.

The following botanical varieties appear to be endemic in Abyssinia and northeastern Africa:

1 var. *gracilius* Koern.,
2 var. *schimperianum* Koern.,
3 var. *latiglumatum* Koern.,
4 var. *atrispicatum* Koern.,
5 var. *eurylepis* Koern.,
6 var. *platylepis* Koern.,
7 var. *contractum* Koern.,

8 var. *heterolepis* Koern.,
9 var. *glabro-heterolepis* Vav. (with a compact spike of type *erectum* and with smooth awns),
10 var. *steudelii* Koern., (rarely met with also in Arabia)
11 var. *atterbergianum* R. Reg. (syn. *subglabrum* Beaven.)
12 var. *africanum* Vav. (belonging to the group of *deficientes* f. *erectum* with yellowish spikes)
13 var. *copticum* Vav. (distinct from the preceding varieties by having black spikelets)
14 var. *abyssinium* Sér.,
15 var. *macrolepis* R. Reg.,
16 var. *leiomacrolepis* R. Reg.,
17 var. *melanocrithum* Koern., and
18 var. *deficiens* Steud. (frequent also in Arabia and India).

Algeria, Tunisia, Morocco, Tripolitania, Arabia and Mesopotamia yield to Abyssinia as far as the diversity of the barleys is concerned and come closer to the composition of the varieties in the European countries. Persia, Turkestan, Bokhara, Khorezm and India are characterized by a small number of common varieties. As far as our investigations of the composition of the strains are concerned, and in spite of the abundance of wild barleys and the diversity of conditions for cultivation, only four to five varieties of barley are grown in Afghanistan, while in Abyssinia all the European varieties such as vars. *pallidum*, *nigrum*, *pyramidatum*, *nutans*, *erectum*, etc., are met with. As far as can be judged from the material available to us, Abyssinia appears to be the single most important one of the major centers where the forms of barley developed.

As shown by the comparative–geographical study of the strains in the Old World, southeastern Asia, China, Japan and the area adjacent to Tibet (and, perhaps, also Nepal) appear to represent a second center of varietal and racial diversity with respect to the barleys. There, varieties with naked grains, short awns, no awns or varieties where the awns are replaced by three-lobed appendages (different kinds of *trifurcatum*) occur. The racial composition of this group of varieties is always variable and includes both extremely late-maturing and extremely early-maturing races. Judging from all the material not yet published and the collections available to us, eastern Asia appears, in general, to have a concentration of naked-grained, furcate, no-awned or short-awned forms. On the other hand, Abyssinia appears to be the main center of the awned, hulled forms.

The following varieties turned out to be endemic in and known only from southeastern Asia:

1 var. *brachyatherum* Koern.,
2 var. *tonsum* Koern.,
3 var. *nigritonsum* Koern.,
4 var. *japonicum* Vav. (hexastichous, hulled; the palea of the central spikelet carries a short awn, the lateral spikelets are almost completely without awns, the spike is dense and yellowish)
5 var. *revelatum* Koern.,
6 var. *nudipyramidatum* Koern.,

7 var. *asiaticum* Vav. (hexastichous, with an open spike and naked grains; the paleas carry short awns; it is cultivated in Japan and India),
8 var. *brevisetum* R. Reg., and
9 var. *trifurcatum* Schlecht.

The ecological type of the Asiatic group of naked-grained barleys differs from that of the previous group. In Asia the races of barley are grown mainly on irrigated soils. In the elevated mountain areas of the southwestern area naked-grained barley is, as a rule, also grown on irrigated soils but, at the same elevation, hulled barley often succeeds on non-irrigated soils.

The New World adopted the cultivation of barley from Europe and is of no importance for the solution of the problem concerning the origin of this crop. The varietal composition of barley in Canada, the U.S.A. and Argentina reminds us, in general, of that in Europe.

Thus, the application of the differential, taxonomic–geographical method proved the presence of two basic centers where forms of cultivated barley developed (Fig. 3).

The existence of two independent centers could also be confirmed by data from experimental hybridization between the Asiatic and the African forms. Between 1915 and 1918 we made a large number of crosses between different varieties of naked-grained and hulled barleys. Studies of the F_2, F_3 and F_4 generations revealed a definite presence in such combinations of a noticeably expressed interrupted spike. Although all the races and varieties of the wild and cultivated barleys are easy to cross and not as clearly isolated from each other as hard and soft wheats, some differentiation between the east Asiatic and the east African groups is indicated by a display of partial sterility in the form of an under-development of the reproductive organs of many plants or in the form of a so-called interrupted spike. In the case of the hybrids between the races within a single group, the interrupted spike is more or less completely absent or only weakly developed.

The striking fact of an isolation between the east Asiatic and the African group was revealed to us in 1925 while we were investigating the second generation hybrids between *H. vulgare* var. *pallidum* f. *coerulescens* (no. 3316 from Abyssinia) and *H. vulgare* var. *coeleste* (no. 3426, a low-growing type from Japan). The first generation in 1924 was normal, hulled, tall-growing and reminded us of the Abyssinian type. The first generation was also similar to the wild barley, *H. spontaneum*, because of the brittleness of the rachis. In the second generation, out of 301 seeds that germinated, 12 plants turned out to be completely sterile; half of the remaining seedlings were generally weakly developed in spite of absolutely normal growing conditions in the greenhouse, while the other half developed normally with thick spikes, which were, however, nevertheless completely sterile. These plants stood with open glumes for a long time (several days) during anthesis, reminding one of the common type of behavior of sterile hybrids. On 46 plants, grains developed in some spikes only; a part of the spikelets of these plants were completely sterile; 197 plants displayed more or less of interrupted spikes, while 46 had normally fertile spikelets (K.V. Ivanova calculated the figures).

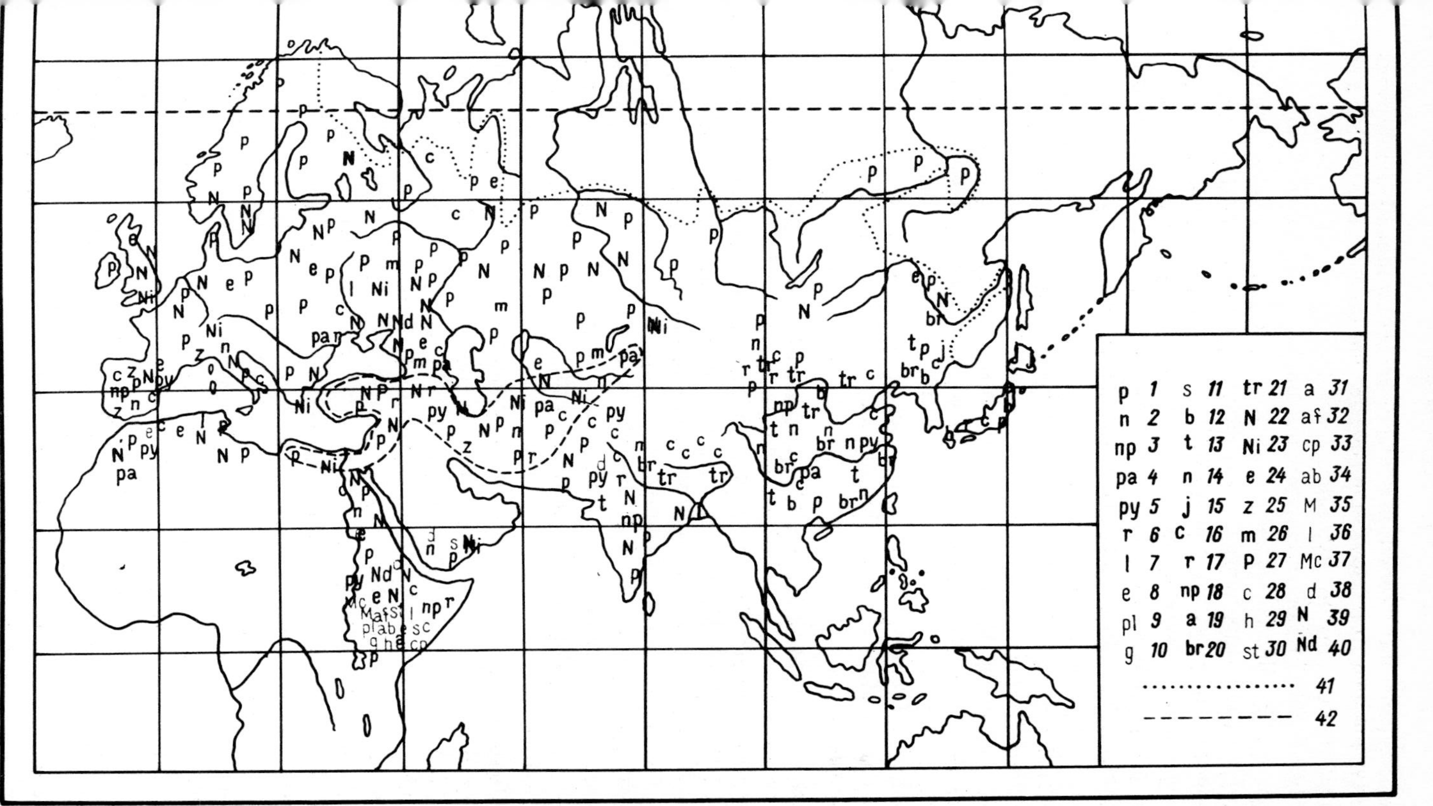

Fig. 3. Geographical distribution of the varieties of cultivated barley and *Hordeum spontaneum* C. Koch. 1. *Hordeum pallidum* Sér; 2. *nigrum* Willd.; 3. *nigropallidum* R. Reg.; 4. *parallelum* Koern.; 5. *pyramidatum* Koern.; *ricotense* R. Reg.; 7. *leirrhynchum* Koern.; 8. *eurylepis* Koern.; 9. *platylepis* Koern.; 10. *gracilis* Koern.; 11. *schimperianum* Koern.; 12. *brachyatherum* Koern.; 13. *tonsum* Koern.; 14. *nigritonsum* Koern.; 15. *japonicum* Koern.; 16. *coeleste* L.; 17. *revelatum* Koern.; 18. *nudipyramidatum* Koern.; 19. *asiaticum* Vav.; 20. *brevisetum* R. Reg.; 21. *trifurcatum* Schlecht.; 22. *nutans* Schuebl.; 23. *nigricans* Sér.; 24. *erectum* Schuebl.; 25. *zeocrithum* L.; 26. *medicum* Koern.; 27. *persicum* Koern.; 28. *contractum* Koern.; 29. *heterolepis* Koern.; 30. *steudelii* Koern.; 31. *atterbergianum* R. Reg.; 32. *africanum* Vav.; 33. *copticum* Vav.; 34. *abyssinicum* Sér.; 35. *macrolepis* R. Reg.; 36. *leiomacrolepis* R. Reg.; 37. *melanocrithum* Koern.; 38. *deficiens* Steud.; 39. *nudum* L.; 40. *nudideficiens* Koern.; 41. northern limit of cultivated barley; 42. distribution area of *H. spontaneum* C. Koch.

The facts stated above concerning the appearance of completely sterile plants following hybridization between representatives of barley from Abyssinia and Japan clearly show that the two basic geographical groups of barley are quite well isolated from each other. In spite of the common opinion that *H. sativum* is a single species, different characteristics of it are, as shown above, entirely isolated due to the appearance of disharmony during the development of the reproductive organs after hybridization. It could be that in a remote past, the different groups of cultivated barley originated from a single center, although we can only guess about that. The present isolation of the groups is so obvious to the plant breeder that he is able to draw corresponding practical conclusions from it.

From the simple hypothesis of De Candolle, we have now arrived at some very complicated ideas. The problem concerning the origin and the evolution of cultivated barley has, over the past 40 years, turned out to be much more involved and intricate. The distribution area of *H. spontaneum* is shown in Fig. 3. Just as could be expected, it does not coincide with the two centers, where cultivated barleys developed, but corresponds more closely to that of the African group of the cultivated, hulled barleys.

The geographical centers of type-formation of cultivated oats

An even more complicated picture was revealed with respect to the origin of oats. In his agronomical monograph on oats, published in 1918, Zade writes: 'Es gibt wohl keine zweite Getreideart oder sonstige Graminee, deren Verwandschafts und Abstammungsverhältnisse wir so klar zu durchschauen vergmögen wie die des Hafers'. [There is hardly any other kind of cereal or similar grass, the relationship and evolutionary condition of which we can understand as easily as that of oats.] Later arguments have revealed the falseness of this opinion.

De Candolle ascribed a European origin to the oats while mainly basing his opinion on conclusions drawn from historical and philological facts. The cultivation of oats is, apparently, not very old. Neither the early Egyptians nor the early Europeans grew oats. Old data in the literature indicate that, as a European crop, oats were a speciality of the northern countries – ancient Germany, Scotland and Norway (Buschan, 1895). European oats are represented by a large group of quite different races, clearly differing in the color of the glumes, the shape of the grains and leaves, vegetative characteristics and resistance to parasites. A number of forms of cultivated oats are without question at present typical only of northern and central Europe.

In addition to the European group of *Avena sativa*, studies over the last decade (Trabut, 1913) have revealed the presence of a large Mediterranean group of oats, comparatively sharply isolated from *A. sativa*, i.e. *A. sterilis* and *A. byzantina*. Northern Africa turned out to be the center of a large group of cultivated oats, in general characterized by resistance to the European types of smut, *Ustilago avenae*, and rust, *Puccinia coronifera*, and physiologically isolated from the European oats by a considerable degree of sterility after hybridizing with the latter. Apparently a number of wild oats, belonging to the subgenus *EuAvena*, can also be included in the group and are met with in northern Africa.

In northwestern Europe it is also necessary to distinguish another independent group of oat species, in addition to *A. sativa*, i.e. *A. strigosa* and *A. brevis*, half weedy but in part also cultivated and characterized by a low number of chromosomes, $2n = 14$ instead of $2n = 42$ (or close to that), typical of the former group, i.e. *A. sativa* and *A. byzantina* (Nikolayeva, 1920). The major part of the forms of *A. strigosa* and *A. brevis* are, according to our experiments, and as confirmed in America by Reed (1920), characterized by resistance to smut, *Ustilago avenae*, and rust, *Puccinia coronifera* as well as mildew (Vavilov, 1919). The physiological isolation of this group is the reason why it is impossible to hybridize *A. strigosa* and *A. brevis* with forms of *A. sativa* and *A. byzantina* (Zhegalov, 1920).

Marquand (1921, 1922) described a considerable number of varieties of *A. strigosa* from Wales. He distinguished var. *alba* f. *fusca* (ssp. *pilosa*), var. *albida* f. *cambrica* (ssp. *glabrescens*), and vars. *flava, intermedia* and *nigra* (ssp. *orcadensis*); some of them are apparently at present endemic in Great Britain. We have established a number of new forms for the European parts of the U.S.S.R., France and Germany. During our experiments with infection of *A. strigosa* with smut, rust and mildew, it could be demonstrated that there were two clearly different strains within that species: one very susceptible to all the parasites enumerated and another very resistant to them. *A. strigosa* appears to be distributed as a weed in cereal crops in Byelorussia and Estonia. It is found as an important crop on sandy soils in England and France. *A. strigosa* and *A. brevis* are not known as a cultivated species nor as weeds in the southern areas of European U.S.S.R., nor in the Asiatic parts of it (Turkestan, Bokhara, Khorezm, Persia or Afghanistan). The geographical distribution area of *A. strigosa* and *A. brevis*, both when in a cultivated state or as weeds among cereal crops, seems to be related to western Europe only. The center of their variation is restricted to the northwestern and western parts of Europe.

There is no doubt that the question about the origin of *A. brevis* and *A. strigosa* must be singled out just as the problem concerning the origin of einkorn (*T. monococcum*) was in the case of wheats. The presence among oats of two physiological groups, either resistant or susceptible to smut and other diseases, indicates that the genesis of this group is not straightforward.

The widely held opinion that the ordinary cultivated oats, united under the epithet of *A. sativa*, should be a definitely north European crop only is not very convincing to us. Studies in Mongolia and northern China have revealed a large number of varieties and races of *A. sativa*, typical of those countries. There, all kinds of colors are found on the glumes from white to black as well as peculiar ash-colored races (var. *grisea*) and some with a heavy coating of wax on the hulls. Oats are grown in many parts of China and Tibet but have so far not been investigated botanically there.

Transcaucasia (Georgia and Armenia) and parts of Persia are characterized by the presence of a very large number of weedy races of hulled oats belonging to *A. sativa* among emmer and barley crops. There, a multitude of peculiar, endemic forms of *A. sativa* appear, in addition to *A. fatua* and *A. ludoviciana*, which are widely distributed all over southwestern Asia (Turkestan, Bokhara, Persia and Afghanistan). We have also found peculiar races of *A. sativa* in the fields of the Pri-Kama area in crops that no doubt originated from more southerly areas.

Most of all, the large- and naked-grained oats, genetically related to the oats cultivated in Europe and characterized by one of the chromosome numbers ($2n=42$), easily hybridizing with each other and reacting similarly to parasitic fungi, are undoubtedly typical, especially of China. The center of its variation must be there. Europe learned about naked-grained oats from China. The very name of one of the widespread varieties, *A. nuda* var. *chinensis*, demonstrates the Chinese origin of the large-grained, naked-grained oats. It is also known from data in the literature that naked-grained oats were brought from China during the fifth century A.D. (Breitschneider, 1881). An expedition from the Department of Applied Botany (under V.E. Pisarev) in 1922 discovered new varieties of naked-grained oats (e.g. *A. nuda* var. *mongolica*) in some settlements in Mongolia. These have dark spots on the exterior paleas. Naked-grained oats are still widespread in eastern Asian crops. The presence in eastern and southeastern Asia of both hulled and naked-grained oats from which, later, endemics were no doubt developed, speaks in favor of the presence there of a basic center of variation and of the possibility for a transfer of the cultivation of oats in part also from Asia.

The small-grained forms of naked-grained oats, found as weeds and very rare in the crops of northern Europe, must necessarily be sharply distinguished from the Chinese naked-grained oats. As shown by our experiments, these forms are very easy to hybridize with *A. brevis* and *A. strigosa*, but are different in their resistance to the parasitic fungi (Vavilov, 1918).

These forms do not hybridize with the common oats or the large- and naked-grained forms. Morphologically they are close to *A. brevis* and *A. strigosa* with respect to the vegetative characteristics as could be assumed on the basis of the data available. Cytologically (according to A.G. Nikolayeva), they also turned out to belong to the group of *A. brevis* and *A. strigosa* ($2n=14$). All the authors, starting with Koernicke, erroneously united them (because of the naked grains and the many florets) with the large- and naked-grained ones into one basic species, *A. nuda*. We consider it more correct to distinguish the naked- and small-grained oats as a basic species, *A. nudibrevis*. This species forms a special group together with *A. strigosa* and *A. brevis*, the genesis of which developed independently of that of the *A. sativa*. In both groups we can find parallel, homologous series of forms as is common during the evolution of plants. *A. nuda*, the large- and naked-grained oats, corresponds to a similar series of *A. sativa*, while the small- and naked-grained oats, *A. nudibrevis*, correspond to the group of *A. brevis* and *A. strigosa*.

The genesis of cultivated oats is no doubt linked to that of weedy as well as wild species, i.e. *A. fatua*, *A. ludoviciana*, *A. sterilis* and *A. barbata*. The common wild oats, *A. fatua* and *A. ludoviciana*, are associated with the crops of *A. sativa* just like *Hordeum spontaneum* is associated with crops of *H. sativum*. The similarity of their karyograms (chromosome numbers), the reaction to parasitic fungi, the general morphological similarity and the possibility of obtaining fertile hybrids between them all indicates this. All the species of wild oats are extremely polymorphic and they can, in their turn, be distinguished into physiological groups as well. *A. fatua*, *A. ludoviciana* and *A. sterilis* are characterized by 42 chromosomes (or close to that) while a number of races of *A. barbata*, studied by A.G. Nikolayeva and Kihara, occupy a position intermediate between *A. nuda* on the one hand and *A. strigosa* and *A. fatua* on the other ($2n = 32$ according to Nikolayeva, 28 according to Kihara); according to Nikolayeva (1922) *A. clauda* and *A. pilosa* from the Yelizavetopol Governate have 14 chromosomes.

As shown by our experiments, *A. sterilis* and *A. barbata* can, in the same way as *A. strigosa*, be distinguished into two physiological groups. One group is at the same time strongly susceptible to smut, *Ustilago avenae*, rust, *Puccinia coronifera*, and mildew, *Erysiphe graminis*, while the second group is immune to all these parasites. Strongly affected forms of *A. barbata* are typical of Persia. Among them, races are found where the effect of the smut is localized on the anthers only, but in the majority of these forms it affects all the organs of the flower. *A. clauda*, *A. pilosa* and *A. wiestii* proved to be susceptible to smut and rust only. In our experiments with infections we have tested races of *A. wiestii* and *A. pilosa* from the Yelizavetopolsk District and *A. clauda* from the same district as well as from Transcaucasia.

The distribution area of *A. fatua* covers an enormous area from the northern districts of the European parts of the U.S.S.R. all the way to Hindukush. It infests cereal crops in the mountain areas of Afghanistan, Transcaucasia, Ukraine and the Vyatskaya Autonomous Territory. A somewhat less wide but still very large area is typical of the races of *A. ludoviciana*; it is especially widespread in southwestern Asia, Transcaucasia and the southern areas of the European U.S.S.R. The areas of *A. fatua* and *A. ludoviciana* overlap in general with the main areas of the racial and varietal diversity of *A. sativa*. There is no doubt that together with the large-grained *A. nuda* these four species form one single group. It is very hard to tell where the geographical center of their type-formation is found. In any case, it would be erroneous to state that the cultivated oats are definitely linked to Europe only. The existence in China of endemic groups of *A. sativa* and *A. nuda*, the wide distribution of *A. fatua* and *A. ludoviciana* in a wild as well as in a weedy state in Turkestan, Bokhara and Afghanistan, Persia, Armenia and Transcaucasia and the presence there of many peculiar groups of weedy and wild oats all point to the participation of Asia in the development of the forms of *A. fatua*, *A. ludoviciana*, *A. sativa* and *A. nuda*.

In Abyssinia, the cultivated and weedy oats are represented by a peculiar group (*A. abyssinica*, *A. schimperi*, *A. hildebrandtii* and *A. braunii*), so far not very well studied. Specimens furnished us by Harlan proved according to the investigations made by E.K. Emme, to have $2n = 28$ chromosomes.

The African group of cultivated, wild and weedy oats, the investigation of which was initiated by Trabut (1913), needs to be subjected to a special study but at present there is nothing fundamental in favor of uniting them with the European or Asiatic species.

It seems impossible to reduce the origin of the cultivated oats to a single geographical center. The cultivated oats have no doubt a polyphyletic origin.

The different species were very likely taken into cultivation independently. We can definitely speak of five geographical and genetical groups of cultivated oats: 1. *A. sativa* (including *A. orientalis*), 2. *A. nuda*, close to the preceding one, 3. *A. strigosa* (including *A. brevis* and *A. nudibrevis*), 4. *A. byzantina* and 5. *A. abyssinica*.

The geographical development of forms of cultivated oats (see Fig. 4) can, thus, be distinguished into the following types:

I For the group of *A. byzantina–A. sterilis*, the center covers the Mediterranean coast but is mainly in northern Africa;

II for the group of *A. abyssinica* and the closely related varieties, the center is found in Abyssinia;

III for the group of *A. strigosa*, *A. brevis* and *A. nudibrevis*, a center in northwestern and western Europe is typical;

IV no center of type-formation with respect to *A. sativa* and *A. orientalis* can at present be established; the old idea of a total relationship between the *sativae* and northern Europe is not supported by facts; oriental forms of *A. sativa* have been found both in China and Transcaucasus;

V the center of *A. nuda*, the large-grained oats, is apparently – in spite of the fact that this group is genetically close to *A. sativa* – definitely located, with respect to the development of forms, exclusively in China and the countries adjacent to it in the south.

When searching for material of strains useful for plant breeding work it is necessary to turn to these different areas.

The problems of the geographical center of the group most essential for cultivation and most polymorphic, i.e. the one assembled under the Linnaean species of *A. sativa*, is especially confused. In the future we may find an explanation of this geographical riddle in connection with that of cultivated emmer.

The geographical center of origin of millets

Among the species of millet, widely distributed within the Old World, we shall concentrate on the common paniculate millet (*Panicum miliaceum*) and the Italian millet (*P. italicum*), or proso and foxtail millets, respectively.

Although he was unable to obtain information from botanists about the localities of ordinary millet in the wild condition, De Candolle assumed on the basis of historical and linguistic data that millet originated in Arabia and Egypt. Millet was found by Heer in prehistoric layers of pile-dwellings in Switzerland. Some philologists have suggested that the five plants, sown at the beginning of the year 3000 B.C. by the Chinese emperors during special ceremonies under the name of 'shu' were actually millet, but one has not yet succeeded in determining the exact meaning of that word. Breitschneider (1870) believed that it could possibly be sorghum. While taking note of the increased sensitivity of millet to low temperatures, Koernicke (1885) suggested that the native land of millet ought to be found in warm areas and thought it likely that such an area could be found in East India or at its northern borders. Buschan (1895)

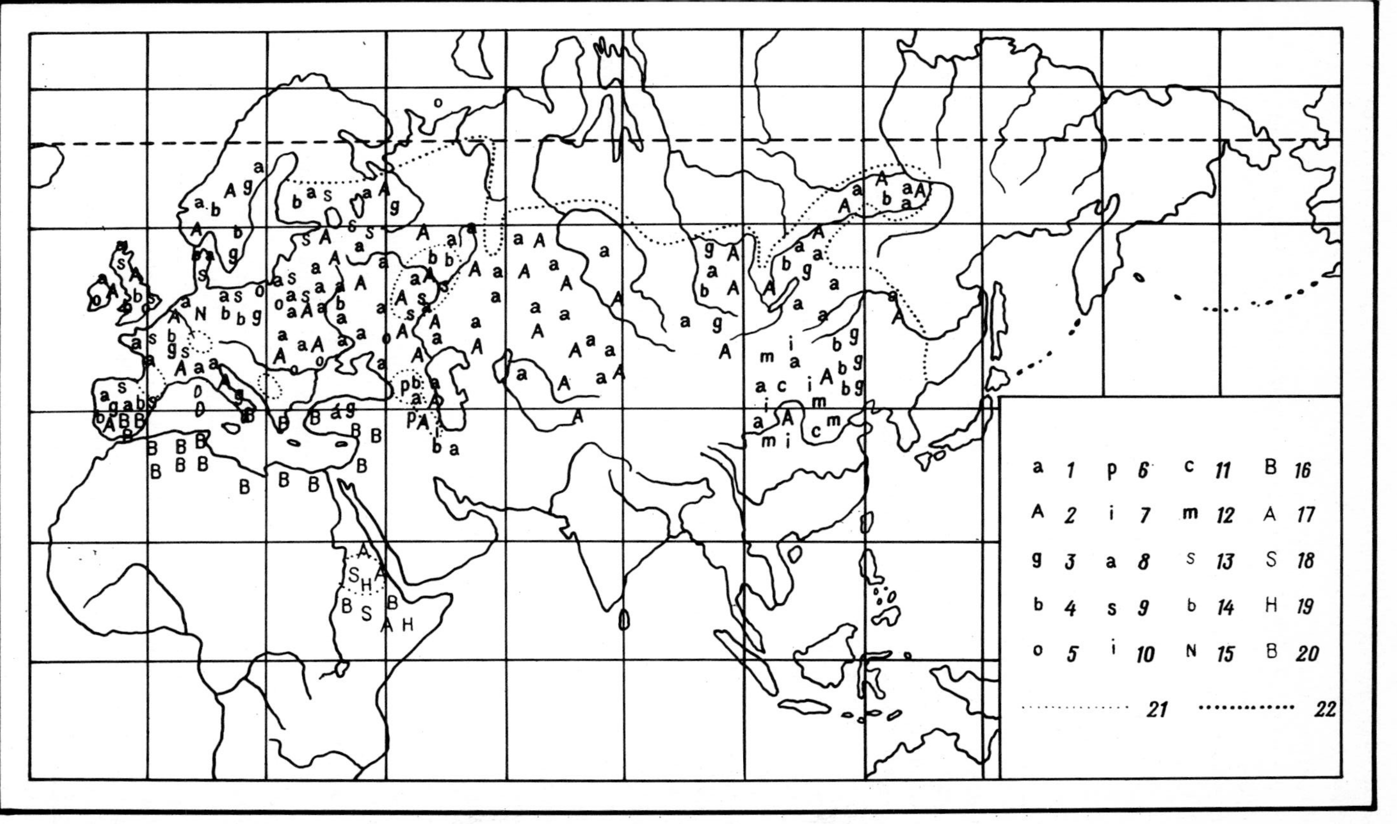

Fig. 4. Geographical distribution of the varieties of cultivated oats. 1. *Avena diffusa* Asch. & Graebn. gr. *alba* Vav.; 2. *A. diffusa* gr. *aurea* Asch. & Graebn.; 3. *A. diffusa* gr. *grisea* Asch. & Graebn.; 4. *A. diffusa* gr. *brunnea* Vav.; 5. *A. orientalis* Schreb.; 6. *A. diffusa* var. *persica* Vav.; 7. *A. diffusa* var. *iranica* Vav.; 8. *A. diffusa* var. *asiatica* Vav.; 9. *A. diffusa* gr. *speltiformis* Vav.; 10. *A. nuda* L. var. *inermis* Koern.; 11. *A. nuda* var. *chinensis* Tisch.; 12. *A. nuda* var. *mongolica* Pissar.; 13. *A. strigosa* Schreib.; 14. *A. brevis* Roth.; 15. *A. nudibrevis* Vav.; 16. *A. byzantina* Koch gr. *aurea* Vav., gr. *nigra* Vav., gr. *alba* Vav., gr. *miltura* Vav.; 17. *A. abyssinica* Hochst.; 18. *A. schimperi* Koern.; 19. *A. hildebrandtii* Koern.; 20. *A. braunii* Koern.; 21. area of cultivated emmer [*T. dicoccum*]; 22. northern limit of cultivated oats.

considered the problem of the native area of millet still open: 'Millet is usually referred to East-India', he writes, 'but there is no proof favoring this assumption'.

Let us now turn to a taxonomic–geographic analysis of the various kinds of millet. At present, the botanists have described more than 60 botanical varieties of *P. miliaceum* (Arnold, 1925). The density of the inflorescens, the color of the paleas and glumes and the roughness of the paleas were used as basis for the division into varieties.

Studies of the racial composition of millet at the Department of Applied Botany together with I.V. Popov revealed even more taxonomic characteristics such as the dimensions of the grains, the pubescence of the leaves, the length of the vegetation period, the height of the plants, and so on.

A study of the complex of millet in various regions of the European and Asiatic parts of the U.S.S.R., Bokhara, Khorezm, Persia, Afghanistan, Mongolia and China demonstrated clearly that the variety of forms increased toward the east in Asia. The maximum diversity was established in eastern and central Asia, where the entire racial and varietal complex of the Linnaean species of *P. miliaceum* is found. So far China has hardly been studied at all but, nevertheless, Mongolia and northeastern China – the areas of the most advanced Chinese civilization – display an exceptional diversity of morphological and biological characteristics.

In the districts of the European parts of the U.S.S.R., an increasing diversity of millet crops can be observed toward the east and southeast. At the same time as in the Voronezh district 22 varieties were found after detailed investigations (by I.V. Yakushina), there were 40 in the Saratov District (Arnold, 1925). In the areas of Bokhara, Chinese Turkestan and the countries adjacent to them, endemic forms not known in Europe are found. There, large quantities of membranaceous and naked-grained millets are grown. In Mongolia, extremely late-maturing kinds of loosely spreading millet, such as never seen in Europe, are found and races with glabrous leaves and large grains have also turned up.

In some parts of Mongolia, races of millet with a characteristic type of rachis, on which the glumes sit, are found. It disarticulates when maturing. This type of formation reminds of the elaiosomes of wild oats. A.K. Holbach established a whole scale of brittleness based on the Asiatic strains. Some cultivated Asiatic races are characterized by the florets dropping off when the grains mature, littering the soil, behaving like wild plants and multiplying as self-sown crops. Some millets actually become weedy.

Apparently, millet dispersed from Asia to Europe together with nomadic people. Among the crops grown by these nomads in Asia, millet appears to be the favorite cereal. It can be sown very late or at different times, and does not tie the nomads down to one place. To seed a 'desyatina' [2.7 acres] only a very small quantity of seeds is required. Millet is extremely easy to transport, universal with respect to its utilization, usually drought resistant and succeeds even on soils where other plants (except for watermelons) do not thrive. Therefore, it is still indispensable to the nomadic economy in the semi-desert areas of Asia and the southeastern European parts of the U.S.S.R.

The mountain regions of Mongolia have an especially large variation of forms with respect to the color of the seeds, the shape of the panicle and the length of the vegetation period.

Studies of the complex of millet strains and their geographical distribution are still in progress, but already now it is possible to definitely outline the general traits of the centers where the forms of this crop developed. The center is not located in Egypt, nor in Arabia, as suggested by De Candolle, but in eastern and central Asia. According to critical studies by Buschan (1985), crops of millet were not found in Egypt in prehistoric times. Persia and Turkey (Asia Minor) cultivated only small quantities of millet in the form of a comparatively poor variety of strains. The variation in Afghanistan was, according to our experiences, also rather small. There, as in Turkey and Persia, it is mainly forms with light-colored grains that predominate. Everything indicates that the diversity increases toward Mongolia, China and eastern Asia. East-Siberian black-colored, brown and dark gray millets come, as shown by the research done by V.E. Pisarev, from the mountain areas of Mongolia. In eastern Europe, the southeastern parts of the European U.S.S.R. and Romania, millet was found also during pre-historic times, as demonstrated by the discoveries by archeologists (Coucouteni) in Romania of millet from the Stone Age. There is also definite evidence of its cultivation there during the fourth century B.C. (Buschan, 1895). Already, in his *Historia Naturalis*, Plinius stated that millet was an important crop among the Slavic tribes (the Sarmatians). Columella (in his *De re rustica*, vol. II) mentioned the preparation of bread baked with millet flour and consumed while still hot.

Italian (or foxtail) millet, *P. italicum* [now = *Setaria italica*], has a very unfortunate epithet since this millet has little in common with Italy. It is grown in large quantities in the mountain areas of southwestern Asia, in Afghanistan, Bokhara and Uzbekistan. Italian millet (also known by several local common Russian names such as 'chumiza', 'gomì' and 'mogare') is often grown in Asia mixed with panicled millet (*P. miliaceum*). In many areas of Bokhara, Turkestan, Afghanistan and Kazakhstan (formerly the Turgay district), the crops of millet grown by the nomads and the settlers consist of both species, *P. miliaceum* and *P. italicum*.

Biologically, both these species are extraordinarily close and it is difficult to separate the species by the seeds. Although these species are sharply delimited genetically as well as physiologically (e.g. in their response to parasitic fungi), the centers where both seem to have formed apparently overlap.

Eastern Asia, including China and Japan, where millet is grown as a cereal crop, appears to be the main center of the Italian millet, *P. italicum*. In Europe, it is grown mainly as a forage crop. Apparently this millet is of alien origin in Transcaucasia, imported there from eastern Asia.

The dispersal of the strains of millet (both *P. miliaceum* and *P. italicum*) into Europe is, as we have succeeded in clearing up, definitely quite regular. Toward the north, the forms cultivated are mainly those branching into loose panicles and having small seeds; toward the south the large-grained and late-maturing forms are increasingly common; in between there are intermediate forms with dense inflorescences, which are drooping or nearly so. Their dispersal is apparently linked to the length of the vegetation period, which correlates with the shape of the inflorescence and the dimensions of the seeds.

The so-called 'mohar' (*P. italicum moharicum*) – a kind of Italian millet with a short, non-branching inflorescence – dispersed toward the north; on the other hand, the branched form with a compound inflorescence is essentially a plant of the south and has a very long vegetation period. The typical forms of millet are grown in Transcaucasia and China. In intermediate areas (e.g. Bokhara) there is a full range of transitional forms from branched Italian millet (*P. italicum maximum*) to non-branching millet with small inflorescences. In the elevated mountain areas the non-branching type of millet predominates, while at lower elevations with a longer vegetative period the branching type is most common.

The ranges of *Setaria glauca* and *S. viridis* [sometimes also called foxtail millet] are very close to that of *P. italicum*, and cover very large areas. These common weeds are widespread in Europe and Asia and, as far as their habitats are concerned, it is necessary to draw conclusions about their centers of type-formation from that of the group close to them, i.e. *P. italicum*.

The geographical centers of type formation of cultivated flax

In his thorough review of the origin of flax and its cultivation in the past, Oswald Heer (1872) suggested that the ordinary cultivated flax originated from the wild species, *Linum angustifolium*, while assuming that winter flax, *L. hyemale-romanum* and *L. ambiguum*, should be the ancestral forms of *L. angustifolium* and *L. usitatissimum*. He considered the native land of the cultivated flax to be located within the Mediterranean area, where *L. angustifolium* still occurs in its wild state. Thus, on the basis of this opinion, Heer referred the discoveries in the pile-dwellings in Switzerland of remnants of stems and seeds to *L. angustifolium* cultivated during the Stone Age.

The questions concerning how the different races of flax originated or how they became cultivated for the sake of the fibers and seeds were not touched upon by Heer (1872) but the establishment of a connection between cultivated flax and the wild forms appeared to him quite important; that would be where the solution of all the problems concerning the origin of the cultivated flax was supposedly hidden.

However, more recent data according to Gentner (1921) compel us to consider Heer's determination of the flax from the pile-dwellings as *L. angustifolium* to be erroneous and instead to assign it to the ordinary winter flax, *L. hyemale romanum*.

De Candolle's opinions changed drastically from 1855 to 1882. In his *Géographie botanique raisonée* (De Candolle, 1855) he assumed on the basis of 1. the existence of many names for flax among various people, 2. the antiquity of the flax crops in Egypt, and 3. the cultivation of flax in India only for linseed but in other countries both for its fibers and its oil that the cultivation of flax was polyphyletic in origin and originated initially from two or three species, which had become united into one single species, *L. usitatissimum*. In 1885, De Candolle began to doubt that the species of flax cultivated in Egypt could be the same as the Russian and the Siberian types of flax. In his *L'Origin des plantes Cultivées* (De Candolle, 1883), he changed his original opinion and leaned toward that of Heer, assuming that the people of the pile-dwellings had cultivated *L. angustifolium*. He looked for the ancestors of the commonly cultivated flax among the wild flax of Asia while basing himself mainly on doubtful discoveries of it in the westerly areas of that continent. In *L'origine des plantes cultivées*, De Candolle acknowledged that *L. usitatissimum* must have arrived in Europe from Asia while assuming that the Egyptian and the Mesopotamian as well as all the presently cultivated kinds of flax came initially from wild flax, originating in localities between the Persian Gulf and the Caspian and Black Seas.

During the last 10 years we have, together with E.V. Ellad', conducted detailed phyto-geographical research on the kinds of flax grown in the U.S.S.R. and other countries. During special expeditions and through the courtesy of correspondents, we have collected more than 1400 samples of flax from all areas where flax is cultivated for fibers or seeds, thus gaining a comparatively complete understanding of what represents cultivated flax as such in relation to the variation of the different kinds thereof. Taxonomic–geographical studies of the kinds of flax grown in various parts of the U.S.S.R. at the experimental stations of the Department of Applied Botany revealed that the first solution of the problem according to De Candolle in 1885 was closer to the truth than the opinion to which he arrived at in 1882. Studies of the botanical complex of the cultivated types of flax and their geographical distribution in Eurasia have to some extent revealed the presence of two basic centers of variation with respect to flax and two major geographical groups of cultivated flax.

According to all the data available, ancient areas for cultivation of flax appear to be found in Asia, i.e. in India, Bokhara, Afghanistan, Khorezm and Turkestan, and along the Mediterranean coasts in Egypt, Algeria, Tunisia, Spain, Italy and Asia Minor. In remote prehistoric time, flax was cultivated also in Central Europe (Heer, 1872; De Candolle, 1883; Buschan, 1895). Among the areas at present covered by cultivated linseed flax, India takes so far first place in the Old World, only recently having a competitor with respect to the cultivation of linseed in Argentina, where its cultivation has developed during the last couple of decades. Just as in Bokhara, Afghanistan, Turkestan and Khorezm, India grows flax exclusively for linseed oil both at present as well as in earlier times (Howard and Khan, 1924). According to the information now available, flax is grown in Asia Minor both for fibers and oil. Ancient Egypt, too, grew flax for both purposes.

Studies of the many specimens of flax collected in southwestern Asia revealed a concentration there of an enormous variety of hereditary forms. In river valleys and lowland habitats, characteristic, low-growing, late-maturing, densely leafy, branched and typically bushy types (gr. *brevimulticaulis*) are concentrated and occasionally also forms with a spreading growth (gr. *prostrata*). Among the typically bushy forms (gr. *brevimulticaulis*) there are races with brownish, umber or pale-yellow colored seeds and endemic races with narrow, crimped petals (var. *angustipetalae*). Entire areas, such as the Khivinskiy oasis, are characterized mainly by the cultivation of yellow- (or white-) seeded forms. In the mountain areas of Fergana, Bokhara and Afghanistan, races are grown that occupy a position, with respect to the branching and the length of the vegetation period, intermediate (gr. *intermediae*) between the typical North-Russian tall- growing early-maturing types and the South-Turkestani bushy types with a long vegetation period. At the altitudinal limit of cultivation in Pamir and Badakhshan one occasionally meets with typical, early-maturing, tall-growing flax (gr. *elongatae*), cultivated for the seeds. In India alone Howard and Khan (1924) have determined 26 botanical varieties and 123 races of linseed flax.

Comparisons between the Indian types described in detail by Howard and Khan, and the races known to us from Turkestan, Afghanistan, Bokhara and Persia revealed a number of characteristics among the former that make us consider them endemic in India. Side by side with races from other countries, we increased experimentally the crops of 20 samples obtained from different areas in India. Excluding the vars. *bicolor*, *vulgatum* and *commune*, described from India but found by us in other countries as well, 23 varieties remain, which were described by Howard and Khan but are not found anywhere else among our enormous material. These varieties are: *luteum*, *lutescens*, *indicum*, *cyaneum*, *purpureum*, *albidum*, *album*, *albocaeruleum*, *officinale*, *bangalense*, *agreste*, *meridionale*, *gangeticum*, *laxum*, *praecox*, *pratense*, *minor*, *herbaceum*, *pulchrum*, *caesium*, *campestre*, *tinctorium* and *sativum*. The general peculiarities, typical of the Indian forms but not found in the races of Turkestan, Bokhara, northern Afghanistan, Persia and Transcaucasia are firm and hard capsules, difficult to thresh (of type *rigidum*) and white flowers not opening fully during anthesis. While in some of the races from Turkestan and Bokhara and those of the African and European types of flax, the petals fall off after anthesis, in the case of the many Indian forms they remain adhered after flowering. Under the conditions in the territory of Kuban' (where the flax was tested) the strong development of anthocyanin in the shoots and the capsules when ripening appears to be a characteristic trait of the Indian forms. Such intensely colored races have been found by us only in Asia Minor. In India there are also typical and peculiar races with light-colored seeds and narrow-petaled flowers, and also typical races with large seeds (up to 6 mm long), which at the same time have white flowers. Forms with very small seeds (from the western parts of India), measuring 3.5–3.75 mm in length, are also found.

During the studies of flax in Asia and Europe, the general impression was that in the direction toward India the composition of the types became more variable. The complex of characteristics distinguishing the races and varieties in Europe and in Asia was especially variable in India, where the maximum variability of the types is concentrated. However, at the same time, the flax of Bokhara and Afghanistan and the countries adjacent to them display a number of endemic characteristics and combinations, not typical of the Indian flax and, perhaps not of those of Europe either. In the areas on both sides of Hindukush and in the mountain areas of Bokhara a number of peculiar types are hidden away; low-growing, branching and bushy races with open flowers and peculiar white-seeded races are found in Khorezm and Persia, although they are distinct from the white-seeded Indian types. The capsules of the Turkestani, Afghani and Bokharan types are easier to thresh than those of the Indian races. In southern Afghanistan, adjacent to India, we found the typical Indian forms growing together with the Afghani races.

In the composition of the ordinary European races, large- and white-seeded races with large flowers, narrow petals and a typical bushy growth and races with capsules difficult to thresh are absent. With respect to the seeds, the flowers and the type of open flowers as well as the leaves, the European races are linked by a gradual transition to the types of flax in southwestern Asia (Afghanistan and Bokhara). In the mountain areas of Bokhara and in Badakhshan, different sites can be found with crops of types similar to fiber flax and impossible to distinguish from the present European types of flax.

In general, in all of Asia (except for along the Mediterranean coast, i.e. Palestine and Syria) and in Europe, including the Caucasus and the European parts of the U.S.S.R., mainly small-seeded (or medium-seeded), small-flowered and small-capsuled races of flax are typical, with tall-growing and bushy as well as intermediate growth-forms. Although an increase in the size of seeds can be observed toward the south of Europe (fiber flax seems to have smaller seeds than the bushy linseed flax), all the larger-seeded European and Asiatic races of the bushy type cede with respect to the size of seed to the typical Mediterranean forms (Kappert, 1921). The center of variation of the large group of small-seeded flax is confined to southwestern Asia.

In the future, investigations of the northern areas of India, Kashmir, Asia Minor and China will allow us to establish more exactly the centers of concentration of the maximum variability of that group of flax.

The Mediterranean coast can be sharply distinguished from Central Europe and southwestern Asia with respect to the complex of flax cultivated there. Mainly large-seeded, large-flowered and large-capsuled forms of linseed flax are typical of the countries extending along the Mediterranean coasts. Egypt, Algeria, Tunisia, Morocco, Tripolitania, Palestine and Syria grow races with large flowers (25–30 mm wide), large seeds (5–6.1 mm long), and large leaves. The European and the Asiatic races have flowers 15–18 mm wide (fibre flax) and up to 18–24 mm (linseed flax). The common Russian fiber flax has seeds 4–5 mm long, but only 3.5–3.75 mm long in the case of the small-seeded types. This group is clearly distinct from the common Asiatic and European races. Large flowers, large capsules and large seeds appear, as a rule, to be genetically linked characteristics, determined by a few genes (Tammes, 1911).

Judging from the eight samples we obtained from Abyssinia, peculiar small-capsuled, low-growing, small-leaved and small-flowered forms are grown there. The dimensions of the seeds are either medium or very small (3.75 mm) in size. Endemic forms were found in Abyssinia, linking a yellow color of the seeds to a blue color of the corolla at the same time as there are also ordinary forms with yellow-colored seeds linked to white flowers.

The large-seeded races of the Mediterranean coasts are, in turn, represented by different varietal types. According to our investigations of 26 samples (from Morocco, Algeria, Tunisia, Egypt and Palestine), there are two types: Morocco, Algeria, Tunisia and parts of Egypt grow large-seeded forms with very large flowers (up to 30 mm or more in diameter), large capsules and large leaves. In Egypt, large-seeded linseed flax with small flowers (16.3–21.7 mm wide), very small and rarely vernate leaves, single stems, similar in type to fiber-flax are also found. The capsules of the latter type are extremely hard, difficult to thresh and have an intense anthocyanin coloring when ripe. The shoots are also of an anthocyanin color. This second type approaches the Indian forms with respect to the small flowers, the threshing difficulties and the anthocyanin coloration of the capsules and the shoots.

When comparing data concerning the dimensions of the seeds of cultivated flax, found during archeological digs at Dra-Abu-Negga, Asserif and Scheich as well as Abd-el-Qurna in Egypt, with those of the African large-seeded forms of linseed, it could be seen that the large-seeded forms were grown in Egypt in the remote past as well. Thus, in his review of the archeological sites of flax, Buschan (1895) furnishes the following measures of the ancient Egyptian seeds: 4.5, 5.0 and 5.5 mm. The present large-flowered linseed flax is characterized by similar dimensions, ranging from 6 to 6.1 mm in length, while Asiatic forms of flax from northern India, Turkestan, Afghanistan and Bokhara, as well as the European forms associated with them, are characterized by seed length from 3 to 4.5 mm. Various races of *L. angustifolium*, in which Heer and De Candolle were inclined to see the 'ancestors' of cultivated flax and which seem to have been grown in prehistoric times, have seeds measuring from 2.5 to 2.9 mm in length.

Taxonomic studies of cultivated flax force us, therefore, to distinguish, as far as possible, between two basic groups: the large-seeded and large-flowered one, genetically associated with the Mediterranean coastal area, and the small-seeded and small-flowered one, belonging to southwestern Asia. It is possible that more thorough studies of flax in Abyssinia will reveal still another group or form.

'Cracking flax' (*L. crepitans*), a kind of flax with capsules that crack when they ripen, is presently grown on small plots in Ukraine (e.g. in the Chernigov district), in the Primor'e area, and in the Alps, the Black Forest and Würtemberg in Germany, as well as in Switzerland, mainly in mountain areas. It is also found as an admixture to ordinary flax with closed capsules. As far as the type is concerned, this flax approaches the Asiatic–European groups. The seeds are medium-sized (4.5 mm), brownish with a weakly developed 'beak'; the capsule is not hard, the flowers are small and blue, the height of the plant and the branching remind one of the intermediate types (gr. *intermediae*) and the capsules are few in number. According to the length of the vegetative period, it can be distinguished into early- and late-maturing forms. The capsules crack wide open when ripe, throwing out the seeds. It is interesting to note that this flax, in essence characterized by 'wild' traits, should still remain in cultivation in spite of the fact that a large amount of its seeds is lost at the time of harvesting or even before that.

Winter flax, *L. hyemale romanum*, is grown on small plots in Spain, northern Italy and the Alps, and in Carinthia and Krain in the mountain areas of Austria (Kramer, 1923). In growth type, this flax approaches the Asiatic kinds with a spreading growth (gr. *prostrata*), being very leafy and characterized by a prostrate habit.

Of all the many species belonging to the genus *Linum*, the wild flax, *L. angustifolium* with its small seeds and narrow leaves comes closest to *L. usitatissimum*. Only this species produces fertile hybrids with ordinary flax. The geographical area of *L. angustifolium* reaches from the Canary Islands to western Persia and Asia Minor, including all the Mediterranean coastal area (Fig. 5). This species is also represented by a large number of forms (cf. Ascherson and Graebner, 1914), easily hybridizing with the cultivated flax and, no doubt, genetically close to the common type of flax (Tammes, 1923). It could perhaps be correct to unite *L. angustifolium* with *L. usitatissimum* into a single Linnaean species *L. usitatissimum s. lat.*, as suggested by Heer (1872). However, there is no objective reason for considering this narrow group of plants as the ancestors of all the European and African kinds of flax, just as there is little reason for considering the distichous barley as the ancestor of all the cultivated types of barley. In our opinion, *L. angustifolium* represents a group of types, closing the cycle of varieties within the species of *L. usitatissimum*. It is no doubt very close to it and genetically related, but just barely so. It is also possible to fit *L. crepitans*, with its capsules that crack open, into the morphological series between typical *L. angustifolium* and typical *L. usitatissimum*. Just like wild barley or wheat, *L. crepitans* can easily persist in the wild state or in the form of a weed. The different types of *L. crepitans* and *L. angustifolium* occupy the same positions within the system of *L. usitatissimum s. lat.* as races of *Hordeum spontaneum* within the system of *Hordeum sativum s. lat.*

It seems likely that the two geographical groups of flax that exist at present were once united and had a common center of origin and that, physiologically, these groups of flax were not sharply delimited. However, in a very remote time, these groups apparently became isolated, of which the discoveries of exclusively large-seeded forms in the Egyptian diggings bear witness. The ancient agricultures of Mesopotamia, India and the countries adjacent to them grew crops of special kinds of flax, different from the Egyptian ones, from which, in turn, an independent kind of crop arose.

In Table 3, we shall make a short review of the geography of the diversity of the kinds of flax in the Old World, compiled according to the most variable characteristics by myself and E.V. Ellad'. From the tables one can see that the most variable characteristics, and consequently also most of the forms, are concentrated within the Indian area. From India we have data from Howard (40 samples) and our own observations of 20 samples, obtained from there. These studies of a small number of samples have already revealed a more or less definite concentration of the largest diversity of types in this country. An important variability characterizes also the areas of Afghanistan and Bokhara, adjacent to India. In spite of detailed investigations of a very large number of specimens (over 1000) in comparison with the small number of samples studied from India, and in spite of the widespread cultivation of fiber flax in the north, in Europe and Siberia, the diversity there is not large. The number of forms definitely decreases from southwestern Asia in the direction toward the Euroasiatic northlands. In

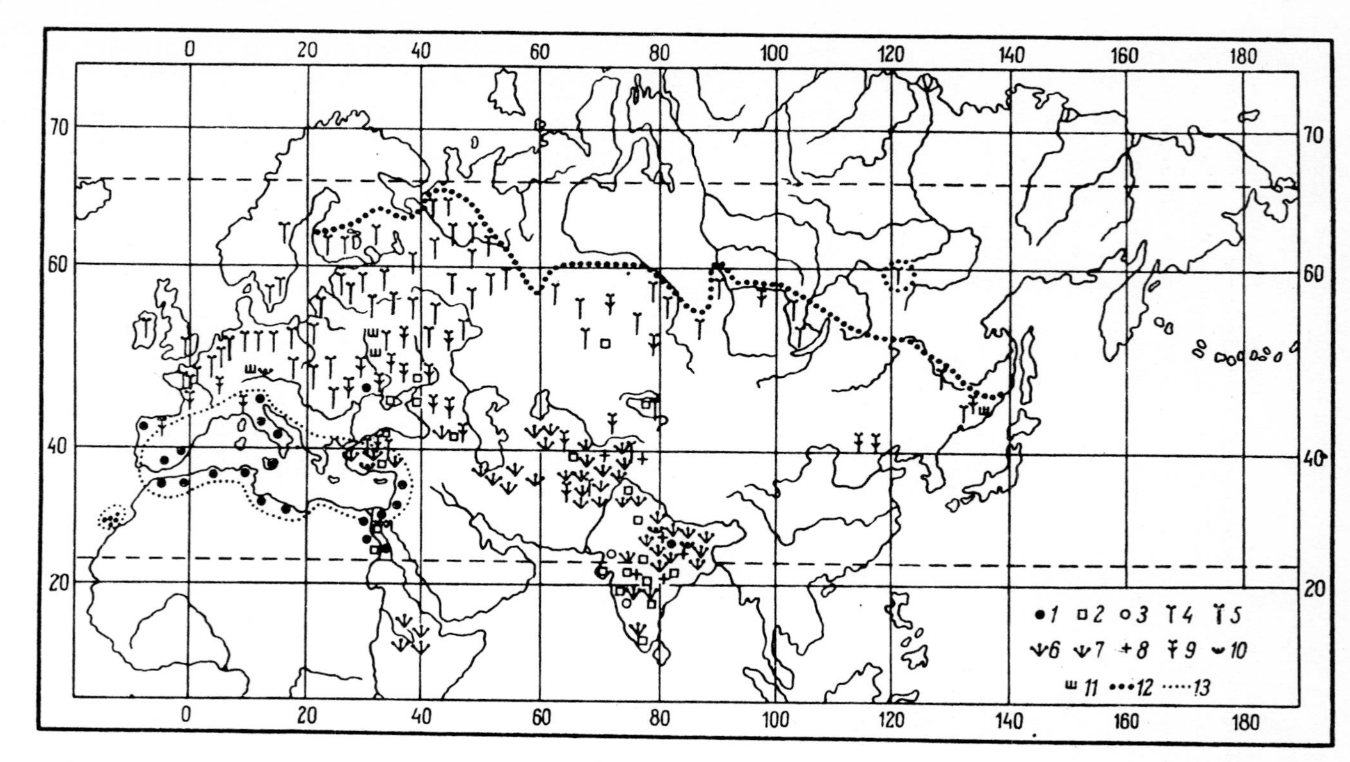

Fig. 5. Geographical distribution of the varieties of *Linum usitatissimum* L. and *L. angustifolium* Huds. 1–3: The *macrospermae* group: 1. var. *grandiflorae*; 2. var. *parviflorae*; 3. var. *leucanthae*; 4–10: the *mesospermae* group: 4–5. *elongatae* (4. var *coeruleae*, 5. var. *albiflorae*); 6–8: *brevimulticaules* (6. var. *brunnea*, 7. var. *leucospermae*, 8. var. *angustipetalae*); 9. *intermediae*; 10. *prostratae*; 11. the *microspermae* group: var. *crepitans*; 12. northern limit of flax cultivation; 13. the distribution area of *L. angustifolium* Huds.

Table 3(a)–(d). *Geography of the type-formation of flax in the Old World*

(a)

	Group of forms				Characteristics					
					Shoots			Flowers		
Area	*elongatae*	*intermediae*	*brevimulticaules*	*prostratae*	With much anthocyanin	With little anthocyanin	Without anthocyanin	Violet	Blue	Pale blue
India	–	+	+	+	+	+	+	+	+	+
Afghanistan	+	+	+	–	+	+	–	–	+	+
Turkestan and Khiva	+	+	+	–	+	+	+	–	+	–
Bokhara	+	+	+	–	–	+	+	–	+	+
Persia	–	–	+	+	+	+	–	–	+	–
N. and Central parts of RSFSR	+	+	–	–	–	+	+	+	+	+
Ukraine and southern RSFSR	+	+	–	–	+	+	–	–	+	–
Transcaucasia	–	+	+	–	+	+	–	–	+	–
Siberia	+	+	–	–	–	+	–	–	+	–
N. Africa	–	+	–	–	+	+	+	–	+	+
Abyssinia	–	–	+	–	+	+	–	–	+	–
Palestine	–	+	–	–	+	–	–	–	+	–
Asia Minor	–	–	+	+	+	+	–	–	+	–
Italy	–	+	–	–	–	+	–	–	+	–

(b)

Area	Characteristics of the flowers									
	White	Open	Convoluted	Large	Medium	Small	Broad-petaled	Narrow-petaled	Blue anthers	Yellow anthers
India	+	+	+	+	+	+	+	+	+	+
Afghanistan	+	+	+	−	+	+	+	−	+	+
Turkestan and Khiva	+	+	−	−	+	+	+	−	+	+
Bokhara	+	+	−	+	+	+	+	+	+	+
Persia	+	+	−	−	−	+	+	−	+	+
N. and Central parts of RSFSR	+	+	−	−	−	+	+	+	+	+
Ukraine and southern RSFSR	−	+	−	+	+	+	+	−	+	−
Transcaucasia	−	+	−	−	+	+	+	−	+	−
Siberia	+	+	−	−	+	+	+	−	+	−
N. Africa	−	+	−	+	+	+	+	−	+	+
Abyssinia	−	+	−	−	−	+	+	−	+	−
Palestine	−	+	−	+	−	−	+	−	+	−
Asia Minor	−	+	−	−	+	+	+	−	+	−
Italy	−	+	−	+	+	+	+	−	+	−

(c)

Area	Characteristics of the capsules									
	Closed	Cracking open	Large	Medium	Small	Septa glabrous	Septa pubescent	Hard to thresh	Medium hard to thresh	Easy to thresh
India	+	−	+	+	+	+	+	+	+	−
Afghanistan	+	−	−	−	+	+	+	+	+	−
Turkestan and Khiva	+	−	−	+	+	+	+	−	+	−
Bokhara	+	−	−	−	+	+	+	−	+	−
Persia	+	−	−	−	+	+	+	−	−	−
N. and Central parts of RSFSR	+	−	−	−	+	+	+	−	−	+
Ukraine and southern RSFSR	+	+	−	+	+	+	+	−	+	+

(c) (*cont.*)

	Characteristics of the capsules									
Area	Closed	Cracking open	Large	Medium	Small	Septa glabrous	Septa pubescent	Hard to thresh	Medium hard to thresh	Easy to thresh
Transcaucasia	+	−	−	+	+	+	+	−	−	+
Siberia	+	+	−	−	+	+	+	−	−	+
N. Africa	+	−	+	+	−	+	+	+	−	−
Abyssinia	+	−	−	−	+	+	−	−	+	−
Palestine	+	−	+	−	−	−	+	+	−	−
Asia Minor	+	−	+	+	+	−	+	−	+	−
Italy	+	−	−	+	+	+	+	+	+	−

(d)

	Characteristics of the seeds						
Area	Brown	Umber-colored	Yellow	Large	Medium	Small	Number of samples
India	+	+	+	+	+	+	33
Afghanistan	+	−	+	−	+	−	25
Turkestan and Khiva	+	+	+	+	+	−	24
Bokhara	+	−	+	−	+	−	23
Persia	+	−	+	−	+	+	19
N. and Central parts of RSFSR	+	−	−	−	+	−	21
Ukraine and southern RSFSR	+	−	−	+	+	−	22
Transcaucasia	+	−	−	+	+	−	19
Siberia	+	−	−	−	+	−	18
N. Africa	+	−	−	+	−	−	21
Abyssinia	+	−	+	−	+	+	16
Palestine	+	−	−	+	−	−	13
Asia Minor	+	−	−	+	+	−	19
Italy	+	−	+	+	−	−	19

India we could record 33 variable characteristics (34, if Afghanistan is included), whereas we know of only 21 variable characteristics within the R.S.F.S.R. and 18 within Siberia.

The countries around the Mediterranean are adequately studied and, while putting together a special morphological and genetical group, it was necessary to treat it differently from the preceding groups, originating from southwestern Asia.

There are definite regularities with respect to the distribution of flax over the Old World, which are important for an understanding of all the different groups of flax and for the interpretation of the plant breeding work with flax.

While studying flax of different origin by means of crops grown under uniform conditions we were able to reveal that northern types, mainly fiber flax, are as a rule characterized by a short vegetative period. Investigations of about 1500 samples from different regions indicated that there is an inverse correlation with respect to flax between the vegetative period and the height of the plants: the shorter the vegetative period, the taller the flax. The closer it comes to the ordinary type of fiber flax, the less the plants are branched and the lower is the production of seeds. With an increasing vegetative period, the branching increases, just like the number of stalks and that of capsules but, correspondingly, the height of the stalks is reduced.

The distribution of hereditary forms of flax in the direction from south toward north is definitely linked to the length of the vegetative period. Toward the north there are mainly early types with a short vegetative period and tall stalks, not branching but used for the long fibers; toward the south the branching, low-growing races, producing many capsules, are used for their seeds.

The fiber flax of the northern parts of European U.S.S.R. belongs to early types. Linseed flax – low-growing, many-stalked types producing many capsules and having a prolonged vegetative period – turns out properly to be typical of southwestern Asia. The southern races of flax grown in the Ukraine, northern Caucasus and the southeastern portion of the European U.S.S.R. belong to types with a long vegetative period but are still closer to fiber flax than the present types of linseed flax (f. *brevimulticaules*).

In Tables 4–6, the characteristics of the types of flax in different areas of origin are listed. A comparison between the inherited differences between these forms was carried out under uniform conditions by means of crops grown at the North-Caucasian Branch of the Institute of Applied Botany, belonging to the Institute of Applied Agronomy (the Kammenaya Step' in the Voronezh District), and at the Central Plant Breeding and Genetical Station (i.e. Detskoye Selo in the Leningrad District).

A pronounced abbreviation of the vegetative period can be clearly seen from these tables to be an inherited characteristic of flax when it disperses from Afghanistan toward northern Europe. From the center of its variation and the actual focus of its origin in southwestern Asia, flax differentiated according to its vegetative period when dispersing northward. Those races that have a long vegetative period remained in the south while the early-maturing forms were dispersed northward because of natural selection.

Table 4. *Height of the plants and the vegetative period of strains of flax of different geographical provenance. The North-Caucasian Branch of the Otrada Kubanskaya Station, 1925*

Provenance of the strains	Latitude (°N)	No. of samples	Mean height of the plants (in cm)	No. of days from sprouting to maturity (vegetative period)
Northern and northeastern regions (districts of Arkhangelsk, N. Dvina, Perm, Vyatsk and Vologod)	65–56	36	75	87
Northwestern region (districts of Leningrad, Novgorod and Pskov)	61–55	95	83	88
Districtsof Tversk, Rybin and Yaroslav	59–56	25	74	86
Districts of Moscow and Vladimir	57–54	20	71	88
Altai	55–46	1	66	87
Byelorussia	56–51	35	66	88
Districts of Kaluzh, Tul′, and Ryazan	55–52	6	63	88
Districts of Orlov, Tambov and Penzensk	55–51	19	63	89
Districts of Samara and Saratov	51–48	27	61	95
Districts of Kursk and Voronezh	54–49	10	61	93
Districts of Chernigov, Poltava and Kharkov	52–49	26	61	93
Districts of Volyn and Podolsk	52–47	18	63	94
Districts of Kiev and Yeakterinoslav	51–48	26	62	96
Crimea	46–44	8	56	95
Northern Caucasus	46–40	61	58	95
Turkestan (at different elevations)		28	58	97
		19	44	99
Khorezm	43–40	4	42	102
Afghanistan (at different elevations)	38–30	57	48	100

Table 5. *Height of the plants and the vegetative period of strains of flax of different geographical provenance. The Central Plant Breeding Station at Detskoye Selo, 1922*

Provenance of the strains	Latitude (°N)	No. of samples	Mean height of the plants (in cm)	No. of days from sprouting to maturity (vegetative period)
Northern and northeastern regions (districts of Arkhangelsk, Perm, Vyatsk and Vologod)	65–56	23	86	85
Northwestern region (districts of Leningrad, Novgorod, Pskov and Vitebsk)	61–55	24	91	86
District of Tversk	59–56	11	77	85
Byelorussia	56–51	5	83	87
Districts of Penzensk and Simbirsk	56–53	3	68	105
The south-east	51–48	16	68	104
Turkestan (different strains collected at different elevations)	43–55	8	64	100
Khiva	43–40	1	55	Did not ripen
Pamir and Bokhara (different strains, collected at different elevations)	43–40	69	47	98
Afghanistan	38–30	3	58	Did not ripen

Table 6. *Height of the plants and the vegetative period of strains of flax of different geographical provenance. The Central Plant Breeding Station at Detskoye Selo. Means of 1921, 1922 and 1925*

Provenance of the strains	Latitude (°N)	No. of samples	Mean height of the plants (in cm)	No. of days from sprouting to maturity (vegetative period)
Northern and northeastern regions (districts of Arkhangelsk, N. Dvina and Vologoda)	65–56	50	83	89
Northwestern region (districts of Leningrad, Pskov and Novgorod)	61–55	61	87	93
District of Yaroslav	59–56	6	83	92
Byelorussia	56–51	22	80	97
Districts of Kaluzh and Ryazan	55–52	2	62	105
Districts of Orlov, Tambov, Penzensk and Simbirsk	55–51	15	71	110
Districts of Kursk and Voronezh	54–49	2	73	103
Districts of Poltava and Voronezh	52–49	2	70	111
Districts of Volyn and Podol′	52–47	6	75	110
Districts of Samara and Saratov	51–48	34	68	112
Districts of Kiev and Yekaterinoslav	51–48	5	70	114
Northern Caucasus	46–40	2	67	118
Turkestan (at different elevations)	43–55	25	61	112
Pamir and Bokhara (at different elevations)	43–40	151	51	106

The distinction of flax into fiber flax and linseed flax is the result of natural selection in the direction from south toward north. The concentration in the north of crops of flax grown for their fibers and in the south of bushy and intermediate forms of flax, grown for the seeds, can be simply explained by the action of natural selection, reflected toward the north in the short vegetative period required by early-maturing flax and, consequently, also in long-stalked flax especially suited for obtaining a satisfactory yield of long fibers. In the south, there was of course a concentration of those races that were able to utilize a longer vegetative period, and consequently, grow more stalks of a lower stature but bear a large number of capsules and, thus, produce more seeds when ripe.

When dispersing northward from southwestern Asia, the crops of flax grown in their native land mainly for linseed oil, became naturally converted into crops of long-staple flax in the north. As demonstrated during the investigation at our biochemical laboratory (by Ivanov, 1926), the percentage of oil in the seeds of some pure-bred strains of flax did not change when transferred from the south to the north or vice versa. The alteration affected the quality of the oil, however (the iodine index has a tendency to increase in the northerly direction). Geographical experiments with crops of 12 different pure-bred lines at 58 sites in the U.S.S.R. demonstrated that in the south there was, in general, a reduction in the height of the flax plants but an increase in the number of stalks and in the number of capsules. The reverse phenomenon was observed when southern types of flax were transferred to the north. Fiber flax grew more stalks of lower stature when moved from the north toward the south. Linseed flax grew taller in northern Europe. The individual variability went in the same direction as the selection of inherited characteristics from south toward north. It is necessary to distinguish the individual variability of the pure-bred lines, which depend on geographical effects, from the inherited differences, typical of the geographical races.

As our research demonstrated, common flax in the high mountains has a tendency to increase the height of the stalks but to decrease the branching. Typical fiber flax from the Pskov district underwent a sharp reduction in height and an increase in the number of branches when moved from the north to the south but in the mountain areas of Transcaucasia, e.g. in Bakurian at 1760 m.s.m., and at an elevation of 2000 m south of Tashkent, it can hardly be distinguished from the typical fiber-flax of Pskov.

Flax is very successfully grown for its fibers at an elevation of 7000–9000 feet in Kenya, right on the equator. Tests at the Belfast College of Technology have shown that this flax surpasses the Russian long-staple-flax in quality (Wigglesworth, 1923).

During our investigations among the high mountains of Bokhara (Roshan and Shugnan) we found in some secluded localities at an elevation of about 2700–3000 m.s.m., settled by Tadzhiks – the ancient inhabitants of Turkestan – some isolated plots of fiber flax, grown for the seeds. During tests, this flax turned out to be almost identical with the present Russian fiber flax of the northern type. In the majority of cases in southwestern Asia and at high elevations, the ordinary low-growing flax is cultivated for its seeds. However, the existence of isolated plots of fiber flax, preserved in southwestern Asia, supports the accuracy of our conclusions concerning the origin of the cultivation of flax in these areas. As can be expected a priori, all the initial types both of the future northern crops of early-maturing fiber flax and the southern, branched and late-maturing types, are concentrated here thanks to the exceptional diversity of habitat conditions. The mountain areas of southwestern Asia favored both the formation and the preservation of the different morphological and physiological types of flax. It is possible that, during detailed studies in southwestern Asia, crops of flax in remote areas of the high mountains will turn out to produce such types of flax that, possibly, became utilized for cultivation of fiber flax in northern Europe. In Asia Minor, flax is still grown for its fibers in mountain areas (data from the expedition led by P. M. Zhukovskiy).

In southern and central areas of India, in North Africa and Palestine, we have discovered low-growing, comparatively rapidly maturing linseed flax that is ephemeral. Such races are often distinguished by hard capsules with seeds that are difficult to thresh out. The intensely hot climate has forced a selection of races with a very short vegetative period, able to utilize the moisture of the winter and spring months and to avoid the summer heat. Early ripening during climatic conditions characterized by drought often results in good quality, just as in the north in the case of a short vegetative period. These races represent a special ecological type with a particular kind of root system (Howard and Khan, 1924) and it is necessary to distinguish them from the relict crops of flax in southwestern Asia and Europe.

Thus, under the conditions in northern Caucasus (at Otrada Kubanskaya Experimental Station), crops grown in 1925, belonging to a number of different lines of flax (Table 4) from Morocco, Algeria and Tunisia (17 samples), reached a mean height of 48 cm and had a mean vegetative period of 98 days. Flax from Abyssinia (8 samples) reached a mean height of 29 cm during a mean vegetative period of 86 days and Egyptian flax (5 samples) grew 41 cm tall after an average of 90 days, while flax from India (15 samples) grew an average of 32 cm tall after a mean of 92 days. At the same time, a series of races, common in Turkestan, Afghanistan and Khivinsk, had a vegetative period of 100 days.

Detailed, systematic and zonal investigations of flax from the mountain areas of India, Asia Minor and northern Africa will no doubt turn up much interesting data, which can be utilized for the cultivation of flax on the European lowlands.

In many cases, it is definitely clear from the facts mentioned above that the center for this type formation of *Linum usitatissimum* can still be traced to southwestern Asia and northern Africa, where plant breeders should go for material of strains for practical purposes. In the case of seed crops, no doubt interest should focus on the large-seeded types of the Mediterranean coastal areas, distinguished by a high percentage of oil, as indicated by the research done by Howard and ourselves.

In the Russian northlands and in Siberia, as well as the northern parts of European U.S.S.R., the cultivation of flax has selected out races with a shorter vegetative period due to both natural selection and the efforts made by Man through the centuries. As far as the presence of a direct correlation between short vegetative period and long stalks is concerned, the Russian North appears long since to have been the possessor of valuable races of fiber flax that force us to pay special attention to a study of the North-Russian flax in areas of more ancient cultivation at the limits where flax can mature and where discoveries of especially practical and interesting types of flax can be expected.

We have examined in detail the cultivation of flax, wheat, barley, oats and millet, in order to demonstrate that, at present, the problem of the local origin of crops and the homeland of any given plant can be solved by means of differential, phyto-geographical methods. Using flax as an example, a number of geographical regularities could be revealed for the dispersal of races and to a great extent for determining the work of plant breeders. It may be possible to find many more examples of complex and more complicated origins of cultivated plants. Investigations at the Institute of Applied Botany have led us to draw the conclusion that the majority of legumes have *two* centers of origin. Cultivated peas, lentils and horse beans come initially from two centers, one of which seems to be concentrated in southwestern Asia, the other, apparently, in northern Africa, or in general, within the Mediterranean area. The same can be said for the castor bean, *Ricinus communis* (Popova, 1926). By means of methods that use detailed analyses of the racial composition, it was possible to establish the centers of type-formation of many cruciferous plants.

The problem of the origin of cultivated plants has turned out to be more complicated than was believed at the time of De Candolle, Heer and many other authors. The discovery of plants in a wild state, which are close to those cultivated and which clearly belong to a single Linnaean species, still tells too little about the local origin of the cultivated plants to form a basis for plant breeding work with them, and to indicate where to look for the initial types of these plants. Wild oats are very close to cultivated oats, but this does not mean that the latter originated directly from the wild oats. The distribution areas of wild and weedy oats do not determine the centers of type formation of the cultivated oats. The wild oats (*Avena fatua* and *A. ludoviciana*) are met with in enormous quantities in southwestern Asia in such wild and weedy conditions, as were never known by the cultivated oats.

The differential taxonomic–geographic approach to the problems concerning origins offers a possibility for utilizing the solutions thereof for practical plant breeding, by means of which it was previously impossible to arrive at an even approximate solution.

Although by means of artificial selection, a plant breeder located far from the native land of the plants can, by crossbreeding of types, produce a race with new characteristics, not revealed by immediate study in their native land, it does not, in essence, change the general concept about the geographical distribution of the variety and its connection with the center of origin. By inbreeding, it is possible to single out a number of recessive types (e.g. in the case of maize, non-ligulate types or a large number of abnormalities). But all of these recessive types are, nevertheless, in their basic genetics, associated with the initial centers of origin and on the basis of these primary varieties one can look for the geographical center of a given crop.

CHAPTER 3 WEEDY PLANTS AS ANCESTORS OF CULTIVATED PLANTS

The origin of cultivated rye

During the study of the geography of to the varietal diversity of rye we ran into an interesting group of facts which, subsequently, turned out to be of general importance.

The rye cultivated in Europe, the European parts of U.S.S.R. and Siberia is botanically very uniform and is, in essence, represented by a single botanical variety, *Secale cereale* var. *vulgare*, with straw-colored and short-awned spikes and an admixture of var. *monstrosum*, a branched type. Enormous areas of northern Europe and Asia, many millions of acres, are sown with these ordinary varieties of rye. Although within the limits of these varieties it is also possible to distinguish forms according to the color of the grains or the more or less pronounced density of the spikes, these characteristics are not localized and do not have particular distribution areas, typical of the varieties of rye where such characteristics are not found.

Figure 6 outlines the northern limits of the cultivation of rye in Europe and Asia and a line south of which crops of wheat predominate over those of rye, but north of which crops of rye are more common than those of wheat.

Investigations by V.F. Antropopov and myself on specimens of rye from various countries demonstrate that all the variation and all the wealth of types of rye are concentrated in those areas where rye is either of secondary or no importance at all as a cultivated plant. The main botanical variation of rye is found in Afghanistan, Persia, Transcaucasia (Georgia and Armenia), Asia Minor and Turkestan. There, rye is known mainly as a weed, infesting wheat and barley crops, in particular the crops of winter wheat. It is also met with among winter barley. The crops of barley are particularly strongly infested by rye in northern Persia and the Transcaucasian areas, where races of winter barley are grown in large quantities.

Among these weedy types of rye an exceptional variety can be established that is not represented within the European agriculture. In Afghanistan alone, where rye is known mainly as a weed, 18 botanical varieties have been found. In southwestern Asia (in a wide sense), red-spiked types and varieties with long awns and pubescent hulls are known. In Armenia, Georgia and eastern Persia as well as in Asia Minor, floral variations are typical: red-spiked, brown-spiked and, occasionally, even black-spiked forms occur. White-spiked forms are typical mainly of Afghanistan, Bokhara, Khorezm and Turkestan. In Badakhshan and the mountain areas of Bokhara adjacent to it (Shugnan and Roshan), we have found a peculiar variety without ligulas or auricles, var. *eligulatum*. In Asia Minor, on the other hand, races with a very large kind of auricles (var. *auriculatum* Vav.) are established. Races with rough spikes and rough awns, with or without a waxy coating or with shoots almost without any anthocyanin (var. *viride* Vav.) are also found. In Shugnan (in Pamir), we succeeded in 1916 in finding forms of cultivated, brittle spring rye. When threshed, the spikelets disarticulate from the upper portion of the spike (Vavilov, 1917). A large

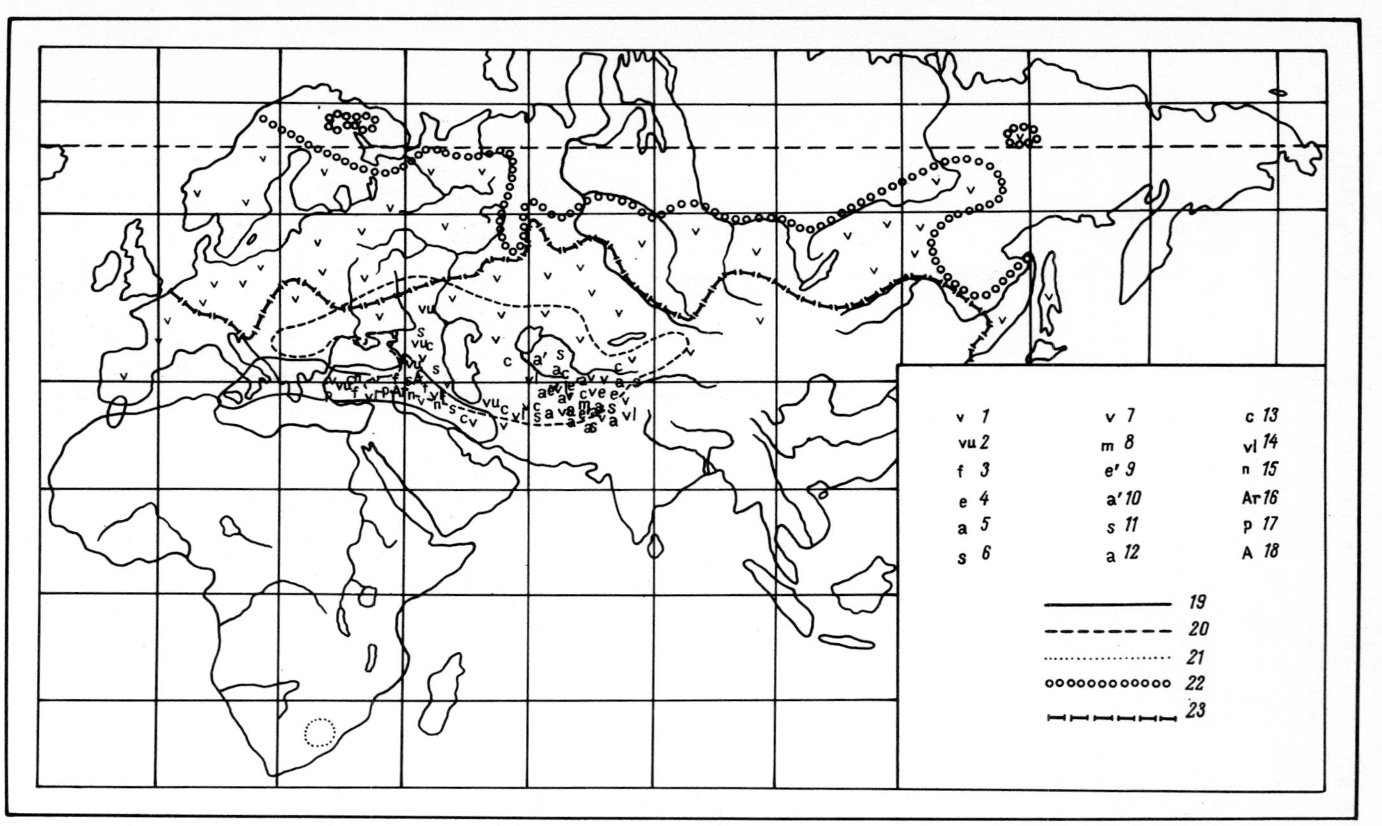

Fig. 6. Centers of origin of cultivated rye. Geographical distribution of the botanical varieties of *Secale cereale* L. and the distribution areas of other species of rye. 1. *vulgare* Koern.; 2. *vulpinum* Koern.; 3. *fuscum* Koern.; 4. *eligulatum* Vav.; 5. *afghanicum* Vav.; 6. *scabriusculum* Vav.; 7. *viride* Vav.; 8. *muticum* Vav.; 9. *epruinosum* Vav.; 10. *articulatum* Vav.; 11. *subarticulatum* Vav.; 12. *asiaticum* Vav.; 13. *clausopaleatum* Vav.; 14. *velutinum* Vav.; 15. *nigrescens* Vav.; 16. *armeniacum* Zhuk.; 17. *persicum* Vav.; 18. *auriculatum* Vav.; 19. distribution area of *S. montanum* Guss.; 20. area of *S. montanum* M.B.; 21. area of *S. africanum* Stapf; 22. northern limit of rye cultivation in the Old World; 23. line south of which crops of wheat predominate over those of rye and north of which rye predominates over wheat.

Fig. 7. Fig. 8.

Fig. 7. Spike of *Secale cereale* var. *afghanicum* Vav. Drawing based on a green, not yet ripe and still not disarticulated spike.

Fig. 8. Spikelets of *Secale cereale* var. *afghanicum* Vav. Drawing by M. Lobanova.

diversity of forms (14 varieties according to our own estimate) has been found in Georgia and Armenia by P.M. Zhukovskiy (1923 a, b), N.A. Maysurian (1925) and E. A. Stoletova. Among the rye infesting the fields, especially in Afghanistan, Persia and Turkestan, races predominate that have enclosed, non-shedding grains and adpressed awns, while in Transcaucasia (Armenia) and Asia Minor such races of weedy rye are found, in considerable quantities, that have spreading awns and florets that do not enclose the grains, which are easily dislodged and remind one of the type found in cultivated rye.

In other words, in southwestern Asia, including Transcaucasia, the investigator encounters, in fact, a striking diversity of rye, reminding one of the varietal complex of soft wheat and, in addition, rye is also found in different forms there, mainly as weedy plants among crops of soft and club wheat as well as winter barley.

A general picture is furnished of the distribution of the botanical varieties of rye, studied by us, from various parts of Eurasia (Fig. 6).

During the basic classification of the variation, according to which the diagram of the geographical distribution is drawn up, we touched upon such characteristics as the presence or absence of ligulae and auricles on the leaves, the brittleness of the rachillae, the color of the spikes, the pubescence of the hulls, the enclosure of the grains, the type of awns and the presence or absence of a waxy coating on the plants. These characteristics are definitely geographically differentiated and make it possible to establish exactly the centers of the inherited variability. For the sake of convenience we excluded characteristics concerning the colors of the grains, listed by N.A. Maysurian (1925) in his classification of the varieties of rye, since they are geographically indeterminate and present within all the areas.

A somewhat abbreviated review of the information available to the Institute of Applied Botany illustrates, nevertheless, that the variation is concentrated in the direction toward southwestern Asia.

BRITTLE *SECALE CEREALE*

In 1924, when sent out to Afghanistan specially to study rye and wheat, i.e. crops that, according to our geographical hypothesis, should display a maximum variability there, we encountered rye that, with respect to morphological and biological characteristics, reminded us of wild oats. In Afghanistan, rye occurs not only as a common weed among soft and club wheat, but most of all it is found (in addition to the common non-disarticulating, Turkestanian type of rye) in the form of extremely brittle types, definitely disarticulating when threshed. With respect to the character of the rachillae, this rye reminds us of wild barley, *Hordeum spontaneum*. Its spikes disarticulate when threshed and behave like the florets of wild oats. Together with wild oats, this weedy, brittle type of rye often covers the ground in enormous quantities after the wheat has been harvested. Biologically, it is in no way distinguishable from wild oats. In northern Afghanistan, this rye is not even distinguished by name from wild oats. The inhabitants of the Herat Province call both brittle rye and wild oats 'tak-

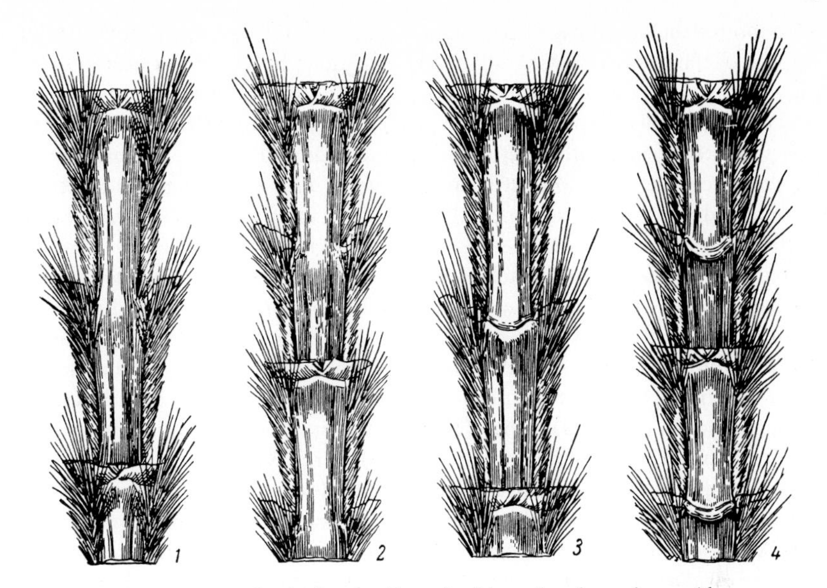

Fig. 9. Types of articulated spikes of cultivated and weedy rye (the rachis of the spikes is illustrated). 1. Ordinary European rye; the rachillae are fused with and delimited by the rachis of the spike; 2. type of north-Caucasian non-disarticulating weedy rye; 3. type of weedy rye from Turkestan; 4. type of disarticulating weedy rye, *Secale cereale* Var. *afghanicum* Vav.

tak'. An enormous quantity of brittle rye is, in particular, found in Hindukush, around Kabul, among *Triticum compactum* [shot wheat] which is frequently cultivated in southeastern Afghanistan. There, after the wheat is harvested, the ground is often literally covered by spikelets of brittle rye, constituting a serious scourge for the cultivation of wheat. In order to clean the fields, the farmers sometimes sweep the spikelets of this rye from their fields with brooms. I have named this variety of weedy, brittle rye *Secale cereale* var. *afghanicum*, although in other respects its characteristics (i.e. the dimensions of the grains, the morphology of the spike, etc.) do not differ from those of ordinary rye.

A Latin diagnosis of this new variety of rye follows below:

Secale cereale L. var. nova *afghanicum*, spica lineari elongata, rachis fragilis, spiculae 2-3-floris, aristae 2 vel 3 cm longae; caryosides clausae. Valde affine varietatii *vulgares Secalis cerealis*. Hab. frequens in segetis *Tritico vulgari* hiberno et *T. compacto* hiberno. Ar. geogr. Afghania, imprimis im parte meridionale prope Herat. (Figs. 7,8)

Thus, there is now a complete morphological series for brittleness and type of rachilla, ranging from the typical, brittle Afghanistani kind all the way to the forms that are definitely not brittle, and are cultivated in the north. The brittle races are characterized by segments jointed to the rachis at right angles; the segments of the rachillae are in the case of this variety easily dislodged (Fig. 9). In the case of the races cultivated in the north, and as the result of an unconscious selection for the durability of the rachillae (i.e. a characteristic promoting a full reaping of the harvest) the segments of the rachillae are jointed so that they can only be disarticulated with difficulty. It is actually possible to line up a whole series of transitional forms from the typical, in essence, wild forms all the way to the cultivated forms with securely joined rachillae. Precisely these forms with brittle spikes – the common attribute of the wild forms and of the wild ancestors – are found just within the general center of the variation of rye, *Secale cereale* (Table 7).

It must, consequently, be acknowledged phytogeographically that within the basic center of its variation and its type-formation, rye appeared in south western Asia just within the center of type-formation of soft and shot wheats. The distribution areas of soft and shot wheats and rye overlap in striking detail. Thus, we found, for instance, non-ligulate rye in exactly the same areas as non-ligulate wheat, i.e. in Badakhshan (N. Afghanistan) and in the adjacent areas of Shugnan and Roshan (in the mountains of Bokhara). Kinds of rye not shedding the grains, i.e. with enclosed grains (var. *clausopaleatum*), and with rough awns, which are widely distributed in Turkestan, Persia and Afghanistan, are at the same time found within the regions where non-shedding wheats of the types *inflatum*, *rigidum* and *speltiforme* occur (Vavilov, 1923).

In general, two major groups of variation can be outlined. One is typical of Afghanistan, Bokhara, Turkestan and eastern Persia; this group consists mainly of single-flowered, straw-colored rye with a presence of all kinds of transitions with respect to brittleness and pubescence; non-ligulate forms are also found there. The second group is typical of Armenia, Georgia and Asia Minor as well as northwestern Persia. There, a great variation is concentrated with respect to the florets of the spike and the shape of the hulls. Although red-spiked varieties are often met with also in Afghanistan, they constitute a considerable portion of the complex of weedy rye within the second group. There, black-spiked varieties, too, can be found. Northern Caucasus is linked to Transcaucasia as far as cultivated and weedy rye are concerned and displays a considerable number of red-spiked types. The presence in these areas of a noticeable brittleness of the rachilla links them also to the weedy rye of Afghanistan.

The local populations of Turkestan (Uzbekistan and Turkmenistan), Persia and Afghanistan know rye mainly as a weed. Rye is not grown in Syria, Palestine or in India, nor in China. S.I. Korzhinskiy launched the hypothesis that the weedy rye of Turkestan and Afghanistan had been a cultivated and widely grown grain, which somehow became replaced by wheat and barley. In my paper, 'On the Origin of Cultivated Rye' (Vavilov, 1917), I cleared up the falseness of this hypothesis and its non-conformity with the historical facts. Most of all, the epithets of rye among various peoples such as the Persians, the Tadzhiks, the Uzbeks and the Turkmeni of Turkestan, the Turks and the Afhgani as well as the Arabs, are 'gandum-dar', 'chou-dar' or 'chou-der' as well as 'dzhou-der', which literally mean 'the plants that infest barley or wheat' (for details, see *On the Origin of Cultivated Rye*). 'Gandum' in the Persian language means 'wheat', 'chou' or 'dzhoud' means barley. 'Dar' with a long 'a', as it is pronounced in some parts of southwestern Asia, corresponds to the present

Table 7. *Distribution of the main forms of rye* – Secale cereale

Varieties	Afghanistan, incl. Badakhshan	Bokhara	Khiva	Turkestan	Persia	Transcaucasia (Georgia, Armenia)	Turkey	N. Caucasus	S. part of European RSFSR	Ukraine	N. and central parts of European RSFSR	Siberia
vulgare – the ordinary form	+	+	+	+	+	+	+	+	+	+	+	+
monstrosum – branched	+	+	+	+	+	+	+	+	+	+	+	+
vulpinum – red-spiked	+	+	+	+	+	+	+	+	+	–	–	–
fuscum – brown-spiked	–	–	–	–	+	+	+	–	–	–	–	–
nigrescens – black-spiked	–	–	–	–	+	+	–	–	–	–	–	–
mediusculum – red-awned	–	–	–	+	+	+	–	–	–	–	–	–
eligulatum – non-ligulate	+	–	–	–	–	–	–	–	–	–	–	–
afghanicum – spike brittle from base up	+	–	–	–	–	–	–	–	–	–	–	–
articulatum – upper $\frac{2}{3}$rd of spike brittle	+	+	+	+	–	–	–	–	–	–	–	–
asiaticum – upper $\frac{2}{3}$rd of spike brittle, pubesc.	+	+	+	+	–	–	–	–	–	–	–	–
subarticulatum – upper $\frac{1}{3}$rd of spike brittle	+	+	+	+	+	+	–	+	–	–	–	–
velutinum – white, pubescent	+	+	+	+	+	+	–	–	–	–	–	–
armeniacum – red, pubescent	–	–	–	–	+	+	–	–	–	–	–	–
persicum – brown, pubescent	–	–	–	–	+	+	–	–	–	–	–	–
scabriusculum – top of palea scabrous, inflated	+	–	–	–	+	–	–	–	–	–	–	–
clausopaleatum – grains hulled	+	+	+	+	+	+	–	+	–	–	–	–
laxum – open-spiked	+	–	–	–	+	+	–	–	–	–	–	–
compactum – with a dense spike	+	+	+	+	+	+	–	–	–	–	–	–
longiaristatum – long-awned (> 5 cm)	+	+	+	+	+	+	+	+	–	–	–	–
muticum – awns 0.8–1 cm long	+	–	–	–	–	–	–	–	–	–	–	–
epruinosum – without waxy coating	+	+	+	+	–	–	–	–	–	–	–	–
viride – shoots green	+	–	–	–	–	–	–	–	–	–	–	–
auriculatum – with long auricles	–	–	–	–	–	–	+	–	–	–	–	–
unauriculatum – without auricles	+	–	–	–	–	–	–	–	–	–	–	–
Total	18	11	11	12	15	14	6	6	3	2	2	2

participle of the verb 'dashtan' ('to be', 'to be found in'); 'der' with a short 'e', as pronounced in other areas, is the present participle of the verb 'derun' (torment) and gives us the literal meaning, 'the plant tormenting barley or wheat'. The Persian epithet is widespread all over southwestern Asia from Turkey to India. The philological analysis of the epithet and its wide distribution all over the East bear witness to the fact that the local inhabitants of southwestern Asia, Persia, Afghanistan and Tadzhikistan from time immemorial have known rye as an annoying, weedy plant, hard to get rid of since it often completely infests the crops of wheat and winter barley or even displaces them, at present sometimes with disastrous results.

When I traveled around Bokhara, Persia, Turkestan, Khorezm and Afghanistan, I had a chance to listen to the complaints of the local inhabitants concerning the difficulties they met with in their fight against rye. Aitchison (1881, 1888) also heard such remarks in Afghanistan. Rye occurs as a weed of ill-repute also in Transcaucasia and Asia Minor (according to P.M. Zhukovskiy) and in Syria as well as in Palestine (according to Aaronsohn, 1910). The grains of wheat are difficult to distinguish from those of rye even for the best contemporary specialists in the sorting of grains, and therefore the complaints heard by the travelers concerning rye in Afghanistan, Persia, Turkestan and Asia Minor can be understood. In northern Caucasus – the granary of winter wheat – such an infestation of wheat by rye is a total disaster, causing a reduction in the value of the crop. An intense campaign is made by the land management for mowing the plots of land at the time when the rye flowers. At that time the contaminated fields have, literally, two tiers: the upper one is represented by the spikes of the rye, the lower one by those of the wheat. It is possible to mow down the rye carefully and not damage the wheat. An intense campaign is being made for this action in the war against rye infesting the wheat crop. In 1925, the land owners were free from taxes if they took such preventive measures.

THE APPEARANCE OF RYE AS AN INDEPENDENT CROP

However, rye does not occur as only a weedy plant in southwestern Asia. In the high mountains of Bokhara, Afghanistan and Asia Minor, it is possible to observe how, when ascending the mountains, rye gradually turns from a weedy plant into an independent crop. In the high mountain areas of Afghanistan, around Kabul and along the Khazariyskaya Road, in the mountains of Bokhara, in Shugnan and Roshan and in Badakhshan, rye has become a cultivated plant. The same can also be seen in southern Fergan. Also, its name changes there. The Tadzhiks of Badakhshan call this rye 'kal'p' or (in Roshan) 'lomak'. At the same time the inhabitants of the valleys still use the epithets 'chou-der' or 'gandum-der', i.e. 'the weed of wheat and barley'.

It is possible to draw a picture of the gradual displacement of wheat by rye and of the replacement in the high mountain areas of winter wheat by rye. According to our own observations, this substitution occurs especially intensely in Afghanistan on slopes facing north, but also on those facing south in

Hindukush at an elevation of 2000 m.s.m. and more. Pure crops of winter rye are associated with elevations of 2000–2500 m.s.m. G.A. Balabayev (1926) established for Zeravshan and other areas that a gradual displacement of winter wheat occurred from 2000 m and upward. Above 2500 m, the areas of spring cereals begin and types of winter rye, both the cultivated and the weedy ones, disappear.

G.A. Balabayev calculated the percentage of wheat infested by rye at various elevations and at different localities; accordingly, a clear picture was obtained for all the areas studied concerning the replacement of wheat by rye in relation to the progress toward more elevated sites. Thus, for instance, the picture shown in the first table developed with respect to the district of Dzhizak.

Settlement	*Elevation (m.s.m.)*	*Percentage of rye in wheat crops*
Yam-Dzhizak	345–457	3.1
Zaamin-Rabat	457–1000	8.9
Sanzar	350–1535	11.9
Turulyash	1535–2330	39.0

Among the settlements in the mountains of Zeravshan the increase toward the more elevated zones showed the proportions shown in the second table.

Zone	*Elevation (m.s.m.)*	*Percentage of rye in wheat crops*
Lower	1130–1448	13.3
Median	1535–1860	25.0
Upper	1886–2439	41.7

G.A. Balabayev and I observed that the process of replacement of wheat by rye in the upper zone (in Central Asia at elevations of 2000 m.s.m. or more) was faster than in the lower zone (from 300–1200 m.s.m.). In relation to the ascent into the mountains, the extent to which rye is distributed is noticeably increased.

As is well known, rye is biologically hardier, tougher and more frost resistant and it is therefore natural that, when the crops are transferred into more severe conditions, rye starts to replace the less hardy plants of winter wheat and, particularly, winter barley. Schindler (1923) wrote: 'The modest demands of this cereal is evident already from the geographical distribution of rye crops. It could actually be said that among the cereals it is the most modest one . . . The minimum temperature tolerated by rye when growing is 1–2 °C lower than that tolerated by all the other cereals. At a soil temperature of 4–5 °C rye can start to sprout in 4 days, while wheat under the same conditions requires 4–6 days . . . Neither is rye sensitive to superfluous moisture . . . Dry, sandy loam is considered as the true soil for rye, providing a steady yield. Even such sandy soils, where only lupines thrive, can be utilized for growing rye in combination with the lupines.'

Concerning its biology, rye has specific characteristics, conditioning it as a wild plant and correlating its growth with the more severe conditions on barren soils under severe winter conditions. In comparison to barley or wheat, its root system possesses a greater capacity for assimilation; it tolerates a higher acidity and is able to utilize less soluble compounds. Physiological investigations by Stoklase and others have cleared up the lesser demands of rye on the substrate in comparison with those of wheat or barley.

Factors not favoring the growth of wheat can be favorable for the cultivation of rye. The replacement of wheat by rye under severe conditions depends, naturally, on this factor, just as happens in the mountains of Central Asia and Transcaucasia. Such a process evidently also occurred during the progress of the cultivation of wheat toward the north. Gradually during the transfer of the crops from their basic focus, i.e. from southwestern Asia and Transcaucasia toward the north, northeast and northwest, the rye infesting them replaced them and became, itself, a pure crop. With much thanks to the wishes of Man, the former weed became a cultivated plant. On poor soils and under severe conditions in the north, rye was, according to a statement by A.D. Thaer 'a better gift from God' and, thus, the people began, without knowing it, to sow rye together with the wheat so that, at present, the farmers of the northern and central districts voluntarily sow only rye and not wheat.

If the above facts are taken into consideration, the history of the origin of cultivated rye becomes straightforward and very simple. The ancient crops of winter wheat and winter barley, when transferred from the south toward the north, east and west (the reduction in the varietal diversity of wheat proceeds in that direction and, consequently, the dispersal of the crops must also have gone in the same direction) brought with it the rye in the form of a weed. When cultivated under more severe conditions with colder winters and on poor, podsolic soils, rye began to overpower the weaker wheat and barley types. Barley was the first one to be 'knocked out'. Barley crops could only grow under the conditions such as in Transcaucasia, the Transcaspian region and the southern valleys of Turkestan. Gradually, during further progress toward the north, wheat, too, began to diminish, and, finally, in Siberia and the European parts of the U.S.S.R. and Germany, rye became a pure crop.

In central Persia as well, rye was tougher and hardier than wheat when grown under uniform conditions and rye, which had been constantly and intentionally weeded out by the farmers elsewhere, turned there into a pure crop. Due to the lesser hardiness of the wheat, rye was allowed to replace winter wheat toward the north, with much thanks to the efforts of Man.

It is interesting that, at the border of the 'struggle' between winter rye and winter wheat, the farmer has long since, and even now, purposely sown a mixture of rye and wheat. Not counting on success with winter wheat every year, the farmer willingly sows a mixture of grains of rye and wheat in the hope that, should the wheat fail during an unfavorable winter, at least the rye will survive and produce half a harvest. The peasants in Normandy and the Russian farmers in the Kuban and Terek River areas and the Stavropolis Governate sow both intentionally and involuntarily a great amount of so-called 'surshy' [mixed crop, D.L.], i.e. a mixture of rye and wheat. In France, they often grow 'le méteil', the flour of which is more valued than that of rye alone. According to information from Pallas, mixed crops of wheat and rye are common also in the Crimea.

It is still possible to trace in detail the entire process of the emergence of rye as a crop. In order to understand the origin of cultivated rye, it is necessary to associate it with the cultivation of wheat. Cultivation of rye originated from the cultivation of wheat and, in part, that of barley; therefore, it is natural that the areas overlap where the greatest diversity of soft wheat and of rye occur.

Thus, rye in the form of a field weed and the varieties thereof appear to be the ancestors of cultivated rye. The plant breeder, interested in the selection of varieties of rye, must turn his attention to the weedy rye that infests the fields of Transcaucasia, Asia Minor, Turkestan, Persia, Bokhara and Afghanistan.

Still, very little has been achieved in this direction. Plant breeders have worked mainly with the uniform west-European rye populations. Although these are suitable for the humid conditions of western Europe and the western districts of the European parts of the U.S.S.R., there should definitely be a greater interest in obtaining seed material from the areas of weedy rye for our southern and southeastern regions.

Our research has demonstrated that it is actually possible to find exceptional varieties, e.g. among the weedy rye of Persia, not only from the point of view of morphology but also that of physiology. In Persia, forms can be found that are especially tolerant to drought and which have, apparently, been selected for this for centuries in these drought-stricken areas (Vavilov, 1922). The southeastern races of rye, more tolerant to drought, are known to us from a complex of weedy rye in southwestern Asia. With respect to its productivity, the dimensions of the grains, its spikes and its sturdiness, weedy rye does not yield to the ordinary, cultivated rye.

Apparently, winter rye became cultivated in Europe and in Asia along two basic pathways: one from Transcaucasia and the other from the area of Turkestan, Afghanistan and areas adjacent to them. The difference between the two geographical groups of weedy rye indicates this fact.

During the dispersal of cultivated rye from the south toward the north, some regularity, similar to that concerning the geography of flax crops, could apparently be observed as demonstrated by the investigations made at the Institute of Applied Botany by V.I. Antropopov and V.F. Antropopova. Toward the north (in the Vyatsk, Vologod and Olonetz districts) races occur that are characterized by tall culms and open spikes, which are easy to thresh. Races with shorter culms and denser spikes that shed their spikelets are more frequent toward the south (Kuban, Crimea and Ukraine). A spreading form of growth is typical of many southern races of cultivated rye, such as the so-called 'bushy' type in contrast to the so-called 'tall-one' of the north.

SPRING RYE

The origin of cultivated spring rye was somewhat different. Studies of cultivated vegetation in Afghanistan allowed us a closer approach to the understanding of that process.

In the mountain areas of Badakhshan (northern Afghanistan) crops of winter wheat and weedy winter rye give way to those of spring cereals, i.e. barley and spring wheat, at an elevation of 2700–3000 m. Spring rye is a common field weed among spring wheat and, more rarely, barley. When studying a considerable-sized area of Badakhshan, from Faizabad to Ishkashim (in Pamir), we were able to demonstrate that the replacement of spring wheat by spring rye was particularly intense in areas with sandy soil. It was evident that such soil conditions worked as a factor of natural selection in favor of the spring rye. It is well known that rye does better than barley and wheat on light, sandy and only slightly loamy soils. Thus, for instance, cultivated spring rye is often grown on sandy soils in the southeastern parts of the U.S.S.R.

The spring rye of eastern Siberia, the Transbaikalian area and the Far East (locally called 'yaritsa') had, evidently, also developed into pure crops from weeds among spring wheat and barley. The expedition to Mongolia, led by V.E. Pisarev, found that spring wheat and barley in northern Mongolia are heavily infested. According to information from the expedition sent out from the Department of Applied Botany, the Siberian and Far East field crops had been adopted from Mongolia. As shown by experiments made by agriculturalists in Transbaikal and eastern Siberia, spring rye is hardier here than spring wheat. The winter cereals do not succeed without a snow cover during the winter months.

The development of spring rye into an independent crop can be followed in the high mountains of Bokhara, in Roshan and Shugnan. There, along the river Gunt and farther north, we were able to prove the presence of pure crops of a peculiar, tall-growing, large-grained type of rye with exceptionally large anthers, almost twice the size of those belonging to ordinary European spring and winter rye. Among this kind of rye, we discovered endemic, non-ligulate forms as well as races with semi-brittle spikes. We also found strange, low-growing rye, grown as pure crops on the border to Fergana (in the mountains of Bokhara) at the Pakshif pass. There, as well as in other areas, it is possible to find a complete transition from pure crops of spring wheat to pure crops of spring rye in every kind of proportion. It is interesting that in these high mountain areas of Badakhshan there are large thickets of wild, brittle rye, *Secale fragile*, growing on sandy soils at elevations of 2600 m.s.m.

All the stages in the evolution of cultivated rye can, in general, be followed in minute detail and the information can also be utilized for practical purposes within plant breeding.

The local origin of rye crops was without doubt polyphyletic. As can still be seen, the process of adopting rye into cultivation occurred simultaneously and independently in several localities. The diversity of the rye populations in different isolated areas bears witness to this. The 'yaritsa' [spring rye] with light-colored (mainly yellow) grains and short spikes, found in Siberia, is very different from the rye with large anthers, thick spikes and large grains found in Pamir or from the low-growing spring rye in the Karateginskiy Range. The spring rye of the Astrakhan district is distinguished by long awns and a strongly developed waxy coating on the spikelets, etc.

The distribution area of the wild, montane rye, *Secale montanum* – a species still considered by botanists to be an ancestor of cultivated rye – does not help solve the problem of the local origin of cultivated rye, *S. cereale*, with an accuracy that is satisfactory to plant breeders. According to Flaksberger (1913), the distribution area of *S. montanum* embraces Morocco, southern Spain, Sicily, Dalmatia, Serbia, Greece, Asia Minor, Persia, Turkestan, Central Asia, the Turkish portion of Armenia, northern Caucasus and Transcaucasia (Abkhazia and Armenia).

It appears to us that placing the beginning of the cultivation of rye in Turkestan (S.I. Korzhinksiy, Koernicke, A. Schulz), Serbia (Kerner) or the Balkans (Engelbrecht) is unfounded; it seems much more likely that in this case Persia, Asia Minor, Afghanistan, Bokhara and Transcaucasia, should also be included.

Growing rye crops appears rather to be the result of natural selection when allowing the weed, winter rye, to replace winter wheat or winter barley or letting the field-weed spring rye displace spring wheat on light soils toward the north and high up in the mountains or in the harsh climatic conditions of eastern Siberia. It is quite natural that such a selection among the populations of rye should occur when they appear in the company of wheat crops and barley as a specialized weed acting similarly to corn cockle [*Agrostemma githago*], Italian rye-grass [*Lolium perenne*] or bachelor buttons [*Centaurea cyanus*]. When those hereditary types that were most fit for cultivation were singled out, the people had only to multiply them.

This fact is in complete agreement with the origin of cultivated rye from weedy crops of wheat and barley according to the data available from linguistic, archeological and historical information, disclosing that the cultivation of rye developed much later than that of wheat and barley. This could be expected since the cultivation of one led to the cultivation of the other in crops of the two main cereals. All the information available to us concerning the cultivation of rye indicates that the beginning thereof took place at the start of the Christian era and no later than during the first centuries of that era.

Thus, the problem concerning the origin of cultivated spring rye can be solved to an extent such as needed by plant breeders and biologists for determining the concentration of the sources where the type-formation of a given crop occurred. It is evident that such a solution is only a first approximation toward the elucidation of the problem concerning the origin of the Linnaean species of *Secale cereale*. That problem affects a more general and more difficult area, i.e. that concerning the origin of Linnaean species in general.

In its genesis, the Linnaean species *Secale cereale* is apparently linked both to the perennial rye, *S. montanum* (in a paper by Engelbrecht, 1917, there are hints at a link between the cultivated and the weedy forms of rye, however, like other authors, Engelbrecht was led astray by the identification by A. Regel of the weedy rye of Turkestan as *S. montanum*) and the annual, brittle rye, *S. fragile*, of sandy soils. The latter is found in large amounts in the former territories of Ural, Turgay, Semireche, Syr-Darya and Transcaspia, in northern and north-eastern Afghanistan (Badakhshan) as well as, toward the south, in the Saratov and the Astrakhan districts. Recently, A.A. Grossheim (1924) discovered an annual type, new with respect to its spike, that reminds one of that of *S. montanum*; he named it *S. vavilovii* Grossh. and considered it a link between *S. cereale* and *S. montanum*. As studies by E.K. Emme in the cytology laboratory of the Institute of Applied Botany demonstrate, all these species are characterized by the same chromosome numbers (normally $n=7$, exceptionally $n=8$). *S. montanum* is able to cross with *S. cereale* (according to E. Tschermak). Thus we have a comparatively narrow group of types, which are genetically close to each other. In order to widen the approach to the genesis of these species, a detailed differential–geographical study of the wild species of rye is necessary, similar to that carried out in the case of cultivated rye. Asia Minor is apparently of exceptional interest for solving the problem concerning the genesis of the *Secale* species since, according to investigations made by P.M. Zhukovskiy, all the species of wild rye are found there in great amounts as well as a great variety of weedy *S. cereale*. According to observations by Zhukovskiy, *S. montanum* – a species ecologically different from *S. cereale* – occurs in the fields of wheat there and has become a weed in the vicinity of Yuzdar, displaying a number of races that show how much this species approaches *S. cereale*.

The problem concerning the origin of the species of *Secale* is, apparently, quite complicated, as demonstrated by a discovery in South Africa – in the Cape colony – of large amounts of a species close to both *S. montanum* and *S. cereale*, i.e. *S. africanum*.

According to investigations carried out by E.K. Emme, this species also has $n=7$ chromosomes which are impossible to distinguish from those of the species discussed above. The discovery of *S. africanum* in South Africa indicates – as an example of the geographical dispersal of species of *Secale* to the most isolated areas – a considerable age for the *Secale* species, the origin of which must go back to long before our present era.

Thus, the weedy rye of southwestern Asia appears to be the direct ancestor of cultivated rye. The wild species of *Secale* are no doubt close to *S. cereale* proper. Coherent forms (including brittle as well as non-brittle races) uniting these were possibly in the remote past distinguished into groups of populations, developing into the complex of specialized weeds infesting ancient wheat and barley crops. The discovery of brittle rye, *S. cereale* var. *afghanicum*, in Afghanistan and of a whole series of transitions with respect to brittleness in *S. cereale*, the establishment by A.A. Grossheim of a new, wild, annual rye, *S. vavilovi*, with a spike of the same type as the perennial *S. montanum*, and finally, the discovery by P.M. Zhukovskiy in Asia Minor of an exceptional concentration of species, including a large variety of weedy as well as cultivated kinds of rye, make it possible to reconstruct, morphologically as well as physiologically, a well-composed evolutionary series of rye. A detailed study of the multiformity of the wild species of rye in areas where their variation is concentrated promises to lead the investigator on to the problem concerning the development of the Linnaean species of *S. cereale*. However, at present, we do not believe that there is any reason for the suggestion that the ancestors of cultivated rye should be *S. montanum* and *S. fragile*, such as has been considered acceptable up until now. In this case, the differential–geographical method has allowed us to come closer to a solution of the problem concerning the ancestors right up to the establishment of the sources for the diversity of the types. So far, these sources have not been utilized by plant breeders because of the distraction on the part of *S. montanum*, which – although a closely related species – does not have any direct relationship to the genesis of cultivated rye.

OATS AS WEEDS AND THE PROBLEM CONCERNING THE ORIGIN OF CULTIVATED OATS

Studies of other plants have revealed that the manner in which cultivated plants become grown from weeds is fairly common.

In 1916, during our travels around Persia, we happened to come across a number of villages in the vicinity of Khamadan [W. Iran] where large amounts of emmer (*Triticum dicoccum*) were grown. In general, the cultivation of emmer was not known in Persia, except for in the Armenian settlements established there about 300 B.C. by Abbas the Great. In that area, it was still possible to see a few isolated fields of this crop, which was imported from Turkish Armenia. Studies of these emmer fields revealed a heavy infestation by oats, *Avena sativa*. This is that much more remarkable since oats were not known in cultivation in Persia, nor in Afghanistan, Bokhara, India or Turkmenistan and since, in Persia, oats appeared to be an unavoidable attribute only of emmer crops. In some fields, it was even possible to observe that the oats were displacing the emmer.

During detailed studies of the crops of these Persian oats and a comparison between them and the kinds known in European crops, it turned out that what we had seen of oats among emmer was represented by independent races, which until then were not known to be cultivated. They were distinguished by peculiar, unilateral, atypical panicles, so far not described by botanists in the case of cultivated oats. Both the paleas and the glumes of this kind of oats are much longer than those of ordinary oats. A full range of glume color from whitish to dark brown could be observed. I singled out the forms discovered as special botanical varieties under the names of *A. sativa* var. *iranica* Vav. (white-grained), var. *persica* Vav. (yellow-grained) and var. *asiatica* Vav. (brown-grained; see Figs. 10 and 11).

The fact that we had found peculiar races of oats in the form of weeds among crops of an almost extinct kind, attracted our attention and made us also investigate the weeds among emmer in other countries.

The studies of emmer (*Triticum dicoccum*) in the Pri-Kama region, in the upper Kazan, Ufa and Simbirsk Governates revealed similar facts. There, too, cultivated emmer was heavily infested by oats, and there, as in Persia, we found a multitude of new, peculiar races and a species, *Avena diffusa*, previously not known in crops. Thus, for instance, it was strange to find *A. sativa* with leaves covered by a dense pubescence, f. *pilosiuscula*, a trait definitely rare among ordinary oats.

In addition to the common forms, characterized by florets not firmly articulated to the panicle and falling apart (as grain) when threshed, there were also peculiar races with firmly articulated spikelets that did not fall apart when

Fig. 10. Weedy oat from a field of emmer in Persia.

Fig. 11. Spikelet of weedy oat from a field of emmer in Persia.

Fig. 12. Different races of weedy oats (*Avena sativa* L.) From a field of emmer along the Volga. 1–3: Typical races with firmly articulated flowers and short pedicels, found among emmer; 4–6: ordinary races or weedy oats with brittle, disarticulating florets (separating into different florets when threshed); the races illustrated differ from each other with respect to the extent of pubescence at the base of the florets.

threshed (Fig. 12). The rachis, on which the two florets sit in the case of such forms, is short and sessile; because of this the articulation of the florets becomes rather firm so that, if you try to separate the florets from the panicle, they do not disarticulate from the rachis as usual in the case of cultivated oats, but only a portion of the hull is torn off. When passing through the threshing machine, such kinds of oats always leave pairs of spikelets, i.e. the florets (grains) remain united (*firme coalitae*). Among the races of oats in the Volga area we discovered, in general, races with panicles morphologically somewhat reminding one, in their compact structure, of the spikes of *Triticum spelta* or *T. dicoccum*. The separation of such spikelets and florets is more difficult than in the case of ordinary, hulled oats. Such forms of oats are found exclusively among crops of emmer. Among these peculiar forms of oats in emmer we also found a number of varieties, distinguished by the color of the glumes and by the awns: white-grained ones with awns (i.e. var. *kasanensis* Vav.) and without awns (var. *volgensis* Vav.) and yellow-grained ones with awns (var. *bashkirorum* Vav.) and without awns, (var. *segetalis* Vav.). The names given refer to those of the localities and the areas where emmer is grown (around lower Volga and in the Kazan Governate).

Detailed investigations of oats among emmer, made by A.I. Mordvinkina, led to the distinction of a multitude of transitional forms ranging from the races typical among emmer to common forms impossible to separate from the usual kind of cultivated oats with respect to the articulation of the spikelets. Among all the varieties growing among emmer, there are special races with short bristles at the base of both the grains (florets), and among var. *kasanensis* Vav. some peculiar forms with long bristles at the base of both florets. So far these forms are definitely not known among cultivated oats.

A number of races were discovered among emmer grown in Dagestan. These have unusually broad glumes. In the Volga area, a new race of *Avena strigosa*, with short awn-like appendages also occurs as a weed.

Studies of emmer brought to northern Caucasus (Dagestan, Ossetia) from Transcaucasia (Armenia, Georgia and Azerbaidzhan), Asia Minor, Bulgaria, Crimea and those grown by the Basques in the Pyrenées, as well as some from Abyssinia, invariably revealed the presence among them of oats as a constant attribute; in these cases, the emmer crops actually turned out to be the 'preserver' of a diversity of oats, and of a peculiar concentration of kinds of oats not known in cultivation. At the same time, ordinary types, such as those well known in cultivation, were also found there. We began to study these weedy oats in detail just as we had done in the case of emmer itself. It turned out that the centers in the world where emmer still remains in cultivation were actually at the same time also the centers of variation in the case of oats. In Transcaucasia, where there are no cultivated oats, there is a large variation of *Avena diffusa*, characterized by a series of glume colors and by firmly articulated florets, although not as clearly expressed as in the case of the group around the Volga. In Transcaucasia, there are also constant forms of var. *transiens* with dark-brown bristles covering the glumes. This type was so far known only as rare hybrid forms, intermediate between oats and the wild oats of the Balkan peninsula. Among winter emmer, obtained with great difficulty from the Basques in the Pyrenées, we found peculiar, late-ripening races of *Avena sativa*. In Abyssinia, the emmer is, just like other cereals, accompanied by peculiar forms of oats, *A. abyssinica* and so on. Emmer from Bulgaria was infested by an enormous amount of oats, represented by different varieties.

Table 8 provides a brief review of the diversity of the varieties and the races ('jordanons') of oats found by us so far among emmer crops. By now, we have studied about 100 samples of emmer and we do not doubt that, in the future, still more new races of oats will be found when investigating emmer from Transcaucasia, Bulgaria and other countries.

Oats appear to be common weeds also among einkorn [*Triticum monococcum*], the cultivation of which is preserved by the Tatars on small plots in the Crimea, in Asia Minor, Bulgaria and in the northern Caucasus. We were able to confirm the presence of weedy oats in samples of einkorn from Bulgaria, Transcaucasia, N. Caucasus and Crimea studied in our laboratory. In the northerly areas (e.g. the Vyatskaya Governate), oats often appear to infect crops of barley (the local epithet of such a crop is 'soritsa', [i.e. the 'weedy one'; D.L.] just as observed in Transcaucasia.

In the Povolzhe area, the amount of oats in emmer takes on such dimensions that, statistically, these crops can be distinguished from those of emmer by the epithet 'supolby' ['emmer companion'; D.L.].

In this case, we are talking only of oats belonging to the cultivated type and sharply distinguished from wild oats, *A. fatua* and *A. ludoviciana*, which frequently, together with ordinary oats, infest emmer, einkorn and barley crops (in Transcaucasia and Bulgaria).

THE DISPLACEMENT OF EMMER BY OATS

Farmers sowing emmer around the Volga and in Crimea and Transcaucasia complain constantly about oats infesting their emmer. The oats lower the quality of the groats and the flour, spoiling them somewhat, and therefore oats are considered an undesirable admixture. Emmer heavily infested by oats is ordinarily not used for groats or flour but only as forage for cattle. According to all information available to us, oats are considered a hardier plant than emmer in the north. Observations by local agronomists in the Pri-Kama region unanimously indicate that the emmer there is being displaced by oats. In a paper from the first half of the last century (Belikov, 1840) we can read as follows: 'Who among us does not know that oats overpower emmer so that from an almost imperceptible number of grains of oats, falling by chance among grains of emmer, after a few harvests, the field, even the entire desyantina [1.2 acres; D.L.] appears to have been sown with oats alone, among which a few specimens of emmer are by chance mixed in'. This means that in a given case we can observe a development similar to the penetration of rye into wheat and barley crops. According to some folklore, collected by E.A. Stoletova from the Kazan territory, there is, for instance, a statement used by the local inhabitants: 'Ask any old man, ask any young man and all will tell you that emmer degenerates into oats' (Stoletova, 1925). Frequently there are instructions to the peasants that they should stop sowing emmer because the wild oats will stifle it.

The ancient authors knew oats only as weeds among crops of other cereals (Theophrastes). Hehn explains the German epithet 'Hafer' as 'Bockskraut', i.e. a weedy grass [actually, 'Devil's grass'; D.L.].

THE ORIGIN OF CULTIVATED OATS

In view of the above-mentioned facts, the problem concerning the origin of cultivated oats has suddenly become clearer and many intricate data concerning the geography of the varietal diversity of oats and examined by us with respect to the center of origin, can now be understood and make full sense when cultivation of oats is associated with the cultivation of emmer, which is about to die out. The link between the cultivation of oats and the oldest cereals in the world, emmer and einkorn, becomes at the same time definitely obvious. Just as rye was brought along with soft wheat, when the latter dispersed toward the north, the ancient crops of emmer during their dispersal over the Old world took along a collection of weedy oats. When the crops were transferred to the north or into much more severe conditions, the weedy oats began to displace the main crop. Thanks to these conditions, more favorable for oats, they turned into crops in their own right. In that manner, the cultivation of oats originated within a number of areas. It is most likely that this led to the origin of the crops of oats in the Pri-Kama region. In the Vyatskaya Governate, many kinds of

Table 8. *Varieties of the kinds of oats that infest emmer (*Triticum dicoccum*)*

	Area					
	Around Volga	Transcaucasia	Persia	Crimea	Altai	Spain
Botanical variety						
Avena diffusa						
var. *mutica*	+	+	–	+	+	–
aristata	+	+	–	+	+	–
aurea	+	+	–	+	–	–
krausei	+	+	–	+	+	–
grisea	+	+	–	+	–	+
cinerea	+	+	–	+	–	–
brunnea	+	+	–	+	–	–
montana	–	+	–	–	–	–
transiens	–	+	–	–	–	–
setosa	–	+	–	–	–	–
iranica	–	+	+	–	–	–
persica	–	+	+	–	–	–
asiatica	–	+	+	–	–	–
volgensis	+	–	–	–	–	–
kasanensis	+	–	–	–	–	–
segetalis	+	–	–	–	–	–
bashkirorum	+	–	–	–	–	–
A. orientalis						
var. *tatarica*	+	–	–	–	–	–
obtusata	+	+	–	–	–	–
A. strigosa	+	–	–	–	–	–
Racial characteristics						
Shoots						
Pubescent	+	–	–	–	–	–
Glabrous	+	+	+	+	+	+
Growth form						
Caespitose	+	+	+	+	+	–
Semicaespitose	–	+	–	–	–	+
Leaf blade						
Pubescent	+	–	–	–	–	–
Glabrous	+	+	+	+	+	+
Culms						
Tall	+	+	–	+	+	+
Short	–	+	+	–	–	+
Nodes						
Glabrous	+	+	+	+	+	+
Pubescent	+	+	+	+	+	–
No. of culms						
Many	–	+	+	–	–	–
Few	+	+	–	+	+	+

Table 8. (*cont.*)

	Area					
	Around Volga	Transcaucasia	Persia	Crimea	Altai	Spain
Panicle						
Spreading	+	+	+	+	+	+
Unilateral	–	–	+	–	–	–
Single-crested	+	+	–	–	–	–
Articulation						
Firm	+	+	–	–	–	–
Not Firm	+	+	+	+	+	+
Pubescence at base of grain						
Only on lowest floret	+	+	+	+	+	+
On both florets	+	–	–	–	–	–
Rachilla						
Long	+	+	–	+	–	–
Short	+	+	+	+	+	+
Dimensions of hulls						
(a) Very long	–	+	+	–	–	–
Long	+	+	+	+	–	+
Short	+	+	–	+	–	+
(b) Wide	+	+	–	–	+	–
Narrow	+	+	–	+	+	+
Awns						
(a) Long	+	+	+	+	+	+
Short	+	+	–	+	+	–
(b) Coarse	+	+	+	+	+	+
Fine	+	+	–	+	+	–
Shape of grain						
Elongate	–	+	+	–	–	–
Ovate	+	+	–	+	+	–
Type of cultivated 'Probshteyskiy type'	+	+	–	–	+	–
Type intermediate between needle-like and the cultivated 'Probshteyskiy' types	+	+	–	+	+	+
Vegetative period						
Long	–	+	–	–	–	+
Intermediate	+	+	+	+	+	–
Short	+	+	–	–	+	–

locally cultivated oats cannot be distinguished from the kinds of oats found in emmer. The type called 'chernoviy' ['the black one'; D.L.] recently produced by the Vyatskaya Experimental Station, and based on local oats, is characterized by firmly articulated spikelets, which do not separate into florets (grains) when threshed, which, as we have seen, is often typical of a number of the weedy oats among emmer in the Pri-Kama region.

In Abyssinia, Kostlan (1913) observed the displacement of barley crops by oats in the mountain areas; in the case of late barley crops (ripening in July) and during cold and harsh weather, half the barley crop when harvested consisted of oats. Such grain, sown in the following years, eventually gave rise to a pure crop of oats.

Just like rye, oats are, according to all historical and archeological data, a younger plant than wheat, barley or emmer, the information about which dates back to thousands of years before our present time. As in the case of rye, the first information about crops of oats does not go further back than to the first centuries of the Christian era (Schulz, 1913). The comparative youth of the cultivation of oats can be understood, since these plants were brought along by other ancient crops when dispersing from the south toward the north.

The reason for the dying out of emmer cultivation remains to be understood. At the same time as it was transferred into areas where it could be cultivated, the crops of naked-grained wheat, grown rather than the hulled emmer, which is less suitable for grinding, were to a considerable extent responsible for the reduction of the acreage under emmer, which also, because of natural selection, had been replaced by oats. Just as in the case of rye, oats were taken into cultivation by Man voluntarily or even against his will.

The absence of a single center of origin of oats can also be understood from this. Oats, like rye, were taken into cultivation simultaneously and independently in various places. This happened not only in the case of different Linnaean species but also within the limits of the single Linnaean species, *Avena sativa*, itself.

As we have seen, the centers of emmer in Transcaucasia, the Pri-Kama region, Spain, Abyssinia and Germany are, to a great extent, represented by different morphological and physiological types, accompanied by corresponding groups of weedy oats. In the past, the weedy oats associated with early-ripening emmer grown around the Volga, turned out to be early-maturing. It was natural that different lines of such oats among emmer should initially give rise to different groups of cultivated oats, as shown by the strains still preserved in plots of cultivated emmer. When considering the very large amount of material of oats among emmer, preserved by primitive people in Eurasia and Africa, there is no doubt that it is still possible to interpret the details of how oats became cultivated and to trace the different lines or groups of oats, which were associated with the different groups of emmer. However, this process cannot be completely retraced since the original cultivation of emmer still occupies the stage to a considerable extent and much of the history about how oats became cultivated cannot be fully traced either. In this respect, the most interesting fact is the petering out toward the north of emmer cultivation, which by now is almost extinct. This is even more obvious in the case of the other ancient, cultivated plant, einkorn, now practically extinct as a crop. Displaced in the north by oat crops, emmer and einkorn themselves died out. Spikes of emmer (*Triticum dicoccum*) are only rarely recorded, demonstrating the link between the oats grown at present and the basic crops of emmer that are now dying out.

It is impossible to understand fully the geographical diversity of oats cultivated in the Old World without associating them with isolated plots of emmer and, perhaps, also with such plots of barley. For instance, in Abyssinia, the plots of emmer still act as 'carriers' of an exceptional variety of oats. New types, suitable for introduction, should be sought by plant breeders in such basic centers of variation. There, chances exist for finding new races, which could be valuable for this or that reason. Experiments, testing strains at our Steppe Station, have shown that some races of oats from emmer crops in the Kazan area, surpass the west-European races with respect to productivity under the conditions present on the Russian steppes.

The cultivation of emmer, the origin of which extends, as we have seen, along the coastal area of the Mediterranean, where an exceptional diversity of forms is concentrated in the mountain areas and where there are areas of multiformity adjacent to it has, thus, taken a close part in the creation of the crops of European oats. During the advance to the north, the cultivation of oats developed in the form of different strains from the weeds in the ancient crops. Without any knowledge of emmer crops, it is impossible to understand the genesis of cultivated oats, just as it is impossible to understand the origin of cultivated rye without associating it with the more ancient cultivation of wheat and winter barley.

Hence, it is not such an incredible fact that the roots of cultivation of oats in the north must – just as in the case of the other northern crop, rye – be searched for in the mountains of the more southerly areas. In the case of *Avena sativa*, this may be a question of the countries extending along the northern part of the Mediterranean coast from the Balkan peninsula to the Pyrenées.

We still have not enough reason to convince ourselves that all the cultivated oats developed in this manner. As we have seen in the chapter on the centers of type-formation, cultivated and wild oats have a polyphyletic origin. The cultivation of oats could also belong to the same type as that of *A. byzantina*, which originated along the Mediterranean coast, perhaps in Africa, where it still constitutes an important crop. Apparently, *A. brevis* and *A. strigosa* came from weeds among barley and other cereal crops in northwestern and northern Europe. They are still, to a considerable extent, weedy plants in Byelorussia and adjacent areas. Their hardiness on light and sandy soils, on which they are still grown in England and France, served no doubt as a favorable factor, operating to the benefit of the selection of these oats into cultivation.

Further studies are necessary, but we do not doubt that the establishment of a connection between the cultivation of oats and its development from that of other crops will be of decisive importance for a detailed elucidation of the genesis of cultivated oats.

When retracing the history concerning the origin of cultivated oats, we must necessarily take into consideration the considerable complications that caused the disappearance of the basic crops, that brought oats with them. As far as the fragments of emmer cultivation are concerned, i.e. that which once occupied a broad belt from the Pyrenées to the Caucasus (as demonstrated by the isolated plots preserved of the now almost extinct races of emmer, *T. dicoccum*), it is necessary to restore the entire picture of the complicated history of that crop and of the species of oats associated with it. The polyphyletic origin of oats is an aggravating complication. In this respect, the matter of how cultivated rye was generated is, however, in every sense much simpler. In the Asiatic countries and within the main center of soft wheat, it is still possible to retrace the process of type-formation and to follow the history of rye, in detail, during its dispersal from the south toward the north.

EXAMPLES OF THE ORIGIN FROM WEEDS OF OTHER PLANTS

No doubt whole groups of oil-producing cruciferous plants developed from weeds infesting other, more ancient crops. As is well known, the cultivation of flax originated from a multitude of specialized weeds – *plantae linicolae* – the seeds or fruits of which are, as far as their dimensions are concerned, similar to those of linseed, so that they cannot easily be distinguished from the latter when sorting and cleaning the seeds. Some of these weeds are at the same time used as cultivated plants. The garden rocket (*Eruca sativa*) is such an oil-producing plant, grown in Persia, Afghanistan, Bokhara and India (Sinskaya, 1925). This plant, common in Central Asia, is invariably a companion of cultivated flax; under harsher conditions it frequently displaces the flax. Crops can be seen where it is difficult to determine whether flax or garden rocket was sown. The suitability of garden rocket as an oil-producing plant turned it from a weed into a culivated plant, when grown in areas where flax succeeds badly.

In the Caucasus (especially in Armenia and Georgia) and in Asia Minor as well as the Altai, but also in northern areas, where flax is grown together with garden rocket, the false flax, *Camelina sativa* and *C. linicola*, occur also as specialized weeds and at the same time as cultivated plants, and behave similarly to garden rocket (Tsinger, 1909). The cultivated, large-seeded corn-spurries (*Spergula linicola* and *S. maxima*) have a similar relationship to flax.

In his excellent paper, N.V. Tsinger (1909) investigated in detail the picture of how this group of *plantae linicolae* was selected from flax.

In Transcaucasia (Armenia and Georgia) it is possible to observe clearly the displacement of flax by false flax and colza (*Brassica campestris*; [most likely var. *silvestris*]) in the high mountain areas. There, as hardier and early ripening plants, the false flax and colza displace common flax and become, thanks to the wishes of Man, grown as oil-producing plants. According to the observations made by E.A. Stoletova among the highest mountains of Armenia, the false flax and the winter cress are successful as oil-producing plants, while lower down, in the median belt, flax grows well, while still lower down, in the lowland, sesame (*Sesamum*) and castor beans (*Ricinus*) are preferably cultivated.

A whole series of cultivated species of wild mustard and rapeseed (*Brassica* and

Sinapis) are linked to corresponding weedy plants in the crops of various cultivated plants. Tatarian buckwheat (*Fagopyrum tataricum*) is a bad weed in the crops of ordinary buckwheat (*F. esculentum*). In Siberia it occurs as a common weed among the spring wheat. However, at the same time, the Tatarian buckwheat replaces the common buckwheat in the elevated mountain areas of Altai and Kashmir.

The origin of some cultivated leguminous plants is similar: vetches (*Vicia sativa* and *V. pannonica*), possibly also the vetchling (*Lathyrus sativus*), and peas (*Pisum arvense*) originated from admixtures to cereal grasses in the mountain areas of Asia. Narbonne vetch, *Vicia narbonensis*, is a common weed in Spain and in Transcaucasia but is a cultivated plant in Italy.

In the Crimea, einkorn (*Triticum monococcum*) infests, together with oats, the crops of emmer (Barulina, 1925). It is also found in France (Schulz, 1913).

Coriander (*Coriandrum sativum*) occurs as a weed among cereal crops in Transcaucasia and in Asia Minor while being cultivated in the same areas as well as in Central Asia.

In Asia Minor, P.M. Zhukovskiy noted that a common weed among wheat, *Cephalaria syriaca*, was taken into cultivation. When displacing the wheat in the mountain areas, this plant is locally utilized as an oil-producing plant. Wild melons (*Cucumis trigonus*) occur as serious weeds on tilled fields in northern Afghanistan and in Turkestan. Crops of Chinese jute (*Abutilon avicennae*), a common weed in many crops, and hemp-mallow (*Hibiscus cannabinus*) have a similar origin. Wild carrots (*Daucus carota*) occur as common weeds in vineyards and vegetable gardens in Afghanistan and Turkestan, where they practically invited themselves to be cultivated by the local agriculturists.

The number of such examples can, no doubt, be increased considerably if cultivated plants are studied closely. However, we are already able to state that a large group of plants, known to us in the mountain areas and in the northlands and Central Europe as cultivated plants, occur in the south and in the centers of their type-formation as more or less aggressive weeds. Thus, in essence, a direct and natural pathway can be outlined for the provenance of this secondary group of cultivated plants. When advancing into areas with severe conditions, up into elevated mountains, or toward the north, these weeds displace to a considerable extent the original crops, both in the fields and in the vegetable gardens. It could be said that the European lowland is at present covered to a great extent by plants that were once weeds. The two main European crops, rye and oats, are typical representatives of such an origin.

PRIMARY AND SECONDARY GROUPS

Thus, it is possible to distinguish two groups of cultivated plants. The first group comprises the basic, ancient cultivated plants, known to Man only in their cultivated state. These we shall call *primary crops*: they are, e.g. wheat, barley, rice, soybeans, flax and cotton. The second group, the *secondary crops*, is no less extensive and comprises all the plants that derive from weeds infesting the primary main crops, especially rye, oats, false flax, garden rocket, large-seeded spurries, Tatarian buckwheat, an number of leguminous plants, American hemp and so on. The natural pathway for being taken into cultivation independently was, in the case of this secondary group, the transfer of the basic crops from the valleys into localities high up in the mountains, to sites in more northerly areas, into more severe conditions or onto poorer soils. At the same time as the transfer of the crops toward the north, into worsening climates or soils, a natural differentiation took place, selecting the more frost-resistant, earlier ripening, hardier and more tolerant plants. This process was assisted by the wishes of Man. The agriculturalists counted on this process as a fact. Hence both the practical and the scientific interest that should be given such species of field weeds at the main sources of their type-formation and variation can be understood. However, all these have not yet been fully explored by the investigators.

CHAPTER 4 MOUNTAIN AREAS AS CENTERS OF AGRICULTURAL CROPS

Representatives of the origin of crops have, in general – just like the basic, major civilizations – been associated with major river systems. In his book, *Civilizations and the great historical rivers*, the geographer Lev Menchikov developed a detailed hypothesis for a link between the major river systems and civilizations. The main civilizations of the Old World were situated mainly in the basins of the Nile, Tigris, Euphrates, Ganges, Indus, Yangtze and Hwang-Ho rivers and, consequently, the idea developed that the cultivation of plants also started in the valleys of these great rivers.

However, during the last couple of years, a closer study of southwestern Asia, Asia Minor and northern Africa has demonstrated that all the varietal diversity of field and vegetable crops is hidden mainly within mountain areas. The mountain regions proved to be where the varietal and racial multiformity of plants is concentrated and, consequently, the hypothesis concerning the beginning of agriculture in the valleys of the great rivers appears to be basically erroneous.

The mountain areas present, of course, optimum conditions for explaining the varietal diversity, the differentiation into varieties and races and for the preservation of all kinds of physiological types. At the same time, the mountains are excellent as isolated areas for the preservation of the varietal wealth. Just as in the case of the wild flora, where the Caucasus, the mountains of Bokhara, mountainous Turkestan, Afghanistan, Asia Minor and Abyssinia, as well as the Cordillera of South America, they appear as 'collectors and guardians' of specific and generic diversity; they are also the 'guardians' of the racial variation of many cultivated plants.

However, it would also be a great mistake to think that the concentration of racial variation in the mountains of southwestern Asia, Asia Minor and Abyssinia is the result of the diversity of ecological conditions only. No doubt, it can also to a great extent be explained on the basis of historical and geographical circumstances, which concentrate the process of type-formation of various Linnaean species just to some particular mountain area.

However varied the conditions for growing plants in Afghanistan are –

ranging from that at the limit up to which crops can be grown in Hindukush to that of the subtropical climate in the areas adjacent to India – some important species of wheat (e.g. *Triticum durum* and *T. dicoccum*) are definitely not found there, although they are common in the mountains of Abyssinia. For historical reasons – mainly proximity to the center of type-formation – Afghanistan is the area where the greatest variety of soft and club wheats is exclusively concentrated, while at the same time the uniformity of barley is striking there. The mountain areas of the Alps and the Pyrenées do not – as far as is known – display any concentration of varietal diversity. The naked-grained oats seem for similar reasons to be concentrated in the mountains of China. An infinite number of such facts can be cited and they all bear witness to the fact that a decisive role in the development of a particular mountain area into a center for type-formation is played by historical circumstances and not only by the diversity of the environment.

When contemplating the process of how agricultural crops developed, we must inevitably acknowledge that the period of the great civilizations, uniting a society of many tribal people, was, of course, preceded by a period when people were isolated as tribes and small groups, inhabiting secluded areas. The mountain areas may be considered as excellent refuges for such purposes. The control of the great rivers, the regulation of the Nile, Tigris and Euphrates and the other major rivers, required an ironhard, despotic organization, the building of dams and the regulation of irrigation; it needed organized mass operations such as the primitive agriculturalists could not even dream of. It is, therefore, so much more likely that, just like the centers of varietal diversity, the centers of the first agricultural crops should be found in mountain areas. The regulation of water for irrigation does not require great effort there. Mountain streams can easily be diverted for irrigation by gravity. The areas of the high mountains often provide opportunities for non-irrigated crops as well because of the large amounts of precipitation in the elevated mountain belts. In the agricultural mountain areas of Bokhara it is still possible to observe various primitive stages in the evolution of agriculture, actually preserved unchanged over thousands of years and still illustrating the different stages of the agricultural civilization.

The differentiation of cultivated plants into races was, no doubt, also favored by the mixture of ethnic societies in the mountain areas of southwestern Asia and northern Africa. Ethnological maps of the Caucasus, the mountains of Turkestan, Afghanistan, Bokhara and northern India can be said to also reflect the variety of the racial composition of the plants cultivated in those areas. The mountain areas mentioned represent not only centers of variation of different kinds of cultivated plants but are also foci of diversity of human tribes.

Vertical belts and horizontal zones in relation to the differentiation of cultivated plants into races

The interest in mountain areas as centers of varietal diversity is especially important in connection with the fact that the dispersal of varieties in relation to vertical belts in the mountains coincides locally, to a great extent, with their local horizontal dispersal.

In the high mountains of the Caucasus, Bokhara and Badakhshan we found Siberian and Archangelian types of spring wheat and barley, characterized by soft awns, easily threshed grains, low stature and narrow glumes. The people of Pamir call the Siberian and North European strains of sweet peas 'pelyuski'. Flax in the elevated mountain areas approaches the long-staple flax of the northlands with respect to the early maturation and the few, tall stems. Northern strains of cultivated plants are also currently preserved in the Asiatic Cis-Pamir, in areas adjacent to the center of type-formation of a number of plants. In the valleys and the foothills of the mountains, races that are definitely different but typical of extensive territories of southwestern Asia and northern Africa are often concentrated. When descending from the mountains, it is possible to observe an interesting change in types from the extremely early-ripening high-mountain races to late-ripening lowland types, frequently associated with irrigated cultivation.

The maximum amplitude of racial variability is usually concentrated in mountain areas, however, not at the extreme limit of cultivation but considerably lower down, e.g. in the case of soft wheat in southwestern Asia according to our own observations at elevations of 800–1800 m.s.m. When ascending from there, the diversity of varieties and races becomes noticeably reduced with respect to types and kinds of crop. Such a reduction in the variety of plants and races is, as a rule, also characteristic during the advance of the crops from the south toward the north.

The mountain areas of southwestern Asia, Asia Minor, Caucasus and those of northern Africa still display the full range of variability of those types of cultivated plants that are able to populate the enormous territory from Mesopotamia all the way to the Siberian taiga [boreal forest; D.L.] and to the limit of agriculture in northern Europe. In the case of cultivated plants, knowledge of the global geography of type-formation leads to the acknowledgement of the exclusive roles played by the Asiatic, African and South American mountain areas. These act as 'granaries' of different strains and need detailed investigation. Many primitive forms, the cultivation of which was initiated long ago can still be found there and, hence, the enormous interest that they represent both for practical and theoretical investigations of these areas can be understood.

CHAPTER 5 THE ORIGIN OF CULTIVATED HEMP AND OF CROPS BELONGING TO THE PRIMARY GROUPS OF PLANTS

The origin of cultivated plants that belong to the secondary group is perfectly clear to us. The dynamics of how different plants were taken into cultivation can be traced in minute detail. The problem concerning the origin of the cultivated plants, referred to by us as 'primary', is considerably more difficult. How did the basic crops of wheat, barley, flax, rice and the many leguminous plants become cultivated? The solution of this problem is much more complicated and we may never succeed in solving it completely. In the case of many plants belonging to this group, the link between the wild and the cultivated forms has already been lost. We are faced only with the result of a clearly expressed selection, executed over thousands of years, and to trace all the historical links requires a lot of more or less believable imagination. However, as demonstrated by the investigations described below, in some cases it is not hopeless and, while not resorting to unnecessary fantasies, the historical process can, in essence, be traced.

During the studies of the botanical complex of hemp, *Cannabis sativa*, in the southeastern parts of European U.S.S.R. and in Asia we were led to pay attention to the problem concerning the origin of this crop as well. Nevertheless, we do not know when, by whom or where hemp was taken into cultivation; however, the natural process of the onset of hemp cultivation turned out to yield to a detailed study and, perhaps, there is no other ancient plant but hemp for which the dynamics of how it became cultivated can be successfully traced with such certainty.

Wild and naturalized forms of hemp

The wide distribution of wild hemp in the Old World is well known by botanists. Statements concerning observations of such types of hemp are constantly seen in botanical literature.

Until recently, the majority of authors considered this wild hemp to be a form that had escaped from cultivation, i.e. had become naturalized. No essential difference could be observed between the cultivated forms and those not cultivated and all can be referred to the Linnaean species *Cannabis sativa*. The assumption of a wild origin of the thickets of naturalized hemp appeared to be self-evident.

The mass reproduction of the form of wild hemp – in particular under conditions excluding the possibility that it had been sown by man – involuntarily led some authors to reason that, perhaps, this hemp was genuinely wild. However, until recently, there was no definite proof supporting either opinion. Direct observations of hemp that had escaped from civilization indicated the possibility that it had become naturalized. In areas where hemp is grown, it can go completely out of bounds, for instance on vacant land and fallow fields; in other words, it is relatively easily naturalized. Even in America, where hemp was brought by European settlers, it is frequently met with as 'an escape from cultivation' (Britton, 1889). I, myself, was able to collect 'escaped' hemp around St. Paul, Minnesota.

Enormous thickets of 'wild' hemp are known both from European and Asiatic parts of the U.S.S.R. Especially large thickets of hemp occur in ravines in the valleys of northern Caucasus, for instance in the area of Nal'chik, and to the south of Rostov, where harvesting of 'wild' hemp has reached practical importance. Ledebour (1846–1851) mentions hemp in a '*quasi-spontaneus*' state all over the southeastern as well as the central and western portions of European Russia, Transcaucasia (at Lenkoran'), in Crimea and along the Don and Ural rivers in the Podoliya area of the Ukraine. Wild-growing hemp is known within the Aralo-Caspian territory as well as the districts of Turgay and Ural

(Bogdan, 1908). S.I. Korzhinskiy mentioned in his 'Tentamen florae Rossiae orientalis' (1898) that hemp occurs in the eastern belt of European Russia in an '*inquilinus*' state, i.e. 'dwelling in a place not its own', and was found in the southeastern areas to be '*spontanea videtur*' [seen as spontaneous]. It is found in enormous quantities east of the Volga in the Samarskaya and Ufa Governates districts as well as in the Saratov and Astrakhan districts. B.A. Keller observed hemp as a forest plant in the oak–maple forests of the Astrakhan and Saratov districts, especially below Sarepta (Dimo & Keller, 1907). In the Buzuluk woodland, it has long since locally formed almost pure, thick and tall thickets of stalks, although there is no cultivation anywhere in the Yergeniy highlands (Vysotsskiy, 1915). It is found there naturalized as well as 'artificial plantations' and in untilled habitats. Dioecious nettles [*Urtica dioica*], black nightshade [*Solanum nigrum*] and thistles [*Cirsium* sp.] are associated with it.

On the Balkan peninsula, 'naturalized' hemp is mentioned from Bulgaria (Velenovsky, 1989) and Serbia (according to Adamovich); it is also found in Hungary and the Banat area (Yanishevskiy, 1924).

In the Asiatic parts of the U.S.S.R. 'wild' hemp is found in enormous quantities as a weed at the Kirgizian camps in Altai, on fallow land, in vacant lots and in vegetable gardens. In addition, weedy hemp grows all over the mountainous area of Altai, reaching up to an elevation of 1440 m.s.m. and, in exceptional conditions, even to 2000 m (at Rakhmanovskiye Springs, according to Sinskaya, 1925). According to a report by A.A. Khrebtov (1926), 'wild' hemp is also found as a distressing weed in fields of spring wheat, barley, spring rye and oats among the mountains of Altai.

According to observations made by E.N. Sinskaya in Altai and by A.A. Khrebtov in western Siberia (in the Semipalatinsk district of the Zaysan Territory, in the Akmolin, Omsk and Tobolsk districts, especially the southern parts, as well as in the Tomsk district, etc.) hemp is often seen on steep riverbanks, in valleys and in out-of-the-way corners where, perhaps, man has never set foot but where the thickets of hemp often appear to be of a primary type.

In an economic essay about southern Altai, N.N. Oganovskiy (1922) also mentions 'wild hemp' when speaking of the utilization of natural resources of that territory and, particularly, of its wild flora. According to his calculations, 'the fibers and the seeds of the wild hemp can easily furnish a profit of up to 150 000 rubles a year' in this area. V.F. Semenov (see Krylov, 1909) reports on a distribution of hemp along the right-hand bank of the river Irtysh in the northeastern part of the Omsk district. 'Wild' hemp is also noted from the Semipalatinsk (B.A. Keller) and Syr-Daryan districts (Knorring & Minkovitz, 1912).

As far as the weedy/naturalized hemp is concerned, it was observed in the former Manchuria by V.L. Komarov, K.I. Maksimovich and L.I. Litvinov (cf. Komarov, 1903). It was noted by V.E. Pisarev also in northern Mongolia.

E.L. Regel (1892) mentions hemp in a 'naturalized state' in the Ussuri territory around houses, in weedy hollows and in sites where yurts [nomad felt-tents; D.L.] have stood, but also secluded along river banks.

Boissier (1879) indicated that hemp in the Himalayan region was 'subspontaneous'. Hooker (1890) called the hemp in northeastern Himalaya a wild plant. We have found 'wild' hemp in northern Persia (at Mazanderan), in Kafiristan and Afghanistan in areas where hemp is definitely not cultivated. Belts of 'black hemp' emanating from crops of maize and other cereals stretch along the Kunar river (on the border between Afghanistan and India) from Chekhosarai to Djelalabad) for a distance of 150–200 verst [100–125 miles; D.L.].

The very abundance of hemp found independent of crops of other plants has made botanists suspect that it has a 'wild nature'. However, prudently and in view of the lack of any differences between it and the cultivated form, the authors prefer to call it 'quasi-spontaneous' (Ledebour), 'subspontaneous' (Boissier), 'seemingly spontaneous' (S.I.Korzhinskiy) or 'erratic' (Erndtelio).

Studies of wild hemp

While studying hemp cultivated in the Saratov and Astrakhan districts in 1920, we turned our attention also to the wild, naturalized or weedy hemp. During a close study thereof, it was revealed that the wild or weedy hemp represented forms sharply different from the cultivated ones as far as the seeds are concerned and, most of all, with respect to how the seeds are shed when ripe.

For a long time we did not succeed in collecting a sufficient quantity of seeds from this kind of hemp. Mature plants of 'wild hemp' habitually do not hold on to their seeds. After careful examination, it turned out that the majority of the plants belonging to this kind of hemp had seeds with characteristic, prolonged formations at the base of the seeds, reminding one of the shapes of the 'horseshoes' [actually, elaiosomes] of wild oats. The disarticulation and shedding of the seed when ripe occurs along this formation. The cell membranes covering the seeds of wild or weedy hemp turned out to be thicker and to have an appendix at the base. The 'wild' seeds are also considerably (ca. $1\frac{1}{2}$ times) smaller than those of the cultivated hemp. Unexpectedly, it was revealed that, just as in the case of wild cereal grasses, the weedy or naturalized hemp had peculiarities of the flowers that favor the self-pollination and self-dispersal of the seeds.

In particular, germination of the ripe seeds of weedy hemp is extraordinarily slow and uneven, a fact reported by V.F. Antropopov and T.Ya. Serebryakova in the case of crops on our experimental fields. After being harvested and sown during the fall, the seeds of wild hemp rest for some weeks, even several months, without germinating, while, under similar conditions, those of the cultivated hemp germinate very nicely within a few days of being sown. During normal spring seeding, the germination of the wild seeds is also peculiarly slow and exceptionally uneven. Out of 100 seeds, usually a few dozen will germinate and, occasionally, only individual seeds of different plants.

In other words, this kind of hemp has peculiar biological characteristics common to many wild and weedy plants and is sharply distinguished from those of the corresponding cultivated plants, e.g. the rapid and even germination of the seeds of the latter. The attributes revealed with respect to the wild plants made us bravely consider the weedy and naturalized hemp as a genuinely wild plant, similar, e.g., to wild barley, *Hordeum distichum* var. *spontaneum*. We suggested therefore that it should be likewise distinguished as a special variety, *Cannabis sativa* var. *spontanea* (Vavilov, 1922).

In 1924, D.E. Yanishevskiy published an interesting paper on the wild hemp in southeastern Russia for which he suggested the name *Cannabis ruderalis* when providing it with a Latin diagnosis. He studied the fruits thoroughly and it was revealed that the essential peculiarity of the fruits of wild hemp was to be found in the membrane covering the seeds and developed from the perianth. It remains on the seed after it is shed. In the case of the cultivated form, no perianth remains on the seed, only traces of it hang on as occasional scraps of a membrane. In the case of the weedy (wild) hemp, the perianth covers the seeds as a thick membrane and makes them appear multicolored or marbled and very hard, because of the pericarp covering them. Due to the development of the pericarp there is, according to D.E. Yanishevskiy, a certain similarity to the same structure of the seeds of hops, *Humulus lupulus*, where the perianth also completely envelops the ripe fruits.

D.E. Yanishevskiy described in detail the characteristic of the seeds of the wild hemp called 'horseshoes' [elaiosomes]. Anatomical studies revealed the development of a special, thick tissue with cells rich in oils and inclusions. This tissue develops between the site where the perianth is attached and the site of the carpels. The observations made by Yanishevskiy soon made it clear that this thick tissue attracts the attention of a red bug, *Pyrrhocoris apterus*, to the seeds shed by the weedy hemp, particularly when the bug is at the stage of young individuals. The bug carries off the seeds, piercing its proboscis into the appendix at the base of the hemp seed, where the thick tissue is opened. *Pyrrhocoris* is able to carry the seeds of the weedy hemp over long distances and to bury them. Since they live around fences, hedges and barns, these bugs bring the hemp seeds there, due to which, perhaps, the frequent occurrence of hemp around dwellings can be explained. It was also revealed that other insects are attracted to this thick tissue at the base of the seeds of the weedy hemp. The basal part of the seeds of the wild hemp, thus, has much in common with the so-called elaiosomes, studied by Sernander in the case of various other plants, the dispersal of which is due to their seeds or fruits being carried away by ants (myrmechory).

The marbled or patchy dark color of the seeds is advantageous for wild hemp and makes them, according to the observations by Yanishevskiy, barely visible on the ground. Indeed, it plays a role of a 'homochromatic' color.

D.E. Yanishevskiy was also able to establish the length of the rest period necessary for weedy hemp seeds in order to favor germination only during the spring; with such a type of germination, weedy hemp does not risk being subjected to winter frosts.

Thus, according to the old botanical scheme for understanding the genesis of cultivated plants alongside the typical cultivated forms, which are deprived of the characteristics of self-fertilization and shedding of seeds, such genuine forms of what is indisputably wild hemp can be acknowledged, in which we can also of course see the ancestors of cultivated hemp.

FORMS OF WILD AND CULTIVATED HEMP

Investigations made in our laboratory by V.F. Antropopov and T.Ya. Tsingerling-Serebryakova and also field tests of a large number of samples of cultivated, weedy and wild hemp have revealed a great diversity of various forms.

Cultivated hemp is represented by a complicated mixture of strains with a wide range of variation and is regularly distributed in the Old World from the Equator to the Polar Circle. The length of the vegetative period in relation to branching increases toward the south. The northern forms in the Archangelsk and similar districts differ considerably by their short (ca. 35–50 cm tall), non-branching stalks, small fruits, small flowers and anthers half the size of those of hemp from the Orlov and Kursk districts, as well as a short vegetative period (Serebryakova, 1927–28). They ripen a month earlier than hemp in Central Russia. The late-maturing Chinese and Far Eastern races do not ripen even in the black soil belt of the European U.S.S.R. but reach enormous proportions (3 m and more tall), are strongly branched and differ by big leaves with broad leaflets, large flowers and seeds three times heavier than those of the small-seeded northern races.

The Central Russian hemp (from Orlov or Kursk) occupies an intermediate position with respect to their characteristics and differs from the northern ones by taller growth, more branches, a medium–long vegetative period and medium-sized seeds. Toward the south and southeast of Asia, the large-seeded and much-branched forms increase, while in the Khibinskiy Oasis, Turkestan and Afghanistan hemp, growing lower than in eastern Asia also occurs, although branching types are found as well.

Under uniform conditions at the Steppe Station (in the Voronesh district) the north Yakutian and Archangelsk forms ripen in 80 days, those from Tambov and Tul' require 105 and those from Ukraine 111 days, while the races from Persia and Bokhara never ripen under the conditions there. The northern races have narrow and ordinary leaves, the southern ones larger leaves with many more leaflets when grown under the same conditions. The geographical regularities in relation to the distribution of the races of hemp are just as evident as those of flax.

There is a wide variation with respect to seed color in the case of cultivated hemp. Besides the predominant type with light-colored pericarp, we know of races with brown and almost black perianths. As a rule, the cultivated forms so far studied are without a pericarp, although the presence thereof is typical of wild hemp.

As could be expected, cultivated races are on the whole characterized by seeds that are not easily shed but there are striking differences in the case of various strains.

The wild and weedy forms, which in essence cannot be distinguished from each other are, apparently, represented by no less complex races than cultivated hemp, although the former have not yet been adequately studied. The different forms can be distinguished by the speed with which the seeds germinate, by the dimensions and colors of the seeds, and by the length of the vegetative period.

The common European and Asiatic wild and weedy hemp has well developed pericarps, giving the seeds a mosaic of colors.

When, in 1924, we traveled along the Kunar River between Chekhosaray and Djelalabad in Afghanistan, we discovered among the wild hemp a peculiar race with light-colored, small seeds, thin membranes, slightly splitting and transparent pericarps (f. *afghanica* Vav.). The seeds of this race are very small (1000 seeds weigh 2.1–2.7 g), ca. 10 times smaller than those of the Far Eastern large-seeded races (1000 seeds – 26.0 g); 1000 seeds of the ordinary, central Russian races (from Orlov and Kursk) weigh ca. 17–19 g.

All the forms of wild hemp, collected along the Kunar River, approach the cultivated type with respect to seed color and the slightly splitting pericarp but differ by shedding their seeds and in the development of 'horseshoes' [elaiosomes; D.L.]. When sown, the seeds germinate very slowly and unevenly, i.e. these plants display attributes of wild plants as well.

The wild races of hemp in eastern Afghanistan have even more new characteristics. The leaflets, of which the compound leaf is composed, differ from those of the ordinary wild and cultivated hemp by having a narrow, obovate shape such as so far not observed by us among the European, Siberian or Turkestani forms. Just as in Afghanistan itself, these races differ when sown in our experimental fields at the Steppe Station (in the Voronezh district), by medium–tall growth and having many branches, which is typical also of the common Turkestani forms.

The wild Afghani races, with light-colored and easily splitting pericarp, from the areas adjacent to India (they have, indeed spread into northern India as well), thus turned out to constitute a morphological link between the wild and cultivated races of hemp with respect to the most important differentiating characteristics.

Characteristics of wild and cultivated hemp

A number of forms of cultivated and wild hemp, including all the diversity of the strains so far studied and listed in Table 9, demonstrate clearly a gradation of characteristics approaching each other. The 'cultivated type' of the wild, Afghani, small-seeded hemp with thin, transparent and slightly splitting pericarp appears to be a clear example of such an intermediate. Some races of wild hemp from Altai and Lower Volga have only slightly developed 'horseshoes' [elaiosomes; D.L.], which in this respect approach those of the cultivated form. As far as the rapid germination of the seeds and their dimensions is concerned, there is also a gradation of types. With respect to the dimensions, the small-seeded Yakutian cultivated form yields to the wild Altaian and Saratovan hemp. As far as the quantitative and qualitative characteristics are concerned, whole series of intermediate forms can be described that link the cultivated types to the wild ones. Only the extreme variants differ sharply from each other, e.g. as the Far Eastern cultivated hemp does from the small-seeded Afghani race with slightly splitting pericarps.

Hybridization between cultivated and wild forms

Although typical wild and cultivated forms of hemp show a gradation of characteristics, by their series of races, this phenomenon becomes so much more evident when studying the commonly occurring hybridization that takes place in areas where cultivated and wild forms occur together. It is possible to observe a mixture of typically intermediate forms in the vegetable gardens in Altai, in addition to the cultivated forms with light-colored seeds, no 'horseshoes' [elaiosomes; D.L.] and no perianth, as well as comparatively few branches. According to the observations made by E.N. Sinskaya (1925) in Altai, all the stages of transition from wild to cultivated hemp can be seen there. In comparison with typical hemp, the intermediate forms in the vegetable gardens of Altai have mostly dark-colored seeds; when compared with genuinely weedy forms, the seeds are larger and the 'horseshoes' less well developed. Among the weedy hemp in Altai, plants are occasionally seen with weakly developed 'horseshoes' but such forms with light-colored seeds have not been found. Some Caucasian forms of cultivated hemp remind one of the intermediate Altaian forms. A hybrid origin of these strains in highly likely since the main types occur together in the Caucasus. The occurrence of male and female flowers on different hemp plants makes it very easy to hybridize the races. The morphological borders between the hemp appearing as a field weed and completely wild plants are hard to draw; the forms can hardly be distinguished from each other as far as the characteristics of the seeds are concerned.

The ecology of wild, weedy and cultivated hemp

As far as cultivated hemp is concerned, it displays an exceptional requirement for well-manured soil. No other plants exhaust the soil as hemp does. It is only necessary to look at crops of hemp in order to see how well they react to changes in the fertility of the soil. A shortage of manure reduces the growth sharply and makes the green color of the plants paler; on the other hand, when increasing the amount of manure, stalks and leaves develop vigorously.

In practice, hemp is sown on special plots, 'hempfields', provided with large amounts of manure. Hemp prefers in particular manure rich in potassium. Because of this biological peculiarity, cultivated hemp is a plant of well-fertilized soils.

Genuinely wild and weedy hemp has the same preference for well-fertilized soils. According to our observations, wild and weedy hemp in Asia and in the European parts of the U.S.S.R. grows mainly along hedges and in ravines and hollows as well as, between harvests, on rich fallow soils. When looking at several examples of habitats of wild and weedy hemp, it can be seen that hemp has a tendency toward selecting fertilized soils and colonizing only sites where there is an accumulation of humus and manure.

The nomadic camps around the lower Volga and in Altai appear to be typical sites for an abundant and splendid growth of hemp. The cattle manure the soil

Table 9. *Series of forms of cultivated hemp*, Cannabis sativa, *and wild hemp*, C. sativa *var.* spontanea (C. ruderalis)

Characteristics	Cultivated hemp *C. sativa*			Wild hemp *C. sativa* var. *spontanea*		
	RSFSR	Yakutia	Far East Primor'e	Altai	Lower Volga (Saratov, Samara, Astrakhan)	Afghanistan
Vegetative						
Stalks						
Height						
Tall (> 150 cm)	–	–	+	–	–	–
Medium (60–150 cm)	+	–	–	+	+	+
Short (35–50 cm)	+	+	–	–	–	–
Diameter						
Thick	–	–	+	–	–	–
Medium	+	–	–	+	+	+
Thin	+	+	–	–	–	–
Branches						
Many	+	–	+	+	+	+
Few or none	+	+	–	–	–	–
Leaves						
Dimensions						
Large	–	–	+	–	–	–
Medium	+	–	–	+	+	–
Small	–	+	–	–	–	+
Leaflets						
3–5	–	+	–	–	–	–
5–7–9	+	–	–	+	+	+
9–11–13	–	–	+	–	–	–
Shape of leaflets						
Ordinary European, spool-shaped	+	+	–	–	–	–
Broad, spool-shaped	–	–	+	–	–	–
Narrow, spool-shaped	–	–	–	+	+	–
Narrow, obovate	–	–	–	–	–	+

Generative						
Seeds						
Dimensions						
Large (4–5.5 mm)	–	–	+	–	–	–
Medium (3–4 mm)	+	+	–	+	+	+
Small (2.7–3 mm)	–	+	–	+	+	+
Color						
Light	+	+	+	–	–	+
Dark	+	–	–	+	+	+
Shape						
Roundish	+	+	–	–	–	+
Spherical	–	–	+	–	–	–
Round-ovate	–	–	–	+	+	+
Elaiosomes						
Present	–	–	–	+	+	+
Absent	+	+	+	+	+	+
Mosaic (presence or absence of perianth)						
Present	–	–	–	+	+	+
Absent or slightly cracked	+	+	+	–	–	+
Flowers						
Dimensions						
Large	–	–	+	–	–	–
Medium	+	–	–	+	+	+
Small	–	+	–	–	–	–
Biological						
Vegetative period						
Short	–	+	–	–	–	–
Intermediate	+	–	–	+	+	–
Long	–	–	+	–	–	+
Germination of seeds						
Rapid, even	+	+	+	–	+	–
Fairly slow, uneven	–	–	–	+	+	+

around the camps during the winter. Weedy hemp thrives especially well in vegetable gardens. When driving around the villages in the Astrakhan district, one can see from afar the thickets of weedy hemp around dump sites, in backyards and around small patches of trees. In Altai and northern Caucasus hemp frequently grows also in gorges, which are hard to penetrate and far from any dwellings; there it mainly occupies depressions and ravines where there is an accumulation of dung and excrement of wild animals, or habitats receiving something like natural manure (Sinskaya, 1925).

The ecological aspect of wild and weedy hemp is determined mainly by its infestation of rich, manured soils, which are not compacted and lying fallow. This is typical of ruderal plants.

Genesis of cultivated hemp

Common wild and weedy hemp appears to be a 'camp-follower' of nomads. Because of its biological peculiarities, wild hemp has, thus followed Man and with his knowledge accompanied the camps, seeding itself on dumps and manured sites.

When considering all the facts mentioned above, the multitude of forms of wild and weedy hemp and the very process by which it became cultivated becomes eminently clear.

A mixture of strains of wild hemp, represented by a multitude of morphological and physiological types ranging from genuinely wild ones with elaiosomes around the small seeds, which germinate slowly and carry pericarps, to the typically cultivated, large-seeded ones with thin seed-membranes, rapid germination and non-shedding seeds, accompanied Man during his wanderings and migrations throughout the Old World. Hemp must have caught the eye of the primitive inhabitants of the Old World, when it followed in their heels and amassed around their dwellings at the same time as it was useful.

During times of hunger, when turning to collection of seeds and fruits, Man naturally selected the races of hemp with large seeds that were not easily shed. Unconsciously, just the 'cultivated type' with less easily shed seeds and rich in fats was selected from the mixture of races of wild hemp; thus, the cultivation of hemp developed, in essence, by the will of Man.

Such a picture of a gradually developing cultivation of hemp can still be observed in Altai. According to reports by E.N. Sinskaya from Altai in 1925, it was at that time still possible to observe the following stages: 1. plants in an entirely wild state; 2. an initial colonization by these wild plants of habitats on dump sites and around dwellings; 3. the utilization of the weedy hemp by the inhabitants; and 4. intentional cultivation of hemp.

The inhabitants of Altai usually collect seeds for their crops from hemp in their vegetable gardens but, frequently, when there is a shortage of seeds, they take them also from wild plants; consequently, one can occasionally even now witness the initial stages of an introduction of hemp into cultivation.

It seems that the cultivation of hemp arose for the sake of the seeds; the process of introduction into cultivation was a natural and unconscious development.

The transition toward cultivation of hemp for the sake of its fibers was more complicated. However, it did not demand special inventiveness: in Turkestan, in the Khibinskiy Oasis, we were able to observe a direct separation of bundles of bast from hemp stalks by grinding the dried plants without first retting them.

The utilization of hemp for hashish did not even require special creative efforts on behalf of Man. When burning stalks and leaves over a wood fire, the humans could certainly not avoid noticing the stupefying effect of the hemp. The quantity of the narcotic substance increases toward the south, where the utilization of hemp for hashish is concentrated.

While not representing a major discovery, the utilization of hemp seeds for their fat content apparently took place during a comparatively late historical epoch. With respect to the wild plants, i.e. a mixture of hereditary strains, there are indications of a widespread colonization associated with humans, who brought the hemp along with them from the Equator to the Polar Circle. As demonstrated by observations concerning the segregation of hybrid forms, many characteristics of 'cultivated' hemp, such as the lack of a perianth and the thin seed-covering, associated with it, as well as the lack of elaiosomes, are apparently recessive characteristics, the selection of which was intentionally correlated with the cultivation of hemp.

It is extremely hard to tell when the cultivation of hemp was initiated and what people were instrumental in this. There is no objective information concerning its cultivation in the Nile Valley (Buschan, 1895). During the investigation of the tombs in ancient Egypt, no traces have been found of hemp fibers. Hemp is not mentioned in the Bible. All the information available, both historical and botanical, indicates that hemp is a plant of Asia. Its maximum variation is concentrated there, including the above-mentioned peculiar form which is endemic in Afghanistan. Large-seeded, late maturing races are typical of Manchuria, the Far East and China. Hemp is known as cultivated from time immemorial in China and India. Epithets for hemp exist in Sanskrit. De Candolle's suggestion that hemp was brought to Europe by Scythian nomads and their camps is not at all unlikely. Herodotus (from the 5th Century B.C.) mentioned that hemp was not known in the ancient Greco-Roman world before 484 B.C. Varro, Columella, Plinius and Dioscorides furnish information about the cultivation of hemp for its fibers. [Herodotus was a Greek historian; Marchus Terentius Varro (127–116 B.C.) was a Roman author, among others of books about agriculture; Columella published books about agriculture, *Rei rustics libri XII*, in 65 A.D.; Plinius (23–79 A.D.) was a Roman scholar making lots of observations concerning daily life; and Dioscorides (the 1st century, A.D.) was a physician and herbalist, whose methods were used well into the Middle Ages: D.L.]

It is more likely that the cultivation of hemp arose simultaneously and independently at several sites. The diversity of the geographical races of cultivated hemp indicates this as does the introduction of wild hemp at present taking place under our own eyes in Altai and northern Caucasus and, most of all, the ecological peculiarity of hemp making it a 'camp-follower' of nomads and leading to its cultivation.

CHAPTER 6 ECOLOGICAL PRINCIPLES FOR THE ORIGIN OF CULTIVATED PLANTS

Using hemp as an example, it was possible to see with one's own eyes that ecological peculiarities could determine the introduction of plants into cultivation. Engelbrecht (1916) justly pointed out the intensified requirement of many kinds of cultivated plants for increased fertilization and, consequently, the link associating these with human inhabitation. When studying cultivated plants and the wild species closely related to them and the varieties thereof in Turkestan, Persia and Afghanistan, we were able to point out a number of facts that, in our own opinion, to some extent clear up the very process toward the introduction of primary plants into cultivation.

When considering a number of primary plants and the wild species known to be close to them and when comparing their ecological peculiarities with each other, some very important facts can be noticed, i.e. that in general the closer the wild species or varieties are related to cultivated plants, the more often they belong to similar ecological groups.

We shall consider some of the examples we studied. Let us start with barley.

Species of *Hordeum*

Enormous stands of wild barley (*H. spontaneum*) can be found in northern Afghanistan, in the Transcaspian district (Turkmenistan) and in the Syr-Daryan district (Uzbekistan) as well as in Bokhara. Already, a quick glance shows that this group of wild types, which are close to cultivated barley, thrives on cultivated soils, along borders between fields, around water and on loose, uncompacted soils. For instance, in the Syr-Daryan district, wild barley (*H. spontaneum*) very often invades fallow fields. Such fallow land occasionally looks as if it had been purposefully sown with wild barley. According to the ecological nature of *H. spontaneum*, it is in all its forms a plant that has been essentially pre-adapted to cultivated conditions and to cultivated soils.

In Turkestan and Afghanistan, large quantities of the genetically somewhat more distant species of the genus *Hordeum* can also be encountered: *H. crinitum*, *H. maritimum*, *H. murinum*, *H. bulbosum* and *H. brevisubulatum*. The first of these grows on dry slopes and the last one, *H. brevisubulatum*, on wet soils; *H. maritimum* grows in saline habitats. This means that these three species are characterized by specialized ecological requirements, sharply different from those of *H. spontaneum* and *H. sativum*.

H. bulbosum and *H. murinum* grow along the banks of irrigation ditches in foothills, thus, ecologically approaching *H. spontaneum*. They all belong to a special ecological group.

In other words, the closer a group is to wild barley (*H. spontaneum*) the closer it correspondingly approaches to cultivated conditions and to conditions found on cultivated fields, and the easier is the introduction into cultivation by Man of the non-brittle forms associated with him.

Species of *Aegilops*

A still more obvious picture of the differentiation into ecological types, paralleling the differentiation into genetical types, is revealed by a number of wild species of *Triticum*, referred by some authors to the genus *Aegilops*, a group very close to wheat. The following species are know from Turkestan and Afghanistan: *T. triunciale* (L.) Gren. & Godr., *T. crassum* (boiss.) Hackel., *T. aegilops* P.B. and *T. cylindricum* Cesat.

Triticum triunciale constitutes a special group genetically, as demonstrated by our investigations concerning its resistance to parasitic fungi (Vavilov, 1918). The other three species, *T. crassum*, *T. cylindricum* and *T. aegilops*, are, on the other hand, close to soft wheat, hybridize comparatively easily with it and are characterized by similar reactions with relation to specialized parasites. In conformity with this, the four species are ecologically differentiated in such a manner that *T. triunciale* grows mainly on more compact soils and on dry slopes, while, on the other hand, the other three species, infesting wheat and barley crops, thrive along boundaries between fields, and remind one of the ecological aspects of the common wheat.

Other species

Wild oats, *Avena fatua* and *A. ludoviciana*, hybridize easily with ordinary oats, *A. sativa*, and thrive under conditions close to cultivated ones, around water, and on uncompacted soils. As weeds, they invade fields of wheat, barley and oats, and show a tendency for being associated with cultivated oats.

Wild lentils, *Ervum orientale*, bitter vetch, and *E. ervilia*, which is close to the cultivated forms of *Ervum*, grow in Uzbekistan on loose, uncompacted soils, around fields and not far from water.

Ecologically, wild melons (*Cucumis trigonus*), which hybridize with ordinary melons, are weeds of rotating crops such as corn, melons and cotton. In the ecology of these plants, which are genetically close to cultivated melons, there are characteristics associating them with loose soils, closeness to water and rotating crops.

The wild carrot (*Daucus carota*) usually invades gardens, vineyards and vegetable plots, as well as borders between fields, just as if inviting itself to be cultivated.

With respect to the ecological aspect of wild forms close to cultivated species, characteristics can, in general, be perceived, that pre-adapt a given plant to cultivation and to tilled soils. Consequently, the introduction into cultivation of various primary plants can be understood. Elements already existed in the wild flora that were attractive to the farmer and that, to a great extent because of the farmer's wishes, became introduced into cultivation.

Hypotheses (cf. Aaronsohn, 1910; Cook, 1913) suggesting that mountain forms of *T. dicoccoides*, growing on slopes in Syria and Palestine, should be the ancestors of cultivated wheat, correspond little to the truth just like that hypothesis, which is far from the truth, that the ancestor of cultivated rye should be the perennial mountain rye (*Secale montanum*). Although some stands of *T. dicoccoides* are found in Syria and Palestine in mountain ravines and in fissures in the rocks on soils rich in lime, humus and adequate amounts of manure are at the same time available there. Statements by Aaronsohn concerning the companions of wild emmer (e.g. *Hordeum spontaneum*) and the description by Cook of local observations already indicate the character of the habitats of the distichous wheat. Just like its companion, *H. spontaneum*, *T. dicoccoides* is, in essence, a plant pre-adapted to cultivated conditions.

The closest wild and ancestral forms of plants are necessarily found in conditions with habitats similar to those of the cultivated ones, as our direct observations of the wild ancestral types of many cultivated plants have demonstrated. This fact explains to a great extent how primary plants became cultivated. We do not doubt that in the case of the original species of wild plants and in mixtures of races which represent ancestral groups, an ecological tendency is already found in Nature that compelled Man to utilize a given plant species. It is quite obvious that Man took what came easily to him. In the case of many plants, both secondary and primary ones, the process of introduction into cultivation occurred to a great extent thanks to the will of Man.

CHAPTER 7 GEOGRAPHICAL REGULARITIES IN THE TYPE-FORMATION OF CULTIVATED PLANTS

During the geographical analysis of hereditary forms of flax, hemp and millet, we discussed above a number of regularities concerning the dispersal of the races from the south toward the north.

The data concerning the general direction of the type-forming processes, typical of different groups of plants in different regions, are even more interesting.

Non-ligulate rye and non-ligulate wheat

Above, we could see that non-ligulate rye was found just where the non-ligulate varieties of soft wheats are established, i.e. in Badakhshan and the mountains of Bokhara. The relationship between the geographical centers of origin of both these groups through some stages explains, in the given case, the coincidence between the habitats of these similar types.

Naked-seeded races

The differential–geographical method makes it possible to establish that the center of type-formation of naked-grained barley appears to be eastern and southeastern Asia, i.e. China and the areas adjacent to it. However, what is especially significant is that at the same time southeastern Asia seems to be the center of type-formation also of the large- and naked-grained oat, *Avena nuda*.

What is even more remarkable is that races of panicled millet, *Panicum miliaceum*, with narrow, easily disarticulating glumes corresponding to those of naked-grained oats and barley, are also concentrated in southeastern Asia as far as type-formation is concerned.

We do not know what causes such coincidences in the provenance of naked grained oats, barley and millet from eastern Asia; in any case, the similarity of the landscapes alone cannot explain the existence of similarities as far as type-formation of these very different genera of grasses is concerned.

Regularities of type-forming processes in southwestern and southeastern Asia and in the Mediterranean Region

As regards the centers of type-formation, we drew the conclusion that the cultivation of many plants originated initially within two centers: Asia and the Mediterranean region.

Wheat, barley, flax, peas, lentils, horse beans and castor beans have such origins. A detailed investigation of numerous samples from the Mediterranean region and from Asia unexpectedly revealed a regularity in the type-forming process, which, after a detailed scrutiny, turned out to be a general phenomenon. It demonstrated that while there is a tendency toward formation mainly of large-fruited, large-seeded and large-flowered types within the Mediterranean regions, the areas of southwestern Asia – Afghanistan, Turkestan, Bokhara and India – are characterized by a corresponding formation of mainly small-seeded, small-fruited and small-flowered forms.

Let us consider a number of concrete examples:

As we could see above, southwestern Asia seems to be where flax is concentrated, both in the form of linseed and fiber flax, but mainly the small-flowered, small-fruited and small-seeded types. On the other hand, Spain, Algeria, Tunisia, Egypt and Palestine produce flax with seeds 2–3 times heavier than those of the other European and Asiatic kinds and, similarly, with very large flowers and capsules.

The tendency to small-seeded leguminous plants in the Indian area is particularly strong. On the other hand, within the Mediterranean area, the corresponding kinds of legumes appear gigantic with respect to their flowers, pods and seeds. The horse beans of coastal Asia Minor, Smyrna, Sicily, Tunisia and Egypt have beans up to 3 cm long while the horse beans collected by us in Afghanistan (in the vicinity of Djelalabad) remind us of small peas. Large, disk-shaped lentils are typical of the Mediterranean coasts, while on the other hand, we found very small-seeded races with small leaves, low growth and black seeds, similar to those of peas, with respect to their dimensions, in India and southeastern Afghanistan.

The same can be no less clearly outlined as far as green peas (*Pisum*), sweet peas (*Lathyrus sativus*) and chickpeas (*Cicer arietinum*) are concerned.

We have compiled some tables illustrating the striking differences with respect to type-formation within the two different areas (Tables 10–14).

Table 10. *Lentils (*Ervum lens*)*

Provenance	Length of pod (cm)	Diameter of seed (mm)	1000-seed weight (g)
The Mediterranean coastal area			
Algeria	1.7	8.0	88.0
Italy	1.8	8.0	82.5
Spain	1.7	7.8	90.0
Tunisia	2.0	7.8	64.0
Southwestern Asia			
Mountains of Bokhara	1.2	4.5	26.0
Persia	1.1	4.7	37.0
India	1.1	3.6	20.0
Afghanistan (Mazar-i-Sharif)	1.2	4.0	26.0
Afghanistan (Djelalabad)[1]	0.7	3.0	7.5

Notes:
[1] Black-seeded.
Remark: Statistics by E.N. Barylina

Table 11. *Horse beans (*Vicia faba*)*

Provenance	Length of pod (cm)	Diameter of bean (mm)	1000-bean weight (g)
The Mediterranean coastal area			
Tunisia	9.4	2.4	171
Spain	9.9	2.3	150
	9.3	2.4	200
Italy	15.2	2.1	130
Southwestern Asia			
Persia	5.8	0.8	30
Bokhara	5.5	0.9	38
Pamir	5.7	0.9	32
Afghanistan (Herat)	5.6	1.1	40
Afghanistan (Kabul)	4.6	0.7	20
India	4.9	0.7	20

Note:
Remark: Statistics by V.S. Muratova

Table 12. *Chick peas (*Cicer arietinum*)*

Provenance	Length of pod (cm)	Diameter of pea (mm)	1000-pea weight (g)
The Mediterranean coastal area			
Spain	2.7	12.0	470
Italy	3.0	9.5	369
	2.3	8.4	280
France	2.2	8.5	331
Southwestern Asia			
Persia	1.7	7.7	110.0
Afghanistan	1.7	7.5	113.5
Turkestan	1.8	7.6	127.0
Bokhara	1.8	7.8	114.0
Pamir	1.8	7.8	128.2
India	1.7	6.4	94.3

Note:
Remark: Statistics by K.G. Prozorova

Table 13. *Peas (*Pisum*)*

	Length of pod (cm)		Diameter of pea (mm)		1000-pea weight (g)	
Provenance	Min.	Max.	Min.	Max.	Min.	Max.
The Mediterranean coastal area						
Italy	4.5	6.5	0.65	0.95	18.0	40.0
Spain	5.5	6.0	0.75	0.83	21.7	—
Africa	5.9	7.3	0.7	0.89	24.0	38.0
Mean	5.3	6.6	0.7	0.89	21.2	39.0
Southwestern Asia						
India	3.0	5.2	0.65	0.65	5.2	23.0
Afghanistan	2.9	4.5	0.4	0.6	5.0	26.5
Khiva	4.9	—	0.5	0.5	6.5	8.5
Pamir	3.2	5.6	0.5	0.7	8.5	22.0
Mean	3.5	5.1	0.47	0.65	6.3	20.0

Note:
Remark: Statistics by V.S. Fedotov

Table 14. *Vetchling (*Lathyrus sativus*)*

Provenance	Length of pod (cm)	Diameter of pea (mm)	1000-pea weight (g)
The Mediterranean coastal area			
Africa	4.3	1.0	24
France	3.9	1.4	60
Southwestern Asia			
India	2.9	0.4	6
Pamir	4.0	0.6	12
Afghanistan	3.8	0.5	15
Mountains of Bokhara	2.9	0.7	18

Note:
Remark: Statistics by K.G. Prozorova

The differences in type-formation of leguminous plants within the two areas affect both the vegetative characteristics (dimensions of stems and leaves) and the generative ones. In Asia, small dimensions are typical, in Africa large ones.

The castor beans (*Ricinus persicus*) of Persia and Afghanistan are distinguished from the African varieties by small beans, low growth, small leaves and small inflorescences (Popova, 1926).

The same is the case also with respect to major field crops: wheat, barley and oats. Both oats and wild oats in the Mediterranean coastal regions are characterized by a tendency toward the formation of large grains and long paleas and glumes. Some African races of *Avena sterilis* are astonishing with respect to the size of their spikelets, which exceed by two and a half times those of ordinary wild oats.

Avena byzantina, a species cultivated in some Mediterranean countries, is distinguished from *A. sativa* by the large dimensions of its spikelets and the lengths of its paleas and glumes in all its races. Small-grained oats, *A. sativa*, and wild oats, *A. fatua* and *A. ludoviciana*, are typical of Transcaucasia, Afghanistan and northeastern Persia.

Similar low-growing, small-grained and small-spiked forms of wheat are typical of India and Afghanistan. The centers of type-formation of *Triticum sphaerococcum* and *T. compactum* are located there. On the other hand, gigantic dimensions, long awns and large grains become increasingly more common toward the Mediterranean coastal areas. In Spain, Morocco, Tunisia, Algeria, Sicily, Egypt and Palestine *T. turgidum*, *T. polonicum* and *T. durum* are grown, the many strains of which can without exaggeration be called gigantic in comparison with the typical compact forms found in southwestern Asia.

In China and Japan, low-growing races of naked-grained barley are typical; even under the best of conditions they do not exceed 40–50 cm in height (e.g.

var. *japonicum* and some forms of var. *coeleste*). All the naked-grained Asiatic races of barley differ in general by their low stature when compared with the African races of hulled barley.

A comparison between the type-formation of Asiatic millet, *Panicum miliaceum*, and its gigantic analog in Africa, the sorghum, *Andropogon sorghum*, practically suggests itself.

It is quite possible to suggest similar differences with respect to type-formation within other genera as well. However, the facts presented are adequate for describing the general character of the regularity that clearly demonstrates that it can be utilized for practical purposes.

In southwestern Asia, the center of type-formation of soft wheat and many leguminous plants, such characteristics that approach those of the small-seeded wild forms are in general typical. On the other hand, the Mediterranean areas are characterized predominantly by crops of large-fruited species and races that undoubtedly are of great interest for practical utilization as far as the European crops are concerned.

Species of *Phaseolus* in the Old and the New Worlds

A similar kind of regularity can be found when comparing the species of *Phaseolus* in the Old and New Worlds. The Asiatic species, *Ph. aureus*, *Ph. radiatus* (*Ph. mungo*), *Ph. calcaratus*, *Ph. acutifolius* and *Ph. angularis*, are, as far as type-forming processes are concerned, characterized by small seeds, small flowers and small, cylindrical pods, a strongly developed pubescence, broad stipules and long, linear bractlets. The American species, *Ph. lunatus* and *Ph. acutifolius* var. *latifolius*, differ by their extreme diversity of large seeds, large flowers and short bractlets. (This problem was worked out in detail at the Institute of Applied Botany by N. R. Ivanova, who prepared it for publication under the title *The Peculiarities of the Type Formation Among the Species of* Phaseolus *in the Old and the New Worlds*; it will appear in 1928.) As demonstrated also by other studies by G.M. Popova and N.R. Ivanova at the Institute of Applied Botany, the variability of the American and the Asiatic species of *Phaseolus* is at the same time definitely subject to the law of homological series.

It was especially interesting to compare the process of type-formation of some plants cultivated in the Old World with species of the same genus grown in the New World. There are rather many such genera. It is enough to mention the cottons of the Old and the New Worlds, *Gossypium herbaceum*, *G. hirsutum* and *G. barbardense*, or the many species of cultivated fruits and berries belonging to such genera as *Vitis*, *Ribes*, *Rubus*, *Malus*, *Prunus* or *Fragaria*, typical of both the Old and the New Worlds.

Just as in the case of *Phaseolus*, studies of the genus *Gossypium* made by G.S. Zaytsev revealed a striking parallelism of type-formation concerning the Asiatic and the American cotton species.

The reason for the different tendencies with respect to type-formation is still not completely understood. A correlation with the landscape hardly constitutes a condition since, when grown under similar conditions, the Mediterranean races remain sharply different from the races from southwestern Asia. To relate them to the great antiquity of agriculture in the Mediterranean area and the more primitive agriculture of southwestern Asia did not appear correct either. We have no definite or uncontestable data indicating that the crops of the Mediterranean coastal areas should be older than those of Mesopotamia. Corresponding differences affect not only the cultivated races but also the wild species close to them just as could be seen, e.g. in the cases of wild peas, oats and wheats (e.g. the dimensions of the grains of *Triticum dicoccoides*).

GENERAL CONCLUSIONS – BASIC GLOBAL CENTERS OF ORIGIN OF CULTIVATED PLANTS

During the evolution of the variety of cultivated plants, general characteristics, i.e. parallelism, definitely occurred. A number of forms of wild and cultivated races can be united into Linnaean species – i.e. whole series of hereditary forms, ranging from typically wild ones with brittle inflorescences and seeds with elaiosomes to such with dehiscing capsules, repeat themselves within different genera and even families. We can find similar formations in the cases of wheat, oats, millet, barley, rye, buckwheat and hemp. Analogous cycles are displayed within the type-formation of many specialized weeds (Thellung, 1925).

The general regularity allows us, in essence, to construct a system of type-formation and to predict the whereabouts of some particular form, whether cultivated or not, and, at the same time, to simplify the problem of its origin.

General centers of type-formation of cultivated plants

There can be no doubt that a detailed study of the centers of type-formation of a large number of cultivated plants will very likely also establish centers common for whole groups of cultivated plants, and help us succeed in approaching in earnest the recognition of universal centers of origin.

Considerable efforts will be needed in order to bring such a work to completion. However, as a result of preliminary investigations of some dozens of plants, it is already now possible to outline five basic centers with respect to the most important field crops, vegetables and garden plants (Fig. 13):

1 *Southwestern Asia*, comprising India, southern Afghanistan and the adjacent mountain areas of Bokhara and Kashmir, Persia, Asia Minor and Transcaucasia; this center initially gave rise to soft and club wheat, rye, small-seeded flax, small-seeded vetches, lentils, horsebeans and other beans, vetchlings and chickpeas as well as a number of vegetables and, in addition, to Asiatic cotton (*Gossypium herbaceum*, *G. arboreum*), etc.

2 *Southeastern Asia*, comprising the mountain areas of China, Japan, Nepal and adjacent areas; this is the center of type-formation of naked-grained oats, naked-grained barley, millet, soybeans, many cultivated cruciferous plants and a number of endemic species of fruit trees.

3 *The Mediterranean focus*, comprising all the coastal areas of the Mediterranean including northern Africa (Egypt, Algeria, Tunisia), Palestine and Syria,

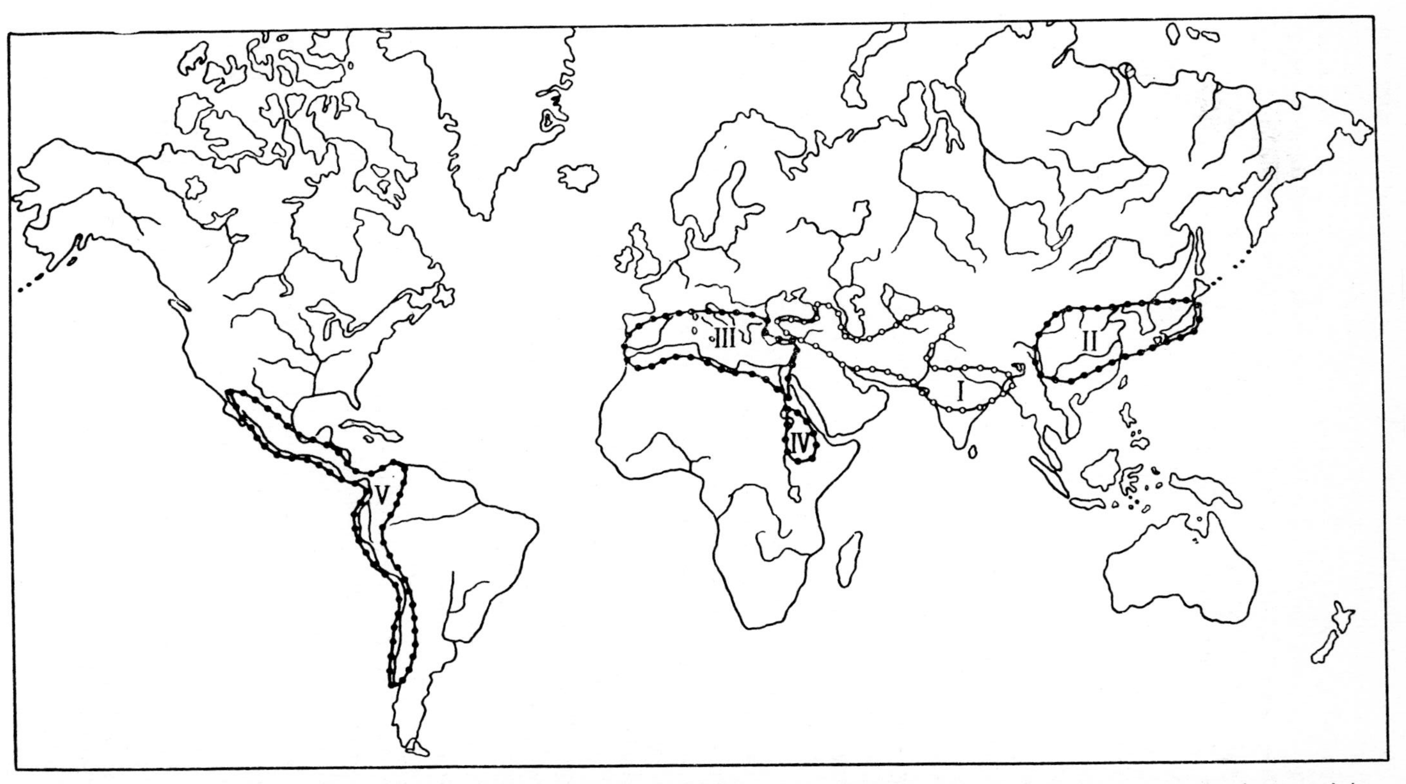

Fig. 13. Basic centers of origin of the most important cultivated plants in the Old and the New Worlds. 1. Southwestern Asia; 2. southeastern Asia; 3. the Mediterranean area; 4. Abyssinia and Egypt; 5. mountain areas of South America and Mexico.

Greece and its archipelago, Spain and Italy and a portion of Asia Minor; the centers of origin of a large number of cultivated plants are concentrated there, such as hard wheats (*sensu latu*) including all the group with $2n=28$ chromosomes, a number of cultivated strains of oats belonging to *Avena byzantina*, large-seeded flax, large-seeded vetches, vetchlings, horsebeans and lentils as well as sugar beets and many vegetables in addition to fruit trees.

4 In northern Africa, an independent center can be observed in *Abyssinia* and the areas adjacent to it; in that area unusual formations of the types of many cultivated plants are concentrated. Globally, the Abyssinian focus stands out by the formation of such types as hulled barley, peculiar races of leguminous plants and oats and a number of other cultivated plants.

5 As far as it is possible to judge from the data available, it is possible to distinguish a focus of primary agriculture in the New World and a center of type-formation in *Mexico and Peru* and the mountain areas adjacent to them. The type-formation of potatoes, maize, species of *Phaseolus*, tobacco, sunflowers, American cotton and many other species and even specific genera of cultivated plants and fruit trees is concentrated in those areas.

The first two Asiatic foci are, apparently, adjoined toward the south by an independent island focus, including both the Philippine Islands and the so-called East Indian islands (e.g. the Sunda islands). In order to single out this focus we can refer to a number of endemic groups of cultivated plants found only there: the basic group of Philippino rice, *Coix*, waxy types of maize, etc. The plants cultivated within this focus are still badly studied from a botanical point of view and therefore we have not yet included this area as a basic center of origin of cultivated plants.

In addition to the above centers, we will, in the future, certainly be able to outline a number of secondary centers and to specify more exactly the actual geographic centers; at the same time, it is necessary to keep in mind that a number of plants such as hemp and rye were introduced into cultivation simultaneously in many places.

As can be seen, the areas of origin and type-formation of the most important cultivated plants which, at the same time, are the foci of a wealth of types, belong mainly to the mountain areas of Asia (Himalaya and its system), the mountain systems of northeastern Africa and the mountain areas of southern Europe (the Pyrenées, the Apennines and the Balkans), the Cordilleras and the southern spurs of the Rocky Mountains. In the Old World, the original areas of cultivated plants belong mostly within a belt between 20° and 40° N. latitude.

These mountain areas are bordered by the deserts of Central Asia and by the Sahara and, with respect to their diversity of climate and soil, they offer optimal conditions for explaining type-forming processes. In those areas there is a gradation with extreme variants in the amounts of precipitation, temperature and soil types.

The diversity of conditions – from those of deserts to those of oases, from soil lacking in humus to those rich in nourishment – promoted a concentration and formation of an exceptional variety of vegetation in the upper and middle belts of the mountains.

The abundance of water for irrigation from melting snow and ice, the ease with which this water can be utilized for irrigation by means of gravity, the opportunity for cultivation even of imperfect crops in areas with high amounts of precipitation, the isolation of the fields and their defense from attacks, all contributed to the development in those areas of a primary kind of agriculture.

So far, the mountains of Asia and Africa represent globally just the kind of areas where people would settle. This fact has recently become even more evident. More than half of the world's people (ca. 900 million) inhabit such mountain areas, which constitute a very small portion of the total area of the world's surface (less than 1/20th).

To a traveler in southwestern Asia, the unusual utilization of every inch of soil and the crops in remote areas are astonishing. If the infertile deserts and waterless mountain areas, slopes inaccessible for crops, rocky screes and areas with a permanent snow cover in Persia, Afghanistan and Bokhara were disregarded and if the density of the population in relation to the accessibility to crops were taken into consideration, a population density exceeding even that of the cultivated areas of Europe would result.

Centers of origin of cultivated plants and foci of human civilizations

The elucidation of the centers of type-formation and the origin of cultivated plants allow us to approach objectively the establishment of basic foci of agricultural civilizations. Arguments about whether the Egyptian civilization is autonomous and has not adopted elements from Mesopotamia and vice versa or the questions about the independence of the autonomy of the Chinese and Indian civilizations can be objectively solved by studying their kinds of crops. Plants and their varieties are not easily dispersed from one area to another; in spite of the many thousands of years of wandering about by peoples and tribes, it was, as we have seen, not difficult to establish the basic foci of type-formation of the majority of cultivated plants. The presence in northern Africa and southwestern Asia of large groups of endemic plants, both species and varieties of cultivated plants, on the basis of which independent agricultural civilizations arose, can solve the problem of autonomy of these civilizations also from an historical point of view.

The history of the origin of human civilizations and agriculture is, of course, much older than the documentation in the form of pyramids, inscriptions and bas-reliefs or tombs can tell us. A close acquaintance with cultivated plants and with the multitude of types and their differentiation into geographical groups as well as their frequently sharp physiological isolation from each other compel us to refer the very origin of cultivated plants to such remote epochs, where periods of 5–10 000 years such as concern archeologists represent but a brief moment.

When studying the foci of type-formation in detail with respect to the most important cultivated plants, the botanist is in a condition to introduce essential corrections into the hypotheses launched by historians and archeologists.

On the map surveying the primary civilizations, compiled by Smith and Perry (1925; cf. chapter VIII, 'Anthropology', written by G. Elliot in the book, *Evolution* . . ., 1925) on the basis of all kinds of archeological, anthropological and historical data, the following areas appear as the basic foci of human civilizations: Egypt, Italy, the Balkan Peninsula, Asia Minor, Mesopotamia, Persia, Afghanistan, a portion of Turkmenistan, Uzbekistan, Transcaucasia and the areas of the ancient Scythians. We do not in the least doubt that, based on the botanical studies of cultivated plants, also India, the mountains of China, Mexico and Peru and the areas adjacent to these should be included among the number of initial areas in addition to those mentioned.

A very large number of endemic, cultivated plant species and varieties in India, China, Mexico and Peru appear to have been the objects of ancient agriculture, as has been demonstrated for these countries. In spite of the lack of corresponding archeological documentation (because of which these areas were not included by Smith among those with primary civilizations), these areas should be included among the ancient centers of agricultural civilizations.

Thus, a phytogeographical analysis of cultivated plants allows us to introduce additional evidence for solving the problems of history concerning human civilizations in general and those of agricultural civilizations in particular. The most recent discoveries in Punjab, the Sunda islands and China of archeological documentation contemporary with the ancient Mesopotamian civilization, support our phytogeographical opinion.

Apart from the spontaneous and utilitarian importance of mastering the sources of a multitude of plant types, the ultimate objective of the investigations discussed is to try to approach, in earnest, the general biological problem of speciation. Evolution occurs both in space and time. It seems to us that only by earnestly approaching the geographical centers of type-formation and by establishing all the links connecting the species will it be possible to find the way to the establishment of a system of Linnaean species and to understand the latter as a system of forms. The geneticist can consciously approach the selection of forms for hybridization and the solution of the problem of experimental phylogenetics only by understanding taxonomy and geography. Hence, the very problem of speciation becomes a problem concerning the evolution not only of the different races, which, according to the Darwin's hypothesis, were isolated into basic species, but also of the origin of the complicated system within which the present Linnaean species developed.

As a result of all that has been said above, the solution of the problem of speciation lies in a synthesis of a far-reaching investigation of various groups of plants, while using the methods of differentiating taxonomical phytogeography for the purpose of establishing centers of type-formation, and the methods of genetics and cytology. The way to grasp the integrity of the species can only be found in a synthesis of taxonomy, differentiating geography, genetics and cytology.

BIBLIOGRAPHY

Aaronsohn, A. (1910). *Agricultural and botanical explorations in Palestine*. Washington, D.C.: U.S. Dept. Agr., Bureau of Plant Industry, Bull. no. 180.

Arnol'd, B.M. (1925). K voprosu o klassifikatsii *Panicum miliaceum* L. [Toward the problem of classification of *Panicum miliaceum* L.]. *Tr. po prikl. botan. i selek.* [*Papers on applied botany and plant breeding*], **14** (1).

Aitchison, Y.E.T. (1881). On the flora of the Kuram valley. *J. Linn. Soc. Bot.*, **18**.

Aitchison, Y.E.T. (1888). The botany of the Afghanistan delimitation commission. *Trans. Linn. Soc. Bot., London, Ser. 2 (Bot.)*, **3**.

Ascherson, P. and Graebner, P. (1914). *Synopsis der Mitteleuropäischen Flora*, [*Synopsis of the Central European flora*], Part 84 and 85, *Linaceae*. Leipzig.

Atabekova, A.I. (1925). Materialy k monograficheskomu izucheniyu novogo vida kulturnoy pshenitsy, *Triticum persicum* Vav. [Material toward a monographic study of the new species of cultivated wheat, *Triticum persicum* Vav.]. *Tr. po prikl. botan. i selek.* [*Papers on botany and plant breeding*], **15** (5).

Bailey, L.Z. (1906). *Sketch of the evolution of our native fruits*, 2nd edn. New York.

Balabayev, G.A. (1926). O pasprostranenii sornoy rzhi, *Secale cereale* L., v gorakh Sredniy Azii [On the distribution of wild rye, *Secale cereale* L., in the mountains of Inner Asia]. *Tr. po. prikl. botan. i selek.*, **16**.

Barylina, E.I. (1925). O zasoranii posevov khlebov odnozernyakoy v Krymu [On the infestations by weeds of the einkorn crops in Crimea]. *Tr. po prikl. botan. i selek.*, **14** (1).

Belikov, V. (1840). *Perezhozdenie rasteniy i vozrozhdenie ikh cherez obrazovanie novikh razvidnostey* [*Degeneration of plants and their regeneration by the development of new varieties*]. Moscow.

Bogdan, V.S. (1908). *Rastitel'nost' Turgaysko–Ural'skogo pereselencheskogo rayona* [*The vegetation of the Turgayan–Uralian nomadic area*]. Orenburg.

Boissier, E. (1879). *Flora orientalis*, **VI**., Geneva.

Breitschneider, E.V. (1870). The study and value of Chinese botanical works. *Chin. Rec. Mission J.*, **3**.

Breitschneider, E.V. (1881). Botanicon Sinicum. *J. North-China Branch Roy. As. Soc. London*, **16**, art. III.

Britton, N.L. (1889). *Catalogue of plants found in New Jersey*. Trenton, New York.

Bukasov, S.M. (1925). Kartofel': Sortovedenie i selektsiya [The potato: control and selection of types]. *Tr. po prikl. botan. i selek.*, **15** (2).

Buschan, G. (1985). *Vorgeschichtliche Botanik der Cultur- und Nutzpflanzen der alten Welt auf Grund prähistorischer Funde*, [*Prehistoric botany of cultivated and utilitarian plants of the Old World on the basis of prehistoric discoveries*]. Breslau.

Carrier, L. (1923). *The beginnings of agriculture in America*. New York.

Clark, J.A., Martin, J.H. and Ball, C.R. (1923). *Classification of American wheat varieties*. Washington, D.C.: U.S.D.A., Bull. no. 23.

Cook, O.E. (1913). *Wild wheat in Palestine*. Washington, D.C.: U.S.D.A., Bureau of Plant Industry, *Bull. no. 274*.

Cook, O.E. (1925). Peru as a center of domestication. *J. Hered.*, **1925** (2,3).

De Candolle, A. (1855). *La géographie botanique raissonée* [*Sensible phytogeography*]. Paris.

De Candolle, A. (1883). L'origine des plantes cultivées [*Origin of cultivated plants*]. Paris.

Dekaprelevich, L.L. (1925). K vyyasneniya persidskoy pshenitsy [Toward elucidation of the area where Persian wheat is grown]. *Tr. po prikl. botan. i selek.*, **15** (1).

Dimo, N.A. and Keller, B.A. (1907). *B oblasti polypustynii*, [*In the region of semideserts*]. Saratov.

Engelbrecht, Th. (1916). Über die Entstehung einiger feldmässige angebauter Kulturpflanzen [On the development of some cultivated plants for field crops]. *Geogr. Z.*, **22**.

Engelbrecht, Th. (1917). Über die Entstehung des Kulturroggens [On the development of cultivated rye]. In *Festschrift Eduard Hahn*, Stuttgart.

Evolution in the light of modern knowledge. (1925). London.

Flaksberger, K.A. (1913). O mestonakhozhdeniyakh *Secale montanum* Guss. na Kavkaze [On the localities of *Secale montanum* Guss. in Caucasus]. In *Yubil. cb. v. yest' 25-letiya nauch. deyat. prof. N.N. Kuznetsov*, [*Festschrift on the occasion of the 25-year scientific activities of Prof. N.N. Kuznetsov*]. Yur'ev.

Flaksberger, K.A. (1925). Pshenitsa odnozernyahki [Einkorn wheat]. *Tr. po prikl. botan. i selek.*, **15** (1).

Flaksberger, K.A. (1926). Bezligul'nye tverdye pshenitsu [Non-ligulate hard wheat]. *Tr. po prikl. botan. i selek.*, **16** (1).

Gentner, G. (1921). Pfahlbauten und Winterlein [Pole-dwellings and winter flax]. *Faserforschung*, **1** (2).

Gibault, G. (1912). *Histoire des légumes* [History of vegetables]. Paris.

Grossheim, A.A. (1924). Novaya rasa rshi [A new race of rye]. *Tr. po. prikl. botan. i selek.*, **13** (2).

Heer, O. (1872). Über den Flachs und die Flachskultur im Althertum [On flax and the cultivation of flax in the past]. *Neujahrbl.* publ. by *Naturforschungs Gesellschaft, Zürich*, **84**.

Hehn, V. (1911). *Kulturpflanzen und Haustiere in ihrem Übergang aus Asien*, [*Cultivated plants and domesticated animals and their dispersal from Asia*], 8th edn., with botanical contributions by A. Engler and F. Pax., Berlin.

Helweg, L. (1908). En monografisk Skildring af de dyrkede Gullerodsformer samt et Bidrag til kulturhistorien [A monographic account of the forms of carrots and a contribution to the history of cultivation]. *Tidskrift Landbrugets Plant.* (Copenhagen), **15**.

Höck., T. (1900). Der gegenwärtige Stand unserer Kentnisse von der angebauten Nutzpflanzen [The present status of knowledge on cultivated, utilitarian plants]. Leipzig. (Reprint from *Geogr. Z.*, **5** (6)).

Hooker, J.D. (1890). *Flora of British India.* **V**. London.

Hoops, J. (1905). *Waldbäume und Kulturpflanzen im germanischen Alterthum*, [*Forest trees and cultivated plants during German antiquity*]. Strasburg.

Howard, A. and Howard, G. (1909). *Wheat in India.* Pusa.

Howard G. and Khan, Abdur Rahman (1924). Studies in Indian oil seeds, 2. Linseed. *Mem. Dept. Agr. India*, **12** (4).

Ivanov, N.N. (1926). Izmenchivost v khimicheskom sostave maslichnykh rasteniy v zavisimosti ot geograficheskikh factorov [Variation in the chemical composition of oil-producing plants depending on geographical factors]. *Tr. po prikl. botan. i selek.*, **16** (3).

Ivanov, N.N. (1928). Osobennosti formoobrazovaniya u vidov fasoli Starogo i Novogo Sveta [The characteristics of the type-formation of *Phaseolus* in the Old and the New Worlds]. *Tr. po prikl. botan., genet i selek.* [*Papers on applied botany, genetics and plant breeding*] **19** (2).

Joret, Ch. (1897–1904). *Les plantes dans l'antiquité au moyen âge*, [Plants from antiquity to the Middle Ages]. Parts I, II. Paris.

Kappret (1921). Über den Wert und die Möglichheit einer Tausendkorngewichts-erhöhung der Leinsaat auf maschinellen Wege [On the value and possibilities for increasing the 1000-grain weight of flax crops by artificial methods]. *Faserforschung*, **1** (2).

Krebtov, A.A. (1926). O dikoy konople [On wild hemp]. *Economica Perm.*, **2–3**.

Knorring, O.E. and Minkvits, Z.A. (1912). Rastitel'nost' Aulieatinskogo uezda Syr-Dar'inskoy oblasti [The vegetation of the Aulieatin district of the Syr-Daryan region]. *Tr. pereselench. upr.*, St. Petersburg.

Koernicke, F. (1885). *Handbuch des Getreidebaus* [Manual of cereal cultivation]. Berlin.

Komarov, V.L. (1903). *Flora Man'chzhurii, I, II*, [Flora of Manchuria]. St. Petersburg.

Korzhinskiy, S.I. (1898). Tentamen Florae Rossiae [Attempt at a flora of Russia]. *Zap. Akad. nauk. po fiz.-mat. otd.*, No. 1.

Kostlan, A. (1913). Die Landwirtschaft in Abissinien [Agriculture in Abyssinia], Part 1. *Acker- und Pflanzenbau*, suppl. *Tropenpflanzen*, vol. **XIV** (3).

Kramer (1923). Beiträge zur Kentniss des Winterleins [Contribution to knowledge of winter flax]. *Faserforsch.* **3** (3).

Krylov, P.N. (1909). *Flora Altaya i Tomskoy gubernii* [*Flora of Altai and the Tomsk district*]. No. 5. Tomsk.

Laufer, B. (1919). *Sino-Iranica.* Chicago.

Ledebour, C.F. (1846). *Flora Rossica.* III. Stuttgart.

Marquand, C.V. (1921). *Avena strigosa segregates. Bot. Soc. and Exchange Club of Brit. Isles, Report.*

Marquand, C.V. (1922). Varieties of oats in cultivation In *Welsh Plant Breeding Station,* University College of Wales, Aberystwyth.

Maysuryan, N.A. (1925). Opyt klassifikatsii vida *Secale cereale* L. [An attempt at a classification of the species of *Secale cereale* L.]. *Zap. Nauch.-prikl. otd. Tiflis botan sada,* No. 4.

Miège, E. (1924). Les formes marocaines de *Triticum monococcum* L. [The Moroccan forms of *Triticum monococcum* L.]. *Bull. Soc. sci. natur. et. phys., Maroc,* No. 7.

Nikolayeva, A. (1920). Primenie tsitologicheskogo metoda pri reshanii nekotorykh voprosovi genetiki: Raboty s ovsami Selektsionnoy Stantsii pri Timiryazevskoy sel'sko-khozyastvennoy akademii [Application of cytological methods for solving some problems in genetics: the work with oats at the Plantbreeding Station of the Timiryasev Agricultural Academy]. *Tr. III Vseros. c'ezda po selektsii i semenovodstvu* (Saratov), **I**.

Nikolayeva, A. (1922). Zur Kentniss der Chromosomenzahlin der Gattung *Avena* [Toward knowledge of the chromosome numbers of *Avena*]. Autoref. *Z. indukt. Abstammungs- u. Vererbungslehre,* **29** (3/4).

Nikolayeva, A. (1923). Tsitologicheskoe issledovanie roda *Triticum* [Cytological study of the genus *Triticum*]. *Tr. po prikl. botan. i selek.,* **13** (1).

Oganovskiy, N.N. (1922). *Yuzhnyy Altay,* [Southern Altai]. Moscow.

Orlov, G.M. (1923). Geograficheskiy tsentr proiskhozhdeniya i rayony vosdelyvaniya tverdooy pshenitsy [The geographical center of origin and the distribution area of hard wheat]. *Tr. po prikl. botan. i selek.,* **13**, (1).

Percival, J. (1921). *The Wheat Plant.* London.

Popova, G.M. (1923). Vidu *Aegilops* i ikh massovaya gibridizatsiya s pshenitsey v Turkestane,' [The species of *Aegilops* and their mass-hybridization with wheat in Turkestan]. *Tr. po prikl. botan. i selek.,* **13** (1).

Popova, G.M. (1926). Kleshchevina i eyo kul'tura v Srednei Azii [The castor bean and its cultivation in Inner Asia]. *Tr. po prikl. botan i selek.,* **16** (4).

Reed, G. (1920). Varietal resistance and susceptibility of oats to powdery mildew, crown rust and smut. *Missouri Agr. Exp. Sta. Columbia, Bull.,* **37**.

Reed, G.M. and Melchers, L.E. (1925). *Sorghum smuts and varietal resistance in sorghums.* Washington, D.C.: U.S.D.A., Bull. no. 1284.

Regel, R.E. (1917). K voprosu o proiskhozhdenii kul'turnykh yachmeney [On the problem of the origin of cultivated barley]. *Tr. po prikl. botan. i selek.* **16** (1).

Regel, E. (1862). *Opyt flory Ussurskoy strany,* [*Attempt at a flora of the country of the Ussuri*]. St. Petersburg.

Reinhardt, L. (1911). *Kulturgeschichte der Nutzpflanzen* [*Cultural history of cultivated plants*]. Vol. IV. Munich.

Safford, W.G. (1925). The potato of Romance and of Reality. *J. Hered.* Nos. 4–6.

Sax, K. (1921). Sterility in wheat hybrids. *Genetics,* **6**.

Sboyev, V.A. (1862). *Issledovanie ob inorodtsakh Kazanskoy gubernii* [*Study of heterogeneity in the Kazanskaya district*]. Chuvashi.

Scharfetter, R. (1922). Klimarhytmik, Vegetationsrhytmik und Formations- rhytmik. Studien zur Bestimmung der Heimat der Pflanzen [Climatic rhythm, vegetation rhythm and formation rhythm. Study of the identification of the homeland of plants]. *Oesterr. Bot. Z.,* **71**.

Schiemann, E. (1922). Die Phylogenie der Getreide [The phylogeny of cereals]. *Naturwiss.,* **10** (6).

Schindler, F. (1923). *Handbuch des Getreidebaues auf wissenschaftlicher und praktischer Grundlage,* [*Manual of cereal cultivation on a scientific and practical basis*], 3rd edn. Berlin.

Schulz, A. (1913). *Die Geschichte der kultivierten Getreide* [*The history of cultivated cereals*]. Halle.

Schweinfurth, G. (1872–1873). On the African origin of cultivated Egyptian plants. *Bull. l'Inst. Egyptien*, No. 12.

Schweinfurth, G. (1922). Was Afrika an Kulturpflanzen Amerika zu verdanken hat und was es ihm gab [What Africa owes America with respect to cultivated plants and what it gave back]. In *Festschrift E. Seler*. Stuttgart.

Serebryakova, T. Ya. (1927–28). Rannaya konoplya [Early-maturing hemp]. *Tr. po prikl. botan., genet. i selek.*, **18** (1).

Sinskaya, E.N. (1924). Zakonomernosti v izmenchivosti sem. Cruciferae [Regularities in the variation within the family Cruciferae]. *Tr. po prikl. botan. i selek.*, **13** (2).

Sinskaya, E.N. (1925a). Indau: Maloizvestnoye maslichnoye i salatnoye rastenia (*Eruca sativa* Lam.) [The garden rocket: a little known oil-producing and salad plant (*Eruca sativa* Lam.)]. *Tr. po prikl. botan. i selek.*, **13** (2).

Sinskaya, E.N. (1925b). O polevykh kul'turnakh Altaya [On the field crops of Altai]. *Tr. po prikl. botan. i selek.*, **14** (1).

Solms-Laubach, H. (1899). *Weizen und Tulpe* [Wheat and tulips]. Leipzig.

Stoletova, E.A. (1925). Polba-emmer *Triticum dicoccum* Schrank [Emmer, *Triticum dicoccum* Schrank]. *Tr. po prikl. botan. i selek.*, **14** (1).

Sturtevant, R.S. (1919). *Notes on edible plants*. Albany.

Tammes, Tine (1911). Das Verhalten fluktuirend varierender Merkmale bei der Bastardierung [The behavior of fluctuating and variable characteristics during hybridization]. *Rec. trav. bot. Neerlandais*, **8** (3).

Tammes, Tine (1923). Das genotypishce Verhältniss zwischen den wilden *Linum angustifolium* und *L. usitatissimum* [The genotypic relationship between *Linum angustifolium* and *L. usitatissimum*]. *Genetica*, **V**.

Thellung, A. (1918). Neuere Wege und Ziele der botanischen Systematik erläutert am Beispiele unsere Getreidearten [New methods and objectives for the botanical systematics explained by means of examples from our cereal species]. *Naturwissenschaft Wochenschr.*, **17** (32,33).

Thellung, A. (1925). Kulturpflanzen – Eigenschaft bei Unkräutern [Characteristics of cultivated plants found among weeds]. In Festschrift Carl Schröter, *Veröff. Geobot. Inst. Rübel in Zurich*, No. 3.

Traubut, L. (1911, 1913). Observations sur l'origine des avoines cultivées [Observations regarding the origin of cultivated oats]. In *Confér. Intern. genetique*. Paris.

Tsinger, N.V. (1909). O zasoryanyushchikh posevy l'na vidax *Camelina* i *Spergula* i ikh proiskho denii [On the infestation of crops of flax by species of *Camelina* and *Spergula* and their provenance]. *Tr. Botan. Muziya Acad., Nauk.*, No. 6.

Vavilov, N. (1914). Immunity to fungus diseases as a physiological test in genetics and systematics as exemplified by cereals. *J. Genet.*, **IV** (1).

Vavilov, N.I. (1917). O proiskhozhdenii kul'turnoy rzhi [On the origin of cultivated rye]. *Tr. Byuro po prikl. botan.*, **10** (7–10).

Vavilov, N.I. (1918). Immunitet rasteniy k infektsionnym zabolevaniyam [Resistance of plants to infectious diseases]. *Izb. Petrovskoy C.-Kh. Akad.*, Nos. 1–4.

Vavilov, N.I. (1920). *Zakon gomologicheskikh ryadov v nasledstvennoy izmenchi vosti* [*The law of homologic series in the case of hereditary variation*]. Saratov.

Vavilov, N.I. (1921). O proiskhozhdenii gladkoostykh yachmeney [On the origin of smooth-awned barley]. *Tr. po prikl. botan. i selek.*, **12** (1).

Vavilov, N.I. (1922a). Polevye kul'turny yugo-Vostoka [Field crops of the South-East]. *Tr. po prikl. botan. i selek.*, Suppl. **23**.

Vavilov, N.I. (1922b). The law of homologous series in variation. *J. Genet.*, **12** (1).

Vavilov, N.I. (1923). K poznaniya myakikh pshenits [Toward an understanding of soft wheat]. *Tr. po prikl. botan. i selek.*, **13** (1).

Vavilov, N.I. (1924). Zakonomernosti v izmenchivosti rasteniy [Regularities in the variation of plants]. In *Selektsiya i semenovodstvo b SSSR, [Plant breeding and agriculture in the U.S.S.R.]*. Moscow.

Vavilov, N.I. (1925). O mekhdurodovikh gibridakh arbuzov, lyn' i tykv [On intergenetic hybrids of watermelons, melons and gourds]. *Tr. po prikl. botan. i selek.*, **14** (2).

Vavilov, N.I. and Yakyshkina, O.V. (1925). K filogenezy pshenits [Toward a phylogeny of wheat]. *Tr. po prikl. botan. i selek.*, **15** (1).

Velenovsky, J. (1898). *Flora Bulgaria*, Prague, Suppl. **1**.

Vysotsskiy, G.N. (1915). Yergenya: Kul'turno-fitologicheskiy ocherk [Yergenya: a cultural–phytological essay]. *Tr. po prikl. botan.*, **8** (10–11).

Wigglesworth, A. (1923). The new era in flax. *J. Text. Inst.*, **XIV**.

Willis, J. Ch. (1922). *Age and area*. Cambridge.

Woenig, E. (1886). *Über Pflanzen des alten Aegypten* [*On the plants of ancient Egypt*]. Leipzig.

Zade, A. (1914). Serologische Studien an Leguminosen und Gramineen [Serological studies of Leguminosae and Graminae]. *Z.Pflanzenzücht.*, **2** (2).

Zade, A. (1918). *Der Hafer*, [Oats]. Jena.

Zade, A. (1921). *Der Werdegang der kultivierten Pflanzen* [The evolution of cultivated plants]. Leipzig, Berlin.

Zhegalov, S.I. (1920). Iz nablyudeniy nad ovsyanymi gibridami [From observations of wild oat hybrids]. *Tr. III Vseros. c'ezda po selek. i semenovodstvu*, Saratov, No. i.

Zhukovskiy, P.M. (1923a). Issledovanie krest'yanskogo semennogo materiale Vostochnoy Gruzii [Studies of the seed material of the peasants of E. Georgia]. *Zap. Nauch.-prikl. otd. Tiflis botan. sada*, No. 3.

Zhukovskiy, P.M. (1923b). Materialy po izucheniyu pshenits Vostochnoy Gruzii [Material for a study of the wheat of E. Georgia]. *Zap. Nauch.-prikl. otd. Tiflis botan. sada*, No. 3.

Zhukovskiy, P.M. (1923c). Persidskaya pshenitsa v Zakavkaz'e, [The Persian wheat in Transcaucasia]. *Tr. po prikl. botan. i selek.*, **13** (1).

Yanishevskiy, D.E. (1924). Forma konopli na sornikh mestakh v Yugo-Vostochnoi Rossii [Types of hemp in ruderal habitats in Southeastern Russia]. *Uchen. Zap. Sarat. un-ta*, **2** (3).

Geographical regularities in relation to the distribution of the genes of cultivated plants

WHEN STUDYING THE varieties and races of cultivated plants all over the world and trying to locate the centers of cultivated agricultural crops, we approached an establishment of geographical centers of accumulation with respect to a variety of phenotypes. By means of a detailed study of the racial composition of various Linnaean species, a system of characteristics was revealed which, to a certain extent, corresponds to the geographical regions where they are concentrated and about which scientists were not aware until recently.

In my book, *Centers of Origin of Cultivated Plants*, I drew the general outlines of these. In the future, new facts and truths will enhance, widen and refine these centers of type-formation.

Concentration of dominant forms – the increasing number of dominant genes in the direction toward the center

Direct studies *in situ*, by the expeditions sent from the Institute of Applied Botany to the centers of type-formation of cultivated plants in the mountain areas of Asia, Africa and the countries along the coast of the Mediterranean and in Transcaucasia, have revealed not only the presence in those areas of a great diversity of forms but also of a *principal accumulation of dominant forms*, i.e. varieties characterized by dominant genes. A significant number of plants studied by us has revealed this fact. Let us furnish some examples.

(a) As demonstrated in particular by the results of the last expeditions from the Institute of Applied Botany to Asia Minor (led by Prof. P. M. Zhukovskiy), Armenia (led by E. A. Stoletova) and Azerbaidzhan (led by Prof. N. N. Kuletov), the most likely centers of origin – both as far as species of cultivated rye, *Secale cereale*, and, actually, the entire genus of *Secale*, are concerned – appeared to be in eastern Asia Minor and Transcaucasia, since all the races of rye and all the varietal characteristics distinguishing the different varieties and races are concentrated there. However, it is especially important that there are found not only a large number of forms, but also many dominant characteristics such as red spikes, brown spikes, and even black spikes, as well as varieties with a

First published in *Priroda* [*Nature*] no. 10, pp. 763–74, 1927; see also *Tr. po. prikl. botan., genet. i selek.* [*Papers on applied botany, genetics and plant breeding*], 17 (3): 411–28, 1927.

strongly developed pubescence. The colors of the spikes and the pubescence of the hulls of weedy rye in Asia Minor and Transcaucasia appear to be characteristic traits and distinguish them from the rye found in Europe, Afghanistan and Tadzhikistan (the mountains of Bokhara, Uzbekistan and Turkestan) – i.e. areas competing with Asia Minor and Transcaucasia, with respect to the number of characteristics, and known mainly by the recessive forms such as leaves without ligulae (the fact that this is a recessive trait has been demonstrated by studying the process of segregation of the heterozygous forms), straw-colored spikes and glabrous or only slightly pubescent paleas.

(b) Abyssinia appears, according to our investigations, to be the center of type-formation of the cultivated varieties of all the large group of hard wheats, more exactly, of the entire group of cultivated species of wheat characterized by $2n=28$ chromosomes. All the diversity of hulled barley, including a large number of endemic forms, is concentrated there as well. However, it is also interesting that we have met such a wealth of dominant forms in Abyssinia as exists nowhere else in the world. A group such as the black-spiked 'deficientes' of barley is, in fact, dominant as was demonstrated by our experiments and other investigations following hybridization with the common European and Asiatic forms of *tetrastichum*, *hexastichum* and *nutans*. A large number of forms with strongly developed anthocyanin in the plants or the spike at the time of ripening are also concentrated here. Even races of naked-grained barley, such as are usually not correlated with any particular color, are black-grained there.

As far as the hardwheats are concerned, we succeeded in finding in those areas not only all the varietal characteristics already known in Europe and Asia but also many peculiar endemic forms, the majority of which are races with clearly dominant traits not yet known in Europe. Among these are the large group of Abyssinian violet-grained wheats, forms of hard wheat without awns and forms with hulls that have colored edges; pubescent varieties are also widespread there. Experimental investigations should soon reveal in detail the genetic composition of what was collected by the last expedition from the Department of Applied Botany, but even now – based on the research of some forms already known to us – the facts concerning the wide distribution of dominant characteristics in Abyssinia have definitely attracted our attention.

(c) Abyssinia is the main center of the Linnaean species *Avena abyssinica*, a field weed usually infesting emmer (*Triticum dicoccum*) and barley. As far as we know, this species belongs exclusively to Abyssinia and the mountains of Egypt. Studies of the races of this kind of oats revealed the existence of a large number of forms with dominant characteristics: pubescent, brown or grey hulls, and so on.

(d) Abyssinia is the true center of type-formation as far as cultivated leguminous plants are concerned. An investigator will, indeed, be astonished at the diversity of their forms and his attention will be drawn involuntarily to the presence among them of black-seeded races, not known in Europe.

Among the chick-peas (*Cicer arietinum*) and the fava beans (*Vicia faba*) black-seeded races are found in Abyssinia; these are unknown or very rare in Turkestan or the Caucasus.

(e) In parts of southeastern Afghanistan, adjacent to India, and in the southern mountains of Bokhara, a great diversity of soft wheats, shot wheats and club wheats (*Triticum vulgare*, *T. compactum*, and *T. sphaerococcum*) is found. In fact, the basic center of type-formation of the entire group of soft wheats is located there. We found a large number of forms with dominant characteristics there as far as the compact types of the spike, the colors, the pubescence of the glumes and the spikelets as well as the inflated type of glumes are concerned.

(f) With respect to the distribution of its forms, the potato shows a clear reduction in dominant traits in the direction toward the periphery of its area of distribution. Mexico, Peru and Chile – without question the original focus of its cultivation – display a multitude of races characterized by tubers and stolons with an intensely violet color, i.e. with anthocyanine penetrating all the way to the center of the tubers. Hybridization experiments have revealed that this type of anthocyanin coloration appears to be definitely dominant.

(g) As demonstrated by studies made by the expeditions from the Institute of Applied Botany, Mexico appears to be one of the centers of type-formation of maize. According to observations by S.M. Bukasov, there is a reduction of dark-colored forms (i.e. of colored cobs) in the direction toward the Atlantic and the Pacific oceans.

Old World, but the main diversity of its forms is concentrated in the eastern mountains of Africa. This was revealed by our latest expedition to Abyssinia and Egypt and was based on subsequent studies by growing crops of the samples collected (by M.S. Shchenkov). Because of this, it is perfectly clear to us that Abyssinia and Egypt are characterized particularly by the presence of mainly dominant forms (black-seeded varieties and such with sectioned leaves, etc.).

Many more such examples can be shown.

In short, when summing up the data, we arrive at the conclusion that the basic centers of type-formation, the foci of diversity – i.e. those of prime interest for plant breeding – are characterized not only by the presence of a large number of types but also – and this is no less important – by the *presence of a large number of dominant characteristics.*

On the other hand, secondary centers of type-formation are characterized by varieties with mainly recessive traits.

Many European crops of imported garden plants, which have been subjected to breeding for a long time, are represented mainly by recessive types. Due to isolation and inbreeding, maize has, during the last couple of years, developed a large number of new characteristics: various albino types, races without ligulae and a multitude of 'freak' traits. Such types are, of course, recessive.

Dominant characteristics of animals and humans

The studies of the genesis of cultivated plants led us to the conclusion that the basic centers of type-formation of vegetable crops are, to a great extent, associated with the distribution of the basic foci of human civilization and, to a similar extent, with the centers of diversity of domesticated animals.

When traveling around in Africa and Asia, we were unintentionally drawn to the fact that there is an unusual diversity with respect to the racial composition of horned cattle, goats and sheep in Abyssinia, Egypt and southeastern Afghanistan, in the very centers of type-formation of many plant crops. The artificial selection of types concerning the shape and color of domesticated animals is revealed even more clearly than in the case of the plants, but the diversity still clearly preserved is no less striking. Those colors that, according to the results of genetic investigations, are dominant (as far as can be judged superficially at least) are fairly often widespread.

This regularity seems even more applicable to humans. The centers of the main cultivated plants in the Old World (i.e. barley, wheat, leguminous plants and flax) seem to have a special predominance of dark-colored human races as well. As far we know from the results of hybridization, the characteristic of black coloration is determined, in the case of negroid races, by a type of genes that, after crossing with whites, frequently results in a complete absorption of the whiteness.

It is interesting that even the ectoparasites of the humans in Abyssinia have a tendency to display dominant traits: the lice in Abyssinia are represented not only by light-colored types but also to a great extent by black varieties.

If we compare a map showing the basic centers of origin of the most important crops in the Old World with the distribution of the coloration of the human races up to the time of modern migrations, it is impossible not to notice the coincidence in general terms between the concentration of the resident dark-colored races and the centers of agricultural crops.

General picture of the establishment in different areas of races of cultivated plants, domesticated animals and humans

Thus, the general picture of the colonization of the different areas by humans and their races of domesticated animals and plants can be understood as a *decrease of dominants from the center toward the periphery of a distribution area.*

The genetical bases of domestic animals and of humans themselves are close to the genetical centers of major cultivated plants, i.e. those that we have already established with definite objectivity.

At a distance from the main geographical/genetical bases toward the periphery, the types of cultivated plants become lighter in color. Europe is characterized mainly by white-spiked rye, wheat and barley; animals, too, become lighter in color as do the people. The northern type appears to be the result of a loss of dominant genes. There is a proportional accumulation of recessive forms resulting from the dispersal and isolation. The recessive boreal types must have developed in that manner.

Due to the physical dispersal when expanding spatially, a differentiation of the basic dominant types took place, isolating the recessive forms. In some cases the isolating factor was an island, in other cases mountains or deserts or, in the long run, even distance itself. The recessive types, spatially isolated while differentiating from the dominant types and multiplying among themselves, initially gave rise to complexes of recessive races and varieties.

When studying the geography of races of cultivated plants we meet with the strikingly illustrated role of geographical factors of isolation in the case of recessive types.

In the isolated areas among the high mountains of Badakhshan and Bokhara, represented by such ideal factors of isolation as, on the one hand, Hindukush, rising to 5000 m.s.m., and, on the other hand, the ridge of Pamir, we have discovered both non-ligulate rye and non-ligulate soft wheat. As shown by our experiments, when hybridized with the ordinary races the non-ligulate character turned out to be governed by at most two genes.

A number of varieties of non-ligulate hard wheats has been discovered on the island of Cyprus (by Flaksberger, 1926). In that case, the isolating role was played by an island; the basic center of type-formation in the case of hard wheat is found in Abyssinia.

The oasis of Khiva is an area ideally isolated toward the north by the Aral Sea and toward the southwest and east by the Karakum and Kyzulkum deserts, respectively. In 1925 our expedition found a large number, even entire stands, of white-seeded and white-flowered races of flax, which are rare within the main center of linseed flax, adjacent to India.

In the mountains of Asturia, along the ridges running out from the Pyrenées, we encountered entire isolated areas with stands of white-flowered flax, *Linum angustifolium*.

According to information given us by S.A. Aegiz, white-seeded recessive forms of 'peasant or Aztec tobacco' [*Nicotiana rustica*] have been found in central Russia (the district of Tambov) far from the center of origin of this tobacco.

Under various conditions of geographical isolation, recessive races of animals and plants have, no doubt, been singled out due to artificial selection by man. In essence, the process of geographical isolation and the action of artificial selection are similar: the recessive forms are selected out from the genetic base, i.e. the source of the genes. The Europeans were, so to say, 'relieved' of the prevalence of dominant genes. This picture applies equally to humans, animals and plants. In all likelihood, the northern and polar recessive varieties of animals developed in that manner. Their genetical bases are no doubt found much farther south. Their differentiation was promoted by the environment and by natural selection in conformity with the 'recessive' coloration of the external conditions. Finally, the dominant types could, of course, also reach the limits of the distribution areas but they became eliminated there because of their incompatibility with the environment. The pathway of the geographical development of races leads, thus, to a great extent to a process of isolation of recessive genes.

The original geographic centers include all the genetic elements. However, due to hybridization (especially in the case of cross-pollinating plants and animals of different sexes) and the occurrence of dominant genes, it is mainly the dominant types that prevail there. We know now that recessive types are especially easily produced by inbreeding under conditions that prevent heterogenous genes from being lost.

The fact that no non-ligulate recessive forms are found within the basic foci does not mean that they do not exist as potential genotypes. This reasoning explains very many data that appear to be exceptions from the general aspect prevailing in the centers.

Amendment of the general picture

We established five basic geographical centers of origin for the most important cultivated plants in the book about the *Centers of Origin of Cultivated Plants*:

1. Southwestern Asia;
2. the mountain areas of China;
3. the mountain areas of the Mediterranean coasts;
4. the mountain areas of eastern Africa; and
5. the mountain areas of the New World (Mexico, Guatemala, Peru and Colombia).

After analyzing the process of dispersal of forms throughout the world, it became necessary to consider the polyphyletic origin of many cultivated plants as well. Wheat, barley, oats and flax, many leguminous plants as well as many fruit trees have polyphyletic origins. This circumstance complicated the picture of the distribution and this must also be taken into consideration in the case of the genesis of domesticated animals and that of humans, too.

It is, of course, evident that artificial selection and interference due to the wishes of man could, under certain conditions, disrupt the general scheme. Dominant horned cattle have multiplied especially in the Russian north but are only rarely met with in the south. Dominant awnless wheat is preferred all over the civilized world. Even in Abyssinia, the populations settled in the foothills prefer white-grained wheat and barley (the latter is not used as fodder there but for brewing beer and baking bread).

The discovery of cultivated black oats in northern Europe is, apparently, associated with the special origin of this group. As our studies have shown, cultivated oats have a fairly complicated polyphyletic origin. Red-grained wheat is, with respect to its physiology, obviously correlated mainly with the humid north and is preferred in cultivation there over the recessive white-grained races. The domesticated animals and cultivated plants associated with man are subject to his wishes, and the more civilized a country is, the more man's wishes predominate.

These exceptions do not ruin the general picture, the concept of which reveals the essence of the evolutionary process in space and time within the limits of the Linnaean species of animals and plants.

Based on our investigations during the last couple of years, our concept about what is dominant and what is recessive has become more complicated. The same characteristics, externally identical, can in *different* races be either dominant or recessive. Thus, for instance, winter and spring types of plants (cf. Vavilov and Kuznetsov, 1922), the color of the glumes of grasses or the color of the flowers are not always equally dominant. Green cotyledons can, in the case of leguminous plants, be either dominant or recessive.

The hypothesis concerning the presence or absence of genes, developed by W. Bateson, has been subject to criticism. It is impossible to accept recessive mutations and recessive forms as due to a loss of genes. The occurrence of multiple alleles compels us to change the old ideas about the nature of differences that are inherited according to the Mendelian laws. Recessive mutations can be conceived as the result of a corresponding qualitative change of the original gene and not only as the result of a loss of a rudiment of a dominant characteristic.

It is necessary to consider these complications in relation to our ideas on the nature of recessive genes. However, at the same time, experiments and discoveries indicate in the great majority of cases an equal differentiation into dominants and recessives. Whether we understand recessive forms as a loss of a gene (a concept that so far has been found to fit fairly well for the description of the phenomenon) or as a different qualitative constitution of the gene, this does not change the gist of our ideas about the geographical distribution of dominant and recessive forms within the limits of the Linnaean species.

General characteristics of the center of accumulation of genes from a utilitarian point of view

The geographical centers, where the genes are concentrated, and the actual centers of type-formation do not always contain the phenotypes necessary for man. The wheat and barley of Abyssinia, characterized by an extraordinary wealth of genes, are on the whole not of any particularly practical interest to plant breeders in Europe. The enormous multitude of races in Abyssinia and Egypt is the result of natural selection correlated with the short equatorial day; the majority of the races in the mountains of eastern Africa are spring-ripening types and are characterized by less-compact spikes.

It is necessary to single out only those genes that are useful to us from the diversity of the mainly dominant genes. It is true that for the European crops there is a greater interest in just the recessive combination and the recessive genes. The building blocks, the genes, are necessary for the plant breeder, but he himself must build up the combinations. The geographical centers of multiformity are the true foci of origin of cultivated plants and domesticated animals and, in practice, are of interest in the same way as deposits of ore are, i.e. they contain dominant and recessive genes, from which it is proper to retrieve only what is interesting for plant breeding, leaving behind the major portion of the dominant genes, which are prevalent because of the free combination of genes.

The very similarity of the majority of the genes at the centers of type-formation does not always result in harmonious combinations from the point of view of the plant breeder; hence, perhaps, the limited cultivation of the Asiatic and African strains and the low civilizations of the people settled in the ancient foci of agricultural crops. The types cultivated according to this hypothesis are the recessives, i.e. types such as can be clearly seen in plants, animals and in man himself.

Conclusions

Thus, we arrive by a different route to the same hypothesis as launched by my teacher, W. Bateson, i.e. that the process of evolution must be regarded as a process of simplification, as an untangling of a complicated set of original genes (cf. Bateson: 'Presidential Address', *Nature* 1914).

We shall define the geographical process of evolution within the limits of the species as a dispersal from the basic centers of type-formation, a combination of genes decreasing from the center of the dominant genes toward the periphery where there is an accumulation of recessive combinations and a disengagement of a portion of the genes (if Bateson's hypothesis, i.e. his concept of dominance and recessiveness as a presence or absence of genes, holds true; it is of no consequence for our own concept if Bateson's hypothesis is eventually refuted).

From the points of view of general civilization and history, this process was positive, leading initially to the great northern civilizations. Being genetically recessive according to the composition of their genes, they added potential to the richer composition of genes of the original populations which were residing in the original centers of human civilization.

The general inductive picture of geographical regularities is similar and leads to a direct knowledge of the geographical forms within the limits of the Linnaean species in relation to their distribution over the globe. Much effort may still be needed on the part of the investigators in order to establish a more exact course of the process of geographical evolution of a large number of animal and plant species. The investigators and geneticists have, at present, hardly touched upon this area which promises to open up facts of paramount importance for the understanding of the process of evolution of different species.

On board the 'Krispi' in the Mediterranean, April 22, 1927.

BIBLIOGRAPHY

Bateson, W. (1926). Segregation. *J. Genet.*, **16**.

Flaksberger, K.A. (1926). Bezligul'nye tverdye pshenitzy c ostrova Kipra [Non-ligulate hard wheats on the island of Cyprus]. *Tr. po prikl. botan. i selek.* [*Papers on applied botany and plant breeding*], **16** (3).

Vavilov, N.I. (1926). Tsentry proiszhokhdeniya kultur'nykh rasteniy [Centers of origin of cultivated plants]. *Tr. po prikl. botan., genet. i selek.* [*Papers on applied botany, genetics and plant breeding*], **16** (2).

Vavilov, N.I. and Kuznetsov, E. (1923). O geneticheskoy prirode ozimykh i yarovykh rasteniy [On the genetic basis of winter and spring type plants]. *Izd. Sarat. s.-kh. in-ta* [*Publications from the Saratov Institute of Agriculture*], **1** (1).

Universal centers of a wealth of types (genes) of cultivated plants

Linnaean species as a system of forms

RESEARCH ON THE inherited variability and studies of the racial composition of various Linnaean species of different cultivated plants led us to the establishment of those principles concerning type-formation, which we have called the laws of homogeneous series of inherited variability (Vavilov, 1920, 1922).

As demonstrated by this research, the Linnaean species are not represented by random composition of races but by a complicated, although definable system of forms developed into a striking conformity of series of variability of closely related species and genera.

We have recently concluded the compilation of a picture concerning the variation within the limits of the Linnaean species as far as the most important botanical families are concerned, i.e. those to which cultivated plants belong. For the time being, such a picture shows systems of phenotypic variability. Experimental genetics has established only a comparatively small number of characteristics; however, there is already no doubt that these principles apply in their general traits to the system of genes and to genetic variability.

Studies of the variability and the systems of variable characteristics within the limits of the Linnaean species of cultivated plants and those closely related to them have made us search for missing links within these systems and for the foci of type-formation, located somewhere on our Earth. Of course, the question arose about where and in which area we would find the focal genes of a given plant. Do such centers exist on Earth or did perhaps the process of type-formation dissipate into thin air? We were logically led to the old problem concerning the origin of cultivated plants, now, however, associated with new, definitely concrete problems. Not only our concept of the system of variability in relation to Linnaean species but also, to a great extent, the practical work of the plant breeder, depend on the solution of these problems just as the solution of the problem concerning the location of the centers of variability of various species as formulated by us will define the mastering of the original material of forms, i.e. the genes of cultivated plants.

Lecture to the General Assembly of the 5th International Genetical Congress in Berlin, September 1927. First published in *Izd. Gos. In-ta, opyt. agr.* [Publications from the National Institutes of Experimental Agriculture], 5 (5), 339–51, 1927.

It seemed that enough information about the local origin of at least the major cultivated plants should be found among the data in the numerous old historical and archeological as well as the botanical works. However, a more-exact definition of the problems revealed immediately a considerable amount of sketchiness and inaccuracies as far as the old statements are concerned. The historical and archeological, as well as biological, investigations during the last century did not distinguish the Linnaean species adequately, not to mention their races or varieties. In the majority of cases, actual studies of the worldwide geography of the varieties revealed discrepancies concerning the actual location of the centers of variability or the centers of type-formation with respect to cultivated plants, as seen already in the classical statements made by De Candolle. Until recently, the botanist had to deal with entire groups of plants, which were not always strictly distinguished into different species within the limits of different genera. For the solution of this kind of problem the contemporary level of knowledge obstructed primarily a strict differentiation of the material into separate groups, although it was, in part, possible to speak, e.g. of the local origin of oats in general. It is now necessary to split up this problem and to consider the genesis of different groups such as *Avena sativa*, *A. byzantina*, *A. strigosa* and *A. abyssinica*, which are sharply distinguished to such an extent that, when crossed with each other, they cannot produce any hybrids.

A concrete and even utilitarian formulation of the problem necessitates a complete revision of the old definition concerning the origin of cultivated plants.

Thus, we shall henceforth understand the correct solution of this problem as being the establishment of actual centers of type-formation of Linnaean species, as foci of diversity of races and varieties or, more exactly, of variable characteristics and, finally, as geographical centers for the concentration of genes.

The expeditions of the Institute of Applied Botany

For the purpose of actual studies concerning the geography of the genes and in order to collect strains of interest to us, the Institute of Applied Botany has, during the last 12 years, organized a large number of expeditions both within the borders of the U.S.S.R. as well as to all corners of the world. Such studies have brought about essential amendments to our information about the racial composition of the species of cultivated plants and the geography of the botanical varieties and races so that we have at present actual data about the geographical concentration of the species of many plants.

Most of all, the geographical investigations carried out by the Institute of Applied Botany revealed a definitely distinct presence of clearly localized centers of variation. In the cases of many cultivated plants we succeeded in establishing the centers where an abundance of their variability is accumulated. In spite of the antiquity of agricultural civilizations, migrations of peoples and colonization, the location of centers of diversity could be clearly revealed and they are of real and practical importance for control over the resources of genes for practical plant breeding.

The Caucasus, Afghanistan, Turkestan and the mountains of China, some countries along the coasts of the Mediterranean, Abyssinia and Egypt, as well as Mexico and Peru have, as shown by the investigations, a multitude of endemics and varieties, typical exclusively of these areas. In mountain areas, in almost inaccessible areas, and in isolated countries in Asia, Africa, the Caucasus, as well as the Cordillera, an enormous number of peculiar types, unknown to Europeans, are still preserved. Detailed studies of the distribution of the varieties have revealed centers of an astonishing diversity, which frequently include all the types known to Europeans as well as variable characteristics and an additional number of new forms and characteristics that have not previously been seen.

In addition to the studies of cultivated plants, research on the closely related wild species and their varieties and races are of no less interest to us.

Methods for determining the centers of type-formation (centers of origin) of cultivated plants

For the purpose of establishing the centers of type-formation or the centers of diversity we applied a method we called the 'differential phyto-geographical method'. It consists of the following:

1 A strict differentiation of the plants studied into Linnaean species and genetical groups by means of various disciplines such as morphology, taxonomy, hybridization, cytology, parasitology, etc.

2 Delimitation of the distribution areas of these plants and, if possible, also of the distribution areas in the remote past when communications were more difficult than at present.

3 A detailed determination of the composition of the varieties and races of each species, and a general system of the inherited variability within the limits of the different species.

4 Establishment of the distribution of the inherited variability of the forms of a given species as far as regions and areas are concerned, and the establishment of the geographical centers where these varieties are now accumulated. Regions of maximum diversity, usually also including a number of endemic types and characteristics, can also be centers of type-formation.

5 For a more exact definition of the center of origin and type-formation it is necessary to establish the geographical centers of concentrations of species that are genetically closely related as well.

6 Finally, the establishment of the areas of diversity of wild species and varieties that are closely related to the species in question should be used for amendment and addition to the area defined as original when the differential method for studying races is applied to them.

For this purpose it is necessary to distinguish the primary foci of type-formation from the secondary ones. Cases are known where the present maximum of type diversity can be the result of different species meeting and hybridizing with each other. Thus, e.g. Spain has a large number of varieties and species of wheat thanks to its mountainous conditions and its general geographical position. However, the total of a large number of species tells us very little since, as shown by our direct analysis, the number of varieties within the limits of the different species in Spain is very small in comparison with that found at the actual centers of type-formation of the same species. There are frequent examples – especially in the case of cross-pollinating plants – of how isolation and exposure far from the primary centers can lead to the appearance of a great variety of recessive forms. Such a development is promoted, e.g., by the application of inbreeding. Within the basic centers, the recessive characteristics are concealed behind the phenotypic appearance of the dominant genes. In modern times, the maximum variation of many garden plants seems to be found in the horticultural nurseries. This is linked to the results of hybridization and isolation of recessive types. It is well known that, e.g., *Drosophila* varieties are the result of mutations in artificial surroundings. It is necessary to be aware of such possibilities, inherent in some genera, and to distinguish secondary centers of type-formation from primary ones.

However, investigations of a large number of cultivated plants show quite clearly that the process of evolution takes place in space as well as in time.

In order to demonstrate how to establish geographical centers of type-formation we have selected some important plants from the Old World for discussion.

WHEAT

Let us start with wheat, the most important cereal in the world. Not very long ago, at the end of the nineteenth century, one of the great phyto-geographers of our time, Solms-Laubach, considered the native land of cultivated wheat impossible to locate and suggested that the explanation was the disappearance of 'missing' links. De Candolle tried to search for the native land of wheat in Asia. The wild wheat, *Triticum dicoccoides*, discovered in 1906 by Aaronsohn in Syria and Palestine, directed the attention of scientists to that part of the East. It seemed that the clew of Ariadne had been found and that the problem of the origin of wheat was solved. Soon, however, new investigations revealed complications concerning this problem.

Our own experiments with hybridization of wild wheat with different species of cultivated wheat, including such a closely related species as *T. dicoccum*, demonstrated that the wild wheat of Aaronsohn is represented by a distinct Linnaean species. As already known, it is characterized by 28 chromosomes, which sharply distinguishes it from the entire group of soft wheats so that, in reality, it is a species of its own, distinguished even from other wheats with 28 chromosomes. Plant breeders and geneticists will turn in vain to Syria and Palestine in their search for the genes of the cultivated wheat: as shown by our own thorough research, Syria and Palestine, as well as Transjordania, are not distinguished by a variety of wheat types. Rather, Syria and Palestine are astonishing by the uniformity of the composition of types of wheat in comparison with that in other areas around the Mediterranean and in eastern Africa.

Through detailed investigations over the last ten years we have succeeded in demonstrating that the actual centers of diversity, with respect to the variable characteristics of the genes of wheat, are found in two different geographical regions.

Soft wheats – characterized by $2n=42$ chromosomes, as shown by detailed research – have their origin in southwestern Asia. When proceeding in the direction from Europe toward southwestern Asia, an investigator will gradually arrive into a region where a variety of strains are concentrated. The mountain areas of southeastern and northern Afghanistan, the foothills of the southern Himalayas, Chitral and Kashmir preserve an exceptional treasure of races and varieties of soft wheat. All the diversity of types, of which until recently the taxonomists had samples, are concentrated in the areas adjacent to northeastern India. In addition to the 22 varieties described by Koernicke and known for about ten years, we have now found more than 70 botanical varieties in this area, including a number of new variable characteristics. Besides, a significant portion of these varieties remains, so far, endemic within the region in question.

It is interesting and important that the geographical distribution of the variety closely related to soft wheats, i.e. club wheat, *T. compactum*, also belongs to the area where a variety of soft wheats is found.

If we keep in mind that also a third kind of wheat, close to the soft ones, i.e. *T. sphaerococcum*, recently described by Percival, is found in all its variations only in northern India, there is a coincidence between the areas of three species which even more supports the accuracy of the application of the differential phyto-geographical method.

With respect to the hard wheats and the Linnaean species closely related to them, it could, on the other hand, be demonstrated that their centers of variation were restricted to the eastern mountains of Africa, particularly to the area of Abyssinia and the mountainous countries adjacent to it.

Investigations made this year have shown that, in addition to our previous information, Abyssinia appears actually to be a striking center for the concentration of an abundance of strains as far as hard wheat is concerned. All the forms of hard wheat with 28 chromosomes are found there and the majority of these are new varieties. In addition, the divergence of types is not as sharp there, so that it is often hard to distinguish between different species of wheat in the fields of Abyssinia.

As we have just learned, the wild species of wheat, *T. dicoccoides*, is in all its diversity geographically typical of southern Syria and northern Palestine. Einkorn [*T. monococcum*], characterized by $2n=14$ chromosomes, is – as demonstrated by investigations made at the Institute of Applied Botany – centered mainly in Asia Minor and southern Syria.

We have now succeeded in establishing the geographical localization of the genes belonging to the most important cereals by means of the analytical phyto-geographical method. The genetical isolation of *T. dicoccoides* from other hard wheats, revealed by means of hybridization, is confirmed by the isolation of its geographical area from that of the last-mentioned group in general.

The hypothesis about a polyphyletic origin of wheat, suggested by Solms-Laubach in 1899 and then appearing to be a fantasy, has thus turned out to be more than accurate.

BARLEY

De Candolle looked for the native land of barley, where the distichous, hulled barley, *Hordeum spontaneum*, grows in a wild state. The area of wild barley covers northern Africa, Asia Minor and all of southwestern Asia. However, our investigations, applying the differential phyto-geographical method, have demonstrated that the distribution area of wild barley furnishes very little information about the location of the actual center of type-formation of cultivated barley. The maximum concentration of its varieties is found in Abyssinia and includes, consequently, also the genes belonging to the group of hulled barley. An exceptional variety of forms is concentrated there, as well as all the characteristics by which a botanist distinguishes the varieties and races of cultivated barley. In addition, there are a number of endemic characteristics, not known in Europe and Asia, such as a group of *deficientes*. It is interesting that in Abyssinia and Egypt, among the rich diversity of the varieties and races of cultivated barley, wild barley is definitely absent.

The investigations allow us to distinguish secondary centers of variation as well, and the fact that a wealth of naked-grained barley is especially concentrated in southeastern Asia. Peculiar forms, e.g. with a type of glumes called *furcatum*, with short awns and with broad leaves are also included in that area.

We also know about endemic, naked-grained barley in Abyssinia, but it is very different from the Asian type. Thus, the differential method has led to the establishment of two centers of type-formation for cultivated barleys.

OTHER EXAMPLES

Consequently, it is possible to define objectively the centers of accumulation of a variety of forms for other plants as well. We know, e.g., at present, that cultivated flax comes originally from a number of centers. The genes of the multiform, cultivated flax are concentrated within three different areas: large-flowered, large-capsuled and large-seeded flax (and linseed) is typical mainly of the Mediterranean countries; on the other hand, central and northeastern Asia Minor and northwestern India as well as the countries adjacent to them are the areas where small-flowered and small-seeded flax, used for both its fibers and linseed, are accumulated, while Abyssinia turned out to be the center of a peculiar, extremely small-seeded type of flax that is grown neither for fibers or oil but is used as food for the inhabitants of the area.

By means of similar comparative studies of the varietal composition of the leguminous plants from various countries in the Old World it was revealed that Abyssinia and India appear to be the definite foci of the genes of such crops as peas and chick-peas.

Only a few plants have distinctly located single centers, where all the present, variable characteristics and genes are concentrated. Species, e.g. of *Avena strigosa*, *A. brevis*, *A. abyssinica*, *Ervum monanthos* [erse] and some fodder plants such as sulla (*Hedysarum coronarium*), berseem clover (*Trifolium alexandrinum*) and *Ulex* are characterized by narrowly delimited centers. Two or three foci are present in the case of many cultivated plants, often in such cases where they belong to a single Linnaean species as, e.g., in the case of lentils, vetchlings and fava beans [*Ervum lens*, *Lathyrus sativus* and *Vicia faba*].

In general, we arrived at the conclusion that it is still possible to define the geographical centers on Earth comparatively exactly and objectively, i.e. those where an abundance of strains are accumulated in centers of type-formation. The presence thereof significantly facilitates selection and plant breeding and even the problems encountered when studying the genesis of cultivated plants.

Geographical regularities concerning the distribution of a variety of cultivated types of plants

When studying the worldwide distribution of a variety of forms, a number of interesting facts become apparent. As already stated, the primary centers of type-formation are characterized not only by a great diversity and a large number of forms but – what is particularly significant – by the presence of a large number of dominant forms with clearly expressed dominant genes. When examining the cereal market of Addis Ababa in the center of Abyssinia, you are surprised by the presence, in contrast to what you see in a similar market in Europe or Asia, of a large number of types of dark-colored barley and violet-grained wheat, which the local farmers call 'black wheat'. In its ancient and basic agricultural areas, Abyssinia mainly cultivates races of distichous barley of the '*deficientes*' type, which, as we have already seen, when hybridizing with the ordinary tetra- or hexastichous races, produces the dominant type of their characteristics. In contrast to what is found in the Mediterranean area, chick-peas (*Cicer arietinum*) from Turkestan are, to a significant degree, represented by dark-colored strains. In the direction toward northeastern India we have found the same with respect to other plants as well. Not only does variation decline in the direction from the center toward the periphery but – what is even more remarkable – the number of dominant genes also decreases.

The entire process of geographical evolution can, in principle, be considered as a process of 'shedding' dominant genes. Toward the periphery of the geographical distribution area, and at the edges of such areas, races of cultivated plants are found to have mainly recessive genes. Of course, artificial, but often also natural, selection interferes with this picture when we are speaking of a general tendency and of a general regularity with respect to the process of distribution of forms.

Within isolated localities, on islands and in almost inaccessible areas close to a center, there are often isolated recessive forms. In the oasis of Khiva in the deserts far from any basic centers of flax, we found a focus of mainly recessive, white-flowered types of flax. The island of Cyprus is home to recessive, non-ligulate races of hard wheat. In the remote mountain areas of Badakhshan and Pamir we have discovered non-ligulate forms of both rye and soft wheat, and in the mountains of Asturia and northern Spain we encountered enclaves with white-flowered, definitely uniform wild flax, *Linum angustifolium*.

The existence of this kind of regularity appears to be quite essential and allows us to distinguish secondary centers, which often are richer in forms than the primary ones. An analysis of the forms, an elucidation of their genetic composition, can facilitate the definition of the concept of what is primary and what is secondary. The presence of a large amount of recessive types far from their basic center of origin – for instance, in the case of maize – does not appear to be something contrary to the basic idea concerning the centers of type-formation; it only confirms it. By inbreeding and isolation of the recessives, the plant breeder can – as we now know so well – bring out a great variety from the environment of the primary phenotypic uniformity.

Comparative studies of the worldwide geography of races and varieties have allowed us to establish a number of other regularities as well. Thus, for instance, the races of many field and vegetable crops of the Mediterranean area are characterized by a tendency toward production of large fruits, large flowers and large seeds. Large-seeded vetchlings, chick-peas, lentils, horse beans and flax, and large-grained races of wheat and oats, are typical of the countries around the Mediterranean coasts. On the other hand, southeastern Asia, in the direction toward India, is characterized by a tendency toward small fruits and small seeds as far as the same kinds of plants are concerned.

Southeastern Asia turned out to be the focus for naked-grained races of some cereals: oat, barley and millet. The nature of this is still not known to us.

Primary and secondary crops

Our experiments have allowed us to separate all the cultivated herbaceous plants into two groups. The first or basic one consists of the ancient cultivated plants which are known only in a cultivated state. They gave rise to what we shall call primary crops. To this group belong, for instance, wheat, barley, rice, maize, soybeans, flax and cotton. The second, or secondary, group is no less extensive. All the plants belonging to this group originate from weeds infesting the primary crops.

RYE

It could, thus, be demonstrated that cultivated winter rye originated from weedy rye infesting winter wheat and winter barley in southeastern Asia and Transcaucasia. In all its multiformity, which is astonishing to Europeans, rye is concentrated just within those countries where it is not cultivated but where it infests crops of wheat and barley. The very epithets of rye in the Persian language and in Turkey, India, Afghanistan and Turkestan as well, are 'choudar' or 'gandum-dar', which means 'the plant that torments wheat or barley'. Rye is considered a weed in those countries. In Afghanistan it is, indeed, a very serious weed, doing no less harm than wild oats. When it is mixed in the granary with the wheat a wide distribution of the common weedy and non-brittle forms is promoted, just as in the case of the brittle types with disarticulating spikelets (*Secale cereale* var. *afghanicum*), which infest emmer [*T. dicoccum*] in a manner similar to that of wild oats.

As demonstrated by the Institute of Applied Botany, weedy rye has a very large number of endemic forms: red-spiked, even black-spiked ones and ones with pubescent hulls or peculiar vegetative characteristics.

As revealed by detailed observations, weedy rye is able to displace wheat because of its frost-hardiness and to turn into a crop of its own. In the belt between 2300 m and 2400 m elevation in Turkestan, winter wheat disappears and is replaced by rye. All the phases of the 'competition' between wheat and rye can be followed there.

The same phenomenon – although on a larger scale – takes places in the direction from south toward north. During the dispersal toward the north of crops from southeastern Asia, the basic center of type-formation of soft wheats, rye began to displace wheat in Siberia and Europe. Thanks to its naturally inherited frost-resistance and its lesser demands on the substrate, rye began, also by the grace of Man, to replace wheat and became a pure crop in the same manner as has happened in the mountain areas of southwestern Asia. The dominance of rye, which at present can be observed on the Great Russian Plains and in Germany, is the result of a natural selection. At the border where wheat and rye 'compete with each other', the farmer himself often sows a mixture of wheat and rye – called 'surzhy' [mixed crop] in Russian – promising a more reliable harvest than of wheat alone which can fail during a hard winter. The entire process of displacement of wheat by rye can be followed in minute detail. The wheat, thus, helped bring rye into cultivation. The genesis of rye cannot possibly be understood without the presence of the crops of wheat and it is very clear to us that all the abundance of genes must, in the case of rye, be looked for just among the weedy types that infest the fields of wheat and barley in southwestern Asia.

Thanks to data assembled by the expeditions of the Institute of Applied Botany to Asia Minor (led by P.M. Zhukovskiy) and to Transcaucasia, Afghanistan and Turkestan, it has now been revealed that the maximum variation of dominant genes with respect to weedy rye is concentrated in Transcaucasia and the eastern parts of Asia Minor, which indeed appear to be the center of origin not only of cultivated rye but of the entire genus of *Secale*.

OATS

The facts revealed by studies of oats are no less interesting. Cultivated oats are represented by a complex group of varieties. At present, we can speak of four basic genetic groups of oats: that of *Avena sativa* and the closely related wild *A. fatua*; that of *A. byzantina* and its close relative *A. sterilis*; that of *A. strigosa*, close to *A. barbata*; and, finally, the fourth one, that of *A. abyssinica*. The first, the extremely polymorphic group of *A. sativa*, is especially intricate.

It has been demonstrated that however great the diversity of cultivated oats is, they have all originated from weeds infesting other crops; some of these crops appear now to be dying out. The presently cultivated oats were in their genesis associated mainly with crops of wheat and rye but particularly with those of emmer. In the few areas around the world, where emmer (*Triticum dicoccum*) is cultivated in small plots, for instance in Bulgaria, or the Caucasus or in the area around River Kama, there is an enormous variety of oats that are included as specialized weeds in crops of emmer. During the dispersal of such crops toward the north and into more severe conditions, the oats displaced the original crop and developed into special crops, cultivated for their own sake. This process can still be closely followed in Abyssinia in the case of *A. abyssinica*, a common weed of emmer and barley. Farther north, in Portugal and Spain, the same process can still be observed in the case of the group of *A. brevis* and *A. strigosa*, which infest rye and wheat in those areas. We succeeded this year in detecting a whole new group of botanical varieties in that area within *A. brevis*.

Among the crops of emmer in the former provinces of Kazan, Vyatsk and Ufim, we have succeeded in finding a peculiar group of so-called 'emmer-like oats', distinguished by grains that are firmly attached to the spikelets.

When looking for new types and new genes of oats, the plant breeder and the geneticist should turn their attention to such foci of emmer cultivation that seem to be veritable granaries of genes for cultivated oats. Consequently, just as in the case of rye, it is necessary to keep in mind the association of oats with the genesis of other crops in order to get a grip on the system of the oats themselves.

Such a process of infestation of one crop by another and the displacement of the basic crop by weeds has also taken place in the case of crops of lentils. The majority of the cruciferous crops originates from weeds, infesting other crops. Such is the origin of many leguminous plants and that of a number of vegetables, for instance carrots, is similar. In general, we are of the opinion that roughly half of the cultivated grasses appear to have originated from weeds in other crops and are plants that, by the grace of Man, were brought into cultivation. Accordingly, the genes of a given plant can, strangely enough, be looked for in the centers of other plants, as in the case of weedy rye in the foci of emmer, although the latter are of limited dimensions in our world. However, a supply of oat genes can be found there. Studies of field weeds can, from this point of view, open up new horizons for plant breeding.

ANTHROPOCHOROUS PLANTS

Investigations of such plants as hemp, nettles, carrots and poppies have revealed that these species are, with respect to their origin, tied to such circumstances that their initial types appear to have been unfailing 'camp-followers' of Man because of their ecological specificity. As shown by observations made during the last expedition from the Institute of Applied Botany to Peru (led by S.V. Yuzepchuk), tomatoes and potatoes appear to be such a kind of anthropochorous plants in their native homelands. Wild hemp in Asia and parts of southeastern Europe appeared to be an unfailing companion of nomadic tribes. They seed themselves around the camps in fertilized and manured spots just as if they had invited themselves to be cultivated. As far as such plants are concerned – in particular where their adoption into cultivation was not planned – a definite center of origin cannot be located: their genes were, so to say, 'scattered' by the nomads or by people settling over wide areas across the continents. The accumulation of the genes of these plants is just as factual as in the case of plants the centers of which can be directly located. The genesis of the anthropochorous plants, i.e. those allowed by the grace of Man to accompany him, can to a great extent be explained by the same process of origin as that of cultivated plants.

Finally, with respect to the diversity of their genes, such plants as watermelons are still growing in a naturalized state in areas far from Africa, where they originated. In modern times, the genes of watermelons have been scattered across enormous areas of the Dark Continent and to gather them presents considerable difficulties. The same can be observed, for instance, as far as red clover is concerned, which has an enormous area of geographic distribution in its wild state. The number of such cultivated plants with colossal distribution areas over which their genes are scattered, is fortunately not very large; instead, they are rather exceptions.

Localities where basic foci of origin can be found

Summing up our geographical knowledge of the worldwide distribution of centers of accumulation with respect to the wealth of types, or genes, of cultivated plants, we have arrived at a schematic definition of some general centers of whole groups of cultivated plants and to the establishment of basic, general centers of origin (cf. Fig. 13, p. 127).

1 Southwestern Asia, including India, southern Afghanistan and the areas adjacent to them in the mountains of Bokhara, Kashmir, Iran, eastern and central Asia Minor and Transcaucasia: This enormous center gave rise to soft, club and shot wheats, rye, small-seeded flax, small-seeded peas, lentils, chick-peas and a number of other vegetables, Indian cotton and a number of fruit trees such as apricots, peaches, etc.

2 Southeastern Asia is the second center, which includes the mountains of China and Japan and the areas adjacent to them: the centers of type-formation of naked-grained oats and naked-grained barley as well as millet are found there; it is the native land, also, of soybeans, many cultivated cruciferous plants and a number of cultivated fruit trees.

3 The Mediterranean center comprises all the coastal areas of this area, including Syria, Palestine and Greece with its archipelago, the Pyrenéan and Apennine peninsulas, the western and southwestern parts of Asia Minor and Egypt. Apparently Algeria, Tunisia and Morocco are not primary centers of origin of any cultivated plants. The centers of origin mainly of such fruit trees as olives, carob and figs also belong to the Mediterranean focus just as do the species of cultivated oats, *Avena byzantina*, large-seeded linseed and large-seeded leguminous plants – fava beans, lentils and vetchlings (*Ervum monanthos*, *E. ervilia*), etc. – and such fodder crops as sulla (*Hedysarum coronarium*) and berseem clover (*Trifolium alexandrinum*), as well as many vegetables.

4 Abyssinia must be distinguished as the fourth independent center in northeastern Africa, together with the areas adjacent to it, especially Egypt. An exceptional diversity of a number of cultivated plants is concentrated in that area. The Abyssinian center is characterized by various forms of hulled barley, violet-grained wheat, peculiar races of peas and special, weedy forms of oats.

Abyssinia has the greatest morphological diversity of hard wheats as well. It is also typical of a number of endemic forms of plants, not known anywhere else in the world, such as, for instance, ramtil (*Guizotia*) and teff (*Eragrostis abyssinica*) as well as *Rhamnus prinoides*. At the same time, many Asiatic and European cultivated fruits, vegetables and root crops are characteristically absent from Abyssinia.

5 The mountains of Mexico, Guatemala, Colombia and Peru, as well as adjacent areas in the New World, can be distinguished as foci of primary agriculture and centers of accumulation of genes of cultivated plants. The centers of type-formation of potatoes, maize, beans, gourds, tobacco and American cotton and many endemic species and genera of cultivated plants are concentrated there. However, the diversity of the species and varieties of sunflowers seems to be found especially in the central parts of North America only, where they occur mainly as weeds and plants on waste lands.

Apparently there is a sixth independent center adjacent to and south of the first two centers in Asia, i.e. one including especially the Philippines and what is called East India. However, cultivated plants of that focus are still not well enough known botanically.

As is obvious, the areas of type-formation with respect to the most important cultivated plants, i.e. the present geographical centers of genes, are associated mainly with the mountain areas of Asia (Himalaya and its systems of outlying ridges), the mountain systems of northeastern Africa, the mountain areas of southern Europe (the Pyrenées, the Apennines and the Balkans) and – in the New World – with the Cordilleras and the southern portion of the Rocky Mountains.

In the Old World, the origin of cultivated plants is associated with a zone between the latitudes of 20° and 40° North. This mountain area is delimited by the deserts of Central Asia and by the Sahara and, as far as the diversity of the climates and soils is concerned, it represents optimal conditions for the manifestation of the type-forming process. The diversity of conditions, ranging from deserts to oases, from soil deficient in humus to soil rich in nutrients, within the alpine and subalpine belts, has made feasible a concentration in those areas of an exceptional diversity of vegetation.

The fact that the most important of them, as far as cultivated plants are concerned, are concentrated just within mountain areas, more exactly in the areas at an elevation between 500 and 2500 meters above sea level, is absolutely certain. In spite of the commonly held opinion about the association between the initial cultivation of crops and the valleys of the great rivers such as the Tigris, Euphrates, Indus, Ganges, Nile, Hwang-Ho, Yangtse and Amu-Darya, as well as Syr-Darya, we have arrived at the conclusion that the foothill areas, which at present appear to be the actual 'custodians' of a wealth of types, are also the initial areas of agriculture. Studies of the process concerning where human civilizations settled show quite clearly that it is just a number of mountain areas that appear to be the original foci thereof. While representing natural fortresses or refuges, they favored the establishment of primitive civilizations, uniting only a few groups of people.

It can still be clearly seen in Abyssinia, Egypt and along the coasts of the

Mediterranean that the crops gradually migrated from the mountain areas down into the lowlands and the valleys, and not the other way around.

The development of the major valley civilizations, associated with artificial irrigation along the Tigris and Euphrates and in the Nile delta, needed a major cooperative organization, uniting peoples and tribes, developed logically, and, undoubtedly, much later and by more advanced processes and not by early, primitive ones.

An analysis of the composition of cultivated plants demonstrates that the great ancient Egyptian civilization started, initially, in all likelihood at the sources of the Nile, i.e. along the upper White and Blue Niles. It seems also likely that the basic elements of the plant crops grown in Upper and Lower Egypt, as well as the human tribes developing the Egyptian civilization, should be looked for in the mountain areas of Abyssinia and the areas adjacent to it. All data, available to us concerning the composition of plants grown in Mesopotamia, speak in favor of the fact that the Egyptian civilization, too, is secondary.

In all the world, the mountain areas of Africa and Asia still represent their own kind of ideal habitats for settlements. This was even more obvious in the remote past: more than half of the human settlements in the world are still concentrated within this mountain zone which represents but a very modest fraction of the Earth's total surface, according to our calculations not amounting to more than 1/20th thereof.

While traveling in southeastern Asia or in the Mediterranean areas, you cannot avoid being surprised by the extreme utilization of every inch of soil accessible for cultivation. If, in Iran, Afghanistan and Bokhara, the infertile deserts and the waterless mountain areas, any slopes too steep for cultivation and rocky screes as well as areas covered by 'eternal' snow, are subtracted, and the density of the settlements is compared to the area accessible to cultivation in the rest of the world, we arrive at a population density that exceeds even that of the cultivated areas of Europe.

Centers of origin of cultivated plants and human civilizations

It is interesting to note that the centers where the genes of cultivated plants are concentrated are associated with the ancient foci of agricultural civilizations as far, as they appear at present. In South and Central America, especially in areas of mass-accumulation of an abundance of plants, there are remnants of primitive agricultural civilizations such as those of the Aztecs, the Mayas, the Incas, the Chibchas or the Muish.

From the problem concerning the origin of cultivated plants and the worldwide geography of their genes, we have unintentionally arrived at questions concerning the problem of the origin of agriculture, i.e. of human civilizations. Whether he wants it or not, an investigator of plant crops must approach in earnest the problem concerning the autonomy and independence of human civilizations. We do not doubt that, after detailed studies concerning the foci of type-formation of the major cultivated plants, a botanist will be able to make essential corrections to what is presented by historians and archeologists.

The independent genetic foci of cultivated plants appear to be truly autonomous foci of human civilizations as well. The questions about the independence of civilizations can be more accurately answered on the basis of plants than by means of various archeological documents.

Thus, after detailed studies of the centers where the genes of cultivated plants are accumulated as if reflecting the geographical distribution thereof throughout the entire world, the investigator can, at present, determine the genetic centers of cultivated plants by means of the differential phyto-geographical method. By chance, such centers of a majority of plants can still be located and be subjected to exact analysis. This circumstance opens up wide horizons with respect to the utilization of these centers for practical purposes as well as for genetics. An immense area of application has been laid open for scientists.

Unfortunately, the basic centers where the genes of cultivated plants are accumulated are located in mountain areas, accessible to the investigator only with difficulty, i.e. where areas of a lot of interlaced interests are centered. It is possible to approach fully an investigation of these interesting and important foci of genetic concentration only by establishing international contacts and by means of an international organization of scientific research. Let the present international congress serve as a new stimulus toward such a united international scientific work in the interest of all mankind.

BIBLIOGRAPHY

Vavilov, N.I. (1920). *Zakon homologicheskiikh ryadov v issledstvennoy izmenchivosti* [The law of homologous series in variation]. Saratov.

Vavilov, N.I. (1922). The law of homologous series in variation. *J. Genet.*, **12** (1).

The problem concerning the origin of cultivated plants as presently understood

SCIENTIFIC WORK WITHIN genetics and related fields has recently taken on a singular meaning making it attractive to research workers. This is an exceptionally clear and noteworthy statement concerning the problems of genetics and plant breeding. The great progress made within the field of Science in our time and uniting those gathered here is unquestionable and undeniable and is a specification of the great theoretical and practical problems that at the same time, so to say, are the most important of the biological problems: i.e., those of type-formation and speciation. The geneticists and plant breeders of our time know very well what to do and – to a significant extent – how to do it. If a contemporary biologist or geneticist is planning to conduct research in order to study the problems concerning the genesis of species and forms, it seems to us that he should be able, without any particular difficulties, to outline specific problems for very interesting experimental and descriptive research lasting several decades. This pertains similarly to plant and animal materials.

As the general basis of the most important problem within biology, that of speciation – the theme which I have selected as the subject for this speech – it constitutes only a modest fraction of the entire, almost boundless theme. However, the particularity of our times – the direct approach to the problem – is different from that of the past and its formulation of the problem concerning the origin of cultivated plants.

In contrast to what was feasible during the nineteenth century, a scientist can now approach the problem concerning the origin of biota mainly in the capacity of an experimenter or a kind of engineer. The previous solutions to the problems concerning various cultivated plants and domesticated animals both in our Old and in the New World and the establishment of the approximate relationship between wild and cultivated plants or concerning, more or less, the antiquity of cultivated biota are no longer satisfactory. The interesting historical and archeological investigations are only some of the many auxiliary methods applied for understanding the pathway toward the creation of species and forms

Speech given at the All-Soviet Congress of Genetics, Plant Breeding, Agriculture and Livestock Breeding in Leningrad, January 10, 1929. First published in *Dostisheniya i perspektivy v oblasty prikladnoy botaniki, genetiki i selektsii.* [*Progress and perspectives within the fields of applied botany, genetics and plant breeding*], Leningrad 1922, pp. 11–29.

but are far from satisfactory to the geneticist/experimenter who wants control over the building materials and the methods for building species and forms.

According to our present ideas, anybody investigating the problems concerning the origin of cultivated biota must do all he or she can to master the original elements, when determining the type-forming process of different Linnaean species. We are faced with very specific and especially utilitarian problems of how to master the stages of type-formation, i.e. the structural materials, in order to, on that basis, apply the imaginative work of the biologist concerning the creation of species and forms according to arbitrary rules; our objective is to learn how to restore the historical process and how to, by ourselves, create new species and forms. In essence, we want to learn how cultivated wheat arose, how its great diversity of forms developed and along which pathways this occurred and how, from the wild forms that are somehow still preserved, the synthesis of those species that actually exist now came about. We must consider the problems, decisive for the origin, where an investigator must actually be able to master all the material for creating forms or species of cultivated organisms. In other words, as far as the problem pertaining to the origin of cultivated biota is concerned, we must presently not only consider it historically but also dynamically and attempt mainly to conquer it experimentally. All that is of significance for the utilitarian objective of breeding can be found in such a specification of the problem concerning the origin of cultivated plants and domestic animals. Without a solution of these difficult theoretical problems, the practical breeder is to a significant extent forced to work relying on only random combinations. At this Congress of Genetics and Experimental Science we will allow ourselves to pay attention to what, at first glance, seemed to be a question of a historical nature.

In its absolute dimensions, which engulf everybody who comes in contact with a Linnaean species in its actual composition and all its multiformity – an example of which is Man himself – the problem of the origin of cultivated biota is far from being straightforward. In the case of many biota, for instance domesticated animals, we do not even know the real dimensions of the taxonomic composition of the Linnaean species or its racial complex. When earnestly raising the problem relating to the origin of cultivated species of animals and plants we can, perhaps, speak of it as stages during an attack conquering some fortresses after a general assault upon still-inaccessible strongholds that, until recently, represented Linnaean species.

It was not by chance that the International Congress in Berlin last year was opened with a speech by Wettstein, who stated that genetics has actually not yet reached the root of the problem concerning the development of species and genera.

Establishment of foci of primary type-formation

The first stage of such an attack consists mainly of establishing the spatial location of the basic, initial type-forming process, i.e. of determining foci of origin of cultivated species. The Linnaean species as presently understood is represented by a whole system of forms. In order to really master this complex it is necessary to know the geography of the entire system and the geographical distribution of the elements composing the species. The ordinary zoological and botanical distribution areas, which do not take into account the genotypic diversity, do not satisfy us. First of all, in order to master a species not only the general geographical outlines of the distribution of its different forms are necessary, but also a knowledge of the area where the maximum accumulation of its elements of diversity is located, i.e. the genes of the species in question, as well as an establishment of the geographical center, where the type-forming process occurred. Until recently we did not have such data. The problem raised compels us to direct a frontal attack on the basic foci of origin of cultivated biota and, similarly, as indicated by the facts, on the diversity of these organisms.

In spite of the old ideas, developed especially by the botanist Alfonse De Candolle, according to which the native land of cultivated plants should be looked for in areas where they exist in a wild state, a discrepancy between the present areas of wild species that are closely related to cultivated plants and the main genetic basis of cultivated plants could in many cases be established, as demonstrated by expeditions sent out by our Institute of Applied Botany to different corners of the world (Fig. 1). Actual investigations of a large amount of material point to the existence of a geographical disparity between cultivated and wild species.

Wild barley, which is very closely related to the cultivated species, is quite common in the foothills of Turkestan. Its distribution covers a considerable area, at present even reaching Asia Minor, Syria, Palestine and Afghanistan. As demonstrated by our expedition, however, the complex of cultivated barley is at the same time strikingly poor in these areas. Dozens of square kilometers of loess soils in northern Afghanistan are covered by *wild* barley and, what is more, the *cultivated* barley in Afghanistan is, as shown by direct investigations, of an extraordinarily poor diversity there; in addition, the basic centers of type-formation are, in the case of cultivated barley, found far away from Asia Minor, i.e., in the mountains of eastern Africa – Abyssinia and Egypt – and in eastern Asia. Large amounts of wild wheat still grow in southern Syria and Palestine. Cultivated types of wheat in all their amazing variety of which botanists until recently were not aware, have been observed in the foothills of western Himalaya and in the mountains of eastern Africa, mainly in Abyssinia. These kinds of examples may be very frequent: wild carrots have an enormous distribution area, from the Pyrenées to Himalaya, but the actual base of cultivated carrots was discovered in the foothills of Himalaya; wild species of lentils and vetches grow abundantly in the eastern part of the Mediterranean area, while the centers of the cultivated forms are, as shown by indisputable research, found in Abyssinia and the foothills of Himalaya; wild flax, *Linum angustifolium*, which is very closely related to the cultivated one, grows in large stands on the Pyrenéan and Appenine peninsulas, while the original genetic centers of cultivated flax are concentrated in southwestern Asia and northern India.

The solution of the problems concerning the initial origin of cultivated plants

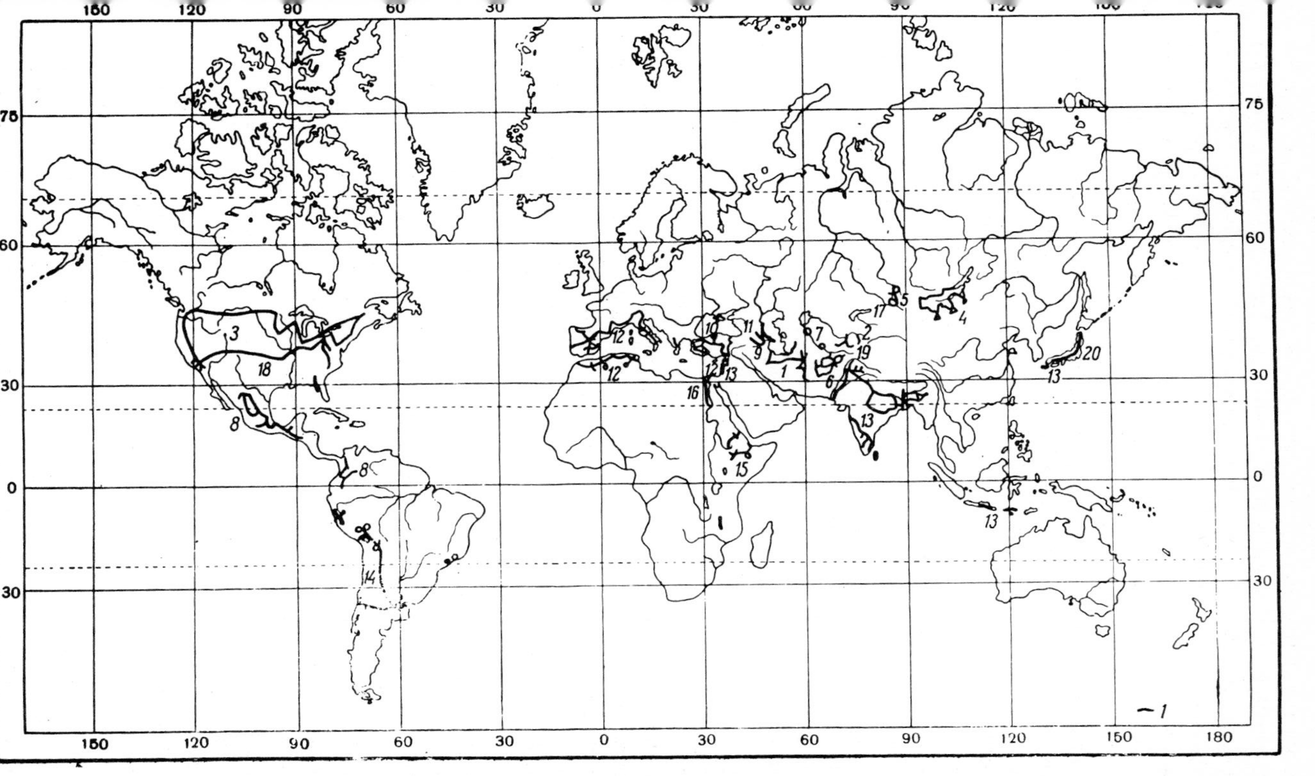

Fig. 1. Map of the expeditions from the All-Soviet Institute of Applied Botany and New Crops and the Department of Applied Botany of the National Institute of Experimental Agronomy. The lines indicate the routes: 1. Persia, N.I. Vavilov, 1916; 2. Pamir, N.I. Vavilov, 1916; 3. North America, N.I. Vavilov, 1921–22; 4. Mongolia, V.E. Pisarev, 1922; 5. Altai, E.N. Sinskaya, 1924; 6. Afghanistan, N.I. Vavilov, 1924; 7. Khiva and Bokhara, N.I. Vavilov, 1925; 8. Central and South America, S.M. Bukasov and Yu. N. Voronov; 9. Armenia, E.A. Stoletova, 1925–26; 10. Asia Minor, P.M. Zhukovskiy, 1925–27; 11. Azerbaidzhan, V.V. Pashkevich and N.N. Kuletov, 1926; 12. the countries along the Mediterranean coasts, N.I. Vavilov, 1926–27; 13. Palestine, India, Ceylon, Java, China and Japan, V.V. Markovich, 1926–27; 14. South America, S.V. Yuzepchuk, 1926–27; 15. Abyssinia and Egypt, N.I. Vavilov, 1927; 16. Egypt, Gudzoni (on behalf of N.I. Vavilov), 1927; 17. Altai, N.P. Gorbunov, 1927; 18. North America, V.V. Talanov; 1927; 19. Pamir, N.P. Gorbunov, 1928; and 20. Japan, E.N. Sinskaya.

by locating the corresponding wild species, which, during De Candolle's time, appeared to be the latest word in Science, seems now, as demonstrated by research, to be only a first approximation toward an answer to this question. We can consider the establishment of the ancient centers of type-formation as the actual determination of regional areas of origin of cultivated plants and domesticated animals. The same solution is associated with the mastering of the original elements and the manifold characteristics composing the Linnaean species of cultivated biota. In turn, a problem arose concerning the exact establishment of the universal distribution of the original genes of cultivated plants and the domesticated animals, i.e., a problem which can only be solved by the identification of the actual sites where the initial type-formation took place.

Recently, when summing up our information about the sites of type-forming processes in the world, we were able to establish six basic main centers in the Old as well as the New World: 1. southwestern Asia; 2. eastern India and the adjacent areas; 3. the mountains of China; 4. Abyssinia; 5. the Mediterranean area; and 6. Central and South America (Vavilov, 1926).

Investigations made by our expeditions during the last couple of years have furnished an enormous amount of material serving as proof of the actual presence in just these areas of centers of basic origin of cultivated plants. For each of these six centers, dozens of different cultivated plants can be mentioned. We consider the problem of an exact establishment of these basic regions to be absolutely essential. When we estimate the large number of cultivated and domesticated biota that are, as a rule, associated with ancient agricultural foci but that are also found in almost inaccessible mountains and foothill areas, for instance of southwestern Asia and the Cordilleras of the New World, it becomes obvious that great effort is required to carry out this immediate task and to master the original material.

Pinpointing the loci of the type-forming process

So far, the pinpointing of the foci of origin with respect to cultivated biota has succeeded on general terms only. The foci agree with the large regions that we established as the six areas mentioned. Now we will start a new stage toward a more detailed and more exact geographical location of the type-forming process.

During the last expeditions and after further research on the material collected, investigations in one country after another revealed, although still inadequately, sharply expressed localization of the type-forming process of a kind impossible to dream of even a few years ago. Research in southwestern Asia – i.e. Iran, Seistan, Afghanistan, India, Kashmir, Asia Minor, Syria and Palestine – and even in the Soviet central-Asiatic and Transcaucasian republics, executed during the last couple of years, has revealed facts about a striking geographical affinity between the process of basic type-formation and certain very limited areas. Thus, a rather small area, enclosed between western Himalaya and Hindukush including the southeastern portion of Afghanistan and north-western India, could be distinguished. There, within a space of not more than a few hundred kilometers in diameter, an isolated focus was observed, where a striking and all-embracing diversity of a number of the most important of our field crops are concentrated, such as soft wheat, leguminous plants, flax, carrots and other plants. Within this peaceful geographical fold formed by the great mountain ridges, an amazingly diverse complex of cultivated plants was observed to be isolated, especially as far as the most important cereal in the world is concerned, i.e. soft wheat. It is interesting that the entire group of cultivated plants is characterized by a number of primitive traits such as small fruits, small seeds, rough types of spikes, low sugar content, etc.

Remarkable observations were also made by the expeditions of S.M. Bukasov and S.V. Yuzepchuk to Central and South America. Each of the ancient civilizations in the Americas had their own specific group of cultivated plants with an abundance of genes. A significant, initial type-forming process is specifically associated in America with Central and South America, particularly the western and eastern slopes of the Cordilleras and their foothills. However, what is even more remarkable is the presence in this large center of several smaller foci, each of which is characterized by an entire group of species and even genera of cultivated plants that are typical of it alone. Thus, for instance: 1. Mexico is typical for the main species of cultivated cotton, *Gossypium hirsutum*, and beans such as *Phaseolus multiflorus*; 2. Guatemala has musk melons, *Cucurbita moschata*, as well as chayote, (*Sechium edule*), cacao and papaya; 3. Colombia – the country of the Chibcha Indians – appears to be the native land of the so-called arrachacha (*Arracacia*); 4. Peru is the homeland of the edible species of *Amaranthus* as well as large squashes; 5. Bolivia appears to be a center of potatoes and quinoa [*Chenopodium quinoa*], *Ullucus* and other tuberous plants; 6. southern Chile represents a second center for potatoes as well as sunchokes [*Helianthus tuberosus*], *Madia* and the plant, now nearly extinct, *Bromus mango*; 7. southern Brazil is the true native land of a number of important plants, such as some other species of cotton (*Gossypium peruvianum*, *G. brasiliense*) and lima beans (*Phaseolus lunatus*) while sunflowers and sunchokes reach all the way up to Canada.

As a result of such detailed investigations, the large center in the New World proved to be composed in general of a number of foci like a row of type-forming 'craters', the mastering of which appears to be a prime task for the solution of the problem concerning the origin of the field crops in America.

When studying the mountainous eastern Africa in 1927, we succeeded in establishing an astonishing localization as far as the type-formation of cultivated barley and wheat are concerned. The awnless hard wheat, interesting from a practical point of view, proved to be definitely associated with the northern part of Abyssinia but absent from its southern part. Abyssinia itself is geographically not very large, but an amazing variety of wheat and barley is typical of it, such as seen nowhere else in the world. A preliminary study in this country has already resulted in more than 200 different varieties of wheat, each of which consists of many forms. The diversity is so great that even the experimental taxonomist feels at a loss trying to classify it.

Interesting facts concerning a narrow localization have also been revealed with respect to the Iberian peninsula. Such species of cultivated plants as *Avena brevis* and *A. strigosa* – oats typical of sandy soils – are as far as their type-formation concerns definitely associated with the western foothills of the Pyrenées, while the eastern Pyrenées and the rest of the Iberian peninsula are alien to these species. The entire genesis of this half-weedy, cultivated group of oats can be followed within this locus. All kinds of facts opened up by such studies can be evaluated while using these oats as an example. Until recently, botanists knew of two or three varieties belonging to these species, *Avena brevis* and *A. strigosa*; now we can distinguish several dozen.

As shown by preliminary research in the Caucasus, the wild relatives of fruit trees proved to have strikingly localized type-formation both from the point of view of Linnaean species and that of its separate elements. In this respect, the investigations carried out by V.P. Ekimov concerning the cherry plum [*Prunus cerasifera*] in Transcaucasia are exceptionally interesting. The same can be said about the research on pomegranates [*Punica granatum*], quince [*Cydonia oblonga*] and some species of pears.

On the map in Fig. 2 we have tried to illustrate the foci within the regions of the basic centers. Orienting studies have been made in the New World, in Africa, Europe and eastern as well as southwestern Asia. Unfortunately, we have still left untouched almost all of southeastern Asia, but no doubt a center of many Asiatic crops will be found there. It is necessary at all costs to explore this untouched part of the world during the next couple of years. We are convinced that the research on the foci as centers of type-formation is of enormous scientific and practical importance. It can, indeed, lead to the establishment of universal gene-banks where many kinds of characteristics, or rather genes, are accumulated.

We are, thus, starting on a new phase of detailed studies concerning the geography of cultivated plants and we are at present actually convinced – in spite of how little has yet been achieved – that enormously interesting perspectives will open up in this respect even in the case of already well-studied materials such as wheat and barley, not to speak of such groups as fruits and vegetables. The latter have, in essence, hardly been touched upon. The same is valid also for domesticated animals. We do not doubt that only when an all-embracing work toward the determination of such foci of major cultivated plants and domesticated animals is completed will it be possible to approach the mastering of their type-formation with certainty.

During this congress of geneticists we can see, among a series of accomplishments with respect to various crops, what kinds of new and important facts have opened up for this type of investigations. The problem concerning the origin of cultivated potatoes, gourds, leguminous plants and melons must definitely be looked into again. Actual studies of material collected by the expeditions of S.M. Bukasov and S.V. Yuzepchuk have already revealed a number of new genetic groups of potatoes (distinguished from each other in the same manner as, for instance, hard and soft wheats), about which, so far, neither plant breeders nor botanists knew anything. Thus, the first stage of the attack on the problem concerning the origin of cultivated biota has rounded a new corner, i.e. that of mastering the foci of the primary type-formation of these biota. The creation of

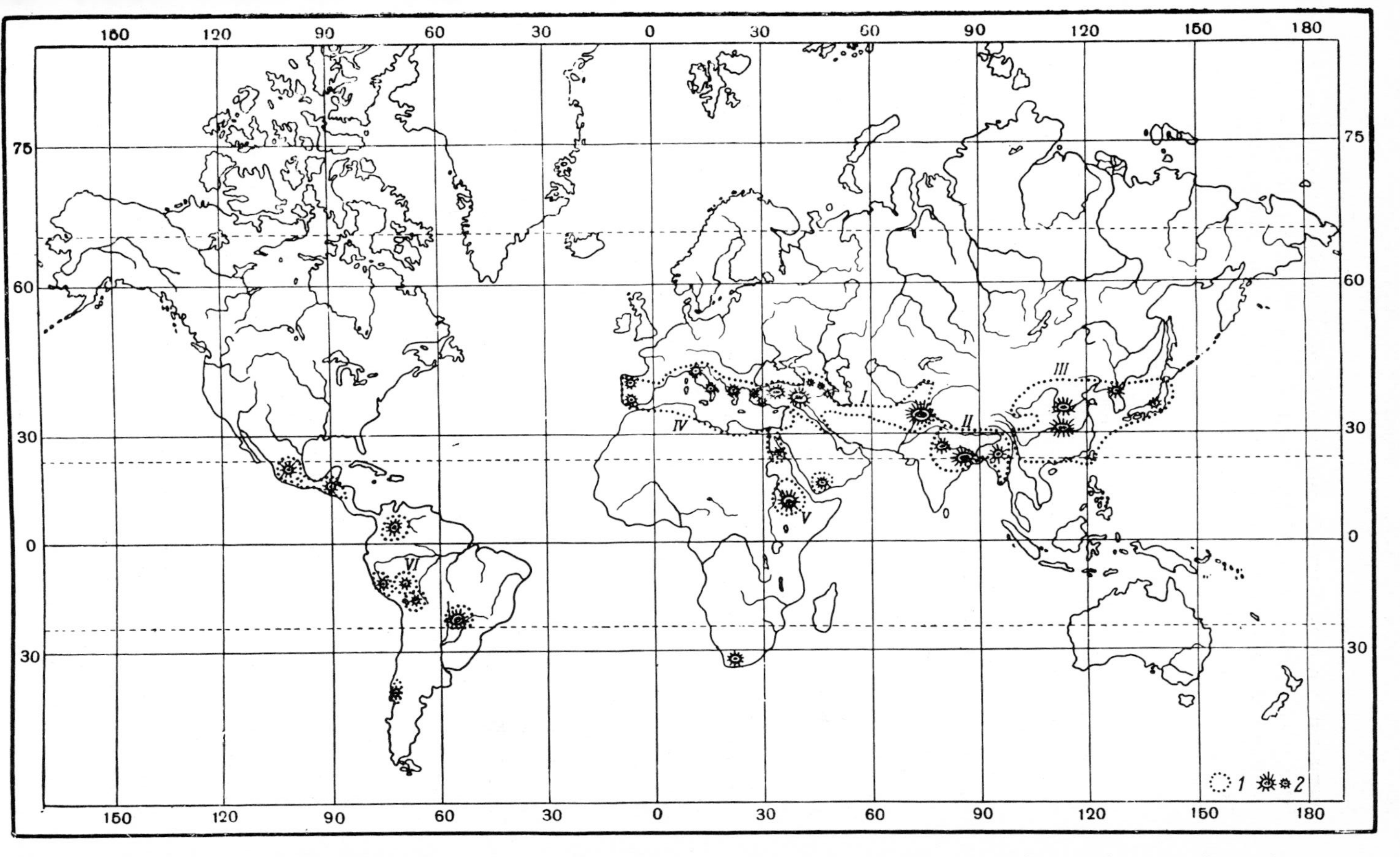

Fig. 2. Universal centers of origin of cultivated plants. 1. Main centers: I. Southwestern Asia; II. eastern India; III. China; IV. the Mediterranean region; V. Abyssinia; VI. America. 2. Sites of type formation foci of different groups of cultivated plants.

nurseries for the treasures of strains in the form of a kind of living granary of genes awaits its turn. This task is, in practice, far from simple, even given the variable conditions of our extensive country. It will require a great diversity of ecological conditions, since such plants as wheat will not find conditions corresponding to those of all its groups even within all of the U.S.S.R.

Genetic analysis of cultivated biota

A morphological study of the foci of type-formation will certainly open up a startling diversity, compelling us to once again work on the system of variability in the case of different species. Physiological research on selected samples, such as we have already begun, has revealed the presence of distinct contrasts. In the forefront stands genetic research using hybridization, which requires an enormous and well-planned work, on which we have just started. If the large amount of material, which belongs to the sphere of agronomy, is taken into consideration and if the hundreds of variable characteristics usually present within the different Linnaean species, not to mention the innumerable combinations thereof, are also taken into consideration, it may be possible to imagine the volume of work involved. So far, genetics has hardly touched upon this almost virgin multiformity. The individual genetics of various species is, in essence, still in its infancy. Characteristics that we only recently considered as simple have turned out to be controlled by a number of genes as, e.g., in the case of wheat awns. By establishing centers of type-formation, the investigator will gain control over almost all the genes of the cultivated biota. As demonstrated by investigations, there is often a large number of genes that determine the varietal and racial differences in spite of a phenotypical uniformity.

Work done by L.I. Govorov has shown that Afghan vetches, although morphologically quite uniform in that area, contain, in essence, all the basic genes for the innumerable cultivated European strains. The establishment of foci of type-formation is essential for us first and foremost for gaining control over a universal store of genes belonging to different species of cultivated biota.

ECOTYPES

Usually the forms of various cultivated biota that are discovered when the foci are opened up are in themselves not very suitable for Soviet ecological conditions. They are represented by too specific combinations of physiological and morphological characteristics. The wheats of Turkestan, Iran and Afghanistan cannot, in spite of their amazing amount of variable characteristics, compete with our common strains, which are the result of centuries of selection, natural as well as artificial. A large amount of collective work is required for a selection of the different elements, for regrouping them and for hybridological analyses in order to distinguish those genes that can be of interest. Perhaps this circumstance is the reason why, so far, the original centers of type-formation have attracted so little attention. Scientists have, so to say, bypassed these deposits of international gene-banks. It is, for example, interesting to note that the universal resources of genes are found in economically badly developed countries; the poorest lands, for instance those of Abyssinia, Afghanistan, Bokhara, Peru, Chile, Mexico and China, can actually be considered as internationally available treasures.

However, the scientists taking part in the last expeditions were also able to reveal that things are not always the same. Samples of barley brought back from the last expedition to Abyssinia proved, after two years of experiments under various conditions, to be exceptionally valuable types, growing beautifully both under the conditions in the Far East and in Byelorussia, competing in productivity and yield with the local strains and representing in their present state an exceptionally valuable kind of material. Also, a number of European ecotypes of field crops and vegetables have been proved to exist in a 'ready-to-use' state in Asia Minor.

Classification of cultivated plants according to the extent to which their genes are localized

Investigations carried out during the last couple of years concerning the spatial localization of the origins of cultivated plants have already allowed us to start grouping their genes according to the extent of localization. Many of the most important and most ancient crops have turned out to be amazingly strictly localized; here belong such important primary plants as wheat, barley, flax, a number of leguminous plants, lentils, chick-peas, even maize and beans. A striking localization has been revealed also for many fodder plants such as Persian and berseem clover [*Trifolium resupinatum*, *T. alexandrinum*] and sulla (*Hedysarum coronarium*), various kinds of vetchling [*Lathyrus sativus*] and fodder lentils. The affinity of these crops shows quite clearly that they can be the means for identifying a number of independent foci of type-formation, which as a rule correspond to the foci of independent, ancient, agricultural civilizations. In particular, many cultivated plants have such a distinctly expressed localization that it facilitates the determination of their foci. But a number of plants are also characterized by areas, so to say, bursting their seams, e.g. species of cultivated cotton which, to a certain extent, are associated with four, perhaps five centers both in the Old and the New Worlds. In the case of such a crop as beets, the area of primary type-formation is apparently very wide, extending over both southern and central Europe so that it is impossible to localize it exactly. Finally, there are cultivated plants, some of which, by chance, have genes that during modern times have attained an enormous distribution, such as, for example, watermelons, the genes of which are scattered over the entire African continent. Our cultivated clover belongs also to this kind of plants, where a determination of the foci of type-formation is impossible. A scientist studying species after species will gradually be able to group such species. That kind of classification is of decisive importance for mastering the process of type-formation.

Geographical regularity in relation to how plants establish themselves in different areas

Indisputable research on foci of type-formation has led to the revelation of a number of regularities concerning the process of how cultivated and domesticated biota establish themselves in different areas. On the basis of a large number of data, the general regularity has been elucidated, mainly in relation to the dispersal of recessive forms from the center toward the periphery of a natural distribution area. As a rule, the primary foci proved to be characterized mainly by the presence of dominant genes and dominant characteristics. In isolated mountain areas, there is an increase of recessive forms toward the periphery. The process of geographical evolution of cultivated biota can actually be considered as an untangling of the original complex tangle of genes. The concept of what is recessive and what is dominant has recently undergone a major change, but with respect to the qualitative characteristics, it is nevertheless usually not difficult to distinguish between dominants and recessives even in nature.

What in practice is particularly essential is the discovery of data concerning establishment within different areas, i.e. the isolation of very valuable strains under various conditions. Thus, for example, according to research carried out by N.N. Kuleshov, the Asiatic kind of maize was, following its dispersal from America to Asia, found to be of an interesting recessive type, called waxy corn. This one is of great practical interest. According to data from the Turkestani Experimental Station, the existence of crops of cotton consisting of an interesting recessive, early-maturing type has been observed. Similar phenomena have been discovered with respect to some crops of flax and peas, which have been very thoroughly studied genetically. Apparently, the ancient Mediterranean countries are represented by a concentration of a number of field and vegetable crops of very valuable recessive, large-seeded forms.

PRE-EMERGENT CHARACTERISTICS

The last expeditions have also discovered a number of facts of great interest to geneticists. In some foci of type-formation there are interesting combinations of genes that represent characteristics that make it difficult to distinguish one species from another. Thus, there are, e.g., many kinds of wheat in Abyssinia that, according to their genetic structure and their chromosome numbers, definitely belong to hard wheats, but the morphology of which cannot be distinguished from that of soft wheats. Therefore, a botanist will – or has at least so far – put them among the soft wheats. In this case it cannot be a question of hybridization between soft and hard wheats. Similar observations have been made in the case of the oats of Abyssinia and Egypt. Professor K.I. Pangalo made a similar observation with respect to pumpkins in South America. These have in some foci pre-emergent characteristics. In other words, studies have suggested an appearance of the absence of divergence between species. This can be understood as a process of type-formation *in statu nascendi* [a condition about to be produced]. For the experimental phyto-geneticist such types are of particularly great interest and can, perhaps, provide us with a clue to some riddles.

The foci that are basic for type-formation can thus be characterized not only by a great diversity such as we have so far assumed but, indeed, also by the presence of a large number of dominant types and a lack of divergence between species as well as a clearly expressed process of producing recessive types toward the periphery of the area of type-formation.

POSITIONING THE FOCI

Analytical work is still in progress toward the determination of the foci of different species: by accumulating facts, a synthesis of the data can, in turn, be accomplished. There are areas of diversity of different plants and foci of different plants superimposed on each other. Consequently, we approach the establishment in the world of a number of foci, from which the racial diversity radiates. Some of these foci embrace type-forming processes of a large number of species. From this point of view we have been able to study fairly well the remarkable site in the foothills of western Himalaya and Hindukush, where the basic type-forming processes of both soft wheat and, to a great extent, the tribes of Vicieae are concentrated.

While simplifying the concept of the geographical type-forming processes, the establishment of such foci led us logically to the problem concerning the establishment of the independent foci of agricultural civilizations and allowed us to further contemplate the history of agriculture. The elementary ideas, which most scientists so far hold on to, i.e. that the onset of an agricultural civilization should be searched for in the areas of Mesopotamia, Syria and Palestine where the wild wheat is found, does not in reality correspond to the distribution of the universal foci of cultivated plants. No doubt, problems linked to the history of agriculture must be thoroughly reconsidered and, in connection with this, we are, obviously, on the doorstep of a general revision of our hypotheses about the history of the civilizations of mankind.

The establishment of the foci led us to facts that force us to subject the entire problem concerning the origin of cultivated plants to a thorough revision. Which of the many small foci in the New World should be considered the oldest one cannot possibly be answered at present. It is certain that the most important cereal in the world, wheat, in its original, basic type-formation was shared by two continents, Africa and Asia. One part of it, comprising the group of soft wheats, developed in the fold between Himalaya and Hindukush, the other on the mountainous plateau of Abyssinia. Each of these groups is represented by typical Linnaean species. In the case of wheat we have a definite divergence into Linnaean species, in that of other crops there is, in relation to their foci, only a divergence of genes and in a third case there is an adaptation of crops going in different directions.

In order to understand the nature of such kinds of divergence which appear absolutely undeniable, the historical process concerning the origin of the major cultivated plants in the world must be pushed back not only into a very remote period of time or into some archeological epoch, but even farther back. Because, in order to understand the process of divergence, e.g. in the case of the species of wheat, we must perhaps go back deep into geological eras. We do not doubt that the rudiments of agricultural civilizations, whether appearing simultaneously or at different times, arose within different regions where the elements for the creation of an agricultural civilization existed (Figs 3–5). When pushed back into such a very remote time, the problem of the origin of cultivated biota becomes, nevertheless, more lucid and more definite, since only when separating it spatially in relation to the localization of the genes is it possible to approach seriously the problem of how to reconstruct the historical process of speciation.

Centers of Origin and Breeding

We must face the problem concerning the origin first and foremost from the point of view of geography, since the origin of cultivated plants and domesticated animals is associated mainly with space and time. The geographical localization of type-forming processes has been given only slight attention. In order to be able to grasp the elements of type-formation, the geographical solution is, in our opinion, of an enormous importance, theoretically as well as practically, from the point of view of mastering the original material for type-formation.

Contemporary genetics approaches the problem of origin mainly from another angle, i.e. by trying to explain the dynamics of type-formation, irrespective of space. There are no doubt great achievements within this field.

Fig. 3. Iran. Working the soil. Photo: N. I. Vavilov.

Fig. 4. Cretan plow. Photo: N. I. Vavilov.

Fig. 5. Sicily. Latin type of plow at work. Photo: N. I. Vavilov.

The method of hybridization, widely practiced by contemporary geneticists and breeders, has opened up very wide perspectives. Recently, our studies have for the most part revealed facts of exceptional importance for the restoration of fertility by means of polyploidization of remotely related interspecific or even intergeneric hybrids. The work carried out by G.D. Karpechenko and S.A. Egiz of the Experimental Stations at Saratov and Odessa have opened up a new era for the actual creation of new species by means of hybridization. A bridge has been thrown across the gulf that, until recently, separated Linnaean species and genera. This kind of research is also of enormous importance for explaining the origin of cultivated biota.

The phenomenon of mutation is, apparently, more extensive than we have believed until recently, but now we can understand the interesting data concerning the frequent mutations of such materials as, e.g., potatoes, not to speak of the perspectives opened up during the past year by American scientists. The frontal attack toward a solution of the basic problem of speciation and type-formation of cultivated or domesticated biota has split up in two directions: one directed toward the mastering of the structural material and the other directed toward a study of the art of creation.

As always, chemistry can furnish us with a close analogy to biology. The contemporary biologist/geneticist must, just like the chemist, simultaneously study both the distribution of the elements on Earth and in the universe, i.e. geochemistry in the wide sense of the word. Simultaneously, he must do research toward a study of the transformation of these elements.

From a short review like the one presented here, concerning general problems, it can be seen what an immense scope of scientific work has been opened up for scientists of our time. There is no doubt that, at the frontier, where we try to master the species and the processes of type-formation, international science will in the near future win great victories.

BIBLIOGRAPHY

Vavilov, N.I. (1926). Tsentry proiskhozhendiya kul'turnykh rasteniy [Centers of origin of cultivated plants]. *Tr. po prikl. bot. i selek.* [*Papers on applied botany and plant breeding*], **16** (2).

The problem concerning the origin of agriculture in the light of recent research

WHERE IN THE world did agriculture first begin? Did it start independently in different places and on different continents? How can the geographical localization of initial agriculture be explained? What plants were first taken into cultivation? What animals were domesticated and where? Where can the initial sources of cultivated plants be found? How can we link the present domesticated farm animals and cultivated plants to their corresponding wild kinds? Along what routes did the evolution of cultivated plants and domesticated animals proceed? What tools did the first farmer use within the different centers?

In light of directly materialistic research, all these historical questions have now become actual, alive and full of meaning for contemporary agriculturalists. In contrast to the past and under conditions of an intensifying opposition to classical interests, today's investigator attempts to find elements, that also existed in the past, for improving the present. In a Soviet country, building up socialism and a socialistic agriculture, problems concerning the origin of agriculture and the origin of cultivated plants and domesticated animals are of interest to us mainly from a dynamic point of view. By knowing the past, looking at elements from which crops developed and gathering cultivated plants in the ancient centers of agriculture, we hope to learn soon how to control the historic processes and how to change cultivated plants and domesticated animals in a manner reflecting the interests of our modern times. It is of comparatively little interest to us that wheat and barley have been found in the tombs of the pharaohs belonging to the first dynasties. The problem of the construction of these tombs, a problem interesting to an engineer, seems more important to us. But it is much more essential to know what distinguished the Egyptian wheat from that grown or found in other countries and what it has of value for improving our own wheat, or to learn how the Egyptian wheat originated and where to find the basic elements, the 'building blocks', from which the present cultivated species and strains were built. This is necessary in order to master the initial material of the strains and for use in practical plant breeding. From the construction of primitive tools, we can find useful hints for manufacturing modern equipment.

In other words, the historical problems concerning the origin of agriculture,

First published in: *Sots. rekonstruktsiya i nauka* [*Socialistic reconstruction and Science*], 1, pp. 34–43, 1931.

the origin of cultivated plants and domesticated animals are of particular interest to us from the point of view of gaining control over plants and animals used within agriculture.

Finally, the results of such research are also interesting to archeologists, historians, naturalists, geneticists, agronomists and plant breeders. Therefore, I want to draw my readers' attention to the fundamental results of research in this direction, carried out and planned during the last couple of years in the land of the Soviet citizens.

While working with practical problems connected with breeding of cultivated plants, we approached the solution of a number of problems concerning the history of cultivated plants, applicable to the program of present research.

During the last couple of decades the All-Soviet Institute of Agriculture (the former Institute of Applied Botany and New Crops) has done extensive, collective research on cultivated plants, grown all over the world, and has executed it according to a very strict plan involving systematic studies of one species after another. It became evident to us that, so far, neither botanists nor agronomists or plant breeders have, in essence, touched upon the inexhaustible wealth of the basic, worldwide resources of even those cultivated plants, the potentials of which exist – as demonstrated by irrefutable research – mainly in ancient agricultural lands. All the plant breeding and all the European and American agricultural crops are based on fragments of the racial composition of such cultivated plants as are derived from the ancient centers of agriculture.

We have started to study cultivated plants all over the world in a well-planned manner. A large number of special expeditions have been directed mainly at ancient agricultural areas in mountains where an enormous amount of material of new strains and data on primitive agricultural technology have been collected. Thus, from 1926 through 1928, we studied all the countries situated along the coasts of the Mediterranean, including Morocco, Algeria, Tunisia, Egypt, Portugal, Spain, Italy, Greece, all of Asia Minor, Syria and Palestine as well as islands such as Sicily, Sardinia, Crete, Cyprus and Rhodes. Abyssinia (in 1927), Egypt, Persia and Afghanistan (in 1924), western China (in 1929), agricultural parts of Mongolia (in 1923), Japan, Korea and Formosa (in 1929) as well as parts of India, Java and Ceylon have also been thoroughly studied. The ancient agricultural work methods in Transcaucasia and Turkestan were given a similar, detailed scrutiny. In the New World, all of Mexico (including Yucatan), Guatemala, Colombia, Peru, Bolivia and Chile, as well as the subtropical areas of the United States were researched from 1925 to 1930.

As a result of these expeditions, an enormous number of seed samples (hundreds of thousands of samples) were gathered and studied by growing crops from them for a number of years at experimental stations. They have already been utilized practically by plant breeding institutes.

These studies have revealed the universal geography of the strains and produced a multitude of strains important to botanists and plant breeders as well as to agronomists. They frequently have valuable and practical characteristics. We have also succeeded in discovering a number of new species of cultivated plants. It is enough to mention that, in addition to the species of potatoes already known to be cultivated (*Solanum tuberosum*), our expeditions found 12 other new species of potatoes in Peru and Bolivia, not to mention hundreds of new strains. We also found new species of wheat, thousands of new cereal grasses and other field crops and vegetables that were so far unknown to Science.

What has been revealed and what is important for the understanding of the entire world-history of agriculture is similarly important – especially the fact that the original sites of the basic strains of potentially important cultivated plants have been located. This was established by direct research. It proved feasible to locate exactly the sites of the original strains of the major cultivated plants such as wheat, barley, rice, maize and many field crops and vegetables. This has led to control over a colossal amount of original material that was previously unknown to botanists.

As shown by our studies, the basic centers of origin of cultivated plants appear, as a rule, to be found where a striking diversity of types is accumulated. Within the small country of Abyssinia alone with its primitive agriculture, where the area under wheat does not amount to half a million hectares, we were able to find a variety of strains of wheat as large as that in all the rest of the world's countries taken together. The same applies to the barley of Abyssinia. As far as the types of maize are concerned, the maximum polymorphism was found in southern Mexico, the basic native land of this plant. An amazing variety of wild fruit trees is concentrated in Transcaucasia, the original native land of many new European fruit trees. In this respect, the area of Transcaucasia and Asia Minor have no equal in all the world.

However, diversity alone does not determine what is an initial center of origin of a particular cultivated plant. A study of closely related wild and cultivated relatives of modern cultivated plants is necessary, just as is a study of where the cultivated species of plants were originally taken into use. We have worked out a method of differential taxonomic phyto-geography, making it possible to determine exactly the native area of a given cultivated plant.

As a result of our studies of some hundreds of cultivated plants, we were led to the establishment of basic universal centers of the most important cultivated plants. As presented by us, facts of exceptional general interest were, thus, revealed.

On the whole, our investigations have led to the establishment of seven basic and independent centers of origin of cultivated plants in the world and, at the same time, to seven definite centers of independently originating agricultural civilizations.

Asia appears to be the basic continent that gave rise to agriculture and to the majority of presently cultivated plants. An enormous number of cultivated plants have their origin in southwestern Asia. It is not by chance that just southern Asia still appears to be home to half the inhabitants of our world. We can distinguish three basic centers of speciation in Asia. The first and most important one is southwestern Asia, comprising the interior and eastern portion of Asia Minor, Persia, Afghanistan, Turkestan and northwestern India. This is the native land of wheat, rye, flax, lupines, shabdar or Persian clover (*Trifolium resupinatum*) and many European fruit trees such as apples, pears, cherry plums

(*Prunus cerasifera*), pomegranates, quince and sweet cherries (*Prunus avium*), as well as grapes and many vegetables. It is not by chance that, according to biblical mythology, the original garden of Eden was geographically located in this area. Forests consisting of wild apples, pears, cherries and cherry plums, all entangled by grapevines, can still be found even now in Transcaucasia and northern Persia. As the most recent research reveals, it is necessary to refer western Transhimalaya and the northwestern portion of India, including all of Punjab and the adjacent western provinces as well, to this center.

In particular, India and the Ganges valley, all of the Hindustani peninsula and the adjacent parts of Indo-China and Siam belong to the second independent main center in Asia. This is the original native land of rice, the most important crop in the world that still feeds half of all mankind. Rice can be found there from the stage of the original wild plants through weeds in other crops to cultivated primitive types, all in an amazing diversity of strains. This is also the native land of an enormous number of tropical cultivated plants such as sugar cane, Asiatic cotton, mangoes and a large number of medical and textile plants.

The third Asiatic center is concentrated within the mountains of eastern and central China. As far as we know now, Central Asia was never of a similar level of importance for primitive agriculture in spite of the immense area over which it is practiced. Neither Mongolia, western China, Tian-Shan or Siberia developed into similar centers of independent agricultural civilizations with respect to the plants cultivated or as far as agricultural technology is concerned.

On the other hand, eastern Asia, especially the upper reaches and the valleys of the major Chinese rivers, i.e. Hwang-Ho and Yangtsekiang (the Yellow River), as well as the adjacent areas, gave rise to the great Chinese – indeed, even pre-Chinese – agricultural civilization. This is the native land of soybeans and many kinds of typically Chinese crops that are little known in Europe. It is also the basic native land of the citrus fruits, jujuba [*Zyziphus*], persimmons, peaches, apricot-plums (*Prunus simonii*), tea, mulberry trees and many tropical and, especially, subtropical fruit trees. Even the agricultural technology is very special there. Manual work predominates; animals are used comparatively little for agricultural work there. Intensive cultivation of vegetables is widespread. Monsoons carry rain into eastern China; a considerable amount of precipitation is characteristic of the main areas of agricultural China. As demonstrated by our studies of Japan and Formosa, these countries have, from an agricultural point of view, adopted their cultivated plants mainly from China. The same can be said of the Philippines and the Malayan islands, where it is mainly cultivated plants of China and India that have been adopted.

In contrast to China and Japan, an extensive use of domesticated animals for agriculture – especially horned cattle, horses, camels and mules – is typical of southwestern Asia (the first main center). The types of agricultural tools from that area are especially diversified.

All the initial agriculture in Europe was concentrated especially in the south. The fourth main center embraces the ancient countries distributed along the coasts of the Mediterranean, including the Iberian, Italian and Balkan peninsulas, coastal Asia Minor, Egypt and the territories of what are now known as Morocco, Algeria, Tunisia, Syria and Palestine.

In spite of the enormous cultural and historical importance of the Mediterranean center, including the major civilizations of Antiquity – the Egyptian, the Etruscan, the Aegean and the old Judaean ones – this center has only a small number of primitive but essential plant crops such as revealed by a detailed analysis of the composition of the kinds of plants cultivated there. Fundamentally, the most ancient agriculture of the Mediterranean countries is based on olives, carob or St. John's Bread (*Ceratonia siliqua*) and figs. The majority of the field crops, such as wheat, barley, leguminous plants, peas and chick-peas, are for the most part adopted from other centers. The composition of types is comparatively poor there in comparison with that of other centers where crops are concentrated. Only a number of forage plants, such as sulla (*Hedysarum coronarium*), fodder lentils [*Vicia ervilia, V. monanthos*], fodder peas (*Lathyrus cicera* and *L. gorgonii*) and berseem [*Trifolium alexandrinum*] definitely arose within the Mediterranean center. Their introduction into cultivation evidently occurred later than the development of husbandry.

The plants cultivated in that area have been subjected to careful selection, favored by the mild climate and the civilization of the populations. As shown by comparative studies, strains of cereal grasses, leguminous plants, flax and vegetables can be distinguished there; these have unusually large fruits, seeds, flowers and bulbs, and a high productivity in comparison with corresponding plants in southwestern Asia. Primitive breeders worked hard to achieve this.

Basic types of tools, both for processing and for harvesting, are typical of primitive Mediterranean agricultural civilizations, e.g. the Latin furrowing plow, the mill set with sharp stones and the stone press. China, India and a major part of the countries in southwestern Asia never knew such types of tools.

The fifth main center is found in the mountains of eastern Africa, mainly in Abyssinia. This small center is unique and is characterized by a number of special and important cultivated plants, appearing there in an amazing variety of forms. The greatest diversity in the world of species such as wheat and barley, and perhaps also of the cereal millet, is concentrated in this area. This is the native land of such peculiar, purely Abyssinian plants as teff (*Eragrostis abyssinica*), the most important cereal in Abyssinia, and ramtil (*Guizotia abyssinica*), a special oil plant. Flax in this area is distinguished by white seeds and it is – unlike what is common in the old Mediterranean countries – grown only as a cereal plant, for flour. The utilization of flax for oil or fibers is still not known by the primitive Ethiopians. Abyssinia is also the native land of the coffee tree and the 'brewing barley', but there are no fruit trees and only a few vegetables. This is mainly a land of field crops.

In spite of the fact that no archeological finds have been made that could indicate the great antiquity of Abyssinia as a center of civilization (except for an old phallic culture recently established in southern Abyssinia), based on the composition of cultivated plants studied and the peculiar agricultural technology (a hoeing agriculture is in part preserved there), this center must without a doubt be considered as independent and very old. We do not doubt that Egypt acquired to a considerable extent its cultivated plants from Abyssinia. All the comparative data concerning cultivated plants and domesticated animals, the life of the farming populations and their native food definitely speak in favor of an independent Abyssinian center. Linguistic data also support this hypothesis.

In the New World, studied by Soviet expeditions during the last five years, it is necessary to distinguish two main centers: the south-Mexican one including a portion of Central America as well, and the Peruvian one including Bolivia. The first one seems to be the larger of the two. From that one originate crops such as maize, highland cotton, cocoa, henequen agave [*Agave fourcroyoides*], musk melons, scarlet runner beans and ordinary beans [*Phaseolus coccineus* and *Ph. vulgaris*, respectively], chayote [*Sechium edule*], papaya [*Carica papaya*] and a multitude of endemic crops of secondary importance, totalling about 70 kinds.

Potatoes, quinine trees [*Cinchona*] and the coca bush [*Erythroxylon coca*] and a large number of secondary crops arose originally in Peru and Bolivia. There, an unusually polymorphic group of soft corn [starchy maize, *Zea mays* convar. *amylacea*] was also isolated.

In spite of having given rise to some crops, the remaining areas of South and North America are not of decisive importance for the history of agriculture in general.

The centers of agriculture in the New World certainly arose independently of those in the Old World. The definitely original, endemic cultivated flora of North and South America indicates this. The ancient civilizations of the Mayas and the Incas did not know of iron and had no plows. The 'foot plow', known in the highland areas of Peru is, in essence, nothing but a spade. Neither Mexico nor Peru used animals for farm work. The llama and the alpaca, as well as the guinea pig, domesticated in Peru, were used mainly for their wool and meat. Only the llama was utilized as a pack animal.

There are, thus, seven basic centers in the world whence originated all the presently used agricultural crops (cf. map, Fig. 1). This map shows that these centers occupy a very limited territory. According to our estimates, the south-Mexican center occupies about 1/40th of the entire, large North American continent. The Peruvian center covers an approximately similar area in relation to all of South America. The same can be said as far as the dimensions of the centers in the Old World are concerned.

The differentiation of the initial centers of origin of cultivated plants corresponds, thus, to the distinction into types of agricultural tools. The mountainous areas of eastern Africa as well as all of primitive Africa still practice a hoeing type of agriculture. According to comparative research conducted by B.N. Zhavoronkov, the tools used in Abyssinia, in China and in southwestern India, as well as in the Mediterranean area, can be distinguished into definite types.

The geographical localization of the primary centers of agriculture is very peculiar. All the seven centers arc found mainly in mountains within tropical and subtropical areas. The centers in the New World are associated with the tropical parts of the Andes, those in the Old World with Himalaya, Hindukush, the mountains of eastern Africa, the mountainous areas of the Mediterranean

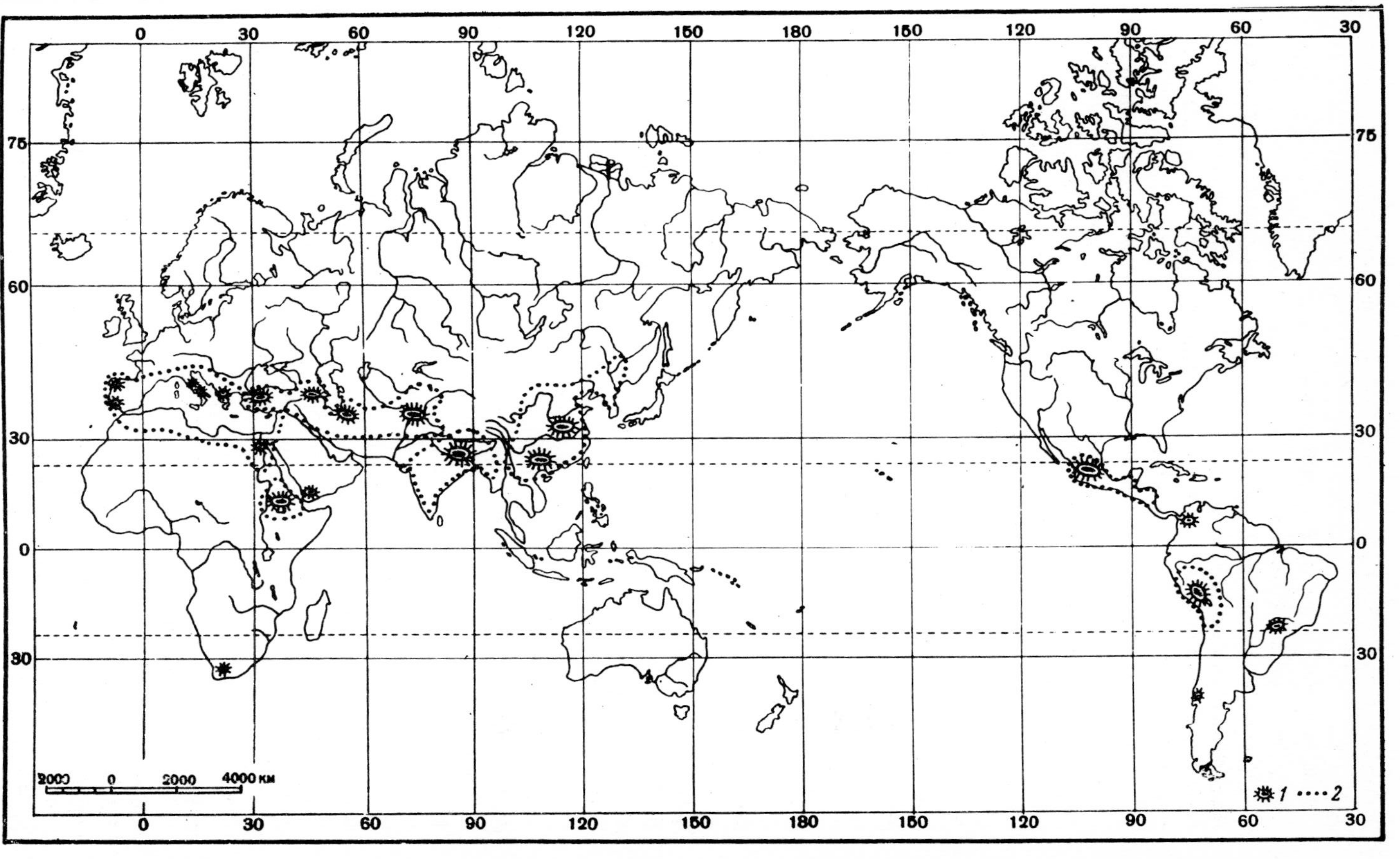

Fig. 1. Universal centers of origin of the most important cultivated plants. 1. main foci of primary type formation of cultivated plants; 2. Basic centers of origin of cultivated plants.

countries and the mountains of China, although mainly within the foothill areas.

In essence, only a narrow belt of the world's continents plays a major role in the history of global agriculture.

Dialectically and in the light of recent research, we have developed a hypothesis of a geographical concentration of the most primitive agricultural crops within these limited areas. The Tropics and Subtropics present optimum conditions for the development of species-forming processes. The maximum racial diversity in the world, with respect to wild vegetation and wild animals, is concentrated towards the Tropics. This is especially obvious in North America, where southern Mexico and Central America, which occupy a comparatively insignificant portion of the total area, have more plant species than all the immense areas of Canada, Alaska and the U.S.A. (including California) taken together. The very small republics of Costa Rica and San Salvador occupy an area about 1/100th of that of the United States but are characterized by an equally large number of plants species as all of the latter. Powerful speciation processes are clearly geographically localized, particularly in the wet tropical areas of the New World.

The same is obvious in the Old World. The Mediterranean countries are very rich in plants. The flora of the Balkans, Asia Minor, Persia, Syria, Palestine, Algeria and Morocco is distinguished by a wealth of species, by far exceeding that of northern and western Europe. According to the estimates by Turill, the Balkans have about 6350 species with Greece being especially rich therein, but India has no less than 14 500 species. The flora of central and eastern China is also very rich. Although we do not know exactly the number of species in the parts of China of interest to us, it definitely amounts, in any case, to many thousands.

There is a wealth of both endemic and ordinary species also in Abyssinia. Thus, the geographical localization of speciation in the case of cultivated plants correlates to a major extent with the localization of speciation processes in general that is typical of all the floras in the world.

Mountain-building processes no doubt played an important role in the differentiation of plants into species, favoring the development of speciation processes. Factors such as isolation and the origin of barriers against the dispersal of species and genera have certainly always been essential for the isolation of different forms and species. Variation in climate and soils, such as those typical of mountain areas, where the main centers of origin of cultivated plants are concentrated, also coincide with the appearance of diversity among the plants and agree also with the racial composition of cultivated plants. On the other hand, the glaciers covering northern Europe and parts of Siberia during the last geological period impoverished the floras in those areas.

Although it is mainly a woody flora that has developed in the wet Tropics, herbaceous species have, on the other hand, developed particularly in those mountains of the Tropics and Subtropics where agricultural crops were initially isolated. To the latter belong the majority of the plants cultivated in our world.

The mountainous tropical and subtropical areas present optimum conditions for humans to settle in. Primitive man feared, and still fights up to this day, the wet Tropics with its unmanageable vegetation and tropical diseases, in spite of the fact that the wet Tropics with their fertile soils occupy one third of the world's landmass (according to Sapper). Man settled, and is still settling down, along the edges of the tropical forests, but the mountain areas of the Tropics and the Subtropics offered more favorable conditions for the original inhabitants in the sense of a warm climate and an abundance of food as well as the possibility of life without the need of clothing. The people in Central America and Mexico, as well as in the mountains of tropical Asia, still use a multitude of wild plants. However, it is always easy to distinguish cultivated plants there from the wild ones that are clearly related to them.

The mountain relief favored the existence of small groups of people: during that stage the development of human societies was initiated. There is no doubt that the mastering of the great river basins such as those of the upper and lower Nile, Euphrates and Tigris, and the Indus was possible only thanks to the efforts of populations that were united into larger groups. This occurred essentially during a much later stage in the evolution of human societies.

Primitive man and primitive farmers lived – and still live – in small, isolated groups. The mountains of the Tropics and the Subtropics offered them exceptionally optimal living conditions.

In contrast to the ordinary presentations of archeologists, our research on the ancient agricultural civilizations has allowed us to draw the conclusion that primitive agriculture was basically a non-irrigated one. An analysis of the composition of cultivated plants in Egypt and Mesopotamia, as well as the irrigated areas of Peru (up to 11 000 ft above sea level) has demonstrated that all had adopted their cultivated plants from outside their own areas. There is no doubt that the ancient agricultural crops of Abyssinia, the mountainous areas of Mexico and Peru (above 11 000 ft altitude), China and India, as well as those of the Mediterranean countries, were not irrigated.

While considering the complicated interaction between opposing factors and basing ourselves on definite facts, which can be verified by irrefutable studies, we have arrived at an exact geographical localization of primitive agricultural crops and at an understanding of the importance of such localization.

It is absolutely clear that these crops were based on different genera and species of plants, developed independently either simultaneously or at different times, and it is necessary to speak of at least seven basic crops or, more exactly, seven basic groups of crops. These are typical of different ethnic and linguistic groups of people. Different kinds of agricultural tools and domesticated animals are also typical of these groups.

Knowledge of the basic function of the universal agricultural centers throws light on the entire history of mankind as well as on the history of common crops.

Our investigations have demonstrated that during the process when cultivated plants were transferred toward the north or up into high elevations, a displacement of the basic crops by weedy associates sometimes took place in such cases where the weeds were of greater interest to the farmer.

Thus, when crops of winter wheat were moved to the north from southwestern Asia, its main center of origin, they were displaced in a number of areas in Asia and Europe by frost-hardy field weeds such as winter rye. Similarly, barley and emmer were displaced by weedy oats, which are less demanding on soils and climate. In Europe, flax was not rarely replaced by false flax (*Camelina sativa*) and in Asia by garden rocket (*Eruca sativa*), and so on.

Consequently, during the resettlement of farmers, a series of crops were developed both due to Man's efforts and by natural selection. When studying rye contamination of wheat in southwestern Asia, we observed an amazing wealth of forms of rye, of which European farmers who grow rye have no representatives. It is peculiar that the cultivation of rye is practically unknown in southwestern Asia and that the very epithet of rye in Persia, Afghanistan and Turkey literally means 'the plant contaminating wheat and rye'.

A number of regularities were observed with respect to changes in the crops during the transfer northwards.

I have furnished only a brief review here of our collective investigations. They led us to an actual control over universal resources of strains and to an understanding of the evolution of cultivated plants as well as to a solution of problems concerning the independence of the basic, original agricultural crops and their interactions.

These results provide a materialistic basis for understanding the initial stages during the evolution of human society. It is evident that the distribution of natural resources of food was a main factor for the original settling of people in permanent quarters.

The data relating to the initial geography of cultivated plants and their wild relatives agree with contemporary knowledge of the evolution of primitive man. Southern Asia and the mountains of eastern Africa were, evidently, the original areas for the creation of permanent human societies, initiating the practice of agriculture. As we have seen, the basic elements were concentrated there, which led to the development of agricultural civilizations.

With this in mind, and in the light of new methods of investigation, the problem concerning the origin of the basic industry of mankind, agriculture, can be sketched out. A direct approach to this problem from the point of view of dialectic materialism and of the actual control over the basic, original potential of plants, leads to a revision of many older hypotheses. What is even more important is that it provides the investigator with an opportunity for controlling the progress of the historical processes in the sense of directing the evolution of cultivated plants and domesticated animals according to his own wishes.

BIBLIOGRAPHY

Vavilov, N.I. (1917). O proiskhozhdenii kulturnoy rzhi [On the origin of cultivated rye]. *Tr. po prikl. botan.* [*Papers on applied botany*] **10** (7–10).

Vavilov, N.I. (1926). Tsentry proiskhozhdeniya kul'turnykh rasteniy [Centers of origin of cultivated plants]. *Tr. po prikl. botan.*, **16** (2).

Vavilov, N.I. (1931a). Meksika i Tsentral'naya Amerika kak osnovoy tsentr proiskhozhdeniya kul'turnykh rasteniy Novogo Sveta [Mexico and Central America as basic centers of origin of cultivated plants in the New World]. *Tr. po prikl. botan.*, **26** (3).

Vavilov, N.I. (1931b). Rol' Tsentral'noy Azii v proiskhozhdenii kul'turnych rasteniy [The role of central Asia for the origin of cultivated plants]. *Tr. po prikl. botan, genet. i selek.* [Papers on applied botany, genetics, and plant breeding], **26** (3).
Vavilov, N.I. (ed.) (1931c). Pshenitzy Abissinii [The wheats of Abyssinia]. *Tr. po prikl. botan., genet. i selek.* Suppl. **51**.

The role of Central Asia in the origin of cultivated plants

Preliminary Review of the Results of an Expedition to Central Asia in 1929.

ON THE BASIS of investigations carried out during the last couple of years in many countries and areas of Eurasia, we have succeeded in elucidating the exceptional importance of the area secluded between southwestern Himalaya and southeastern Hindukush in the history of the origin of cultivated plants. A comparative study of cultivated plants in that area and of the composition of their strains revealed a concentration in the areas of northwestern India and southwestern Afghanistan of basic, primitive type-forming processes of many field crops and vegetables of the Old World. Soft, club and shot wheats (*Triticum vulgare*, Vill; *T. compactum* Host; and *T. sphaerococcum* Perc.), rye, peas, lentils, chickling vetch and beans (*Vicia faba*), flax and linseed, carrots and turnips appear there with an amazing concentration of different varieties such as is not known in any other part of the world (Vavilov and Bukinich, 1929).

So far, the investigations have been carried out mainly in southwestern Asia in such countries as Afghanistan, Iran, the Soviet Inner-Asiatic republics, Syria, Palestine, Transjordania (by N.I. Vavilov), Asia Minor (by P.M. Zhukovskiy), and India (by A. Watt and G. Howard, Shaw and V.V. Markovich).

In turn, a problem arose about how to approach the northern slopes of the Himalayas, the foothills of Kun′-Lun′-Shan and the immense expanses of Central Asia by starting out from Pamir and Kun′-Lun′-Shan.

It was natural to suggest that the type-forming processes could not be limited to the southwestern slopes of the Himalayas but should involve also the areas north of the Himalayas. Could it be that the wide expanses of Central Asia did not to some extent participate in the type-forming processes of cultivated plants? Such ideas were expressed at the end of the nineteenth century by the famous botanist Solms-Laubach who, on the basis of phyto-geographical reasoning, reached the conclusion that the native land of wheat should be looked for in Central Asia. ‘Who knows’, Laubach wrote in 1899, ‘if we shall not succeed somehow in finding traces of crops assumed by us to be the common ancestors of the Chinese and the western crops, known to us, in the mountain valleys of Kun′-Lun′ or Tien-shan (perhaps even in a fossil condition), i.e. just in those areas where most probably discoveries of ancient, tertiary remains of Man

First published in *Tr. po prikl. botan, genet. i selek.* [*Papers on applied botany, genetics and plant breeding*] **26** (3), 1931.

be expected but, in any case, rather there than on the small western peninsula of the large [Eurasian] continent, which we call Europe.' Ancient agricultural crops are concentrated in the foothills of Kun'-Lun'-Shan and Tien-Shan and the suggestion that there was a possibility for discovering elements of a primitive type-formation of cultivated plants there could not be excluded.

After direct studies *in situ* of this problem, a solution might amount to only an approximation or a conjecture. Hence, one can understand the interest that led to our studies of the Chinese Turkestan, i.e., the Sinkiang province of China immediately adjacent to the Himalayas and, as well known, including the extensive ancient oases with agricultural crops such as Kashgar, Yarkand [Soch'e] and Khotan [Hot'ien] as well as the oases along the southern slopes of Tien-Shan. Already a quick glance at the hypsometric map of Chinese Turkestan reveals an exceptionally interesting area in Sinkiang. Chinese Turkestan is situated in the northern foothills of the Himalayas around Pamir and along the southern slopes of Tien-shan; it is separated from eastern and central China by the Takla-Makan desert. Glancing at the map drawn by Aurel Stein, the position of the oasis of Kashgar can be seen. (Cf. the remarkable article by Stein, 1925, in which there is a map summing up the geographical information about Sinkiang.)

As demonstrated by the investigations of Aurel Stein, S.F. Oldenburg and others, there were two historical 'silk roads': the Sogdian and the Bactrian, along which silk was sent from Ancient China to the Roman Empire. A large number of English, German, Swedish, Russian, American and Japanese expeditions have, during the last couple of decades, crossed through Sinkiang in different directions, mainly in search of documentation about the history of the ancient arts.

At the beginning of the present century (in 1910) the United States Department of Agriculture sent the famous student of cultivated plants, Frank Meyer, to Sinkiang. During the winter of 1910–11 (October 1910–March 1911), he made a great tour of Sinkiang, passing over Irkutsk to Kashgar and Yarkand and from there to Ak-Su and on to Kul'dzhe. However, the scientific results of Meyer's work remain unpublished. As far as is known to us, all his work was limited to the collection of interesting seeds and grafting materials of cultivated plants. These are reported on only in the *Inventory of Seeds*, published by the United States Department of Agriculture (1912), where there is a detailed list of the plant material collected by Meyer. Among others, the desert poplar (abele) of Sinkiang attracted the special attention of the U.S.D.A. as did the tamarisk, the willows, the different bushes of the dry areas of Sinkiang (*Spiraea*, *Lonicera*, *Caragana*, *Reaumuria*, *Ribes*, etc.) and the wild apricots and apples of Tien-Shan, as well as the perennial kinds of melons in Sinkiang. The arrival of *Apocynum hendersonii* Hook and *A. venetum* L. was also mentioned in the same inventory.

Last summer (July–November of 1929) I made an expedition to Sinkiang together with M.G. Popov on behalf of the All-Soviet Institute of Applied Botany and New Crops in order to cover the area of Central Asia from Kashgar to Urumchi; the areas of Uch-Turfana, Ak-Su, Kuch'i, Turfana, Urumchi and Kul'dzhe were studied (cf. Figs. 1 and 2). While staying in Kashgar we also

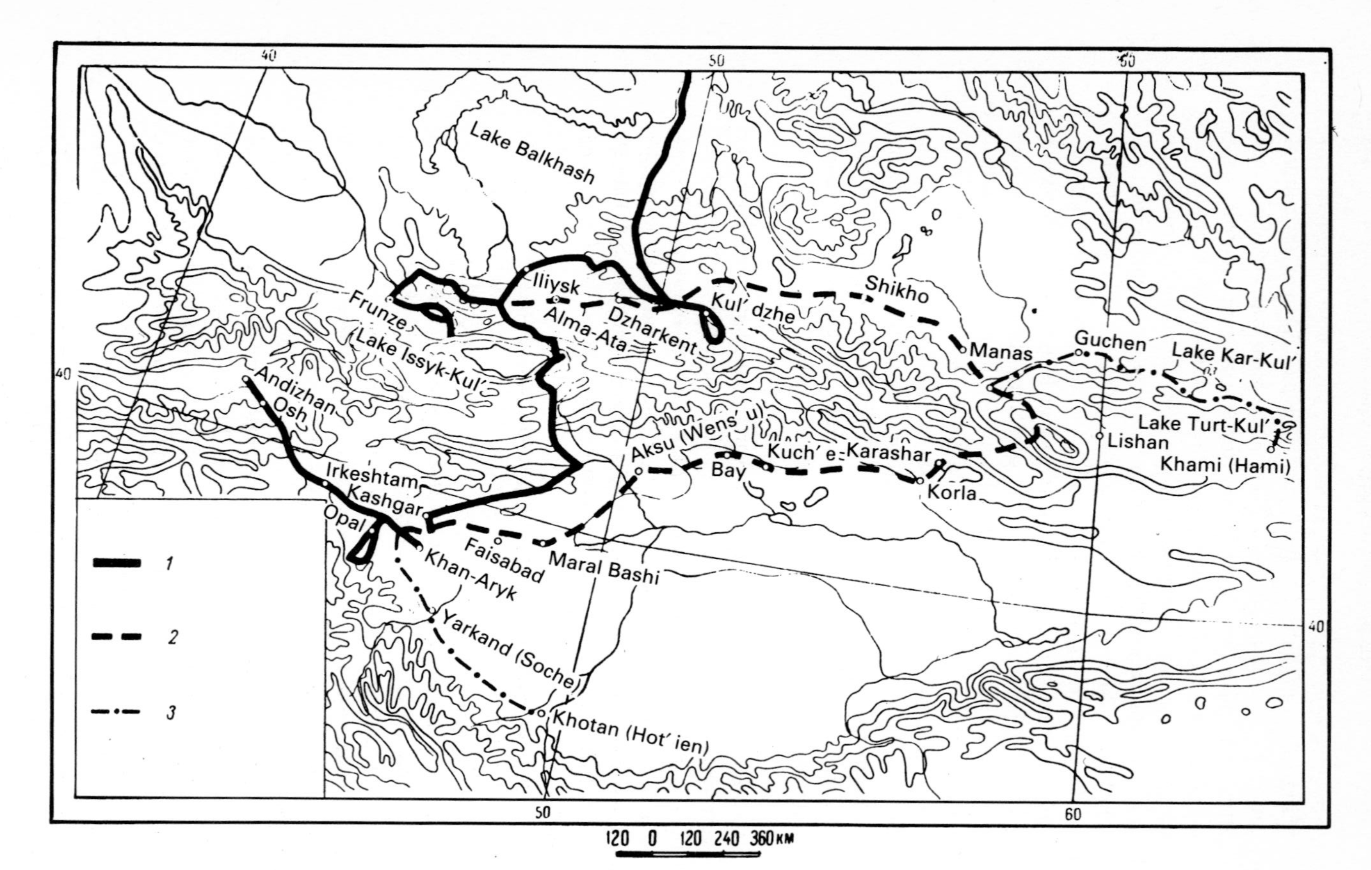

Fig. 1. Routes of the N.I. Vavilov and M.G. Popov expedition to Chinese Turkestan (Sinkiang), Kirgizstan and Kazakhstan in 1929. 1. Route covered by Vavilov; 2. route covered by Popov; 3. seed collected, courtesy of local habitants.

Fig. 2. Talib the interpreter-guide, who accompanied us from Kashgar to Alma-Ata. After a photo by N.I. Vavilov.

organized a collection of seed material in the Yarkand and Khotan areas. In addition, we studied the major agricultural areas of Kazakhstan and Kirgizstan, as well as some areas of western and eastern Siberia and the Far East, for the purpose of elucidating the composition of the cultivated flora of Central Asia. A large amount of material of strains was collected (ca. 5000 seed samples). Last fall (1929) I, myself, set about to study also the east-Asiatic centers of cultivated plants, Korea, Japan and Formosa.

What did the comparative botanical and agronomical studies of the agricultural crops of Central Asia reveal?

By means of decisive samples, it could be demonstrated that Central Asia does not at present correspond to the typical concept of genesis of cultivated plants or to a typical center of origin, in spite of the existing hypotheses by Solms-Laubach, but it does, in contrast, reveal with full certainty that there is irrefutable evidence for the introduction of cultivated plants to that area mainly from Asia Minor and Inner Asia and, in part, also from India and central and eastern China.

1 First of all, as demonstrated by our expedition to the secluded ancient agricultural areas of Sinkiang (Fig. 3), the highly essential fact emerges, when describing the role of Central Asia, that there is only a limited and small number of field crops there in comparison with the number in the adjacent basic centers of agriculture, i.e., in southwestern Asia and eastern and central China. This can be distinctly concluded from the crops of leguminous plants and cereal grasses grown there. In spite of its large and densely settled oases such as Kashgar and Yarkand, Sinkiang hardly grows any crops of peas, lentils, beans (*Vicia faba*) or chickling vetch (*Lathyrus sativus*) and only very little of chick-peas (*Cicer arietinum*), i.e. plants common in Inner Asia. Lentils, beans and peas are grown in an appreciable quantity only in the eastern portion of Tien-Shan, in Dzungaria,

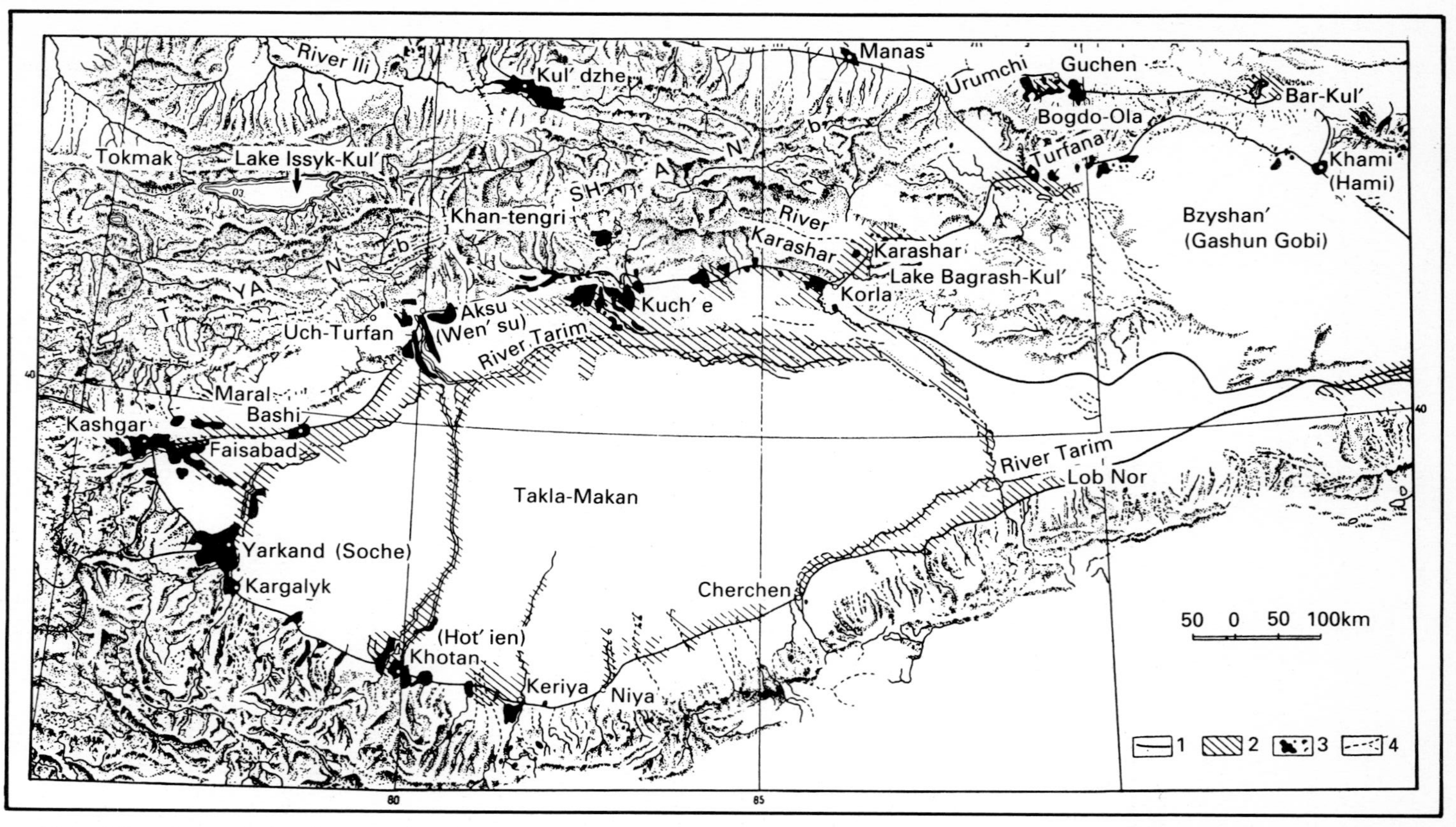

Fig. 3. Map of the agricultural areas within the Sinkiang province of China. 1. Roads; 2. sandy areas with desert vegetation; 3. oases; 4. dry river beds.

and in the area around Urumchi. Toward the west and southwest of Pamir, at the same altitude as Kashgar, peas, beans, chickling vetch and chick-peas are commonly distributed crops; under the conditions for cultivation that exist in Sinkiang there is absolutely nothing unfavorable for these plants, yet there is no Persian clover (*Trifolium resupinatum*) although this is a common forage plant in Afghanistan and Iran.

While rye (*Secale cereale*) is a very widely distributed obnoxious weed beyond Hindukush in Afghanistan, often infesting winter wheat but also spring wheat and sometimes even appearing as a cultivated plant, it is not even a weed in Sinkiang. In spite of a very thorough search for rye among the wheat in that area, we did not encounter a single straw of rye in Kashgar. (M.G. Popov observed rye only in the area of Shikho in Dzungaria.) Even the cultivation of naked-grained barley, a very common plant in the elevated mountain areas of southwestern Asia, is almost unknown in the wide Kashgarian and Yarkandian oases. Naked-grained barley occurs only in Dzungaria.

In other words, when comparing the specific composition of field crops beyond the Himalayas and to the north and east of Pamir with that of Fergana, Iran and Afghanistan, a limited number of crops has to be reported although there are irrefutable ecological conditions for growing the same kinds of crops in both these areas.

2 In Sinkiang the attention of the botanist/agronomist is involuntarily attracted to the absence of the many wild relatives of cultivated plants, which are common in southwestern Asia. Species of *Aegilops* such as *Ae. triuncialis*, *Ae. cylindrica*, *Ae. squarrosa* and *Ae. crassa*, very common in Inner Asia, are definitely missing in Kashgar and Dzungaria. Along the slopes of Tien-Shan, Kun′-Lun′-Shan and eastern Pamir, there is no *Secale montanum*, which is common in Asia Minor and the mountains of Iran. The wild barley, *Hordeum spontaneum*, has also disappeared, although stands of it cover the foothills of northern Hindukush and the Transcaspian area. There is no wild *Ervum orientale* Boiss., nor any *Ervum* [*Vicia*] *ervilia*, *Lathyrus cicera* nor any wild *Allium cepa* or *A. sativum*. Not even *Andropogon halepensis*, a common weed in Inner Asia, was seen there. All these relatives of cultivated plants are missing beyond Pamir and toward the east.

3 Even more important roles for the composition of strains are played by the Himalayas, Pamir and Tien-Shan in the form of mighty barriers or ‘filters’ with respect to the number of botanical varieties, which are represented here by important Linnaean species of cultivated plants. The number is definitely lower in Sinkiang than in northern India, Fergana or Afghanistan. Soft and club wheats are represented by a dozen varieties in Sinkiang with a clear predominance of two such, i.e. *Triticum vulgare* var. *erythrospermum* Koern. and *T. vulgare* var. *ferrugineum* Al. in the form of winter crops and the pubescent, white-grained *T. vulgare* var. *pseudoturcicum* Vav. as a spring crop. In the northerly area, on the northern slopes of Tien-Shan but rarely on its southern slopes, we can find a limited variety of club wheat, *T. compactum*. In the settlements of Afghanistan we found, in 1924, no less than 60 varieties of soft wheat (*T. vulgare*) and 50 of club wheat (*T. compactum*), not to mention an innumerable multitude of minor taxonomic units, the major portion of which appear to be endemic in

Afghanistan. In Sinkiang, the species *T. durum* and *T. turgidum* and the closely related 28-chromosome representatives of wheat are definitely missing.

The rare crops of barley are in Sinkiang represented mainly by *Hordeum vulgare* var. *pallidum* Sér.

The composition of rice, one of the most important crops of Chinese Turkestan, is poor; varieties with brown and black hulls or with black awns, common in northeastern Afghanistan and in India, are missing here. In the Kashgar and Yarkand oases, the coarse-grained asiatic races of rice predominate and these are, no doubt, adopted from southwestern Asia; in Urumchi and in Kul′dzhe typical Chinese cultivated types with light-colored hulls and no awns predominate.

The composition of chick-peas (*Cicer arietinum*) is exceptionally poor in spite of the wealth of varieties in Afghanistan and India (Vavilov and Bukinich, 1929; Howard *et al.*, 1915).

The number of types of grapevines in Sinkiang is very low in comparison with that in the Inner-Asiatic republics; according to the investigations made by M.G. Popov, the kinds there are clearly brought in from the adjacent Fergana and regions close to it.

The racial composition of fruit trees is even poorer, i.e. that of apples and pears, cultivated Russian olives (*Eleagnus angustifolia*), cherry plums (*Prunus divaricata* Ledeb.) and even apricots, which are the richest crops in Fergana. Carrots – usually variable in Afghanistan, where 26 varieties are found, including black and violet ones (Matskevich, 1929) – are represented in Kashgar mainly by uniform, yellow types.

More detailed comparative research on the seed samples collected from other kinds of crop is necessary, e.g. by cultivating them, but their relative uniformity is already now a definite fact.

A similar racial uniformity and poverty of field crops is typical also of Kazakhstan, Kirgizstan and Siberia and all of Central Siberia in a wide sense. The so-called 'yaritsa' rye of Eastern Siberia is homogenous and the local wheat, grown in the Amur region, is amazingly uniform.

4 The following fact, established by our expedition, is even more striking and bears witness not only to the reduced racial diversity but also to the preferential selection here of genetically recessive types that are peculiar of aspect and characteristic of the periphery of basic distribution areas of plants and animals or of isolated areas (Vavilov, 1927). Isolated or occasional recessive types usually soon disappear when they exist simultaneously with the dominant races. Here, however, in this geographically isolated area, they are preserved and have been bred into a pure state.

The Kashgar, Yarkand and Khotan oases – especially the ancient agricultural area of Kashgar – attract one's attention mainly because of the peculiar strain of flax grown there. The cultivation of flax is widespread there and is represented mainly by an extraordinary type with white flowers and narrow crumpled petals in a corolla reminding one of that of *Cerastium*; the anthers of the flowers are yellow. The seeds of this kind of flax are white, not brown as in our kinds of flax. More rarely a variety is met with that has white flowers without the

Fig. 4. The peculiar white-flowered Kashgarian flax with narrow and crumpled petals and white seeds (a typical recessive form). Ordinary blue-flowered Kashgarian flax is also illustrated.

crumpled petals. In spite of a thorough search we never found blue flowers with crumpled petals. Apparently there is a definite correlation between narrow, crumpled petals and white-colored flowers. Normal blue-flowered and brown-seeded types are also grown, but rarely, and are distributed mainly in Dzungaria and north of Tien-Shan. As known on the basis of research executed by Tina Tammes, the white-flowered type with crumpled petals is definitely recessive in relation to the normal blue-flowered types and to races with ordinary petals in the corolla. White flowers, narrow petals, yellow anthers and white seeds are all recessive characteristics. Flax is grown here for oil and is, apparently, in practice represented by a very valuable group of strains. The outward appearance of the Kashgarian white-flowered type of flax with narrow petals is very characteristic and, when flowering, very different from ordinary types of flax (Fig. 4).

Sesame (*Sesamum indicum* L.) is also exceptional in Kashgar by having white seeds and pale-blue flowers. This is, again, a recessive type. As demonstrated at one of our research stations during the study of the seed samples sent to us from Kashgar by the Consul General to the U.S.S.R., M.F. Dumpis, in 1927, this is one of an earlier-ripening group of practical interest to us for our more northerly crops of sesame (e.g. at Kuban).

Black-seeded types of sesame were observed by M.G. Popov only in Turfana; there are no such in Kashgar, Yarkand, Khotan or Aksu. This black-seeded race is low-growing and has ternate leaves, reminding one of Japanese sesame. It has, in fact, been introduced from China.

Among the samples of cotton secured by M.G. Popov in 1927 from Kashgar, the early-ripening races grown at our Turkestani research station are of particular interest to us. According to studies in 1930 of harvests of our worldwide collection of cotton types including new material from our expedition, it could be demonstrated that the cotton that ripened the fastest (*Gossypium herbaceum*) originated from eastern China.

It is particularly interesting that a number of these have pods that open before being fully ripe. In contrast to *G. hirsutum*, the races of *G. herbaceum* appear to be more drought-tolerant; hence, there is clearly an exceptional practical interest for this new group of cotton and for moving it northward into new areas in the northern Caucasus and in the Ukraine.

At Urumchi, the expedition of M.G. Popov established the presence of safflower plants without spines, i.e., recessive forms, sometimes encountered in Inner Asia and in Afghanistan as well.

The discovery of recessive types of cultivated plants in Sinkiang is a common event. There, carrots with yellow-fleshed roots predominate, but sometimes there is a tendency toward white-fleshed ones. A mustard (*Brassica juncea*) called 'kchi', has yellow seeds, i.e. again it is a question of a recessive race. Among the fruit crops, nectarines with smooth-skinned fruits (a recessive characteristic) predominate. The nectarines in Sinkiang (locally called 'togach') are usually bright red. Occasionally, white nectarines, 'ak-togach', are found as well.

Just as in the case of rice, barley is represented mainly by varieties with pale-colored hulls.

It should be mentioned that the wild 'camel's thorn', *Alhagi camelorum*, of Kashgar, is also often characterized by light-colored flowers with clearly expressed flavonid hues, distinct from the more brightly red-colored races of Inner Asia.

In this sense, there is, to an important extent, a manifestation of reconstruction, of a development of characteristic type-forming processes in the direction toward a selection of recessive types, such as, e.g., early-ripening races or white-seeded types, which, under different conditions, can be of a significant practical interest.

Thus, there is a comparatively small number of plant crops in comparison with those of southwestern and eastern Asia, a lack of wild relatives, racial uniformity and the presence of a large number of typically recessive forms. *All this clearly bears witness to the more-or-less absence of indigenous agriculture in Central Asia and of the strange, secondary character of the crops and the plants cultivated there. There is no independently cultivated flora in Central Asia.*

Finally, Himalaya, Pamir and Hindukush appear to be powerful barriers separating Central Asia from the basic centers of type-formation of cultivated plants in southwestern Asia. The genesis of a number of cultivated plants in southwestern Asia is most definitely geographically located within the foothills of southwestern Himalaya, southeastern Hindukush and west of Pamir. Therefore, only fragments of the wealth of the cultivated types of plants were able to filter into Central Asia through the almost impassable mountain barriers.

It is interesting that the role of Himalaya, Pamir and Hindukush as powerful 'filters' applies also to the illnesses of the agricultural population of Kashgar. In Kashgar and Yarkand there are few of the infectious diseases that are so widespread in Afghanistan, in the Soviet Inner-Asiatic republics and in Iran, in spite of the fact that the ecological conditions in Sinkiang are not sharply different from the conditions in Afghanistan, Iran and some areas of Soviet Asia. According to information from Soviet and Swedish physicians, in Kashgar there is no trachoma, the horrible eye disease tormenting the inhabitants of Herat and other cities in Afghanistan, and no typhus or recurrent typhus, malaria, erysepalus, cholera or diphtheria. In this respect, Chinese Turkestan is really a kind of a 'veritable Paradise'.

A floristic investigation of Sinkiang undertaken by M.G. Popov also demonstrated general facts showing a comparatively great poverty with respect to the specific composition of crops in Kashgar and the absence there of any noticeable indigenous speciation processes (Popov, 1931).

When comparing the racial and specific composition of the cultivated flora of Central Asia with that of southwestern and southeastern Asia, it is possible to definitely trace from where, in a wide sense, the present cultivated flora of Central Asia has come.

The influence of southwestern Asia

Comparative botanical investigations in southwestern Asia allowed us to prove that the field crops in Sinkiang are represented mainly by plants native to southwestern Asia. Wheat (*Triticum vulgare, T. compactum*), barley, mung beans (*Phaseolus aureus* (Roxb.) Piper), chick-peas (*Cicer arietinum* L.), flax and linseed, garden rocket (*Eruca sativa*), cress (*Lepidium sativum* L.), fenugreek (*Trigonella foenum-graecum* L.), alfalfa (*Medicago sativa*), cotton and safflower, all these field-crops of Sinkiang clearly derive from southwestern Asia. The rice of Kashgar and Yarkand is hardly different from that in Fergana and is, in fact, introduced from the Soviet Turkestan or from northern Afghanistan. The spices of Sinkiang, coriander (*Anethum graveolens* L.), fennel (*Foeniculum vulgare* Mill.), cumin (*Cuminum cuminum* L.), 'azhgon' (*Ammi copticum* L.), anise (*Pimpinella anisum* L.) and black cumin (*Nigella sativa* L.), and many others were also introduced from the southwest.

Some of the melons (Cucurbitaceae; Figs 5,6) arrived by the same route and, in part, the same applies to a number of vegetables such as carrots, turnips, onions (*Allium cepa*) and garlic (*Allium sativum*).

Grapevines of all kinds, apricots, cherry plums [*Prunus cerasifera, v. divaricata*], pomegranates, almonds and figs evidently also came from southwestern Asia into Chinese Turkestan (Fig. 7). A comparison between the racial composition in Kashgar and that in Uzbekistan clearly indicates this.

Thus, the main flow of cultivated plants runs from southwestern Asia into Chinese Turkestan and the more northerly and easterly areas of Central Asia.

The influence from China

At present it is, however, also quite possible to trace the influence of the east-Asiatic center on Central Asia. The vegetable gardens of the Kashgarians, the Dzungarians and the Mongols indicate (according to studies made by expeditions from the Institute of Applied Botany in 1923) an obvious influence from China. The peculiar Chinese 'asparagus lettuce' (*Lactuca serriola* f. *integrifolia*) is practically unknown in southwestern Asia but is widely cultivated for its thick stalks in the vegetable gardens in Sinkiang. The local name is 'usum' or 'on-sen'.

Fig. 5. Types of melon in Urumchi. The Turkestani type of melon reaches all the way to far eastern Sinkiang. Photo: M. G. Popov.

Fig. 6. Melons of Turkestani type in Urumchi. Photo: M. G. Popov.

Fig. 7. A white kind of Khusayn grapes, common in Sinkiang but originating from southwestern Asia. The Turfana valley. Photo: M. G. Popov.

The long cucumbers (*Cucumis sativus*), the extraordinary, small-seeded melons (*Cucumis chinensis*), different kinds of cow peas (*Vigna catiang* Endl.; syn. *V. sinensis* Endl.), sometimes with pods measuring one half meter in length, and Mongolian turnips of all kinds and varieties (Fig. 9) are widely distributed in the vegetable gardens as are Chinese cabbage (*Brassica pekinensis*) and pak-choi (*B. chinensis*), turban squash (*Cucurbita turbaniformis*; Fig. 8), Chinese onions (*Allium chinense*, an onion that can be used for many years, and *A. fistulosum*, here replacing leeks) and, finally, yams (*Dioscorea batatas* Decne). All these demonstrate the particular influence exerted by China on the composition of vegetables in Central Asia (including Kashgar) (Fig. 10).

In Sinkiang, vegetables are very often cultivated by Chinese, Korean and Dungan peoples (the latter are a group of Chinese Muslims, inhabiting Central China). The influence from China is particularly strong with respect to the organization of vegetable gardens. Here they are remarkable by their neatness and the intensiveness of the crops, utilizing every inch of soil, the intensive application of fertilizers and the diversity of the crops.

As far as field crops are concerned, the influence from the East can be seen mainly with respect to the millets (*Setaria italica* and *Panicum miliaceum*), here called 'kunak' and 'proso', respectively. These two cereal grasses of the Asiatic East are definitely appreciated by the nomadic inhabitants because of their modest demands, their drought-resistance and the relatively good yield. The Koreans and the Chinese consider 'kunak' millet (*S. italica maximum*) a valuable cereal grass (used for a kind of porridge). The Kazakhs and the Kirgizes value it

Fig. 8. Early-ripening types of Khanderlyak melons from Kashgar, locally called 'chil'ga'. The turban-like shape is apparently recessive.

Fig. 9. Pink Chinese turnips. Urumchi. Photo: M. G. Popov.

Fig. 10. Chinese vegetable garden in Kul'dzhe. In the foreground, Japanese leek (*Allium fistulosum*), in the background, yams (*Dioscorea*) on trellises. Photo: V. A. Dubyanskiy.

as cattle feed. Only a small amount by weight of seeds is needed for a crop. Both kinds of millet are characterized by a comparatively wide variation in Central Asia (Kazakhstan and Kirgizstan). During its dispersal into Kirgizstan, the Chinese semi-vegetable millet (*S. italica maximum*), with a large, branched panicle, became associated with *S. italica* gr. *moharicus*, which approaches it in the type of a spike-like panicle. Both 'kunak' and 'proso' millets definitely arrived from eastern Asia, where racial diversity is concentrated. The many samples of millet (*S. italica*) collected by us in Korea and on the island of Kyushu in Japan, clearly demonstrate the concentration of type-forming processes of this crop in southeastern Asia.

The soybean, sometimes found in cultivation also in Central Asia, is obviously an east-Asiatic plant. Its genetic center is situated in Manchuria, Korea and northern China, where all the variable characteristics and genes are concentrated. In the region of Amur a wild species, the Ussurian soybean (*Soya ussuriensis*) is often found as a weed in the form of a type with small beans and a pod that cracks open when ripening. It is interesting that exactly in the Amur region – the leading area with respect to this crop – crops of soybeans are of a striking diversity as far as the dimensions of the beans, their shape and colors, and the pods are concerned, and also as far as early ripening is concerned (information from V.A. Solotnitskiy).

In contrast to what is seen in Kashgar, there is also an effect from the East (but not from southwestern Asia) on rice grown in Semirech'e (in the Chilik river valley), Kul'dzhe and Urumchi. There, contrary to what is the case in the Soviet Turkestan, the cultivation of awnless Chinese varieties or varieties with slender, delicate awns, and faint venation of the hulls predominates. These are sharply distinguished from the coarser, inner-Asiatic types of rice. In the Soviet Far East, an extraordinarily early-ripening Japanese type of rice is grown The most widely distributed strain bears the name of the north Japanese island, Hokkaido. This, no doubt, is a highly bred strain on which primitive selectors started to work long ago. Just as the civilized Europeans selected awnless types of soft wheat and oats and presently grow mainly such awnless crops, similarly the cereal growers in the Asiatic South-East, who had attained a high level of civilization due to population density, selected awnless types of rice and barley.

According to data from the V.E. Pisarev expedition to Mongolia in 1923, the farmers there grow a considerable quantity of typical, Chinese, short-awned and red-grained barley. The cultivation of naked-grained barley has apparently spread as far as to Urumchi.

The opium poppy (*Papaver somniferum*), grown in large quantities in the foothill areas of Central Asia, attracts the attention of investigators due to its unusual diversity with respect to the colors of the corolla and the dimensions, shape and colors of the capsule, the shape and the structure of the seeds and the shape of the leaves. The flowering poppies of Kirgizstan, Kazakhstan and Dzungaria are striking due to their bright colors and the shapes of the corollas. In Dzungaria, forms are often found with bright reddish-violet capsules. In this respect, the opium poppies of Iran and Afghanistan, studied by us in 1916 and 1924, differ sharply by their uniformity and the mainly pale colors of the corollas and seeds. The Arabian myth about 'snowdrifts' of opium poppies in China is in need of serious revision! In any case, in comparison with Afghanistan and Iran, the opium poppies of China are extraordinarily variable: there are many dominant types and it is not even necessary to prove that the basic, ancient center of type-formation of this crop is found in China. However, the strictly enforced periodic prohibition of its cultivation in China makes it difficult to establish the exact main area of its type-formation. The opium-growing areas of Central Asia obtained, in any case, this kind of crop from the Chinese.

Buckwheat (*Fagopyrum esculentum*), often found in cultivation in Dzungaria, is represented mainly by pink-flowered races that are similarly an introduction from eastern Asia.

Thus, 'proso' and 'kunak' millets, soybeans and rice as well as opium poppies – the main field crops of the agricultural areas of Central Asia – can definitely be traced to eastern Asia.

The fruit crops in Central Asia show similar traces of two-sided influences. The main fruit- and grape-growers in Kashgar grow strains from Fergana. The gardens are small and the growing of fruits and grapes is of limited importance in Kashgar. But, even there, it is also possible to see typical east-Asiatic representatives in the gardens: often there are Japanese apricot plums (*Prunus simonii* Carr.) with large red or yellow fruits, and sometimes the east-Asiatic, small-fruited *Prunus tomentosa* Thunb. is found as well; its fruits remind us of cherries. There is no doubt about the influence from the East with respect to the pear (*Pyrus sinensis* Lindl.), with its characteristic sharply spiny leaf edges. The Kuchar pear, famous in Kashgar, belongs to *P. sinensis* Lindl. In the markets of Kashgar, it is sometimes possible to find large-fruited jujuba (*Zizyphus sativa* Gaertn.) introduced from China; it is a typical representative of southeastern China. Also, large-fruited mulberries, *Morus nigra*, i.e. the black mulberry, belong to the introductions from eastern Asia. It is possible that, in the end, the white or ordinary *Morus alba* will also be found to have a similar origin.

In the agricultural areas of Siberia, adjacent to Central Asia, the influence from the East is also great. Under the severe conditions prevailing in the province of Irkutsk and along the Amur river, the ordinary European strains of apples, pears and prunes, as well as grapevines, do not thrive, but perish even during an ordinary winter. However, the native strains from the Ussuri region and even those from northeastern Asia, selected under the severe conditions there, succeed well. *Pyrus ussuriensis* Maxim., the Ussurian pear of Siberia, is, in spite of its astringent taste and low yield, a common, edible fruit in Blagoveshchensk and Irkutsk and even farther north. *Malus baccata* (L.) Borkh., the wild Siberian crabapple with small fruits, is common in the markets of Irkutsk and Tuluma. During the last couple of years, so-called 'rennets' were often found there as well; these are hybrids between *Malus baccata* and the European *M. pumila* Mill. Ussurian plum trees fill some gardens in the area of Amur. This is a very valuable source of material for hybridization with European and other east-Asiatic species and can, no doubt, be a source of future fruit-growing in the north, not only in Siberia but in the European parts of the Soviet Union as well. It is also interesting to note that *Vitis amurensis* Rupr., the small black Amur grape, is occasionally cultivated in the Amur region. It even grows wild there. The forests on the island of Hokkaido in Japan are covered by thickets of this kind of grapevine. In the fall it takes on beautiful colors, covers the coniferous trees and creates the scenic landscapes of northern Japan.

The indigenous species (I have) enumerated can, in northeastern Asia, in essence be traced only as far back as when people started to have an effect on them. Man usually multiplied seeds that were collected directly from wild specimens. The racial diversity was no doubt high, but this question has so far hardly been touched on by any investigators.

The effect of India on what is grown in Kashgar can perhaps be traced to *Momordica charantia* L. (bitter gourd), sometimes called 'Indian gourd' in that area. It is grown in the vegetable gardens of the Chinese and is especially readily consumed by Indian merchants. The strains of small-fruited *Momordica*, observed by us in Sinkiang, is hardly different from the Indian type we collected in Kandahar and in Kabul.

Indigenous crops of central Asia

There are very few indigenous crops in Central Asia but they do exist. Hemp appears to be the most important one among the field crops. Wild *Cannabis sativa* var. *spontanea* Vav. is a very common plant in northern Tien-Shan, especially on its north-facing slopes and valleys, but also north of this range. The vacant lots of Dzharkand are full of hemp thickets. Borders of wild hemp are frequently found along the edges of fields and roads. Weedy hemp is the most common plant in ravines, at the edges of forests and on poor soils and wasteland around the villages. As a weed, hemp reaches as far as the provinces of Irkutsk, Omsk and Amur. Ordinarily, wild hemp is not utilized, but here and there it is made into ropes. It is especially widely used in Altai. The representatives of wild and weedy hemp are, as a rule, types shedding their seeds, which have a so-called 'horseshoe'-formation [elaiosomes] (var. *spontanea*) and seeds of various dimensions up to the size of those of the large-seeded cultivated strains of hemp.

It has not been verified whether here and there in the ancient permanent settlements, e.g. those below Minusinsk, this field-weed hemp may have served as a source of crops. In Altai, E.N. Sinskaya was able to follow all the stages of hemp from typically wild ones to the cultivated kinds (Sinskaya, 1925).

We do not doubt that the introduction of hemp into cultivation from a plant existing under wild conditions and characterized by an enormous distribution area from the southeastern parts of the Soviet Union all the way to the Pacific Ocean occurred simultaneously at different times and in different areas; perhaps this occurred even in the agricultural areas of Central Asia (Vavilov, 1925).

Among other wild, native but also cultivated plants chicory, *Cichorium intybus*, is often found, especially on northfacing slopes of Tien-Shan, although it is doubtful whether it was originally taken into cultivation there; no evidence of that exists and no signs of a gradual transfer into cultivation have been found.

There are a lot of wild carrots in this area, i.e. of *Daucus carota* ssp. *carota* (L.) Thellung. On waste land and along edges of forests this common plant occurs as a weed, but there is no basis for suggesting that it was initially taken into cultivation here. Cultivated carrots are of a definitely introduced type. The distribution area of wild carrots reaches from the Atlantic Ocean all the way to Tien-Shan.

Among the wild fruit trees, the apples of the mountain areas of Central Asia are exceptionally interesting. In translation, the name of the city of Alma-Ata means 'apple city'. In the ravines of Tien-Shan, around Alma-Ata, there is a great variety of wild paradise apples (*Malus pumila* Mill.), constituting a whole chain of links from typically wild, astringent, almost inedible small apples to good, cultivated, comparatively large and sweet types. Some of the wild apples are even grown in the gardens. In the apple forests around Alma-Ata, types can be seen with red, yellow and white longish fruits, but also with roundish, flat or segmented fruits, large as well as small. All this variety deserves a very thorough pomological study. Similar stands of apples are know from Lepsinsk. It is highly probable that apples were transferred into the gardens and produced, perhaps, some initial strains there. In any case, it is still possible to see with one's own eyes whole forests of wild apples and to approach there the problem of the origin of cultivated apples, not to speak also of the significant stock of apple genes that will be of great interest for the future breeding of apples.

When traveling around in the Tien-Shan in March of 1911, Frank Meyer turned his attention especially to wild apples. In the *Inventory of Seeds*, published by the United States Department of Agriculture, the characteristics of wild apples are mentioned. Meyer called attention to the great interest in these forms for hybridization and for selection of drought- and cold-resistant strains, especially for higher elevations (up to 2400 m.s.m.).

Wild apples (*Malus pumila* Mill.) are characterized by an enormous distribution area reaching from the Pyrenées to Tien-Shan. There is a great diversity also in the Caucasus, although more small-fruited forms predominate there. It is quite likely that apples were taken into cultivation at different times and in different areas. Semirech'e may, indeed, be one of such centers of genesis in the case of cultivated apples, although it is on the periphery of the main area of wild apples.

Wild apricots (*Prunus armeniaca* L.) of Semirech'e are also of considerable interest. Thick stands thereof are often encountered at the foot of Tien-Shan. We also found them in the mountains between Frunze and Alma-Ata. Frank Meyer collected them in Tien-Shan around Kichik-Dzhigalan (at 4100 ft. elevation). The variation with respect to the pits, taste, shape and dimensions is quite wide there and, no doubt, some races have been and are being brought into cultivation there. The main center of apricots is concentrated toward Inner Asia. Tien-Shan is on the periphery of the main distribution area of wild apricots.

The wild bilberry bush (*Vaccinium uliginosum*) with large berries, the blueberries (*V. myrtillus*) and lingonberries (*V. vitis-idaea*) as well as bird-cherries (*Prunus padus*), are highly valued in Siberia. The last of these is often taken from the forests into cultivation around farmsteads and the trees beautify the villages while the fruits and a flour made from the dried fruits are considered an essential food product of Siberia. Pine nuts are another important foodstuff in Siberia.

In Central Asia, the farmers themselves are not represented by a single ethnic unity either. The enormous space of Central Asia is occupied by nomadic populations of Kazakhs, Kirgizes and Buryats, who have become farmers only comparatively recently. On the other hand, the inhabitants of the oases of Sinkiang (e.g. the Kashgarians) – referred to by A. Stein as *Homo alpinus* – have an intensive type of settled economy. Immigrants from China – Chinese, Koreans and also Dungans – practice both intensive market gardening and agriculture. However, on the whole, Central Asia was, until recently, a realm of nomads. Even its ancient agricultural areas do not, as we have seen, have any special, indigenous crops, nor any specific composition of cultivated plants or any typical agricultural technology. In Kashgar, the latter is not different from that of Inner Asia as far as the tools, plows or harrows, are concerned (Figs 11–16).

A comparative study of the agricultural areas of Central Asia results most definitely in a confirmation of the geographical localization of the type-forming processes with respect to Asiatic cultivated plants by placing them, on the one hand, in southwestern Asia and, on the other, in eastern and central China. The hypotheses concerning a narrow localization of primary type-formation of important European cultivated plants such as wheat and leguminous plants to the foothills of southwestern Himalaya and southeastern Hindukush, as pre-

Fig. 11. Tilling with a primitive tiller-plow in the Kashgar oasis. Photo: N. I. Vavilov.

Fig. 12. A typical roller for threshing in the Kashgar oasis. After a photo by N. I. Vavilov.

Fig. 13. Threshing wheat by means of animals, a more common method in the oasis at Kashgar. After a photo by N. I. Vavilov.

Fig. 14. Spade for digging irrigation ditches and doing other kinds of agricultural work.

Fig. 15. Pitchfork for removing chaff when threshing wheat.

Fig. 16. Stone mill used for producing flour from wheat. Figs 14–16 are all from the oasis at Kashgar and drawn after photos by N. I. Vavilov.

Fig. 17. Melons for sale in the market of Kashgar.

Fig. 18. Chinese vegetables. Kashgar.

Fig. 19. Carrots, Urumchi. Photo: N. I. Vavilov.

Fig. 20. Camel caravan travelling from Kashgar to the U.S.S.R.
Photo: N. I. Vavilov.

viously suggested by us (Vavilov, 1926; Vavilov and Bukinich, 1929), have been fully corroborated. Southwestern Asia, comprising Transcaucasia, Asia Minor, Iran, Afghanistan and the mountains of Inner Asia, as well as northwestern India, represents in the light of investigations made during the last decades a definite, basic general center of origin of an enormous number of cultivated plants such as field crops, vegetables and fruits. On the other hand, studies of Central Asia led us to a thorough study of the other basic general center of agriculture, i.e. the mighty, ancient east-Asiatic Chinese civilization, so far hardly touched upon by investigators (Figs 17–20).

As established by the recent American expedition, led by Andrew, Central Asia is a basic center where strange forms of animals linking primitive reptiles to modern mammals can be found in a fossil state. Fossil dinosaurs, titanotherians [primitive ungulates] and belutchitherians [primitive, giant rhinoceroses] are found in abundance there. In Mongolia and the adjacent parts of China we can also look for the native land of camels and horses.

With respect to cultivated plants, the genesis of which belongs to a more recent geological period, *Central Asia cannot* – as can be seen from what is stated above – *be considered a primary base of speciation or even of type-formation.*

BIBLIOGRAPHY

Howard, A., Howard, G. and Khan, A.R. (1915). Some varieties of Indian Gram (*Cicer arietinum*). *Mem. Dep. Agr. in India, Bot ser.*, **7** (7).

Matskevich, V.A. (1929). Morkov' Afghanistana [The carrots of Afghanistan]. *Tr. po prikl. botan., genet. i selek.* [Papers on applied botany, genetics and plant breeding], **20**.

Popov, M.G. (1931). Mezhdu Mongoliey i Iranom [Among Mongols and Iranians]. *Tr. po prikl. botan., genet. i selek.*, **26** (3).

Sinskaya, E.N. (1925). O polevykh kul'turnakh Altaya [On the field crops of Altai]. *Tr. po prikl. botan i selek.*, **16** (1).

Solms–Laubach, (1899). *Weizen und Tulpe* [Wheats and tulips]. Leipzig.

Stein, Aurel (1925). Innermost Asia: Its geography as a factor in history. *Geogr. J.*, **65** (5,6).

Vavilov, N.I. (1926). Tsentry proiskhozhdeniya kul'turnykh rasteniy [Centers of Origin of Cultivated Plants]. *Tr. po prikl. botan i selek.*, **16** (2).

Vavilov, N.I. (1927). Geograficheskie zakonomernosti v raspredelenii genov kul'turnikh resteniy [Geographical regularities in the distribution of the genes of cultivated plants]. *Tr. po prikl. botan., genet. i selek.*, **17** (3).

Vavilov, N.I. and Bukinich, D.D. (1929). Zemledel'cheskiy Afghanistan [Agricultural Afghanistan]. *Tr. po prikl. botan., genet. i selek.*, Suppl. **33**.

U.S. Department of Agriculture (1912). Imported seeds and plants. Inventory no. 27. *Bureau of Plant Industry, Bull.*, No. 242.

Mexico and Central America as a basic center of origin of cultivated plants in the New World

Preliminary report on the results of an expedition to Central America in 1930.

DURING THE LAST couple of decades, the necessity of procuring new material for plant breeding has compelled us to pay much attention to the study and collection of various resources of worldwide vegetation. In the U.S.S.R., this work is carried out mainly by the All-Union Institute of Agriculture according to a definite and strict plan for the purpose of maximum utilization within a short time of the areas in the world of greatest interest to us. In addition to studies of the resources in the Old World, during the last couple of years we have also crossed over to the New World which is of particular interest to us because of its commercial crops such as cotton, rubber, potatoes, and so on.

During the fall of 1925, a special expedition, led by S.M. Bukasov (1931), was sent out to Mexico, Guatemala and Colombia. The work of this expedition was continued in 1927 and 1928 by S.V. Yusepchuk, who studied the areas of Peru, Bolivia and Chile. As a result of these expeditions, an extraordinarily valuable amount of various material was collected. Not only the varietal, but also the specific composition of the important crops of the New World such as potatoes, maize, squashes and so on, is of fundamental importance to our own representatives.

In 1921, I acquainted myself with Canada and the northern parts of the United States. When traveling in North America during the fall of 1930, my attention was drawn mainly to the subtropical and tropical areas (Fig. 1). The main task of my trip became to clear up the localization of the primary speciation and the type-forming processes in North America with respect to the most important cultivated plants, which are of specific interest for the U.S.S.R.

Much was accomplished last year in Florida, Louisiana, Arizona, Texas, Mexico, Guatemala and parts of tropical Honduras. Important varietal material of different kinds of crops was collected.

The present preliminary report presents a collection of the basic results concerning the studies made by the All-Union Institute of Agriculture in North America.

For the purpose of establishing the centers of primary type-formation of

Lecture to a session of the Academy of the Sciences of the U.S.S.R., March 1931. First published in *Tr. po prikl. botan., genet, i selek.* [*Papers on applied botany, genetics and plant breeding*] **26** (3), 1931.

cultivated plants, knowledge of which is necessary for mastering the original varietal material, we have worked out a method of differentiating taxonomy and phyto-geography. In short, this one leads to the following:

1 The actual establishment of the primary area on the globe of a given plant;
2 the actual establishment of a system of varietal diversity with respect to Linnaean species;
3 an elucidation of the differentiating geography of the wild plants that are closely related to the cultivated ones, and of the varietal composition of their forms;
4 the establishment of primary centers, characterized as a rule by the discovery of many endemic varietal characteristics; under such conditions where the endemism of a given plant is of ancient origin (paleo-endemism), it can involve not only the characteristics of varieties and species but also those of entire genera of cultivated plants;
5 the fact that varieties of cultivated plants in primary areas are frequently characterized by a lack of divergence of the specific characteristics due to the absence of intraspecific hybridization;
6 the establishment of primary centers, which often contain a large number of genetically dominant factors; as demonstrated by direct studies of the geographical distribution of cultivated plants due to isolation (on islands or in the mountains) mainly recessive forms have developed and become singled out toward the periphery of the basic, ancient areas of species of cultivated plants; and
7 the fact that, in addition to the phyto-geographical method, data from archeology, history and linguistics can be of some use, although on the whole, they are too general for plant breeding purposes, which demand an absolute and exact knowledge of the species and varieties involved.

It is obvious that the application of this complicated method requires very extensive teamwork and a knowledge of all the world. However, we must also master the actual, original potential of the specific diversity of the morphological and physiological characteristics that are necessary for plant breeding and genetics, and approach closely to an understanding of the dynamics of the historical, evolutionary processes.

As can be distinctly affirmed at present, the cultivated flora of the New World was, in pre-Columbian times, definitely independent and represented by species and even genera alien to the Old World. The New and Old Worlds have comparatively few genera of cultivated plants in common, such as *Gossypium*, *Phaseolus*, *Solanum*, *Prunus*, *Vitis*, *Malus* and *Crataegus*. With respect to the overwhelming majority of cultivated plants in America and the Old World, even the genera are definitely different. It is enough to state that before Columbus, all the present Old World cereal grasses and leguminous plants such as *Pisum*, *Lathyrus*, *Vicia*, *Ervum* and *Cicer* as well as foliage plants such as alfalfa and clover, were alien to the New World. The majority of the Asiatic and Mediterranean fruit trees, flax and millets and the enormous multitude of vegetables in the Old World were unknown in America before colonization by the Europeans.

As far as we know, only one plant of the Old World had reached America before Columbus and been grown already by the farmers of the ancient

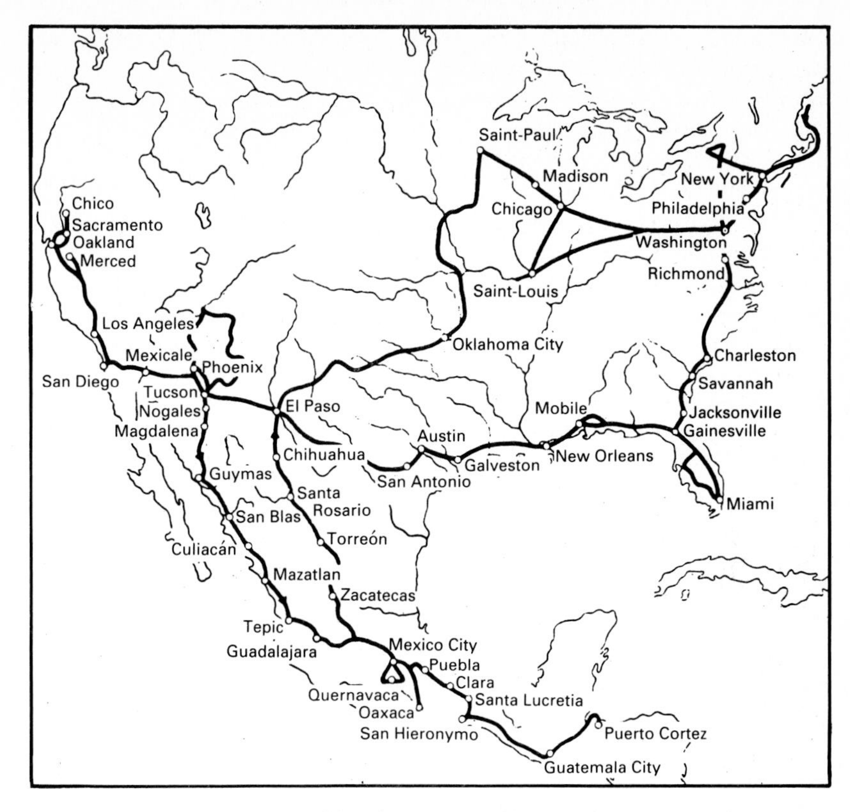

Fig. 1. Route covered by the N.I. Vavilov expedition to North America in 1930. The route is indicated by a heavy line.

civilizations in Central America and, perhaps, also in South America. This is the bottle gourd (*Lagenaria vulgaris* Sér.), the fruits of which are used as pots. However, as demonstrated by our research in Guatemala and Mexico in 1930, in contrast to what is the case of the Old World varieties of *Lagenaria vulgaris*, these gourds are represented in Central America by only a few forms. In spite of its general distribution in Guatemala, we were always surprised by the uniformity of the gourds there in contrast to the multiformity that we were able to observe during our studies of them in the old agricultural countries in Asia and Africa. Perhaps these gourds were transported by ocean currents or other means to the shores of the New World and taken into cultivation by the pre-Columbian farmers there. Their knowledge of it in pre-Columbian times is supported by discoveries of pots made of such gourds in the tombs belonging to the Mayan and Incan civilizations.

The agriculture of pre-Columbian America definitely developed independently from that of the Old World. If, as suggested by the majority of the present scientists (Wissler, 1922; Kroeber, 1923; Spinden, 1928), the people of the New World came from Asia, they arrived without any of the plants cultivated in Asia and their taking of wild plants into cultivation in the New World certainly depended on independent processes, as is shown by the peculiar, endemic cultivated plants of Central and South America.

In contrast to the farmers in the Old World, those of the New World did not keep any domesticated farm animals. Even the llamas and alpacas of Peru were used only as pack animals or, especially in the case of the latter, for their wool and meat. Most peculiar of all, the ancient farmers of Central America did not know of iron tools and they had no plows; copper tools were only fairly recently employed. The ancient agricultural tools were manufactured out of wood, stone and bones (Figs 2 and 3). Something similar to a plow, a so-called 'foot-plow', existed only in Peru (Cook, 1920). Since there were no farm animals (except for turkeys that were not really tame), the ancient farmers of Mexico paid great attention to the wild vegetation, as shown by the great number of edible plants taken into cultivation and the existence of the cultivation of ornamental plants before the arrival of the Europeans.

Our expedition collected an extensive amount of material of the many New World crops. A number of the plants of greatest interest to us, such as maize, potatoes, cotton and vegetables, were studied according to the method of differential taxonomy. Valuable data were also located in works by American scientists, especially in the papers published by the 'Bureau of Plant Industry' at the Department of Agriculture in Washington.

The basic fact that we can now establish as a result of studies of the entire extensive continent of North America is that *there is a striking geographical localization of speciation and type-forming processes with respect to the overwhelming majority of cultivated plants within the continental area of the New World.*

As far as their primary varietal potential, but also that of the wild relatives are concerned, the great diversity of endemic species of cultivated plants that initially originated from North America *turned out to be associated with the exceptionally limited territory of Central America, consisting of southern Mexico,*

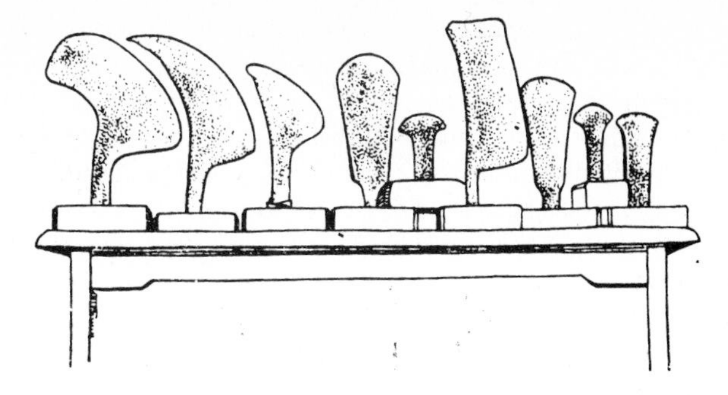

Fig. 2. Agricultural copper tools of ancient Mexico, preserved in the archeological museum, Mexico City. After a photo by N. I. Vavilov.

Крестьянская усадьба
в Mitla
(Oaxaca)

Fig. 3. Typical country house at Mitle (near Oaxaca). Photo and handwritten text by N. I. Vavilov.

Guatemala, Honduras, El Salvador, Nicaragua, Costa Rica and Panama. However, it is necessary to subtract a considerable portion of the large Yucatan Peninsula form this comparatively small territory.

On the whole the area of Central America, from where cultivated plants of North America initially originated, occupies not more than one twentieth of the entire continent. If the areas covered by volcanos and free of vegetation, as well as deserts and swamps, are subtracted, the surface on which the endemic cultivated flora of continental North America is presently concentrated becomes even smaller. Nevertheless, within this fairly limited area an amazing concentration is found of species and of racial diversity of cultivated plants. *A very important part of the plant resources of our globe has its origin just within this part of North America.*

As we know now, maize and cotton, i.e. the most important species now grown all over the world, originally came from there; the cultivation of some species of squash, beans, chayote, cacao, agave, papaya and a number of very widespread ornamental plants such as dahlias, cosmos, zinnias, tagetes and morning glories also came from this area.

We have compiled a list of species of cultivated plants that, according to present knowledge based on the above-mentioned method, originated in Central America and southern Mexico.

1. Maize – *Zea mays* L.
2. Teosinte – *Euchlaena mexicana* Schrad.
3. Upland cotton – *Gossypium hirsutum* L., *G. mexicanum* Todaro and some wild species of cotton (*G. hypadenum, G. patens, G. contextum, G. dicladum, G. morilli*), described by Cook and Hubbard (1926).
4. Garden beans – *Phaseolus vulgaris* L.
5. Runner beans – *Phaseolus multiflorus* Willd., both in a wild and a cultivated state, and in an enormous variety of forms.
6. Small-seeded lima beans – *Phaseolus lunatus* L. gr. *microspermus*, both in a wild and a cultivated state and in an enormous variety of forms.
7. Tepary beans – *Phaseolus acutifolius* A. Gray.
8. Fig-leaved gourd – *Cucurbita ficifolia* Bouché (*C. melanosperma* A. Braun).
9. Pumpkin squash – *Cucurbita moschata* Duch.
10. *Cucurbita mixta* Pang. – a new kind of squash described by K.I. Pangolo.
11. Chayote – *Sechium edule* (Jacq.) Swartz, also found in a wild state in Guatemala.
12. Annual pimento peppers – *Capsicum annuum* L. The exceptionally variable and ancient crops of pepper in southern Mexico bear witness to the independence of the south-Mexican and the Guatemalan centers from the Brazilian center which is also rich in varieties of peppers. Colombia, Venezuela, Panama and Costa Rica are poor in pepper varieties (according to S.M. Bukasov).
13. Perennial pepper – *Capsicum frutescens* L., [tabasco] both in a wild and a cultivated state.
14. Jicama – *Pachyrhizus angulatus* Rich (syn. *Cacara edulis* Kuntze).
15. Maguey – *Agave atrovirens* Karw., the pulque agave. Other species are also used for making pulque. Maguey is both cultivated and found in a wild state.
16. Sisal – *Agave sisalana* Perrine.
17. Ixtle agave – *Agave ixtli* Karw.

Fig. 4. Fence made of cacti (*Cereus*) in a village near Mexico City.

18 Lechuguilla – *Agave lechuguilla* Torr., fiber agave, used mainly in its wild state.
19 Century plant – *Agave americana* L. In Mexico alone, Standley mentions up to 170 wild species of agave. A number of these are used for various purposes.
20 Sotol – *Dasylirion duranguense* Treb. and other species of the same genus, known mainly in a wild state and used for making alcoholic beverages (from the base where the leaves sit); the leaves of some species are used for mats, hats and baskets.
21 'Tuna' – a number of cactus species belonging to the genus *Opuntia*, used both in a wild and a cultivated state.
22 Cactus species belonging to the genus *Cereus* s. lat. (e.g., *Pachycereus marginatus* Zucc., used for fencing (Fig. 4), *Lemaireocereus queretarensis* and *L. weberi*, used as wood, and so on; cf. Britton and Rose (1919–23).
23 Cochineal cactus – *Nopalea coccinellifera* Salm-Dyck. There is a particularly rich flora of cacti in Mexico; nowhere in the world is there so great a diversity of these as in Mexico. There are no less than 54 genera of the family Cactaceae and *Opuntia* alone has 87 species. Many of these species are grown for various purposes (as fruits or ornamentals, for medicines, etc.); cf. Britton and Rose (1919–23).
24 Huautli – *Amaranthus paniculatus* L. [purple amaranth].
25 Mexican tea or apazote – *Chenopodium ambrosioides* L. and *Ch. nuttaliae* Saff.
26 Cacomite – *Tigridia pavonia* Ker-Gawl.
27 Tomatillo – *Physalis aequata* Jacq., both a cultivated and a ruderal plant.
28 Cherry tomato – *Lycospermum cerasiforme* Dun., both a cultivated and a ruderal plant.
29 Chia – *Salvia chia* Fern., an oil plant.

30 Cacao – *Theobroma cacao* L., both cultivated and found in a wild state in Central America (especially in Guatemala and Honduras).
31 Achiote or annatto – *Bixa orellana* L., a dye plant, used both in a wild and cultivated state.

THE FOLLOWING FRUITS CAN BE MENTIONED

32 Anona or cherimoya – *Annona cherimola* Mill., *A. squamosa* L., *A. muricata* L., *A. purpurea* Moc. and Sessé., *A. cinerea* Dun., *A. diversifolia* Saff., *A. glabra* L. and many other species. They are cultivated but are also found in the wild.
33 Sapodilla – *Achras sapota* L. (*Sapota sapotilla* Coville), used mainly in its wild state.
34 White sapote – *Casimiroa edulis* La Llave.
35 Mamey – *Calocarpum mammosum* (L.) Pierre.
36 Incherto – *Calocarpum viride* Pitt. (syn. *Achradelpha virides* Cook).
37 Yellow sapote – *Lucuma salicifolia* H.B.K., used mainly in its wild state.
38 Papaya, melon tree – *Carica papaya* L.
39 Avocado – *Persea schiedeana* Ness and *Persea americana* Mill. (*Persea gratissima* Gaertn.), grown as a crop but used also in its wild state. A number of *Persea* species are found here, e.g., *P. longipes* Schlecht., *P.veraguensis* Sum., *P. podadenia* Blake, *P. liebmani* Mez., *P. chamissonis* Mez., *P. floccosa* Mez. and *P. cinerascens* Blake.
40 Guava – *Psidium guajava* L., used in its wild state but also cultivated.
41 Costa Rican guava – *Psidium friedrichsthalianum* (Berg.) Niedenzu, cultivated in El Salvador but also used in its wild state. *P. sartoranium* (Berg.) Niedenzu is also cultivated.
42 Mombin, ciruela or jobo – *Spondias mombin* L. and *S. purpurea* L. are used both in their wild and cultivated states.
43 Mexican hawthorne or texocote – *Crataegus mexicana* Moc. and Sessé. In addition, *C. stipulosa* Steud. is also found here. Both species are used in their wild and cultivated states.
44 Black cherry – *Prunus capuli* Cav., used mainly in its wild state.
45 Vanilla – *Vanilla fragrans* (Salisb.) Ames, used mainly in its wild state.
46 Pecan nuts – *Hicoria pecan* (Marsh) Britton, mainly in northern Mexico.
47 Guaymochil, Manila tamarind – *Pithecellobium dulce* (Roxb.) Benth. and other species, used in their wild state.
48 Silk-cotton tree or kapote – *Ceiba pentandra* L. Gaertn.
49 Castilloa rubber – *Castilla elastica* Cerv. in southern Mexico, El Salvador and Honduras, used in the wild state but also cultivated. Other species of castilloa are also used.
50 Wax myrtle – *Myrica mexicana* Willd., used mainly in the wild state. *M. cerifera* L. is also found in a wild state in Honduras and Yucatan.

ORNAMENTALS

51 Dahlias – *Dahlia excelsa* Benth., *D. imperialis* Reczb., *D. variabilis* Desf., *D. merckii* Lehm., *D. pinnata* Cav., *D. coccinea* Cav., *D. popenovii* Saff., *D. maximiliana* Host., etc. (ca. 15 species), known from the mountains of Mexico and Guatemala.

52 Cosmos – *Cosmos bipinnatus* Cav., *C. sulfureus* Cav., *C. diversifolius* Otto and *C. caudatus* H.B.K., etc.
53 Marigolds – *Tagetes lucida* Cav., *T. signata* Bartl., *T. patula* L., and *T. erecta* L. are all Mexican species.
54 Zinnia – *Zinnia elegans* Jacq., *Z. mexicana* Hart., and *Z. multiflora* L. are all Mexican plants.
55 Thorn-apple – *Datura candida* (Pers.) Pasquale.
56 Morning glories – *Ipomoea purga* Wender (syn. *I. jalapa* Royle), *I. purpurea* Lam., *I. schiedeana* Ham., *I. heterophylla* Ort., and *I. tyriantha* Lindl.
57 Lanthana – *Lantana camara* L.
58 Mock orange – *Philadelphus mexicanus* Schlecht.
59 Poinsettia – *Poinsettia pulcherrima* Grah. (*Eurphorbia pulcherrima* Willd.)
60 Mirabilis – *Mirabilis longiflora* L.
61 Bomarea – *Bomarea acutifolia* Herb.
62 Tuberose – *Polianthes tuberosa* L.
63 Yucca – *Yucca elephantipes* Regel and *Y. aloifolia* L. are often cultivated around shacks in Mexico and Guatemala.
64 Bouvardia – *Bouvardia ternifolia* (Cav.) Schl., used in the wild state as a medical plant but also cultivated as an ornamental.
65 Orchids (several species) – Just like bromeliads, these were already being cultivated at the time of the Aztecs.
66 Guayule – *Parthenium argentatum* A. Gray. The mountain areas of northern Mexico are the main area of its distribution.

For more detailed lists concerning the different crops and cultures, see Vilmorin's 'Blumengärtnerie' [Horticulture] (1896), Rose (1899), Standley (1920–26), Martinez (1928), Standley (1930) and Bukasov (1931).

As we now know, the commonly cultivated potato, *Solanum tuberosum* L., has 48 chromosomes. It originates from South America and seems to be a Chilean species. However, in the areas of Mexico and Central America, there are more than 30 tuberiferous wild species of potatoes, some of which are genetically comparatively close to the common potato, as demonstrated by research executed by S.V. Yusepchuk and S.M. Bukasov (1929).

According to data from V.A. Rubin (1929), the wild Mexican potato species represent a whole series of polyploids with $2n = 24$, 36, 48, 60 and 72 chromosomes.

Some of the wild, south-Mexican species such as *Solanum antipoviczii* Buk, *S. demissum* Lindl., and *S. ajuscoense* Buk. hybridize easily with the ordinary cultivated potato. Species such as *S. antipoviczii* and *S. demissum* appear to be resistant to *Phytophthora infestans* and some other diseases; consequently, it is evident that they are of exceptional practical interest for the breeding of potatoes, and even more so since *S. demissum* is frost-hardy.

Apart from the widely grown species, a number of other wild species of potato, such as *S. neo-antipoviczii* Buk., *S. candelarianum* Buk., and *S. coyoacanum* Buk., etc., were discovered in Mexico by the Soviet expeditions. Some of the species mentioned are used as food by the indigenous populations of Mexico. In spite of this, there are no definite data on the cultivation of potatoes in Mexico before Columbus.

As far as tobacco is concerned, the use of which was known by the Aztecs and the Mayans, there are no data on its geographical origin. Neither *Nicotiana tabacum* nor *N. rustica* were found by us in the wild or as a ruderal plant in Mexico or Central America. The 1925 and 1926 expeditions from the Institute of Applied Botany did not find any either. In all likelihood, *N. tabacum* is a native of South America.

Sunflowers, *Helianthus annuus* L., and 'Jerusalem artichokes' or sunchokes, *H. tuberosus* L., are the only two North American species which, in their primary type-formation, have spread beyond Mexico and now occur in the areas of the present U.S.A. and Canada. Sunflowers still occur in northern Mexico as a field weed but do not extend further south. We did not once observe any sunflowers in central or southern Mexico. Instead another, closely related genus, superficially reminding one of the sunflower, i.e. *Tithania tubaeformis* Cass. (syn. *Helianthus tubaeformis* Ort.) is widespread there; it is distinguished by the characteristic, conically shaped heads. Since it reminds one so much of the real sunflower, it is also called 'Mexican sunflower'. It is mainly and almost without fail associated with crops of maize in the mountain areas of Mexico and Guatemala.

'Jerusalem artichokes' are still found in a wild state in southern Canada and reach from Saskatchewan far toward the south in the U.S.A. into the Indian reservations in Arkansas and the central parts of Georgia. The 'Jerusalem artichoke' was first taken into cultivation in northeastern U.S.A. and southern Canada.

Transition from wild to cultivated forms

In contrast to some Asiatic and African centers of agriculture (e.g. in Abyssinia or Afghanistan), *the presence of many cultivated plants corresponding to their wild relatives are very typical of Central America and Mexico.* In the case of one half – if not an even higher number – of the endemic species cultivated in Central America and Mexico, all the stages of their introduction into cultivation can be traced. Thus, the fruit trees of Mexico and Central America are directly linked to their wild relatives. When clearing the forests, farmers often leave the wild Mexican plum tree (*Spondias mombin*), the guava [*Psidium guajava*] and the Mexican hawthorne [*Crataegus mexicana*] standing in the fields. We were able to observe this both in Guatemala and southern Mexico. (The same can be seen in the Old World with respect to the wild pears and wild apples in the Caucasus and in Turkestan.) Wild runner beans [*Phaseolus coccineus*] wind around bushes in the mountains of southern Mexico. A number of cultivated plants, such as agave, sisal and maguey [*Agave atrovirens*] do not, in essence, differ from the wild species and it is often difficult to tell whether the people are using the wild plants growing around their fields or if they have voluntarily brought them there. *Physalis aequata* (the tomatillo) is commonly utilized even as a ruderal plant and the same is true in the case of the cherry tomato (*Lycopersicum cerasiforme*). Wild cacao differs little from the cultivated one. It is also possible to trace transitional forms between wild and cultivated avocados and papayas.

Many species of cacti are growing wild at the same time as they are being multiplied by Man.

The farmers of Mexico and Central America still use a multitude of wild plants just as do the Indians on the Indian reservations in the adjacent parts of the United States. In the deserts, they use the fruits of the mesquite (*Prosopis pubescens* Benth. and *P. juliflora* D.C.) as well as of many cacti. Fruits of wild plants can be seen in the markets of small towns with a predominantly Indian population. The fragrant plants and spices in the local Indian markets in southern Mexico are usually represented by wild species. Teosinte, the ancestor of maize, is a common weed in the corn-fields of southern Mexico.

In southern Mexico and Central America, a student of cultivated plants can, in the full sense of the words, 'partake of the horn of plenty'.

The differentiating type-forming processes within Mexico and Central America

There is, thus, an enormous potential of species and primitive forms of cultivated plants in Central America. These are confined to a very small area in relation to that of the entire continent. This fact can still be traced in detail with respect to a number of plants. As demonstrated by direct observations, there is, especially within the borders of this comparatively small area, a sharply defined geographical differentiation as far as the speciation and type-forming processes are concerned.

The greatest wealth of endemic species of cultivated plants can be found in southern Mexico. The greatest racial diversity of maize is still concentrated just within this area and only there can one still see a large amount and variety of teosinte (*Euchlaena mexicana*). The maguey (*Agave atrovirens*) is typical of the high mountain areas of southern Mexico. Jicama (*Pachyrrhizus angulatus*) is distributed mainly in central and southern Mexico. On the other hand, the cherry tomato, *Lycopersicum cerasiforme* is unique to Guatemala. Runner beans [*Phaseolus coccineus*] are also amazingly variable in Guatemala. The large range of variation of maize in the mountains of Guatemala is striking; there are in particular many kinds of so-called 'flint-corn'. In southern Mexico, large ears of maize are rare. According to our own observations, *Cucurbita moschata* (squash), is especially variable in Guatemala (particularly in the area of Antigua; see Fig. 5). Guayule, *Parthenium argentatum*, and its companion, the mariole (*P. incanum*), are, as far as their diversity is concerned, definitely concentrated in the mountains of northern Mexico. Sisal agave [*Agave fourcroyoides*] originated in Yucatan. It is also evident that the typical tropical species such as cacao (*Theobroma*), vanilla and the rubber tree *Castilla* belong to the lowlands of Mexico and Central America.

The geographical localization within the limits of southern Mexico and Central America is to a great extent associated with the exceptional variation of ecological conditions in that environment, reaching from the humid Tropics at sea level to the upper limit for plant cultivation at 3200 m elevation and from the mountain semi-desert to the lowland deserts. During the dispersal of the crops and their varieties, the differentiation of the human populations that is closely linked to past civilizations of South America played, indeed, a quite important role. There is, however, no doubt that the natural-history aspect of the evolutionary process, linked to time and space, also puts its mark on the different areas.

Unfortunately, just like southern Mexico, Central America is fairly poorly known from a botanical point of view. There are still no serious floristic studies available for some areas, such as Honduras and Guatemala. No doubt more investigations concerning cultivated plants and their wild relatives can still result in interesting discoveries.

It is quite essential to keep in mind that the agriculture of Central America and southern Mexico is basically primitive. Therefore, there is no sharply reflected differential localization of it associated with river basins or water reservoirs, such as is found in areas with irrigated agricultures (e.g. in Egypt, Afghanistan and western China). The ancient agricultural settlements in southern Mexico and Central America were scattered and often difficult to find. For detailed studies of cultivated plants and their wild relatives, special long-term expeditions will be necessary here. We have still far from mastered the inexhaustible material for varieties that can be traced with respect to a number of ancient agricultural countries with irrigated agriculture.

How can the localization of the primary speciation and type-forming processes in North America be explained?

How can we explain or understand the striking localization of the primary speciation and type-forming processes of cultivated plants within the extensive area of the North American continent? Why is there such an amazing concentration of species and forms just in the southern part of this continent?

No doubt, studies of the wild flora of the New World can do much toward the understanding of these problems. Contemporary botanists have still far from any exhaustive knowledge of the flora of the New World. Rather, we are astonished by our ignorance in this respect. Whole areas of South America are still not botanically explored; there are still enormous areas of Brazil that have not been penetrated by any botanist and that have still not even been topographically surveyed. On the whole, there is no really exact geographical map of South America. The floras of Honduras and Guatemala are practically unknown. The geographical distribution of the specific diversity within the territories of North and South America and the portions of these areas with the maximum concentration of species and genera still await investigators.

Nevertheless, some basic phyto-geographical facts concerning the New World can now be considered as established. *It is a fact that the concentration of speciation and type-forming processes with respect to cultivated plants are in full agreement with the same regularities observed for wild plants: the flora of the northern parts of North America is very poor in number of species in comparison with that of Mexico and Central America* (cf. Fig. 6).

When speaking of the number of species we should, finally, also consider some conditions for understanding this as well as the more-or-less wide

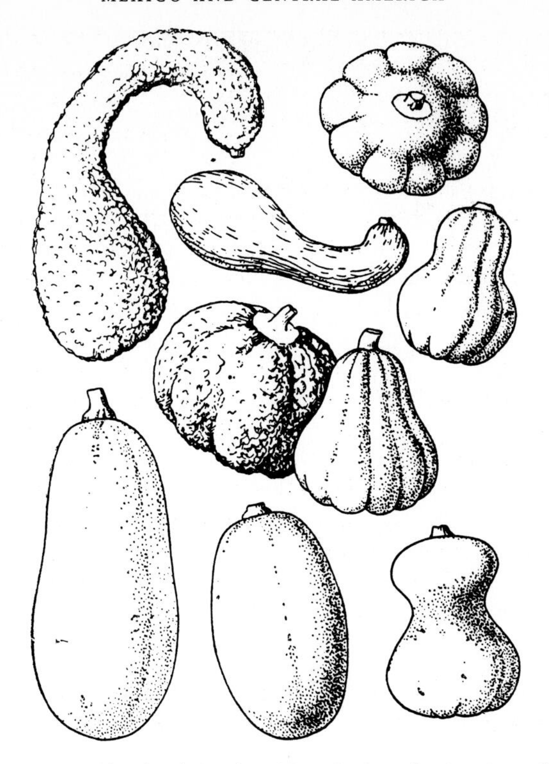

Fig. 5. Various kinds of squashes (*Cucurbita moschata*) in a vendor's stall on the market of Antigua (Guatemala). Drawn from nature.

amplitude used by different taxonomists in relation to the concept of species. However, in the case of a large number of plants or of entire floras, it is generally possible to compare, with some prudence, data from various investigators.

The most reliable estimates concerning the distribution of the specific diversity of plants with respect to the entire continent of North America can be found in the classical work, *Biologica Centralia-Americana* (four volumes), devoted to botany and compiled by Hemsley (1879–88).

For the part of North America including Canada and Alaska and the United States, Hemsley estimated the number of the mono- and dicotyledons at 1513 genera. For Mexico and Central America, he came up with a figure of 1794 genera.

The same difference is obvious with respect to the number of species (see table):

	No. of mono- and dicotyledonous species
North America (Alaska, Canada, U.S.A.)	9403
Mexico and Central America	11626

Fig. 6. Schematic map of North America. The heavy line delimits the basic center of origin of cultivated plants in North America.

Out of 11 626 species in Mexico and Central America, 8193 (70.5%) are endemic, i.e. species typical exclusively of Mexico and Central America.

The most recent data on the various countries and states support the basic opinions held during the last century, although, in the end, they furnish somewhat higher figures with respect to the number of species. Investigations of various parts of the U.S.A., even within the areas that are ecologically the most variable, such as Arizona, New Mexico and California, support Hemsley's conclusions concerning the concentration of the number of species and genera within Central America and Mexico. In his most recent review of Mexico alone, Standley arrives at a number of shrub and tree species amounting to 5450; moreover, he indicates that this figure is undoubtedly underestimated.

How high this figure really is can be judged from the fact that the total number of trees and shrub species in the entire European part of the U.S.S.R., including the Caucasus, is approximately ten times lower.

Within the small area of Costa Rica, Standley estimated more than 6000 species of flowering plants and ferns, although the territory of this republic is 130th the size of the U.S.A. In addition, Standley himself indicated that the number of species in Costa Rica could be 'greatly increased'. In that area, orchids alone number about 1000 species (Standley, 1925).

Although the U.S.A. and Canada are considerably better known from a botanical point of view, the contrast between Central America and Mexico and all the rest of the North American continent is still quite amazing.

Hemsley had already noticed the gradual concentration of species toward the southern part of Mexico. In comparison to all of Central America and Mexico, the southern portion of Mexico stands out because of the number of species found there.

Based on the total number of species known to Hemsley and determined to be 11 626, we have arrived at the figures shown in the table.

Area	*No. of species*	*Percentage of total number*
Northern Mexico	2930	25.2
Southern Mexico	7546	64.0
Guatemala	1337	11.5
Honduras	152	1.3
Nicaragua	843	7.3
Costa Rica	1086	9.3
Panama	1436	12.4

The role of southern Mexico as a center of accumulation with respect to the diversity of species is definitely undeniable and the data on the wild flora agree with the results of the studies of the geography of the endemic cultivated plants.

The number of species established, especially for Mexico, has grown during the last couple of decades. The figures added from the most recent information give an even more striking picture of the contrast between the north and the south. In 1908, Harshberger estimated the total number of species for all of North America at about 22 000 (Harshberger, 1911).

The revelation of such an obvious concentration of the speciation processes toward the south of the continent of North America is, evidently, associated with the approach to the humid Tropics, in the direction of which an increase in specific and generic diversity can be observed all over the world.

The optimal conditions for growth, the absence of winters and drought and the extraordinary fertility of the soil are all factors favoring the development of speciation. In his description of life in the Tropics, Wallace already mentioned the appearance of powerful speciation processes in the humid Tropics.

We cannot here go into a far-reaching examination of the causes for the localization of the multiformity of species and genera, since essentially different processes are involved here: on the one hand, in the south there are exceptionally favorable conditions of moisture, heat and substrates for the development of vegetation, while, on the other hand, in Canada and the U.S.A. the eliminating factor of glaciation occurred during the last geological period, starting from the north and eradicating entire floras over a large portion of the North American territory. It is also well known that mountain areas, just like islands, act in part as preservation areas for ancient species and genera. Mountain-building processes are sharply expressed only in southern Mexico and Central America. The presence there of a large number of endemic genera no doubt bears witness to the great antiquity of their formation and indicates definitely that the localization of many species and genera of presently cultivated plants is a phenomenon pre-dating the appearance of Man in America.

The actual presence among the endemics of Central America and southern Mexico of entire genera such as *Zea*, *Euchlaena*, *Sechium*, *Dahlia*, etc., including all their diversity with respect to both species and varieties, indicates that in all likelihood we have in that area, not only by chance, a center of secondary variation, but most definitely a basic, very old center of origin, as far as cultivated plants are concerned. According to the terminology of Chevalier (in Chevalier and Guenot, 1925), many of the endemic species of plants cultivated in Central America appear at present to be paleo-endemics.

Just as in the case of the study regarding the localization of the type- and species-forming processes in Asia, Africa and Europe, we were able to establish the exceptional importance in the New World of the mountains in the subtropical and tropical areas and their role as preservators of specific and varietal diversity. In Central America, this association between the speciation processes of cultivated plants and the mountain areas of the Tropics is especially obvious and leads us to an understanding of the primary type-forming processes as far as the important plants that are presently cultivated are concerned. This is evidently associated with the development of the flora on the whole.

The affinity of the basic, universal centers of type-formation with respect to cultivated plants (i.e. their centers of origin) to mainly the subtropical and tropical mountain belts, such as already established by our investigations, becomes to some extent dialectically clear since it is associated both with the general evolutionary processes occurring in the plant kingdom and with environmental factors. The Tropics and the Subtropics, with their optimum moisture and temperature and rich substrates, are conditions for the existence of powerful speciation processes. Mountain-building processes, the mountain relief, the diversity of the ecological conditions and the presence of natural, isolating factors there served as new and important factors for the distinction and divergence of species. The presence of mountains made settlement easier for those people who still feared the humid Tropics where the climate is so difficult to endure and where there is yellow fever. The mountains of the tropical and subtropical areas present maximum favorable conditions for the development of a settled and permanent, although primitive, life. They offer opportunities for the establishment of small, isolated groups of people and present optimum conditions for them with respect to temperature, fuel and food. Although the humid Tropics as such are characterized by a predominance of woody species, there are optimum conditions in the mountain areas within their limits for

Fig. 7. Geographical localization of primitive civilizations in the New World (after Spinden and Wissler). The hatched area outlines the area occupied by these civilizations.

species-formation of herbaceous plants and annuals, to which the majority of the most important cultivated species belong.

In Central America and Mexico, an investigator can approach in earnest the establishment of a direct association between the wild flora and the isolation of cultivated plants from the composition of their wild relatives.

Localization of the primitive civilizations in the New World

It remains for us to understand the localization of the great civilizations of North America, the Mayas, the Aztecs, the Zapotecs and the Toltecs, in light of the establishment of a clear geographical localization of the original plant resources used for food. Just within those areas where – as demonstrated by the differential phyto-geographical method – in the past the initial sources of food in America were found, the greatest human civilizations of antiquity were also formed.

In Fig. 7 we present a sketch by Wissler (1926) of the distribution of the civilizations in the New World before Columbus. It is hardly necessary to point out the agreement between data concerning the initial geography of cultivated plants and the geographical distribution of the ancient American civilizations. The same applies to North as well as to South America, and to Peru, Bolivia, Colombia and northern Chile, where the composition of the initial resources of strains is especially rich. Just there the ancient South American crops of the pre-Incan civilizations, the Chibchas and the Araucans, are concentrated.

To the amazement of the present investigator, remarkable monuments of art have been discovered in the ruins, the original structures of which were demolished especially by the Spaniards who were, as is well known, abetted by the Catholic Church. Before us, remarkable buildings, astronomical observatoria and an unusual knowledge of the calendar, as well as the original writings of the Mayans, rise up from these ruins. The hot Yucatan now attracts the attention of archeologists from all over the world and every year new evidence of the genius of the initial inhabitants of Mexico and Central America is revealed.

At the same time, Yucatan no doubt represents only a fragment of the great civilizations, perhaps especially well preserved here due to the conditions of a relatively dry climate. The high level of these ancient agricultural civilizations is astonishing, especially since the people did not know about iron or bronze and had no domesticated animals.

The interest of Central and South America for the rural economy of the U.S.S.R.

The uniqueness of the basic type-forming center of cultivated plants in Central America is also manifest in the extraordinary diversity of conditions in that area. We find both humid and dry Tropics there and conditions for growth of xerophytes and hygrophytes, as well as plants with long or short vegetation periods. Plants can be cultivated from sea-level up to an elevation of 3200 m elevation. The valleys in the mountains of southern Mexico, in which the agricultural civilizations of the Aztecs developed, have become barren. The soil around Mexico City is poor and hardly fertilized at all. In the past, the farmers of Mexico kept no domesticated animals and did not apply any manure. Fertilizers are still fairly rarely and sparingly used. The humid Tropics, on the other hand, are characterized by rich soils with strongly developed biological processes.

The multiformity of the ecological types within the limits of the various species has, until recently, attracted little attention as far as primary type-formation is concerned.

MAIZE

Central America, including Mexico, is of exceptional interest for the Soviet Union since it is the center of origin of maize, which at present is widely cultivated in the U.S.S.R. Studies in Mexico and Guatemala have revealed an enormous potential of strains, especially in southern Mexico. What is particularly noticeable, but still badly recorded, is the striking ecological differentiation of this species in its native homeland.

Fig. 8. A corn (maize) field near Mexico City (Chico), heavily infested by teosinte. The peasant claims that the teosinte hybridizes with the corn and ruins it.

Here, maize can grow under the conditions of the humid Tropics with a precipitation of 2000 mm or more a year. In Mexico, it reaches from sea-level in the coastal zone up to an altitude of 3150 m, almost to the top of volcanos (Fig. 8). At the same time, other strains of corn are grown in the dry areas of northern Mexico and the arid Yucatan, where the natural landscape supports cacti and agaves, i.e., typical desert and semi-desert plants.

There can be no doubt that, together with the exceptional morphological diversity of the plants, such as is unknown in any other country in the world, there has, since long ago, existed a concentration of physiological and ecological strains that are still not adequately utilized for practical plant-breeding purposes.

COTTON

Cotton, *Gossypium hirsutum* L., is cultivated in the United States of America but was originally introduced from Mexico. The best strains at present, such as 'Acala', 'Texas', 'Big Boll' and 'Durango' are directly adopted from Mexico and are, in essence, not different from the original, primary strains. Both the wild species that are closely related to *G. hirsutum*, such as *G. palmeri* Watt., and a number of new species of completely wild cottons, described by Cook and Hubbard (1926), e.g. *G. davidsonii*, *G. hypadenum*, *G. patens*, *G. contextum*, *G. dicladum* and *G. morilli*, are found there. Mexico is definitely of an exceptional interest since development of cotton cultivation is based mainly on the Mexican highland types. So far, the original material in the Soviet Union has been introduced mainly from the U.S.A. and is, in relation to our Soviet conditions, represented by strains that are inadequately early ripening. In order to extend the cultivation of cotton, we need more early types. Therefore, a search in the areas among the high mountains of southern Mexico could result in valuable material. Such a search would not be easy, since the industrial crops produced during the last decades in northern Mexico and the United States have almost destroyed the local cultivation of cotton in southern Mexico and Guatemala. Specialized and extended expeditions will be necessary when searching for primitive crops of cotton, which is far from simple when considering the dispersal of primitive non-irrigated agriculture. It is necessary to look for the primitive cultivated races and the original wild types in those high mountain areas, where it is difficult to gain access, in southern Mexico and Central America. We have only recently started this work. As demonstrated by preliminary acquaintance, the primitive local highland races of southern Mexico are extremely variable. Our expeditions have found types with differently colored fibers, with either pubescent or glabrous leaves, with yellow or cream-colored corollas and with variously shaped leaves, as well as strains with a high production of fibers, comparatively early-ripening races, and so on. In 1930, important new material for the cotton industry was collected from the mountain areas of southern Mexico and from Yucatan.

The endemic assortment of vegetables in Central America and Mexico

There is great interest in the vegetable crops of Central America and southern Mexico (Fig. 9) for Soviet market gardening. The diversity here of strains of beans – both the ordinary ones and the runner bean types – is nothing less than amazing. A new kind of squash, *Cucurbita mixta* Pang., was discovered by our 1926 expedition. Musk melons, successfully grown in the Soviet Astrakhan territory and Inner Asia, are represented here by a large number of forms. It is hard to imagine a greater diversity of peppers than can be seen in southern Mexico. They range from extremely small-fruited Mexican peppers, not more than 1 cm long, to variants with gigantic fruits, 25–30 cm long and 10–15 cm broad. The variation in taste is no less remarkable. The chayote, *Sechium edule*, the fruits of which are widely utilized in Mexico and Guatemala, is of special interest for our Soviet subtropical areas.

FRUIT CROPS

Some of the fruitbearing trees, such as annona or cherimoya, avocado and guava are no doubt of interest to the Soviet subtropical areas and deserve closer attention. During the last couple of years, these crops have been tried out in the subtropical parts of the United States. Finally, we shall point to the great interest to the Soviet Union of a strange rubber plant, the guayule [*Parthenium*

Fig. 9. Jicama (*Pachyrrhizus angulatus*), a local, typical root crop, used in large quantities for food in Mexico.

argentatum], presently grown in Turkmenistan and Azerbaidzhan but originally imported from Mexico (Fig. 10). Its distribution area does not agree with the basic center of origin of cultivated plants in Central America. This plant was introduced into cultivation only 18 years ago; until then only stands of wild guayule were utilized. The main bulk of its diversity is concentrated in the mountains of northern Mexico, in the states of Durango and Coahuila, at altitudes between 1500 and 23–2400 m elevation, and stretches up toward southwestern Texas. The high-elevation, frost-hardy and northernmost races of the guayule are of the greatest interest to us.

The relationship between the endemic cultivated plants in North and South America

The majority of cultivated plants in the New World that have been introduced to the Old World originally came from Central America and southern Mexico. Fewer plants have been adopted from South America.

Among the more important plants introduced from South America, we can mention the potato, tobacco, strawberries, manioc (*Manihot utilissima* Pohl), sweet potato (*Ipomoea batatas* Poir), and giant squash (*Cucurbita maxima* Duch.), peanuts (*Arachis hypogaea* L.), tomato (*Lycopersicum esculentum*) and pineapple (*Ananas sativus* Schult.).

Of the wild plants taken from South America into cultivation in the Old World, we can mention the rubber tree (*Hevea*), the quinine tree [*Cinchona*] and the coca bush [*Erythroxylon*].

Many of the crops endemic in Colombia, Peru, Chile and Brazil do not succeed very well when grown in the Old World, e.g. tuberiferous plants and root crops such as ulluco (*Ullucus tuberosus* Lozano), oca (*Oxalis tuberosa* Molina), magua (*Tropaeolum tuberosum* Ruiz and Pavon.), mangarito (*Xanthosoma sagittifolium* (L.) Schott.), arrowroot (*Maranta edulis* Wedd.), canna (*Canna edulis* Ker-Gawl.), llacon (*Polymnia edulis* Wedd.) or arracacha (*Arracacia esculenta* DC.), although these plants are of essential importance in their native lands. Similarly, we can list the quinoa (*Chenopodium quinoa* Willd.), a grain of the elevated mountain areas of Peru and Bolivia, the Brazilian species of yams (*Dioscorea dodecaneura* Vell., *D. brasiliensis*, etc.), Cape gooseberry (*Physalis peruviana*), Cunningham's lupine (*Lupinus cunninghamii*), tree tomato (*Cyphomandra betacea* Sendtn.), an eggplant (*Solanum muricatum* Ait.), pepino (*Cyclanthera pedata* Schrad.) and cassabana or sicano (*Sicana odorifera* Naud.). Even the South American cotton, *Gossypium barbadense* L., to which 'Egyptian cotton' belongs, succeeds comparatively badly when transferred to the Old World.[1] (Cf. Figs. 11–17).

There can nevertheless be no doubt that South America has its own, ancient and definitely characteristic centers of agriculture, based on its own flora. Just as in North America, the narrow localization of these agree with the geography of ancient civilizations.

Among these, the *Peruvian center* is definitely the most remarkable. It is characterized by a flora rich in endemic cultivated plants. Among its representatives we can mention the potato, *Ullucus*, *Oxalis tuberosa*, *Tropaeolum tuberosus*, *Canna edulis*, *Gossypium peruvianum*, the quinine tree (*Cinchona*), *Chenopodium quinoa*, *Lupinus cunninghamii*, the pepper tree (*Schinus*), the llacon (*Polymnia sonchifolia*) and the ordinary tomato (*Lycopersicum esculentum*), etc.

With respect to the number of endemics, the Peruvian center is comparable to the south Mexican and Central American centers. However, in contrast to Cook (1916, 1925), we consider the latter to be more important as far as cultivated plants are concerned with

[1] [Translator's remark: It seems that Vavilov has mixed up the South American G. *barbadense* aut., now G. *vitifolium* Lam. – 'sea island cotton', with G. *arboreum* L. (syn. G. *barbadense* L.), which originates from the Indus Valley.]

Fig. 10. A stand of wild guayule [*Parthenium argentatum*] in northern Mexico.

respect both to amount and composition. The superiority of Peru apparently depends only on its domesticated animals. At the same time as the Mayas kept only turkeys in a non-domesticated condition, the pre-Incan civilizations already had llamas, alpacas and guinea pigs. There were also civilizations that were familiar with cochineal and the cultivation of cacti. Out of a complex of 70 plants, quoted by Cook as specific for Peru, about one half were no doubt introduced. The entire list produced by Cook is very inexact, and confuses varieties, species and genera. He even considered such a plant as *Lagenaria* to be Peruvian.

It is of interest to mention that the flora of Peru, just like that of southern Mexico, is very rich in endemic species. There is no doubt that both these centers developed on the basis of typical and independent floras, but some species and even genera, many of which belong to typical paleo-endemics, were also taken into cultivation both in Mexico and Guatemala as well as in Peru.

Even before Columbus, there were undoubtedly populations common to both Central and South America. Data from excavations indicate crops, such as maize, beans and cotton, common to both areas. In spite of many differences, all these important civilizations of the ancient Americas had some things in common.

The direction of the migrations of the peoples of Central and South America is still far from fully known, much more so since even small centers, such as those in southern Chile, Bolivia, Colombia and the mountains of Brazil, display characteristics of autonomy as far as the composition of cultivated plants is concerned.

Even the maize in Peru is represented by a typical and special group (basically

Fig. 11. Dr. S. C. Harland in front of a tree-cotton plant. Trinidad.

Fig. 12. N. I. Vavilov by a quinine tree [*Cinchona* sp.]. Bolivia.

Fig. 13. Pineapple plantation, Cuba. Photo: N. I. Vavilov.

Fig. 14. Market place in Antigua, Guatemala. Photo: N. I. Vavilov.

Fig. 15. N. I. Vavilov in a Mexican market place, buying bean seeds.

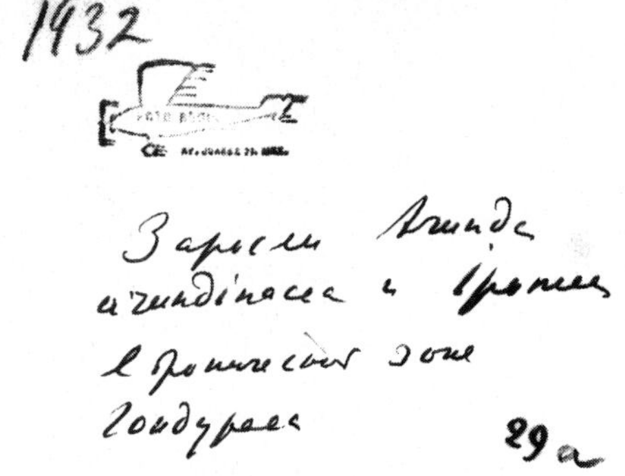

Fig. 16. Stands of *Arundinaria arundinacea* and *Ipomoea* in the tropical zone. Photo and handwritten text by N. I. Vavilov.

a complex of ssp. *amylacea*), although there are many data indicating that southern Mexico and Central America are the origin of this crop. The wild ancestor of maize, teosinte, is known only from Mexico and Guatemala. Maize is grown in Peru exclusively on irrigated plots but not in the non-irrigated mountain belt where – as elucidated by S.V. Yuzepchuk – the endemics are mainly concentrated. The cotton of Peru and Bolivia consists of a basic group of species, not found in the Mexican highland, although all have the same number of chromosomes. Crosses between the upland and South American species display a clear specificity, resulting in disharmony (sterility, deformities, etc.) among the hybrids. *Cucurbita maxima* of South America is a species well isolated from *C. moschata* and *C. ficifolia*, typical of Mexico.

Fig. 17. Varieties of gourds from Central America, grown at the Inner-Asiatic research station of the VIR. In the photo, Oleg, a son of N. I. Vavilov.

In spite of commonly held opinions (cf. Spinden, 1928), we consider, on the basis of our studies, that the primitive, general agriculture was, as a rule, not an irrigated one. Irrigated crops belong to a later stage. The ancient, primitive agricultural crops of Abyssinia, the Mediterranean area, China, India, Central America and southern Mexico were not irrigated, just like those now grown in the high mountains of Peru.

The amazing ancient terraces, used for irrigated crops in Peru and existing at elevations below 11 000 ft, have been well described by Cook (1916; cf. his interesting paper with beautiful illustrations in the *National Geographic*) but were apparently of a later origin than those used for the non-irrigated crops of the alpine areas of Peru (above 11 000 ft). These were characterized by a large number of definitely endemic species of cultivated plants, such as potatoes, oca, quinoa, ulluco, etc.

Egypt and Mesopotamia, characterized by irrigated agriculture, are – as shown by our investigations – distinguished by a poor assortment of strains, which apparently were introduced from permanently settled agricultural areas, in mountains. In spite of their great importance from an archeological point of view, Egypt and Mesopotamia appear secondary with respect to agriculture. The irrigated crops of rice in China and Japan were introduced ones, in spite of their present importance in the life of these countries. A comparison between the complexes of rice in India and those of China and Japan clearly indicates this fact.

The basic ancient agricultural areas in southern Mexico and Central America were not irrigated. Irrigated crops here are concentrated within new, more northerly areas, mainly where European immigrants have settled.

The investigator can only begin the study of plant resources in Central America and South America while, so far, making use of the crops of the primitive farmers; they mainly used plants within the limits of the Tropics and the mountain areas where the first civilizations of ancient America settled.

The enormous stores of plants that are concentrated in the tropical regions of Central America and South America are still not fully known to Man. Some 10 000 species are characteristic of the floras of Brazil, Peru, Venezuela and Colombia. Primitive man avoided these regions; the powerful forces of tropical nature stopped him. He fought and to this day still fights the diseases associated with the terrible elements of the Tropics. However, during the twentieth century and in the future, the intention is to master the Tropics and to reveal the colossal plant resources that can be utilized for the various needs of mankind.

BIBLIOGRAPHY

Britton, N.L. and Rose, J.N. (1919–1923). *The Cactaceae*. Vols. 1–4. Washington, D.C. Carnegie Institution.

Bukasov, S.M. (1930). Vozdelyvaemye rasteniya Meksiki, Gvatemali i Kolumbii [Cultivated plants of Mexico, Guatamala and Colombia]. *Tr. po prikl. botan., genet, i selek.* [*Papers on applied botany, genetics and plant breeding*], Suppl. **47**.

Chevalier, A. and Guenot, L. (1925). Biogéographie. In Martonne, *Traité de géographie physique*. Paris.

Cook, O.F. (1916). Staircase farms of the Ancients' *Nat. Geographic*, **May** 1916.
Cook, O.F. (1920). Foot-plow agriculture in Peru. *Smithsonian Report for 1918*. Washington.
Cook, O.F. (1925). Peru as a center of domestication. *J. Hered.* **16** (2).
Cook, O.F. and Hubbard, J.W. (1926). New species of cotton plants from Sonora and Sinaloa, Mexico. *J. Wash. Acad. Sci.*, **16** (12).
Harshberger, John W. (1911). Phytogeographic survey of North America. In *Die Vegetation der Erde*, ed., Engler and Drude, Leipzig, New York.
Hemsley, W.B. (1879–1888). Biologia Centralia-America of contributions to the knowledge of the fauna and flora Mexico and Central America. *Botany, London*, **1–4**.
Kroeber, A.L. (1923). *Anthropology*. New York.
Martinez, A. (1928). *Plantas utiles de la republica Mexicana*. Mexico.
Martonne (1925). *Traité de géographie physique*. Paris.
Rose, J.N. (1899). Notes on useful plants of Mexico. *Contributions from the U.S. Nat. Herb.*, **5** (4).
Rybin, V.A. (1929). Karyologicheskiy analiz nekotorykh dikikh i tuzemnykh sortov kartofelya Ameriki [Caryological analysis of some wild and native cultivated sorts of potato in America.] *Tr. Vsesoyuz. s'ezda po genet., selek., semenovod* [*Papers of the All-Union Congress of Genetics, Plant Breeding and Seed Propagation*], **3**.
Spinden, H.J. (1928). Ancient civilizations of Mexico and Central America. *Am. Mus. Nat. Hist.*, New York.
Standley, P.C. (1920–26). Trees and Shrubs of Mexico. *Contributions from the U.S. Nat. Herb.*, Pts. I–V.
Standley, P.C. (1925). Orchid collecting in Central America. *Smithsonian Report for 1924*. Washington.
Standley, P.C. (1930). *Flora of Yucatan. Chicago Field Mus. Nat. Hist.*, Publ. **279**.
Vilmorin (1896). *Blumengärtnerei* [Horticulture]. Vols. I–II. Berlin.
Wissler, Clark (1922). *The American Indian*, 2nd edn. New York.
Wissler, Clark (1926). *The relation of nature to man in aboriginal America*. New York.
Yuzepchuk, S.V. and Bukasov, S.M. (1929). K voprosu o proishozhdenii kartofelya [On the problem concerning the origin of potatoes]. *Tr. Vsesoyuz. s"ezda po genet., selek, semenovod*, **3**.

The plant resources of the world and the work performed by the All-Union Institute of Plant Industry toward their utilization

AN ENORMOUS PROBLEM faces the farmer in a Soviet country. In connection with the collectivization of small farms and the State farm system, mechanization and the use of chemicals for agriculture, we have now a powerful lever by which the backward, indigent and broken down rural economy of the past will, within the shortest time possible, be transformed to the level of an agriculture that will be advanced from a technical point of view.

In particular, the area under field crops, vegetables and fruits must be increased. The separate, small individual farms will not be able to control the enormous arable resources in our country. Very extensive, arable areas remain unused. The total acreage under crops in our country during pre-revolutionary time amounted to only 4.9% of all the landmass. According to estimates by our soil scientists, the Soviet country has exceptional possibilities for increasing the acreage under cultivation. The reserve of the singular black soil in the Soviet Union consists – after deducting unusable areas – of 240 million hectares, i.e. twice the area presently used, which in 1930 was determined to be 120 million hectares.

Already last year we were able to increase the area of land cultivated at present by 12% in comparison with that of the past.

In addition to an increase in the crops to be raised, there is an enormous problem concerning new and more rational investments in crops. *A planned and regulated plant industry means: breeding the crops and strains and multiplying them; the exchange of low-yielding crops and strains for more valuable ones; increasing the quality of the raw material of plants; increasing the productivity of our fields and gardens; creating valuable standard strains; increasing all kinds of technical crops; a wide application of new plant crops such as rubber and tannin-producing crops; an extensive development of cotton crops on new, non-irrigated areas in the northern Caucasus and the southern Ukraine; increasing all kinds of forage crops, necessary for the development of the cattle industry; utilization of every inch of soil in our subtropical areas for relieving us from the importation of foreign raw materials; and extending the area of agriculture toward the north while transforming it into areas for reliable crops. These are the problems currently facing the Soviet plant industry.*

Like never before, the Soviet Union is now faced with problems concerning

First published in Semenovodstvo [The Seed Industry], Nos. 13–14, pp. 6–10, 1931.

the most efficient utilization of its plant resources by selection and/or breeding of the most valuable crops and most productive and high-quality strains. For the purpose of utilizing the progress in Western Europe and America as far as field crops, vegetables and fruits are concerned, we have procured an enormous amount of material of strains from plant breeding institutions and seed companies in Germany, France, Denmark, Czechoslovakia, Yugoslavia, Holland, Sweden, Canada, the United States of America and Argentina. Out of the foreign-bred strains of wheat alone, 320 kinds have been tested. West-European vegetables proved to be especially valuable. The most valuable strains were multiplied in great amounts and hundreds of thousands of kilograms of seeds have been multiplied on Soviet soils, according to a definite plan.

We have attempted to obtain all the best that can be found of cereals, forage grasses, root crops, potatoes, vegetables and fruit trees from abroad, in particular from such areas where the climate and soil conditions are similar to those of ours. As demonstrated by research workers at the All-Union Institute of Plant Industry (formerly the Institute of Applied Botany and New Crops), there is a considerable wealth of strains also within the limits of the Soviet Union itself. Thus, as elucidated during the last couple of years, Transcaucasia and Inner Asia have an extraordinary wealth of local strains of fruit trees, in this respect surpassing many other countries in the world.

As is well known, a large number of cultivated plants originally came from different ancient agricultural countries in Asia, Africa and America. The centers of origin of many cultivated plants such as wheat, barley, maize, cotton, and many vegetables have been found in such mountainous countries as Mexico, Abyssinia, India, Afghanistan, eastern China, Peru and Bolivia.

These countries, not important from an economic point of view, turned out to be the wealthiest with respect to the diversity of strains. Indeed, the very beginnings of agriculture arose just in these countries and it still has a primitive character there.

We know, for instance, that we can find a very valuable source of strains for our Black Sea coast in Japan, which, besides being a very old agricultural country, also reminds us of the areas around Batumi, Sukhumi and Chakvi.

For the purpose of a systematic enlistment of material to be used for practical plant breeding, the Institute of Plant Industry has, during the last couple of decades, conducted many investigations toward a study of our planet in order to establish local resources of strains, find new crops and new plants, and collect seeds. In order to solve the practical problems of agriculture, Soviet expeditions have been able, during the last couple of years, to conduct research in almost three fourths of our world, in spite of the many difficulties and obstacles put in the way of Soviet scholars traveling in different countries.

The All-Union Institute of Plant Industry, belonging to the system of the Lenin Academy of Agricultural Sciences, was faced with the difficult problem of mastering, within a short period of time, a very valuable collection of strains of cultivated plants from wherever they were found in the world.

It seemed to us that only by controlling all the original strains and by applying plant breeding methods for gaining control over the necessary stock of strains from all over the world, it would be possible to solve the basic, practical problems of the plant industry.

Starting in 1923, one Soviet expedition after another was sent out to different areas of the world. They studied all the agricultural areas of Mongolia and Afghanistan and spent three years studying Asia Minor and Persia; they researched all the ancient agricultural countries along the coasts of the Mediterranean: Spain, Portugal, Italy, Greece, Algeria, Tunisia, Morocco, Cyprus, Sardinia, Sicily, Syria, Palestine, Transjordania and Egypt. Soviet expeditions also penetrated Abyssinia, French Somalia and the Italian colony of Africa, i.e., Eritrea [Note: In 1962, Eritrea became a province of Ethiopia]. An enormous amount of seeds has been collected from everywhere. Cultivated plants of Japan, Korea, Formosa, western China, northern India, Java and Ceylon have been studied in detail. The Soviet agricultural expeditions have, in particular, traveled around in the New World, in Mexico and Central America (Guatemala and Honduras). They have also gone to South America and carried out studies of Colombia, Peru, Bolivia and Chile, such as had never been done before.

These expeditions, carried out according to a definite plan on the basis of a purely materialistic theory concerning the origin of cultivated plants, have, within a short time, furnished the Soviet State with a colossal, new material of strains such as has never been seen before by agronomists, plant breeders or botanists. At present, the Soviet Union controls an exceptionally complete collection of strains as far as the most important crops in the world are concerned. This is now being multiplied at research stations throughout the Union and shall serve as a basis for practical plant breeding.

Among this material there are strains of exceptional value, such as have so far been unknown to agriculture. It should be especially pointed out that, e.g. in the mountains of Mexico, potatoes are found that are extraordinarily resistant to frost: the tubers can withstand temperatures as low as $-10°$ C. Most of all, there are potato species resistant to the European potato diseases. At present, such strains are being hybridized with our own strains to improve their quality.

We found the world's earliest-ripening cotton in western China. A particularly interesting sample of barley was found in Abyssinia, the native land of 'brewing' barley. Asia Minor furnished very valuable kinds of melons, which surpass our very best Turkestani strains. In northern Africa, we found drought-resistant large-grained varieties of oats, which at the same time are not affected by either smut or rust. There was also linseed, the seeds of which are three times larger than those of our own ordinary flax.

Within the Soviet Union itself, in the Caucasus, the Inner-Asiatic republics and the Far East, expeditions from the Institute of Plant Industry studied wild fruit trees. There, wild apples, pears and cherry plums frequently form whole forests – literally 'fruitful forests'. Based on the research carried out during the last couple of years, there can at present be no doubt that the native lands of the majority of the European fruit trees are found in Transcaucasia and the ancient areas of Asia Minor, but also in Inner Asia and northern Persia. In forests consisting of wild fruit trees, strains of exceptional value can be found, which can be taken directly into cultivation. Some of the wild strains of almonds in Turkmenistan, the apples of the Caucasus and the cherry plum of Georgia are not inferior to but often surpass our best cultivated kinds.

It is necessary to point out that the basic stock of strains turned out to occur in mountains and often in areas of high mountains in the south. Therefore, although the native lands of many cultivated plants are found in subtropical and tropical areas, it is nevertheless a favorable characteristic of these plants that they belong mainly to mountain areas, so that they seem to be comparatively cold-hardy and can be transplanted far toward the north. It is sufficient to state that the native land of the potato is Peru, not far from the equator. However, these potatoes occur at elevations of 3000–4000 m.s.m. and it is, therefore, not astonishing that tropical potatoes can be grown as far north as near Murmansk in our country.

The practical mobilization of the resources of strains from all over the world, the extensive utilization of the original wealth of material for practical plant breeding purposes and the creation of new, more valuable strains appear at present to be the primary task of the Soviet plant industry. For the purpose of extending the research concerning the practical value of new strains and for making comparisons with our ordinary kinds of material, the Institute of Plant Industry has built up a series of major laboratories and research stations, which study the strains from every point of view and investigate the different strains according to new physiological methods, with respect to their resistance to cold, drought and diseases as well as their qualifications as a source of flour and bread.

We have, at present, excellently equipped milling and bread-baking laboratories for studying the quality of bread made from various kinds of wheat and rye. Research conducted by these laboratories during the last couple of years has demonstrated that our best strains of spring and winter wheat are in no way inferior to the best Canadian and Argentinian strains of wheat that compete with ours on the world market. They even surpass them in quality. *Our hard wheat has no competitor on the world market as acknowledged by the milling laboratory in Washington when it investigated strains of wheat from all the countries exporting it.*

Research at our technical laboratories has demonstrated that, as far as the improvement of flax is concerned, much can still be achieved, but also that we have at our disposal high quality flax that can be multiplied within the near future.

On the basis of new material, Professor Pisarev of the Detskoye Selo Station of the Institute of Plant Industry has produced, by hybridization, a strain of wheat called 'Novinka', which is suitable for the Far North and already yields great harvests and grain of a beautiful quality under the conditions of the Leningrad area.

In the Steppe belt at the Kamennostepnaya Station of the Institute of Plant Industry, valuable leguminous plants, at present occupying thousands of hectares, have been produced thanks to new materials from all over the world.

At Sukhumi, where research on subtropical crops is presently concentrated, an enormous variety of all possible kinds of subtropical field crops has been assembled during the last couple of years, such as fruits and technical as well as decorative plants. About 3000 kinds of useful trees alone have been assembled here from all over the world. This is now the richest nursery in all of the U.S.S.R., indeed, in all the world.

For the purpose of systematic multiplication of crops and strains all over the U.S.S.R. and for systematic regulation of seed production, the Institute of Plant Industry has organized a so-called State Bureau for Testing Agricultural Products, run by Professor V.V. Talanov. Competitive comparisons between our best strains, bred during the last couple of years at our plant breeding stations, are conducted in the field at 200 sites distributed throughout the Union. The best standard strains in the world, e.g. the best of American wheat or maize, are compared with them for the purpose of selecting the most valuable among them. In agreement with the conclusions drawn by the State Bureau for Testing Agricultural Products, the most valuable strains are sent out to the state and collective farms for extensive multiplication on millions of hectares. Thus, during the last couple of years, the American line of maize, 'Ivory King', which yields a high quality of flour, has been put into production; last year about 1.5 million hectares were already sown with this strain. The National Seed Bank of the U.S.S.R. now controls some 10 344 000 'double centers' of seeds. Soon, after 2–3 years, all the seed material on the fields of the Soviet Union will be standardized and of high quality. Experiments have demonstrated that the introduction of such highly bred strains will increase production by 25–30%.

Systematic control by the State over the change of strains instead of the spontaneous processes that were relied upon in the past, will allow us to modify, in a short period of time, the composition of our fields and to make our gardens, vegetable plots and fields more productive.

We find ourselves at the very beginning of a reconstruction of the rural economy. We have only just begun the work, but already now we can predict a colossal change for the better, which will take place in the near future within the agriculture of our country as far as our fields and gardens are concerned, and which shall make our rural economy hard to recognize, indeed, make it into the most advanced one.

What is happening at present within the agriculture of the Soviet Union is, as we have demonstrated, of historic importance for all the world.

At the same time as a crisis threatens all the major capitalistic countries and at the same time as the capitalistic countries – in the vice of an anarchy of competition, over-production and the decline of value – attempt to check the growth of their acreage under crops, the Soviet countries steadily increase the production of their soils and produce at an unprecedented tempo a reconstruction of the rural economy on the basis of collectivization, the formation of state farms, mechanization and the use of fertilizers for agriculture.

Out of the two billion inhabitants at present living in the world, more than three-quarters are farmers. At the same time as the capitalistic countries do not know what to do with their over-production or where to put their superfluous grain and other plant crops, India, with its 300 million inhabitants, and China, with its 400 millions, are literally starving.

The most recent research, produced on a global scale, indicates that there are enormous resources for worldwide agriculture. However, only about 5% of all land is used for agriculture. Enormous areas can still be used within the temperate areas of Canada and the United States. A farmer presently 'armed' with science and technology can fearlessly begin to enter the tropical zone with its exceptionally fertile soils, its absence of winters and its abundance of precipitation. Until recently, Man feared the Tropics, its forces, its teeming vegetation, tropical diseases and yellow fever. Now, however, he certainly knows of methods for controlling these diseases. One entire third of the landmass in the world is occupied by the humid Tropics, which are suitable for agriculture and still, in essence, definitely not utilized by mankind.

The opportunities for global agriculture are almost unlimited. Although the inhabitants of the world should increase three to four times, the resources are adequate for an abundant, correct and full utilization thereof.

However, it is definitely obvious that one should take the stand here that only a new organization of humanity will be able to utilize the enormous resources of our world.

The world centers of origin of agriculture and the soil map of the world

THE BUREAU OF AGRICULTURAL GEOGRAPHY of the Lenin Academy of Agricultural Sciences has drawn a soil map of the agriculture of the whole world.

As far as we know, this is the first attempt to illustrate accurately the distribution of the cultivated surface of the entire world. Although some inexactness may have slipped in with respect to the description of badly investigated areas that are still not readily accessible for regular statistics, the map in question clearly shows the utilization of the arable land of the earth.

A comparatively small portion of the landmass is occupied for agricultural purposes; about 4% of all the territory is used for field, vegetable and fruit crops. In the Soviet Union, less than 5% of the soil is tilled; 95% is still not used for agriculture.

When comparing the map of worldwide agriculture with the soil map, we can see that, on the whole, agriculture at present has a tendency toward utilization of a major part of the natural soil resources. Agriculture is in our time concentrated within the most fertile areas, i.e. the black soils of Eurasia and the New World. However, as we shall see below, such utilization of the natural resources seems to be recent and characteristic only of the last two centuries, i.e. the eighteenth, but particularly the nineteenth.

During the last few decades, the Institute of Applied Botany has paid special attention both to solving the problems concerning the origin of cultivated plants and to the related problems concerning the history of agriculture. The first of these investigations, conducted on the basis of a new method, forced us to considerably revise our initial opinion about the history of agriculture and to adopt a new point of view.

When studying the problem concerning the origin of cultivated plants from the point of view of plant breeders and geneticists, we logically decided to approach the problem from a different angle, while considering a solution of the problem of the origin as an exact establishment of the sites where that process initially developed and thanks to which the types of given plants evolved.

We are now able to suggest that such areas where the original potential of types of given species is concentrated seem to be the native lands of cultivated

First published in *Proceedings and Papers of the Second International Congress of Soil Science*, July 20–21, 1930, Leningrad–Moscow, U.S.S.R., Commission no. IV: Soil Fertility, Moscow, 1932.

plants. For the purpose of establishing exactly the original centers of type-formation with respect to cultivated plants, but also to study the initial material of types for plant breeding purposes, the Institute has, during the last ten years, organized a number of expeditions to various parts of the world. New, rich material, collected by these expeditions, has provided us with an opportunity for definitely establishing the basic areas of origin of cultivated plants by means of the so-called differential–taxonomical method, but has also allowed us to furnish plant breeders with a large variety of new strains.

As a result of studies of the most important cultivated plants, we have established six basic centers in the world as the origins of various plants. These are, possibly, also the centers where agriculture was born. It no doubt developed at different times and in different areas.

The first of the basic centers in the world, giving rise to many crops in Eurasia, is situated in southwestern Asia and comprises northwestern India and the adjacent parts of southwestern Himalaya and Hindukush. The center extends over Afghanistan, Persia, central and eastern Asia Minor and Transcaucasia as well as the mountains and foothill areas of the Soviet Turkestan. This is the basic center of origin for many types of important cereals such as soft wheat, rye and many leguminous plants – peas, beans and lentils. There, a large number of vegetable types also appeared – turnips, carrots, etc. – as well as forage crops such as alfalfa, *Trifolium resupinatum*, and so on. It is also the general center of origin for the majority of fruitbearing trees and bushes: grapes, pears, apples, apricots, pomegranates, quince, sweet cherries (*Prunus avium*), etc. It is also the center that gave rise to fruit growing.

Southeastern Asia, in particular eastern and central China, constitutes the second world center of origin. This extensive center of agriculture, characterized by a very intensive kind of agriculture, is still inadequately studied, but sorghum and millet were initially grown there. This is also the center of origin of many vegetables and the focus of origin of the many fruit and citrus crops of eastern Asia.

The third independent Asiatic center is located in northeastern India and the adjacent parts of Indo-China and Burma, and gave rise to many tropical crops, mainly rice. It is, at present, possible to find a great diversity of wild rice in these countries and to follow the transformation from wild into cultivated types of rice.

We consider the countries situated along the coasts of the Mediterranean to be the fourth world center. This was also the center of agriculture of the ancient peoples, the Etruscans, the Aegeans, the ancient Hebrews and the Egyptians. As our research has demonstrated, these were also important areas for fruitbearing trees such as olives, figs and carob trees. A number of original forage crops also developed there: *Trifolium alexandrinum*, *Hedysarum* species and the lentils of Spain.

The fifth center was established in Abyssinia. This small country, not of much interest to archeologists, turned out to be one of the most ancient independent centers of agriculture in which hard wheat, barley and a number of endemic genera such as, e.g., *Eragrostis* and *Eleusine* appeared. This center is still character-

Fig. 1. At the Second International Congress of Soil Science: Yu. M. Shokal'skiy, N. I. Vavilov and G. Schantz in the hall of the Soil Science Institute of the Academy of the Sciences, U.S.S.R. (Leningrad, July 1930.)

ized by a primitive agriculture and a striking variety of crops is grown. Nowhere else in the world can one find such a diversity of wheat and barley as within this small territory. It is also the most basic center of agriculture in the Old World.

The sixth center is in the New World. This center of origin of cultivated plants is situated in Central America but, in particular, in Mexico and in South America in Colombia, Peru and Bolivia as well as Chile and the adjacent parts of the mountainous areas of Brazil.

A remarkable fact was established by our expeditions: the process of type-formation in the New World is comparatively narrowly located. At present we know of not less than eight small, independent centers in South and Central America distributed within the general area. Special genera and species of cultivated plants in all their initial diversity are characteristic of each of these. It is of great interest to mention that such a differentiation agrees with the independent centers of agriculture and civilizations in ancient America that developed later. These are the regions where the civilizations of the Incas, the Chibchas, the Mayas and the Araucarians were concentrated.

The areas listed above are the six basic centers in the world where types of cultivated plants developed and which, simultaneously, appear to be the main regions for an independent development of ancient agricultural civilizations.

In order to establish these centers, considerable efforts were necessary, but also extensive studies of an enormous number – many hundreds – of cultivated plants according to the method of differential taxonomy.

As can be seen from what is stated above, the portion of the world where the centers of origin of cultivated plants are located is very small. According to approximate estimates, the initial centers of type-formation occupy roughly one-thirtieth of the landmass. It should be noted that the areas in question are situated mainly in mountains or in the foothills of mountains. The centers of origin of cultivated plants and, simultaneously, the ones of primitive agriculture are located close to the most important mountain areas of the world, in the foothills of the Himalayas and Hindukush, in the mountain massif of China, in the Cordilleras and in the mountains of eastern Africa. Thus, these areas not only represent potentials of cultivated plants but appear at the same time as centers for the development of agriculture, confirmed not only by the presence of the basic elements of agriculture, i.e. cultivated plants, but also the ancient constructions, built by the farmers and the primitive technology of the agriculture such as the primitive agricultural tools, which are still characteristic of the areas in question.

The hypotheses, launched by some authors, stating that the sources of agriculture should be sought in Syria and Palestine, where wild wheat is found (incidentally, wild wheat was also found in Armenia last year!), are absolutely false. They do not take into consideration the fact that the present level of our botanical and agricultural knowledge clearly indicates that agriculture developed within different areas.

We do not doubt that even countries with ancient civilizations, such as Mesopotamia and Egypt, are not the initial areas where agriculture arose because, with respect to the diversity of the plants cultivated in these countries, there are traits borrowed from settled areas in the mountains or the nearby foothills. As far as the diversity of types is concerned, Egypt and the interriverine Mesopotamia were definitely poorer than the adjacent centers in the mountains of Abyssinia, Asia Minor and Armenia. It is entirely natural that control over the lowlands along the Nile, the Tigris and the Euphrates occurred at a much later stage during the progress of mankind. In order to master the Nile, a collective effort of an entire population was necessary just to regulate the utilization of its water. It is also completely natural that different tribes, speaking different languages and fighting each other, preferred the safety of isolated mountain areas rather than open valleys and plains. People were also attracted to the mountains because of their healthy air. Primitive people, practicing agriculture still avoid settling in lowlands.

From the present state of our research, the role of the mountains and foothill areas as basic initial areas of agriculture has become all the more evident. This is, however, not valid for all mountain areas of the world. In spite of thorough research in Tien-Shan last year, we were unable to discover any independent agriculture in that area. In the mountains north of Himalaya, the original elements of cultivated plants are entirely absent while at the same time the foothills of southern Himalaya are the site of powerful processes of type-formation.

The regions mentioned above are still inadequately investigated but appear to be the most amazing ones on our planet, since that is where agriculture first developed in the remote past.

If, while not looking too deeply back through the centuries, we attempt to draw a map based on the historical data available concerning the distribution of agriculture before the New World was discovered and compare it to a modern map of agriculture, we can see a considerable difference between them. [A map of world agriculture before the discovery of the New World was presented at this point of the speech.] On the basis of this map, it is evident that in the remote past only small areas of arable land were utilized. Now the populations have some 900 000 hectares under crops, each year plowing up about 650–700 million hectares, which amounts to 5% of the land mass; yet, not long ago not more than 1/5th of the area now controlled was utilized. When comparing the maps of the present agriculture with that from the period before the discovery of America, we can see that the basic agricultural arable land of North and South America – Canada, the U.S.A., Argentina and Brazil – as well as that of Siberia, is of a quite recent origin. In the European parts of the Soviet Union, the agricultural control of the northern Caucasus, the steppes below Novorossiysk and the eastern European part of the U.S.S.R., as well as its northern parts, started relatively recently.

When comparing the maps of agriculture during the fifteenth century and those of the twentieth with the soil map put together by Academician Glinka, it becomes evident that the utilization of the main fertile areas in the world started not very long ago.

It can be seen when comparing the soil map with the original centers of agriculture in the world that *these centers were situated far away from the areas with the best soils*. There can be no doubt that the black soil – all this enormously fertile soil at present extending over Ukraine, eastern Siberia, northern Caucasus and parts of North America – was put under the plow only during the eighteenth and nineteenth centuries. A comparison between the soil map and maps of agriculture in the past and in the present confirms that all the potential opportunities for agriculture are still not fully utilized and that only during the last century have the majority of the world's resources become utilized.

We do not doubt that the enormous agricultural area, now concentrated in Canada, the U.S.A., eastern Siberia, the northern Caucasus, Argentina, Brazil and Australia, is the result of Man's activities only during the nineteenth century.

Until recently, soil resources were badly studied on a systematic base. *Up until the most recent time mankind did not rationally utilize the resources of the planet. Even information about soil resources became available only recently and a soil map of the world has existed only for a few years.*

There are enormous opportunities for agriculture. Until recently, the selection of cultivated plants depended mostly on chance. There was no system for the utilization of the wealth of types found in the initial centers of origin. As we have seen, the worldwide potentials for agriculture are concentrated within such countries as Abyssinia, northern India, Peru, Bolivia, Chile and Mexico, but we still use the results of only natural selection rather than a rational breeding of crops.

The soil resources of the world are not sensibly utilized. One of the next tasks facing our country is the study of new regions and the management of new areas: i.e. control over the 95% of arable land not yet utilized for agriculture.

Finally, a very basic fact is that the centers of agriculture were situated in mountains and in foothills of mountains, which, I stress, occur within a broad belt around the world. In the ecological system of mountain areas, on the soils of these sites, we can encounter combinations of conditions approaching those found essentially in the lowland areas of Europe, Asia and America.

A comparison between the soil map of the world and the maps of agriculture of the past and present demonstrates what enormous opportunities and what lofty tasks face the agricultural sciences. Only joint studies by all nations, such as can be done by an international congress, will allow us to approach a solution to the important problems and to enable mankind to utilize rationally the natural resources of our planet and the wealth of the plant kingdom.

Soviet science and the study of the problem concerning the origin of domesticated animals

THE SOVIET COUNTRY, which occupies an enormous area of Asia and the Caucasus, borders upon an important part of the world with respect to the speciation of domesticated animals. It especially includes territories where the basic type-forming processes of the most important species of domesticated animals took place.

The native land of the horse and the camel is in Central Asia. The Caucasus and the mountain areas of Inner Asia and Kazakhstan are exceptionally rich in primitive genera of sheep and goats. Wild pigs exist in many areas of the U.S.S.R. Our peripheral republics border upon the native land of horned cattle. Reindeer husbandry was initiated in our Asiatic and European northlands and was introduced from there to America.

Thus, as stipulated by the geographical position of our country, the responsible role of studying the important evolutionary problem concerning the origin of domesticated animals becomes the duty of Soviet sciences both within our own territory as well as in the Asiatic countries contiguous with it, since no other single country except the Soviet Union offers such exceptional possibilities for studying the dynamics of the evolutionary processes of domesticated animals, i.e. the key to control over these animals and to an establishment of their type-forming processes.

While West European and American zoologists and cattle breeders can approach the study of this interesting problem only by means of odd museum specimens, the Soviet scientists can tackle it in all its width and breadth.

Anybody who has traveled in Transcaucasia, Iran, Central or Inner Asia is more than familiar with the multiformity of the types of the indigenous domesticated animals. A striking diversity of breeds occurs there both with respect to breed, color of coats, shape and form of horns, pelts and constitution. It is more than true that among the characteristics present in this striking diversity, there are many valuable ones that can be utilized for breeding.

Soviet science now approaches the problems of cattle rearing, improvement of breeds, intraspecific hybridization and a very wide application of artificial

Lecture presented at the opening of a conference in the Academy of the Sciences of the U.S.S.R. concerning the origin of domesticated animals, March 25–27, 1932. First published in *Priroda* [*Nature*], nos. 6–7, pp. 539–546, 1932.

insemination on a scale never witnessed before. Last year, the National Commissariate of Agriculture opened an All-Union Hybridological Institute in Ukraine, at the base of Askania-Nova. The task of the Institute is, among others, to carry out extensive experiments concerning a wide-ranging hybridization of animals. From the dilettantic beginnings made by Falts-Fein [a rich landowner in southern Ukraine, who founded the Askania-Nova Natural Reserve], we want to proceed to systematic research for improving the breeds of domesticated animals. On the steppes of Ukraine, with their exceptionally favorable conditions for the domestication of wild animals within the enormous nature preserves that cover hundreds of thousands of hectares that have now been set aside by the Soviet authorities, it should be possible to conduct extensive research on the process of domestication and hybridization on a strictly scientific basis, with participation by zoologists, geneticists and animal breeders.

At present, our animal breeding enters into a new era of extensive and systematic improvement of the breeds. The collectivization of the farming industries and the building up of major state-operated industries allow an organized approach to experiments toward the improvement of breeds on an unprecedented industrial scale. As has never before been possible, Soviet scientists can now execute a colossal experimental work. The diversity of conditions in our country, the necessity for different types of animal within the various breeds that are used for intensive and extensive farming, make the problem of studying the racial composition of domesticated animals in our country and those bordering upon it exceptionally realistic.

Systematic research on the interspecific variability of different species and the evaluation of extreme variants of the most valuable farming characteristics also allow us to develop the breeding work. Studies of the racial complexes of domesticated animals on a global scale will provide us with the basis for our own breeding program. The questions concerning the original material, knowledge of the intraspecific composition and the subsequent intraspecific variability constitute the fundamentals of breeding work.

The advances in genetics during the last couple of years have opened up new possibilities for learning about type-forming processes and for further deepening our understanding of the problem concerning the origin. In this sense, domesticated animals provide us with an exceptionally favorable object, where every new fact and every regularity can be quickly utilized by breeders.

We can approach a scientific classification and taxonomy of domesticated animals only on the basis of a fundamental knowledge of the racial composition both of the present West European and American breeds as well as of the primitive types of domesticated animals.

The problem concerning the origin of domesticated animals is similar to that of cultivated plants and is associated with the history of mankind: it is a part of the history of the materialistic civilization. Many historical problems can be understood thanks to the interaction between people, animals and plants. Agriculture and animal husbandry are the basic professions by which the great majority of people still live. The destiny of the people is interlinked with that of cultivated plants and domesticated animals. We have been able to establish the fact that the geography of the primitive civilizations in the world agrees to a major extent with the geographical distribution in the world of primitive cultivated plants. A number of historical problems can be solved only by far-reaching research on domesticated animals and cultivated plants. These play an enormous role in the history of materialistic civilizations.

The problem concerning the origin of domesticated animals needs extensive zoological scholarship and a knowledge of both taxonomy and zoogeography to understand it. A number of problems can be solved only by means of paleo-zoology.

Archeologists and philologists can offer valuable assistance towards understanding the historical process of dispersal as far as domesticated animals and cultivated plants are concerned. Very important information for research on the problem concerning the origin of domesticated animals can be furnished, e.g. by a study of the arts and ornaments of the Cretan civilization.

What has been stated above illustrates the complexity of questions that are basic for and associated with the problem concerning the origin of domesticated animals. That is why this problem is of particular concern to the Academy of the Sciences of the U.S.S.R. and that is why this problem should be properly studied in all its width and breadth by the Academy of Sciences since only this body includes within its set-up all the basic disciplines necessary for this.

We can approach this interesting problem properly only by mobilizing scientists from the various disciplines. It can certainly be solved by a team of Soviet scientists.

It is hard to accept it as a fully normal event, when an American expedition conducts studies behind our backs on the problem concerning the origin of domesticated animals within our territory and in friendly countries contiguous with ours. Our own scientific collective is adequately capable and powerful enough to solve such tasks. We think this should be a truly and fully cooperative effort by Soviet and foreign scientists together. However, it is also necessary not to forget that the high level of Soviet science, its activities and its major role within the world of science is a powerful political force.

In the past, the Academy of the Sciences was not alien to the problem concerning the origin of domesticated animals. The excellent work done by A.F. Middendorf constituted the beginning of serious studies of domesticated animals. His *Investigations of the Contemporary Conditions of Cattle Rearing in Russia* (1884–1885) is still of major interest. It is impossible not to mention the systematic and taxonomic papers concerning a study of the fossils of domesticated animals, written by D.N. Anuchin. In particular, N.V. Nasonov at the Zoological Museum of the Academy of the Sciences, has paid much attention to a study of the genus *Ovis*.

During the last seven years, the Genetics Laboratory of the Academy of Sciences has conducted investigations on the racial composition of the domesticated animals of Mongolia, Kirgizia, Kazakhstan and Turkmenia. These investigations have covered almost a third of Asia. Thereby, a large material was collected for the first time concerning the variability of the domesticated animals of Asia. The primitive stages could be elucidated. A multitude of facts of great importance were thereby revealed, which need to be generalized and synthesized. In the light of our present knowledge of genetics, we do not doubt that these facts contribute much toward an explanation of the process giving rise to the various breeds.

Professor E.F. Liskun has made great efforts toward an investigation of the breeds of domesticated animals. The collection of skulls that he has assembled is at present located in the Timiryazevskiy Agricultural Academy and represents one of the richest collections in Europe.

We should also mention the valuable work done by A.A. Brauner, dealing with the problem concerning the origin of domesticated animals. His explanation of the role of the ecological factors for the geography of the initial wild species of animals is of particular interest. His development of the interpretation of the natural winter pasture as a geographical factor for past dispersal of the ancestors of the wild horse is of exceptional importance for us.

Very valuable investigations of domesticated cattle have been conducted during the last couple of years in Armenia and Georgia, i.e. countries with ancient civilizations, where the original endemic breeds are still preserved. The research done by Professor A.S. Serebrovskiy concerning the geography of chicken genes in the Caucasus and especially his research on the 'genogeography' of chicken are of great interest.

During the last couple of years, with advances in breeding, genetics and science all over the world, a great interest has arisen concerning the geography of domesticated breeds of cattle. A number of important papers have been produced, devoted to the origin of domesticated animals. We need only mention the books by Klatt, Adamets, Nachtsheim and Steegman. The remarkable publication by Hehn, *The cultivated plants and domesticated animals and their dispersal from Asia into Greece and Italy and the rest of Europe*, has appeared in a new, amended edition. By the way, this Victor Hehn was an Assistant Professor at the University of Dorpat and Librarian at the Petersburg Public Library.

What has been stated above has compelled the Academy of Sciences of the U.S.S.R. to arrange, on its own initiative, this convention of specialists for a conference, the tasks of which are to mobilize an interest in the problem concerning the origin of domesticated animals, to register our forces and to make salient remarks concerning the themes associated with the problems of the origin of domesticated animals. One of the objectives of this conference is to stimulate the necessity for reviewing all the work produced by Soviet scientists and to propose a suggestion for the establishment of a section dealing with the origin of domesticated animals in a future museum of evolution. The creation of such a museum could stimulate the mind, advance matters concerning the research on the evolution of domesticated animals and be of an exceptional pedagogical importance.

It was not by chance that Darwin was a regular visitor at the Museum of Natural History in London, the location of the fundamental collection of museum exhibitions displaying the breeds of domesticated animals. It is hard to present a better picture of the striking amplitude of variability with respect to breeds of dogs than that located in the British Natural History Museum, which demonstrates the enormous possibilities opened up by breeding.

On the whole, the problem of the origin is of interest to us primarily from a dynamic point of view, seen from the angle of socialistic cattle rearing. We want to control the animals, learn to create new types based on existing breeds in agreement with the needs of the economy. However, for this we must approach a solution to the actual problems associated with breeding: knowledge of the history and an intrusion into the historical method are necessary for understanding all the problems. Just as in the case of plants, the breeding of domesticated animals represents, in essence, an experimental kind of evolution, but historical knowledge of the process of evolution is necessary to understand this and to be able to control it step by step.

Soviet scientists must consider our exceptional opportunities for solving this complicated problem and follow the path of complex research while all the time keeping in mind the final objective: the affirmation of a materialistic understanding of the evolutionary process and the creation of a scientific basis for practical breeding.

Problems concerning new crops

OUT OF THE 13.6 billion hectares of land, circa 950 million, or 1/15th (7%) of the world's landmass, are utilized for agriculture. Within the Soviet Union, only about 7% of its landmass were also until recently used for agriculture. Enormous territories are still available for this purpose and even very fertile areas remain unused. It is not more than 100 years since farmers started to grow crops on the fertile plains of North America, Argentina, Australia, Siberia and the northern Caucasus.

A study of global agriculture clearly shows the disarray that is widespread on our earth, even from the point of view of an elementary utilization of its natural resources. The control over territories well suited for agriculture is still at a primitive stage. The period when people settle down in a rational manner in accordance with the existing resources of the Earth is far from over. One half of our globe's inhabitants still live in the southern tropical and subtropical areas of Asia where, in all likelihood, the cradle of mankind was located. Enormous areas of the most fertile land in South America, Central America and Tropical Africa are still hardly used at all. At the same time as the southern areas of China and India are overpopulated and the allotment of land in southern China has reached the extreme limits of what is possible for the existence of a farming population, wide-ranging areas, whole continents in the western hemisphere, are still not utilized by any people. The barriers erected during the last hundred years as a result of imperialistic colonization politics and the competition between the capitalistic countries have halted the farm-steading of people. In spite of the great technical discoveries of the nineteenth and twentieth centuries and the resulting development of new means of traffic and communication, the capitalistic system not only did not promote a speedy utilization of the resources for global agriculture but even impeded and hindered it. The most important scientific discoveries have been utilized primarily for military purposes. The gulf developing between the technical and the capitalistic systems, of which Marx wrote, is growing into a threat to human progress right under our eyes.

World history, in the sense of a systematic organization of the world for the purpose of the interests of the working populations, has only just begun.

Although the process of distribution of populations over the globe and the

First published as *Problema Novykh Kul'tur*, Moscow-Leningrad, 1932 utilization of territories for agriculture is at the starting stage, the plant resources of the world have still hardly been touched upon. Global agriculture is entering a stage when there will be a rational utilization of the riches provided by the plant kingdom.

Out of the 950 million hectares (including fallow land) used for agriculture globally, about 750–800 million hectares are used annually for field crops. A significant portion of this is occupied by cereal crops: wheat, rice, maize, oats, rye, barley, millet and sorghum.

According to worldwide statistics in 1930:

150 million hectares were used for wheat,
120 million hectares were used for rice,
80 million hectares were used for maize,
60 million hectares were used for oats,
48 million hectares were used for rye,
40 million hectares were used for barley,
25 million hectares were used for millet,
20 million hectares were used for sorghum.
Total 543 million hectares

A study of the statistics concerning cultivated plants demonstrates the prominent role played by a small number of these plants. Only 11 crops occupy an area larger than 15 million hectares each. In addition to the eight cereal crops mentioned above, cotton, potatoes and soybeans should be included here. The development of the last three crops dates back only to the beginning of the twentieth century. The specific composition of field crops, both within our own country and beyond its borders, is characterized by monotony. All the 11 crops enumerated (but not including soybeans, potatoes and cotton) belong to the oldest cultivated crops in the world, i.e. plants that were already grown at the dawn of agriculture many thousands of years ago. Important developments in technical, oil-producing and sugar-producing plants as well as so-called 'colonial' crops began only during the last couple of decades.

In addition to the eight cereal crops mentioned above, the following plants occupied the largest areas under crops and were cultivated all over the world in 1930:

Cotton	36 mill. ha.	Sunflowers	6 mill. ha.
Potatoes	20 mill. ha.	Peanuts	5 mill. ha.
Soybeans	15 mill. ha.	Beans	4 mill. ha.
Flax (linseed)	10 mill. ha.	Peas	3 mill. ha.
Chick-peas	8 mill. ha.	Sugar beets	3 mill. ha.
Sugar cane	6 mill. ha.	Rapes (turnips)	3 mill. ha.
Olives	6 mill. ha.	Sesame	3 mill. ha.
Coffee	6 mill. ha.		

All other crops occupy areas less than 2 million hectares each, however, with the probable exception of apples, about which there is no information.

It is not very long since global agriculture had mainly a consumer character, satisfying only the most urgent and elementary needs of nourishment. The level of need limited questions from farmers concerning exceptionally nutritional plants. The composition of the fields in our country and – to a considerable extent – also that of other countries, still reflects thousands of years of routine.

As elucidated by our research, a number of plants, including rye and oats, were taken into cultivation as a result of natural selection. In the south, in Inner Asia and Transcaucasia, rye is not grown but appears as a serious weed, infesting crops of wheat and barley. When crops of wheat were moved northward or into high elevations, the frost-hardier weed – the rye – outcompeted the wheat and became, thanks to natural selection, the predominant plant in the fields. Cultivated rye, oats and a number of oil-producing plants developed from weeds.

The development of industry at the end of the nineteenth century and during the twentieth has revolutionized agriculture and modified the composition of field crops. In addition to the important plants besides wheat, rye and barley, we can now count cotton, sugar beets, sugar cane, flax and linseed, maize, potatoes and several 'colonial' plants.

The unprecedented and rapid development of the socialistic industries and the rural economy in the Soviet Union has, in turn, posed a problem for a general revision of crops and strains grown and for new investments in crops. The reconstruction of the rural economy on the basis of collectivization and the development of a state-controlled management with widespread mechanization and use of chemicals is opening up exceptional opportunities for agriculture in the Soviet country. The Soviet Union now holds a key, of which the rest of the world is so far ignorant. In front of our own eyes, the necessities that are cultivated for our socialistic country now grow at rates incomparable with even that of a short time ago.

New aspects of the history of global agriculture are opening up. The problem of how to utilize the resources of our Earth has been posed anew. A radical revision of our fields, a change of the strains grown, a rational utilization of arable land and an advancement of agricultural crops now take place in a new manner and on a definitely new scale.

The concept of new crops

By the term 'new crops' we understand a wide area of measures associated with new plants and strains. By 'new crops' we mean not only definitely new kinds of plants, not used before, but also old plants, well-known to us or used beyond our borders, although previously only slightly distributed in our country in spite of the fact that, in practice, they deserve a wider introduction. According to our understanding, the problem concerning new crops is definitely associated with a wide application of plant resources from all over the world and with a mobilization of the wealth of plants growing on our planet.

Research conducted during the last couple of years has revealed the presence of enormous resources of plants in various southern mountain areas within old agricultural countries. The lack of communication between peoples and the routines prevailing within agriculture created barriers for the utilization of experiences gained by different peoples and in different countries. A large number of valuable species of cultivated plants grown in various countries were, until recently, extremely badly utilized in the U.S.S.R. This was the case, e.g. with the American maize, African sorghum, Sudan grass, American sweet potatoes, early-ripening Japanese strains of rice, soybeans, tea, ramie [*Boehmeria*], hemp-mallow [*Hibiscus cannabinus*], citrus fruits, and so on.

As part of the problem concerning new crops, we also include the question regarding the introduction of crops into new areas and especially the problem of how to maximally develop crops of cotton in new, non-irrigated areas. Here we also include the problem of how to transfer crops toward the north in connection with the northward extension of agriculture.

The maximum utilization of our valuable areas of both humid and dry subtropical land also belongs to the sphere of problems concerning the application of new crops. Our subtropical areas are limited; therefore we must utilize every square inch thereof in a rational manner when an area is suitable for growing raw materials from plants that are better than those imported by us from abroad.

We should also mention here the utilization of the extensive, presently desert-like areas of 'worthless' land for crops, i.e. the deserts of Inner Asia, Kazakhstan and the southeastern portions of the Soviet Union. Control over desert areas can be associated with the discovery of new crops for them.

A general revision of the species and varieties of our cultivated plants is also included under the term of 'new crops'. The wheat of the future will, indeed, be different from the present kind. Plant breeders have already started a change, not only of the varieties but also the species of wheat used; interesting perspectives are opening up in this direction. The method of 'vernalization', i.e. an abbreviation of the vegetative period, will no doubt make it possible to utilize in a more extensive manner some valuable species and varieties of wheat, which so far could not be cultivated within growing seasons like ours. The experiments with vernalization with respect to the worldwide assortment of wheat and barley, made this year (1932), have clearly demonstrated that it is definitely possible to change the behavior of species and varieties and to make new ones (e.g. *Triticum turgidum*) grow in the southeastern portions of the Soviet Union, i.e. species that are unable to grow (without vernalization) under the ordinary conditions there with hard winters and dry summers. The concept of crops cannot be distinguished from that of strains. The success of a new crop that is introduced usually depends on the selection of suitable strains. The problem of new crops is therefore inseparably associated with highly developed plant breeding and an application of hybridization for selecting strains.

Experiments on Java, trebling the production of sugar cane during the last couple of decades by hybridization of old Chinese strains with wild sugar canes that are resistant to diseases, demonstrate what enormous changes for the better can be made within the plant industry, what transformation many of our crops can undergo and how something new can be created out of something old. The scientific alteration of cultivated plants is, in essence, only in its infancy but the perspectives that already now have opened up for interspecific hybridization promise to produce major changes in the varietal and even the specific composition of our ordinary but important cultivated plants. Contemporary genetics dares to create new species and even new plant genera.

As we understand it, the problem concerning new crops means, thus, to gain control over the global resources of plants, and, if possible, to utilize them for the socialistic agriculture and production, to organize quickly a widely practiced introduction of new crops as well as to exchange less valuable crops for more valuable ones and, finally, to establish new crops in new areas.

Major problems face Soviet scientists concerning the utilization of the wild flora, both within the borders of the Soviet Union and beyond them, in connection with its application as crops of new and valuable plants. It is necessary to develop the work in this direction without delay. As we shall see below, the main attention must be focused on research on new plants with respect to a number of crops.

At the very beginning it is necessary to mention that, as far as a part of the basic group of plants is concerned, our greatest interest is primarily the utilization of all the experiments made worldwide and available from such countries as the United States, Japan, Algeria and the West-European countries but also a broad enlistment of the assortment of material in the ancient agricultural countries of the world that is available for practical plant breeding.

The role of new crops for agriculture in other countries

In the past, the problem of new crops was indeed not as broadly established in any other country as it was in the United States of America. With the exception of some crops such as maize, tobacco and cotton, all the basic crops of the U.S.A. are introduced from other countries. As early as during the seventeenth century the first European colonizers brought with them European cereals and other plants. All the fruit, vegetable and melon crops of the U.S. are based on such plants, which were introduced from Europe, Japan, China and India and not known in the U.S. before the seventeenth century.

In 1898, a special division for the introduction of seeds and plants was established at the Department of Agriculture in Washington in cooperation with the Bureau of Plant Industry. It developed active research on plant resources all over the world. The 'hunters of new plants', sent out from Washington, went to all corners of the world collecting an enormous varietal material and describing not less than 100 000 specimens. Among the American scientists we can mention David Fairchild, Walter Swingle, the Danes Hansen and Bess, the Dutchman Frank Meyer, Cook, Kern, Harland, Rook, Wilson, Popenoy, Dorset, Morse, Ryerson, Westover and Carlton. A whole series of European, Inner-Asiatic, Mexican and South American epithets of plants that are now cultivated in the United States are associated with these names. The history of these expeditions has, unfortunately, not yet been written but would brilliantly fill pages about the research, as far as the plant resources of the Earth are concerned.

Among the new introductions to the United States we can mention strains of spring and winter wheat brought in from Russia and now considered some of the best ones in America. The hardiest strains of American wheat originated from our Ukrainian and Crimean winter wheats. The Washington expeditions brought back valuable strains of barley ('Trebi' and 'Club Mariot') from Asia Minor and Egypt. These are now widely cultivated in dry areas. New types of rice were brought from Taiwan. Sweet clover [*Melilotus*] was obtained from us. From Africa came new black kinds of sorghum – among those, the famous 'Feterita' – brought from Egypt in 1916. Sudan grass was introduced from Southern Sudan in 1909. Lately a forage crop, *Lespedeza* [Japan clover], has been introduced from Korea. A number of valuable cotton types were discovered by Cook in Mexico, the native land of the upland cotton. During the last couple of years, the Americans have also adopted date palms from Mesopotamia, the same kind as grown in the Saharan deserts. They are now widely cultivated in Arizona and southern California. For the greening of American cities, a great role is played by foreign species of trees and ornamental herbs. In this respect, the Americans make really good use of all the world. Wilson's expedition to China gathered the majority of the valuable decorative trees and shrubs now taken into cultivation.

Canada has obtained all her crops from Europe but mainly from Russia. Wheat, rye, oats and barley, the principal crops of that country, were adopted from our country. It can be stated that, with the exception of maize and cotton, all the agriculture of North America is based on 'adopted new species'.

On the whole, the entire agriculture of Australia, Argentine and Brazil is built on 'new' crops. The coffee-trees of Brazil were introduced from Africa and their citrus fruits from southeastern Asia. All the field crops of Australia and Argentina consist of strains brought from Europe or the U.S.A. Such countries as Uruguay and Argentina have no local crops. All their agriculture is based on plants adopted from Europe and the U.S.A.

A French botanist, Trabut, has done modest but exceptionally productive work, bringing in new plants to Algeria and Tunisia; the present Algeria is obliged to him for the majority of its new crops. When entering Algeria, it is hard to recall the aspect of northern Africa not long ago. Its landscape includes at present many 'foreigners', such as various kinds of exotic, fast-growing Australian eucalyptus, acacias, casuarinas, Peruvian pepper trees and Mexican cacti and agaves.

The indigenous flora of eastern Asia is exceptionally rich. Nevertheless, the practical farmers of Japan have introduced the majority of their new crops. In his interesting book, Professor Sirau Kotaro (1929) mentions some 100 new plants, brought to Japan from other countries. Even the basic crops of rice and tea, cultivated at present, were introduced into Japan from China.

Striking facts concerning the introduction of new crops have been revealed on the basis of global experiments. Thus, e.g. the introduction into cultivation of fast-growing Australian eucalyptus trees into Abyssinia at the end of the nineteenth century furnished them with an opportunity finally to create a fixed capital within that country. Until then, they had to move their capital to new forested areas whenever the slow-growing native trees around it had been destroyed.

All the wealth and all the enormous productivity of Java and Malaya is based on new, introduced plants: 90% of the worldwide production of rubber is at present grown in Malaya, Java and Ceylon. This rubber was obtained from plantations that were established 20–25 years ago on the basis of seeds of wild rubber trees [*Hevea*] brought from Brazil.

The quinine trees [*Cinchona*], adopted in Java at the middle of the last century from South America (Ecuador, Peru and Bolivia), presently produce 93% of all the quinine needed for the inhabitants of our world.

The agriculture of Egypt is based on cotton, brought from South America. The so-called 'Egyptian cotton' is no doubt of American origin and was taken into cultivation in Egypt at the end of the eighteenth century.

It should also be mentioned that our own cotton industry is based on American upland strains. Our humid subtropical areas have adopted their cultivated flora (citrus fruits, tea, bamboo, ramie, cryptomerias, etc.) from abroad. The well-known strain of maize, 'Ivory King', presently occupying a large area in the U.S.S.R., originated from Indian villages in the northern United States.

The scientific basis for the introduction of and the search for new crops

BASIC GEOGRAPHICAL REGULARITIES WITH RESPECT TO THE DISPERSAL OF PLANT SPECIES

The research at the All-Union Institute of Plant Industry, conducted during the last couple of years, has revealed a number of regularities concerning the geographical dispersal of the plant resources of the world, which determines to a considerable extent in which direction it is necessary to proceed when looking for new plants, new species and new varieties.

The botanical exploration of the world is still far from complete. Indeed, botanists do not know more than one half of all the plant species existing in the wild. The enormous continents of South America and Africa, as well as India, China, Indo-China and Asia Minor are extraordinarily badly known. There are still large areas where no botanist has ever set foot. Even within the borders of our own Caucasian and Inner-Asiatic republics, hundreds of new species could still be discovered within the next couple of years, according to the opinion of competent botanists. There is an extraordinarily badly studied wild vegetation in the countries contiguous with our own. The wild flora of our neighboring Afghanistan has hardly been touched upon and the same goes for that of Persia and Asia Minor, which are especially rich in plant species. An extensive field is still open to botanists. The most urgent task for Soviet botanists appears to be an exhaustive study of the floras of Inner Asia, the Caucasus, the Far East and the adjacent areas. Only by means of such studies will we be able to create a serious, scientific basis for the search for plants for new crops.

However, even from the incomplete data we have as a result of our recent botanical studies of the world, facts are revealed concerning the great importance of knowing the geographical localization of the type-forming processes that are just established. Phytogeography definitely demonstrates that specific diversity is not regularly distributed over the world. It is possible to distinguish a number of areas in the world, characterized by an extraordinary variety of species. The floras of southeastern China, Indo-China, India, southwestern Asia, tropical Africa, the Cape colony, Abyssinia, Central and South America, Mexico and that of the countries along the coasts of the Mediterranean are all distinguished by an unusual concentration of specific diversity. On the other hand, the northern countries, Siberia, all of Central and northern Europe and North America are characterized by a poor composition of species. Central Asia has a surprisingly low number of species. The number of species in our country definitely increases in the direction toward Crimea, Transcaucasia, the mountain areas of Inner Asia, Altai and Tien-Shan. Although there are up to 6000 species of higher plants in the mountains and foothill areas within the borders of the Soviet Inner Asia – 6000 to 7000 species in the Caucasus and up to 7000 species in Iran (according to estimates by E.G. Chernyakovskiy) – we can usually count the species in the extensive European and Siberian areas only in the hundreds. No doubt, many more species will still be found in Inner Asia, Transcaucasia and Persia.

In some areas of the world, the concentration of the specific diversity appears to be strikingly confined. Thus, e.g. the small republics of Central America such as Costa Rica and El Salvador with a surface of only 1/10th that of the U.S.A. do not yield in numbers of species to all of the rest of North America, i.e. the United States, Canada and Alaska.

During the search for arable species within our own country, the fact that there is a concentration of specific diversity in the mountain areas of Inner Asia, the Caucasus and parts of Altai and Tien-Shan as well as in the Far East must be kept in mind and we must direct our attention mainly to these areas. Thus, e.g. the remarkable rubber-producing 'tau-saghyz' [*Scorzonera tau-saghus*] is found only in the Karatau mountain range of Kazakhstan and nowhere else, in spite of special searches for it in Inner Asia and Kazakhstan.

Our research during the past years has identified, with great exactitude, the center of species formation as far as the most important presently cultivated plants are concerned. The areas where the original species and the varietal diversity of our most important plants are concentrated turned out to be located within limited territories, occupying approximately 1/40th of the landmass of the Earth (Vavilov, 1929, 1931c).

The development of species and the original type-formation of major cultivated plants seems to be located mainly within the foothills and the mountains of subtropical areas. The centers of origin of the majority of the plants that are cultivated at present, belong at the same time to botanical areas displaying powerful speciation processes. It is evident that primitive man went to these areas, which are rich in specific composition of plants and contain a large number of edible plants. Here, in the foothills and in the mountains, where it is healthier and more agreeable for small groups of people to settle, primitive civilizations developed. There, many plants, which are still not well known by Europeans, were first taken into cultivation.

Soviet expeditions have, during the last couple of years, been sent out to such areas as have a maximum concentration of specific and varietal diversity of cultivated plants in order to collect an enormous amount of material according to a definite plan from Asia, Africa and America. In contrast to the expeditions from Washington, Soviet ones proceeded according to a plan, outlined on the basis of the theory concerning the origin of the plants and a knowledge of the geographical centers of their origin. This has made it possible to conduct the research within a considerably shorter period of time and in a more economic manner.

In the first place, it was natural that our attention was directed toward studies of cultivated plants and a collection of an assortment of plants of basic interest to us; however, simultaneously, everything cultivated or utilized for various purposes, including those alien to the U.S.S.R., was also studied. The Soviet expeditions, which were directed according to a plan with definite objectives, opened, in the full sense of that word, a 'global treasure' of cultivated plants and were compelled, in a new manner, to 'draw milk' from many old cultivated plants, too.

In addition to the basic geographical regularities mentioned concerning the evolutionary process, a number of facts with respect to the dispersal of various groups of species and varieties of plants all over the world should also be considered. Different floras are characterized by the presence within them of different groups of species.

While the wild flora of the European parts of the Soviet Union is very poor in species of woody plants, the Caucasus and the mountains and foothill areas of Inner Asia are very rich in species of trees and shrubs. However, in this respect even our Caucasus and the Inner-Asiatic republics are inferior to a number of other southern countries. Thus, while in all the European parts of the Soviet Union (except for the Crimea) there are about 360 species of trees, shrubs and half-shrubs, there are 370 such species in the Caucasus but about 500 in Inner Asia. According to estimates by Standley, there are more than 6000 species (including cacti and agaves) in Mexico. In the Congo there are more than 1000 kinds of valuable trees and in the Philippines about 2000, while the forests of East India have ca. 1550 species. According to research conducted by Wilson, the East-Chinese flora is exceptionally rich in trees that are valuable from a technical point of view.

Bolivia and Peru are not only the native land of potatoes but also of many other cultivated plants, especially wild tuberous plants such as oca [*Oxalis tuberosa*], ulluco [*Ullucus tuberosus*] and the tuberiferous *Tropaeolum tuberosum*. This evidently means that there are corresponding conditions for the development of tuberous species, which are usually rare in other floras.

As our expeditions have demonstrated, the countries surrounding the Mediterranean are distinguished by large-flowered, large-fruited and large-seeded varieties of both cultivated and wild plants. Wheat, oats, flax (linseed), beans and many other plants attract attention because of their large seeds. On the other hand, northern India, Afghanistan and Soviet Inner Asia are characterized by small-seeded kinds of the same species of cultivated plants.

The low mountains of Yemen ('*Arabia felix*' of the Old Romans), which are surrounded by deserts, harbor the earliest-ripening varieties of wheat, barley, flax, lentils and blue alfalfa in all the world.

Many species of trees in Australia are distinguished by fast growth (e.g. eucalyptus, acacias and casuarinas). A similar rapid growth and development of a great vegetative mass within a short time are also characteristic of many species of herbaceous plants in western Siberia and the foothills of the Altai (according to Turesson).

Eastern China turned out to be the native land of naked-grained oats, naked-grained barley and naked-grained millet.

The flora of the mountains and the foothill areas of the Mediterranean countries and southwestern Asia, including Soviet Inner Asia, the Caucasus and Persia, is very rich in plants producing etheric oils.

Knowledge of such facts when searching for particular plants shows where they can most expediently be found.

The established fact that the majority of cultivated plants of the greatest interest to us come from the mountains and foothill areas of subtropical and tropical areas is of a more decisive importance. This facilitates the transfer toward the north of plants from southern countries, since many of the tropical plants found at high altitude in the mountains are distinguished by short vegetative periods. Thus, e.g. barley, collected by us in the mountains of Abyssinia and situated close to the equator, has turned out to be able to ripen not only in the area around Leningrad, but even at the Polar Circle near Hibiny on the Kola Peninsula.

THE VARIETAL AND THE SPECIFIC COMPOSITION OF OUR BASIC CROPS

Systematic studies of the resources presented by the plant kingdom were started by the Institute of Plant Industry and have led to the discovery of a large number of varieties and even species of cultivated plants. It turned out that even the most important and oldest cultivated plants in the world were not well known by botanists and agronomists. As a result of detailed research by means of the available methods, species of many cultivated plants proved to represent complicated and active morphological systems and complexes of varietal strains, which are associated with respect to their origin with definite environments and distribution areas (Vavilov, 1931b). Expeditions have discovered new species of wheat; three such new species were found in Georgia and Armenia and a number of subspecies in Abyssinia. Some of these species are distinguished by a striking resistance to diseases and are of great interest for practical plant breeding.

Soviet expeditions also found a number of new vegetables that were previously unknown to Science. We also succeeded in establishing a large number of new varieties of many cultivated plants, starting with wheat. Thus, e.g. three times as many forms as were known prior to the Soviet expeditions have been recognized during the last eight years.

In the mountains of Bolivia, Peru and Chile, the Soviet expeditions found 13 new species of potatoes, not known to Europeans, which made it necessary to start breeding this plant all over again. Some of the potato species found turned out to be resistant to frost, others to drought; some species were also resistant to diseases. The interest aroused by these discoveries was such that the United States Department of Agriculture and the German Institute of Plant Breeding sent out their own expeditions, following in the footsteps of the Soviet ones to search for these potato species (according to information from the Director of the U.S.D.A., Ryerson, a new expedition was sent out from Washington last year, 1932, to collect samples of potatoes in Peru and Bolivia).

The history of European potato breeding shows that the present assortment in Europe and even in the United States originated from some tubers brought back by some travelers following Columbus. Until very recently, the initial specific and varietal potential from the countries of origin of these plants, i.e. Bolivia, Peru and Chile, had definitely not been used for practical plant breeding. The modest Soviet expeditions, which approached the utilization of the different plant resources of the world according to a definite plan, opened up in this manner a specific and varietal diversity of the potato plant not only for our own country but also for all the plant industries in the whole world. This illustrates that a broad approach to the problem concerning new crops is necessary, since a radical change in the routine assortment of old crops is still feasible. We have not yet reached a stage where studies have been exhausted even with respect to our common crops.

Even as late as 1931, L.L. Dekaprelevich found a new species of wheat in Georgia. It is at present hard to distinguish the problems of new crops from those concerning the old ones. The exchange of one species of wheat for another, of starchy maize for a flinty one, of bitter lupines for sweet and non-alkaloid ones, the breeding of tobacco without nicotine and of sweet clover without coumarine corresponds, in essence, to the breeding of 'new crops'.

When searching for new crops and new plants in Transcaucasia and Inner Asia, it is necessary to take into account the inexact knowledge of these mountainous areas and not only the possibility of finding new plants there, but also , simultaneously, the necessity for attentively studying the so-called 'old crops' for the purpose of discovering new, previously unknown species and varieties.

SELECTION OF CROPS AND STRAINS ON THE BASIS OF CLIMATIC ANALOGY

It is evident that when selecting species and strains for the U.S.S.R. it is necessary to take the climate and the soil conditions at their origin into consideration in order to introduce strains from areas that are more or less similar to those in our own country. Knowledge of the climate of our own country and that of the areas from where we collected the seeds is of great importance.

However, it is also necessary to state that the problem of climatic analogy is at present not regarded in the same simple manner as it was not very long ago.

Studies of the dispersal of cultivated plants clearly demonstrate the complexity of this process. Some crops and strains appear to be surprisingly universal. To these belong many forage crops, especially grasses, and many vegetables. The Danes export seeds of vegetables to many countries all over the world. For many of our areas, which are quite different from the conditions in Denmark, Danish vegetable seeds seem to be the best, according to data from tests on the strains. They grow well here both at the Polar Circle and all over the Soviet Union. Seeds of forage grasses are exported all over the world from southern Sweden. One of the best customers for Swedish seeds is New Zealand. The Swedish 'Victory' brand of oat is successfully grown in western Siberia, the Ukraine and western Europe under climate and soil conditions quite different from those of southern Sweden. The so-called 'Petkus' rye, which originated from northern Germany, grows successfully all over Germany as well as in our own country. In other words, some plants and strains seem to be really cosmopolitan.

On the other hand, the majority of foreign strains of spring and winter wheat, sugar beets and barley are, as a rule, of less interest to us even when they are used under similar conditions. Even the best standard brands in the world such as the 'Marquise' and the 'Kitchener' wheats have not attained any important distribution in the U.S.S.R. in comparison to our own brands. Many strains of this group of plants appear to be too specialized and the interest in these crops is now declining.

However, the Saratov brand of wheat (Saratov 062) has, for instance, turned out to be unexpectedly successful in the Amur region, where the climate is far from similar to that of Saratov. The barley from the mountains of Abyssinia grows excellently under the conditions existing around Leningrad, in spite of essential differences in climate and the length of daylight in our North in comparison with the short days in Abyssinia. The Abyssinian spring wheat, which is grown alongside the barley in the mountains of eastern Africa, will, however, not ripen at Leningrad and does, in general, not succeed anywhere in the European parts of the U.S.S.R. Unexpectedly, the Amur region turned out to be a better match for it. These are the results of experiments.

The amplitude of the varietal difference within the limits of species is of exceptional importance for the introduction of crops. Thus, e.g. many plants show quite a wide amplitude with respect to the vegetative period. In 1929, we found strains of maize in western China that ripened at Kharkov in 60 to 70 days. Under the same conditions, ordinary strains of maize need twice as much time. A number of varieties of maize from Mexico and South America do not sprout ears in our country even under the subtropical conditions at Sukhumi. We have succeeded in obtaining strains of barley, wheat and sweet clover from Yemen that ripen surprisingly fast and in this respect out-do all other stains. For the transfer of cotton northward all our attention was focused on research concerning early-ripening strains. The amplitude of the varietal differences in the case of this plant is exceptionally favorable. For instance, in India, strains are known that ripen in three months, while there are also perennial types.

A complete climatic and soil-related analogy concerning the needs of different varieties does not exist. The meteorological data that are usually taken into consideration with respect to introduction, i.e. temperature and precipitation, are important. However, daylength is also very essential for introductions, since it decreases in the direction toward the equator. Physiological experiments by Allard, Garner, N.A. Maksimov and T.D. Lysenko and other scientists, but also geographical experiments conducted by ourselves, show the great importance of the length of the night (photoperiodism) for the vegetation within different areas. Direct experiments with a number of root crops, especially turnips and beets brought in from the mountain areas of southwestern Asia, have demonstrated that these plants, which are biennial under the short southern day, change into annuals under the long northern days. Among these root crops, the Afghani turnip turned, in the North, into an annual oil-producing plant without forming any thick roots (according to E.N. Sinskaya).

The main native lands producing the majority of cultivated plants are the subtropical and tropical mountains and foothill areas, characterized by short daylength, which – so to speak – prevents the transfer of the plants northwards. However, in reality this is very complicated since, within the same area, much that was unexpected and not within the framework of the usual, elementary scheme awaited us. Thus, direct research conducted during the last couple of years at the Institute of Plant Industry has demonstrated that 'short-day plants' from the tropical mountains such as, e.g. some varieties and species of potatoes from Bolivia and Peru or barley from Abyssinia and Yemen, can be successfully grown as far north as the Polar Circle in areas with long days during their vegetative periods. Fruit-setting and seed-setting, as well as the formation of tubers, were especially successful even in the Far North. In other words, the potato, the origin of which is close to the equator, turned out to be very suitable far toward the north, where the process of tuber-formation and fruit-setting are able to proceed quite well.

Actually, this means that in the case of the potato the development of the vegetative mass and flowering proceed particularly rapidly under a long day while the formation of tubers starts during the long nights. Long days at the start of the vegetative period and short ones at the end of it are characteristic of the Far North. It means that there are favorable conditions both for the former and the latter stages during the development of the potato plant. According to experiments made by Allard and Garner and confirmed at the Institute of Plant Industry, short days are especially favorable for the formation of tubers in the case of both the potato and other South American root crops. The production of seed potatoes with a short vegetative period is of great importance for the crops in the Far North. Most important of all – potatoes are less subject to diseases there.

Apparently, these kinds of facts are general for a whole group of plants.

Usually, the parallelism between the vertical belts of the mountains and the latitudinal zones is unnecessarily exaggerated. The difference in photoperiodism clearly demonstrates that it is not correct to construct an elementary scheme for the parallelism between vertical and latitudinal zonality.

Allowing oneself to be guided by data concerning climate and soils when selecting species and strains in order to arrive at a so-called 'climatic analogy' is something that must be approached with much caution, while not exaggerating its importance. Thus, e.g. the hypothesis that cotton should not be able to grow successfully in the northern Caucasus or the southern Ukraine, since it is not hot enough there, has not been possible to confirm. Under the conditions of the light and sandy soils of the northern Caucasus and the southern Ukraine, which are suitable for a good harvest, cotton reaches a stage of excellent development and produces ripe 'bolls' such as was impossible during experiments under the conditions in Inner Asia, where crops of cotton usually succeed only when irrigated and rarely at regular intervals even then.

Direct experimentation is necessary in order to be able to determine with confidence the suitability of crops, species or strains.

Organization of quarantine

The development of wide-ranging introductions of new plants must proceed simultaneously with the establishment of quarantines for preventing the importation of new parasites and new pests together with the new plants. The organization of an agricultural health inspection constitutes an unavoidable task when introducing plants. Each parcel of seeds from abroad must be inspected by entomologists and phytopathologists. Infected plants must be fumigated. In the case of doubtful material, it must be sent to special nurseries for control. The establishment of special housing for quarantines is necessary. This is why the introduction of plants from abroad must be centralized and strictly controlled.

Plant resources of the world and the search for new plants to be taken into cultivation

It is now possible, up to a certain degree, to characterize the global resources of plants on Earth from the point of view of how to utilize them and how to search for new wild and valuable species on the basis of data assembled during the last 30 years by the United States Department of Agriculture and the All-Union Institute of Plant Industry.

GRAINS AND OTHER FOOD CROPS

We do not expect very much change or much new in the sense of discovering new cereal grasses. Thousands of years of tests and trials by farmers of the Old and New Worlds have apparently taken from Nature what is most valuable. Hunger and the constant search for food compelled primitive people to select what was most valuable as food and most suitable for crops within the temperate and subtropical areas. No doubt, an enormous work was performed in this direction during the course of thousands of years by these unskilled plant breeders. We can only marvel at the diversity of strains of wheat, barley, maize and sorghum that were already known by primitive civilizations. It is not easy to find competitors to the cereals that are known to us at present, and an attempt toward this may never succeed. It is hard to replace wheat and maize with other crops. Botanical investigations made during the last 50 years all over the world have in any case not given any serious hints in this direction. However, it is possible essentially to alter the strains of wheat, to exchange one species for another, to create new species by means of hybridization and to open up wide prospects by means of plant breeding. However, searching for new genera of cereals may never meet with great success.

Nevertheless, the 'old' cereal crops have still not been utilized widely enough. At the end of the second five-year plan, the area under wheat in the U.S.S.R. should be up to 50 million hectares in comparison with the 37 million ha in 1931. The area under maize and sorghum should be considerably increased during the next couple of years. Early-ripening crops of Japanese strains of rice should, together with mechanization of the work, be advanced into new regions. It is necessary to approach the old type of rice cultivation in a definitely new manner within the near future.

To a certain extent, the same is generally valid for other edible plants. Primitive agricultural civilizations, developing in areas with a maximum accumulation of plant species made, indeed, good use of those plants most needed by them. To find definitely new wild beans or new edible wild roots or tubers could be an arduous task. We must first and foremost utilize the experiments made by the old civilizations and adopt what is most valuable from them, since in this respect still far from everything has been done, e.g., towards improving the edible raw material of the vegetation. How much can be achieved in this direction can be demonstrated by the examples of maize and soybeans, from which we have, during the last decades, succeeded in obtaining hundreds of new products.

No doubt the sweet potato, so far almost unknown in our country, is one of the most promising plants for our southern areas. Among the grain crops, sorghum and leguminous plants, such as chick-peas, vetchling and some species of beans can be mentioned. The cultivation of these has been known by Man for thousands of years but they are still not widely enough distributed in our country. Special attention should be paid to grain-sorghum as one of the most drought-tolerant crops for the south. Some types of sorghum from southern Palestine are – as shown by experiments during the last two years – of a most exceptional interest to the southern Ukraine and, indeed, also to Kazakhstan.

For the introduction of sweet potatoes into our country, it is necessary to pay special attention to the problems of storage and the selection of types maturing faster than the ordinary American strains grown in the southern states of the U.S.A. If we succeed in selecting such early types, we can have a very important new root crop that will at the same time be suitable for the areas where potatoes do not do as well. As experiments conducted during the last two years at the Institute of Plant Industry with seed material obtained from Taiwan have demonstrated, the amplitude of variability and the utilization of seeds rather than tubers open up extensive prospects for early-maturing crops of sweet potatoes as well as for the quality and the chemical composition of the tubers.

Next in turn is the introduction of crops of lupines without alkaloids.

Although German scientists, the first ones to find these, kept their discovery secret, Soviet chemists succeeded in discovering methods for quickly distinguishing the non-alkaloid types. We shall, no doubt within the next two to three years, have a considerable amount of the most valuable non-alkaloid types on hand. Non-alkaloid forms can be found among yellow, white, narrow-leaved and perennial types of lupines.

Still more room is open for investigations of crops of new wild leguminous forage plants. Only crops of vetches, fava beans and peas are grown in our country. The latest papers from Italian and Portuguese research stations show that much can still be done here in this direction. The drought-tolerant French lentils, widely grown in Spain, on Cyprus and in other Mediterranean countries deserve attention, just as do crops of chickling vetch and some Syrian strains thereof. The solution to the problem concerning the albumen used as food for animals and found in leguminous forage crops is of great practical interest. It is possible that, so far, not all has been exhausted as far as the wild leguminous plants are concerned. It is particularly necessary to search for them in the wild floras of the mountains and foothills of southwestern Asia (including Soviet Inner Asia and Caucasus) and among the exceptional richness of leguminous species of pulses in the Mediterranean countries.

TECHNICAL AND MEDICINAL PLANTS AND PLANTS PRODUCING ETHERIC OILS

There are almost unlimited possibilities for what can be discovered when searching for wild technical plants to be taken into cultivation. In this respect, the farmers of the past have done comparatively little. Even with respect to fiber plants, which furnish material for primary necessities, there are still – as demonstrated by short-term Soviet experiments – considerable opportunities. For instance, the Indian hemp (*Abroma angustata*), wild plants brought back from Inner Asia and Kazakhstan during the last couple of years, is especially worthy of consideration. The utilization of a plant from eastern Asia, such as ramie [*Boehmeria*], is, however, of considerably greater interest. Here it is necessary to keep in mind that we have still far from mastered the most valuable types of this southern textile plant. During tests at Sukhumi, cultivated types of ramie collected by us in Taiwan turned out to be more productive and to grow faster and taller than the ordinary Chinese strains.

We could, indeed, still introduce a number of textile plants from South America and southern Asia into our subtropical areas. There is a great interest in New Zealand flax (*Phormium tenax*) for our humid subtropical areas.

The field concerning research into medicinal plants and plants producing etheric oils has hardly been touched upon. In the past, nothing was done in this direction within our country. We have only recently begun this work but already now facts of primary importance have been revealed. The ommu [*Trachyspermum ammi* syn. *Carum copticum* Benth & Hook.], a plant that produces an etheric oil and was brought back by expeditions to Afghanistan and Abyssinia in 1927, solved our problem concerning thymol, a very important pharmaceutical medium for external disinfection, which we have so far had to import from abroad. The collection of seed material of this plant from Afghanistan and Abyssinia turned out to furnish a number of types with a high yield of oil. One of the samples produced up to 10% oil, containing up to 50% thymol. Thymol can now be obtained directly in the Soviet Union, whereas it was previously imported from abroad.

A great amount of camphor was found in a basil [*Ocisum kilimanmcharicum*], collected by a Soviet expedition in Northern Africa. It reached a quality hardly less than that obtained from the camphor tree [*Cinnamomum camphora*]. Thus, the problem concerning camphor can also be considered solved.

Already the first collection of plants containing etheric oils, found by the Institute of Plant Industry in the Mediterranean countries, furnished a material considerably surpassing what we had until then. The types of opium poppy obtained from Asia Minor by the expedition led by Professor P.M. Zhukovkskiy turned out to be exceptional with respect to their yield of morphine. Preliminary investigations have already revealed that there are valuable plants producing etheric oils in the wild floras belonging to the families Umbelliferae, Compositae and Labiatae in Inner Asia, Crimea and Transcaucasus.

No doubt there are great opportunities for the application of altogether 'new' plants, such as those producing rubber. These were not used in the past. A shining example is the 'tau-saghyz' (*Scorzonera tau-saughus*). The All-Soviet Conference on Rubber-Producing Plants, just now ending in Moscow, projected an area of 600 000 ha under 'tau-saghyz' at the end of this five-year plan. This should furnish five times more rubber than so far imported annually.

It is of interest to mention that the 'tau-saghyz' is found in the mountain areas adjacent to Inner Asia, i.e. in areas that are at the same time rich in a variety of plants. However, so far, this plant has only been found in the Karatau mountain range between two semi-deserts, although it was searched for everywhere in 1931.

The 1931 expedition to collect rubber plants found at the border with western China, between Uch-Turfan and Kuld'zhe, found a species of dandelion [*Taraxacum bicorne*] yielding up to 28% rubber in its roots. In Armenia and Azerbaidzhan, attention was focused upon a number of wild scorzoneras that could possibly produce rubber. In this respect, there is apparently some interest also in the *Lactuca* species, especially in the Far East. It is quite likely that new rubber-producing plants can be found if one searches far and wide both in Inner Asia, Armenia, Azerbaidzhan and Georgia, as well as in the countries adjacent to them, such as Afghanistan, Persia and Asia Minor, all of which are especially rich in the rubber-producing genera of plants. It is necessary to extend strongly the search for rubber-producers while focusing attention especially on areas rich in species and genera belonging to families known to produce rubber. Searches in this direction should, of course, also be conducted within our own borders.

It is true that valuable rubber plants can also be discovered in the mountains of Mexico and Central America. Systematic research in this direction has still not been undertaken. Even the Mexican guayule [*Parthenium argentatum*] is still not thoroughly studied in detail. It is necessary to organize a collection of these plants at higher latitudes, especially in Texas and the U.S.A., the northern limit to where this plant reaches within its natural distribution area. There is, apparently, also great interest in a Florida golden-rod. According to preliminary experiments at Sukhumi with seeds sent to us from Florida by Edison in 1931, it was revealed that some types of *Solidago* produce a large amount of gutta-percha also under our own conditions. As a herbaceous plant, producing a great amount of vegetative mass that can be cut down annually and which also produces a considerable output of gutta-percha, the Florida goldenrod deserves serious attention. It is also necessary to extend the search for various forms of the Chinese *Eucommia* that produces gutta-percha, the more so since Soviet workers have succeeded in mastering the vegetative reproduction of this plant. The American species of dogbane (*Apocynum*) deserve investigation as a rubber producer but also as a fiber and 'cotton' producer (Vavilov, 1931e). It is necessary to extend the research on dogbanes to Kazakhstan, Inner Asia and western China, where there are forms (e.g. *Apocynum hendersonii*) that are exceptionally rich in salts. The American species, *Chrysothamnus nauseus*, *Asclepias subulata* and also *Apocynum cannabinum*, should be studied under Soviet conditions.

So far, little has been done in the case of tannin-producing plants. The wild 'badan' [*Bergenia crassifolia*], which grows in western Siberia, has been taken into cultivation as a tannin-producing plant. Professor P.A. Yakimov at the biochemical laboratory of the Institute of Plant Industry demonstrated in 1931 that not only can a high percentage of tannin be extracted from its leaves but also a large amount of hydroquinones (from ensilage of badan), which is of great importance to the photochemical industry. It is now possible to consider the badan as a 'hydroquinone-producer' as well.

The first attempts at an analytical approach to the tannin-producing plants have already revealed many possibilities. Thus, e.g. we have found a form of smoke tree [*Rhus* sp.] on Crimea that has up to 40% of tannic matter in the leaves while the common types produce only 15–20% thereof.

Cultivation of the Chinese tung-oil tree [*Vernicia fordii*] is of great interest for our humid subtropical areas. It produces a high quality of industrial oil. The same applies to the extensively tested crops of wax trees (*Toxicodendrum succedaneum*) and tallow trees [*Sapium sebiferum*]. Nothing has been done within the field of dye and 'cotton'-producing plants. It is necessary to take note of breeding in our Subtropics of the north-African 'alfa' [*Epilobium hirsutum*], which produces cotton-like fibers of high quality around its seeds.

Based on experiments in Japan and England, we ought to adopt widely the cultivation of the so-called chamomile-type pyrethrum [*Anacyclus cotula* L.], which appears to be strongly active in the fight against a number of insects. This plant can be easily grown within the U.S.S.R. and succeeds very well in Crimea and in the Caucasus. Apparently, even some of our wild Caucasian chamomiles are of even greater interest than the foreign species but still need to be subjected to studies.

Nicotine is, as is well known, strongly insecticidal, especially in the case of aphids. In America, they have succeeded in making a new product, neo-nicotine that is considerably stronger than nicotine itself. It was discovered recently that our common Turkestani plant called 'itsegek' or 'barnyard millet' (*Anabasis aphylla*), which belongs to the Chenopodiaceae, contains a large amount of neo-nicotine and similar alkaloids. Attempts are now being made to take this plant into cultivation in the United States. The plant, which is very common in Turkestan and Kazakhstan, grows on saline soils in ravines. Cattle do not feed on it since it is poisonous.

FORAGE PLANTS

The choice of forage plants both in western Europe and in the Soviet Union is fairly limited. No doubt many species can be found among the wild vegetation that would not yield less than our ordinary forage plants. Although, with respect to other plants, it is necessary to turn our attention to the southern mountains and the foothill areas – the center of a concentration of species – there are also endless possibilities elsewhere as far as forage plants are concerned. It is still possible to discover valuable forage plants both among the grasses and the leguminous species in the plant associations of the northern and Siberian meadows and forests. A number of winterhardy, wild vetches deserve, no doubt, exceptional attention.

Preliminary, although not very extensive, experiments already indicate great possibilities. It is necessary to stop our inertia and begin seriously to utilize the wild flora as far as more valuable forage plants are concerned. It is necessary to start a collection according to a definite system and to establish seed producing centers.

A small Swedish expedition worked in 1928 in western Siberia and Altai under the direction of the well-known ecologist, G. Turesson. They collected a large amount of seeds from various wild forage grasses. The study of the west-Siberian and Altaian species of *Dactylis*, *Festuca* and other forage grasses by growing large crops thereof has shown that these species are distinguished by an unusually rapid growth and an enormous vegetative mass, in this respect surpassing the ordinary European type by 50–100%. The breeding of forage grasses in Sweden is now based on these grasses from western Siberia and the foothills of Altai. The data produced demonstrate what unlimited possibilities are open to us.

It is necessary to proceed on a broad industrial scale with respect to the questions about drought-tolerant forage grasses while using all experimental methods for this.

All the data relating to forage within the dry zones must be subjected to a radical revision. Many of the wild plants in the deserts, the semi-deserts and the foothill areas of Inner Asia and Kazakhstan are of great interest from the point of view of forage. It is time to digress from the pattern of West-European forage plants that we have so far utilized. Thousands of years of experiments made by the nomadic peoples of Kazakhstan, Inner Asia, Persia, Afghanistan and Mongolia reveal the nutrient value of many desert and semi-desert plants such as

Russian thistle [*Salsola iberica*], sage brush [*Artemisia*], *Gundelia tournefortii* L., species of *Elymus* and *Atropis*, the sedge *Carex physodes*, the bromegrass *Bromus tectorum* L. and species of *Astragalus*, not to mention the so-called camel's thorn [*Alhagi maurorum*]. It is necessary to begin to appreciate the wild forage grasses again, to pay greater attention to their chemical composition and to their value as fodder and to start extensive, large-scale experiments toward taking the most interesting species into cultivation in order to be able to utilize the enormous open semi-desert spaces of Kazakhstan and Inner Asia.

The method of ensilage has opened up possibilities for a great variety of fast-growing plants, starting with sunflowers and corn and ending, really, with many of our common weeds.

In addition to weedy grasses, the utilization of new leguminous plants both for the northlands and for our southern republics is of considerable interest. There are promising tests with such leguminous plants, such as various species of *Crotolaria* L. and *Sesbania* Scop., for the coasts of the Black Sea and for Inner Asia.

FRUIT TREES AND BERRY BUSHES

Immense resources of fruit trees are concentrated within the foothills and the mountains of Inner Asia, in the mountains of Tien-Shan and in Transcaucasus. These areas are the native land of many important European fruit trees as well as of a number of species still not taken into cultivation but nevertheless representing great value. Some types of wild almonds, cherry plums [*Prunus cerasifera*], and walnuts are as good as the best kinds presently grown and could be taken directly into cultivation (Vavilov, 1931a).

Very special attention should be paid to fruits for our humid, subtropical areas. The best kinds of citrus fruits from Japan, Florida and California can be grown there. Preliminary experiments have already shown that mandarin oranges, lemons, grapefruits and oranges succeed very well there. It is necessary to start a systematic acquisition of a better assortment thereof from Japan and China. Extensive hybridization work should also be done in order to obtain new and hardier strains. There is a great interest in the frost-hardy mandarins and lemons obtained from Swingle in the U.S.A. In our humid Subtropics it is possible to cultivate many subtropical American and Asiatic fruit trees as well as feijoas [*Feijoa sellowiana*], persimmons [*Diospyros kaki*], loquats [*Eriobotrya japonica*], avocado [*Persea americana*] and Manilcara sapote [*Manilkara zapota*].

The primitive fruit growers did not do well in this field. The cultivated fruits of Central America and Brazil differ little from those growing in the wild. Much can still be achieved by means of systematic breeding.

The Australian wattle [*Acacia* ssp.] can be grown along the coast of the Black Sea. Nowhere are there better opportunities for utilizing tropical and subtropical plant resources than in our own, unfortunately territorially very limited, subtropical areas. Nowhere else can such big changes be carried out as in our subtropical areas. Indeed, five to ten years from now we will hardly be able to recognize this area. It is especially necessary to introduce systematically new crops into this area while rationally utilizing every inch of the soil for what is most valuable to us. No less than 100 000 hectares of tea, some 10 000 ha of citrus fruits, some 10 000 ha under ramie and wide-ranging plantations of various subtropical plants producing etheric oils – that is the program for the second five-year plan.

TREE SPECIES

There are practically unlimited opportunities for utilizing so-called 'exotic' trees here. It is possible to list hundreds of species of interesting foreign technical and ornamental trees that are able to grow vigorously in our country and that are of use to the plant industry as well as for the greening of our cities from the Subtropics all the way to the Far North. Some of these 'exotic' species can be of great importance for our forestry and rural economy. Among these we can mention a number of conifers, mainly Douglas firs from California, Sitka spruces from Alaska, gigantic Thujas, Virginian junipers, Canadian hemlocks, Lebanese cedars and such from the Atlas and the Himalayan mountains, California's sequoias, and Japanese cryptomerias, all with very valuable kinds of wood. Among the deciduous trees we can mention Chinese *Ailanthus*, American white acacias, and honey locusts [*Gleditschia*], Canadian red oak, American hickory, fast-growing catalpas, decorative magnolias and Canadian poplars. From eastern Asia, the camphor tree and the mulberry tree are of interest for our Soviet tree plantations. The American black walnut, which is more resistant to spring frosts than the English walnut, the Virginian chokecherry [*Prunus virginiana*], which is suitable for the afforestation of the empty, sandy and stony deserts, and the Canadian frost-tolerant pines also deserve attention. Of similar interest are the Rumelian pines from Macedonia, the Canadian sugar maples, the Manchurian walnuts and the oriental plane trees. It is also possible to enumerate some dozens of valuable exotic trees such as camphor trees, eucalyptus and Australian acacias suitable for the coast of the Black Sea. (cf. Kern, 1926, and a large number of papers from 1925 on in *Papers Concerning Applied Botany and Plant Breeding*.)

During the initial, mostly dilettantic approach to matters concerning the naturalization of trees in our country, there was already, in the past, much material that could have been utilized in every possible way for a socialistic reconstruction.

With respect to 'new crops', it is necessary to introduce a great variety of species, corresponding to and necessary for the present, complicated agriculture and industry and embracing all kinds of plant materials.

NEW CROPS OF THE NORTH

So far, mainly the steppe lands and the forest-steppe ecotone are under agriculture both in our own country and beyond its borders, but especially in North and South America, Australia and Africa. This is due to the fact that it was comparatively easy to take the black soil areas of the steppes into cultivation since they did not need fertilization. The farmers followed the path of least resistance. We know, however, very well, that more humid areas without black soil can produce stable crops after proper treatment of the soil, drainage and the use of manure. The industrial development of mineral fertilizers and the new socialistic order, which allow us to make great capital investments, the development of industrial centers in the Northlands and the creation of new means of communications – all this together compels us to turn our attention to farming of the North. At present, a selection of early-ripening winterhardy strains of new crops for the North is the next task facing us. The possibilities for selecting strains of vegetables, fruits and forage crops for the North are much better than usually assumed.

Geographical tests made at the Institute of Plant Industry have shown that conditions in the northern areas are especially favorable for fiber and forage plants, producing considerably larger amounts of vegetative mass and fibers. Even in the Far North, along the Polar Circle at Hibiny on the Kola Peninsula, carrots and sugar beets produce no less sugar than at more southerly localities (data from the biochemical laboratory of the Institute of Plant Industry).

A number of crops such as of flax, barley, potatoes, fodder beets and many vegetables are as good as those in the south, with respect to quality and productivity, and even surpass them. The Northlands hold first place with respect to the production of mass. Considering the widespread development of ensilage, many southern crops can be used in the North, such as sunflowers, vetches and maize (cf. Vavilov, 1932). Fruit crops of a great specific and varietal diversity should succeed very well all the way to the Far North. Even among the wild species in the North itself, there are kinds that can be taken directly into cultivation. Ussurian pears, Siberian apples and Agaric mushrooms can withstand much frost and can be used just as well for northern fruit-growing as can imported materials.

NEW CROPS FOR THE FIGHT AGAINST DROUGHT

The fight against drought is a complicated problem. In order to solve it, it is necessary to resort to major irrigation work and to change over to agrotechnology. However, simultaneously it is necessary to take into consideration that, in spite of the predominance of moisture-loving plants, there is in nature also a great supply of quite drought-tolerant varieties and species that are still used inadequately or not at all in the U.S.S.R.

Among the number of still inadequately distributed new crops that are distinguished by drought-tolerance (Vavilov, 1931d), we can mention sorghum, Mediterranean kinds of oats, negrito-millet (*Pennisetum*), tepary beans [*Phaseolus acutifolius*], fodder lentils, white-seeded Persian lentils, chick-peas, different species of chickling vetch and olives. To the group of drought-resistant plants belong also a considerable number of plants that produce etheric oils, such as coriander [*Coriandrum sativum*] and fennel [*Foeniculum vulgare*], rubber-producers such as 'tau-saghys' [*Scorzonera tau-saghus*], guayule [*Parthenium argentatum*] and *Chondrilla* as well as oil-producers such as safflower [*Carthamus tinctorius*]. Among the melons, the American watermelons, which are used as fodder, deserve to be more widely spread as do figs, pistachio [*Pistacia vera* L.], steppe cherries, carob trees [*Ceratonia siliqua*] and a number of pears among fruits, as well as cork oaks. Among the forage plants, Sudan grass [*Sorghum drummondi*], sweet clover [*Melilotus*], crested wheat grass [*Agropyrum desertorum*], yellow lucern [*Medicago sativa* and foxtail-millet [*Setaria italica*] can be listed.

In the wild flora of the semi-deserts and even the deserts proper of Kazakhstan, the area around the lower Volga and Inner Asia as well as in the countries bordering on ours, there are stores of wild, drought-tolerant forage plants. Among these are various species of *Elymus*, Russian thistles [*Salsola iberica*] crested wheat grass [*Agropyrum desertorum*], (viviparous) meadow grass [*Poa* ssp.], *Atropis* and species of sagebrush and *Astragalus*.

It is also necessary to turn our attention to the drought-tolerant trees of the deserts and semi-deserts of Inner Asia and the wealth of shrubs and trees that are distinguished by a high level of heat and drought resistance. There are among them those that are valuable from an agricultural point of view and produce wood for building and construction and deserve to be taken widely into cultivation. We must distinguish between the kinds of trees that need the presence of water close to the ground and those that do not do as well as those able to tolerate extensive periods of drought during the vegetative period. To the first category belong some species of poplars, mulberries, tamariscs, e.g. Androsov's and Pallas tamariscs, which are excellent for stabilizing sand. Here we can also mention oleaster [*Eleagnus*], white acacias, ailanthus and some willows as well as the haloxylon [*Haloxylon halimodendron*] of salt marshes.

To the woody plants not needing shallow groundwater belong some shrubs and half-shrubs that are scattered about in the sandy and sandy-clayey deserts, e.g. various species of *Calligonum*, *Halimodendron*, *Eremospartum* and Russian thistle [*Salsola iberica*]. These plants represent a great value as fuel in desolate and dry areas.

THE FIGHT AGAINST DESERTS

We are again faced with the task of fighting the deserts and conquering the desert expanses. Half of the world's landmass is occupied by deserts. Hundreds of millions of hectares in our own country consist of sandy, clayey and stony deserts. It is necessary to start advancing upon the deserts in order to utilize a part of them for pasture, for planting drought-tolerant plants and for planting trees. The fight against the deserts is a question of primary importance for Turkmenia, Uzbekistan, Tadzhikistan and Kazakhstan. The deserts there surround both cities and agricultural areas.

However, it is necessary to emphasize that the fight against the deserts is not a simple matter. It is not by chance that natural selection has turned out to be useless for 'controlling' the desert, since the entire plant world is controlled by it. The great hope of developing crops of fodder cacti for the deserts – however much this has been tried experimentally all over the world – has so far been fulfilled only to a limited extent. In spite of all the advertising connected with the 'invention' – or rather discovery – of the spineless cacti by Burbank (more than 20 years ago), it has so far not met with success during the extensive efforts to control the deserts. Spineless cacti [*Opuntia*] have been known for more than 100 years. In 1926, we saw a small plantation of such cacti, which had existed for several decades around Nazareth in Palestine. The statisticians, even those from such countries as the U.S.A., Algeria, Tunisia, Sicily and Australia, where the cultivation of cacti has been primarily introduced, do not consider the area under cacti to be of importance. We have traveled through many deserts in Asia, America and Africa, including the native land of the cacti, i.e., Mexico, but we have not seen any plantations there of importance as forage. There are no such in southern California where Burbank worked, or in the adjacent deserts of Arizona, although these areas could be said to be even more suitable for developing forage of cacti. Utilization of cacti (mainly wild ones) for cattle fodder occurs in the U.S.A., especially in New Mexico, during exceptionally dry years, when there is no other forage. Attempts to build up milk-producing farms based on cactus plantations did not succeed even during the initial attempt. Cacti are used widely in the dry Subtropics, mainly for fences, as ornamentals and for the fruits. This is valid mainly in the Subtropics that do not have dry winters such as we have to deal with in our dry and humid Subtropics. There are species of cacti (*Opuntia*) that are comparatively frost-hardy, but exploratory experiments must be done with them before recommending them for wide application.

The problems encountered during extensive cultivation of cacti for forage have still not been finally solved in spite of the interest existing for this. A lot of literature is devoted to the fight against the thorny cacti as malicious weeds. However, in order to solve the problem of cacti as fodder plants in our southern deserts, it is necessary to make extensive experiments.

In the fight against the deserts there is a great interest in the utilization of the wild vegetation from the desert areas of Inner Asia and Kazakhstan. The psammophytic (sand-loving) native flora of the region is extraordinarily rich and variable. Taking some of the Inner-Asiatic desert and semi-desert plants, including pistachio trees (*Pistacia* ssp.), into cultivation is quite tempting and realizable. Recently, attempts were made in the U.S.A. with north-Chinese *Calamagrostis* for their deserts. The so-called 'barnyard millet' (*Anabasis aphylla* L.) could be used for the saline lowlands of Inner Asia and Kazakhstan, since it produces neo-nicotine. It is necessary to plan broadly laid-out experiments on a new scale and to use an original approach in order to be able to utilize the desert wastelands of the world. The application of new crops and new plants there will be of decisive importance.

List of new crops deserving primary attention for large-scale planning

When taking new crops into cultivation, it is necessary to pay attention both to our own experiments and those made elsewhere in the world in order to be able to distinguish what is essential from what is not and to be able to pick out what is most valuable from the enormous diversity of plants available. In our own literature as well as that from abroad, a multitude of plants are recommended without any exact basis for this, while diverting attention from the really important objectives. Under such conditions usually great – not rarely even fabulous – harvests are claimed on the basis of averages from small experimental plots without any critical evaluation of the actual agricultural conditions. When evaluating new plants, it is necessary to compare them with well-known crops. Any kind of plant can somehow be useful. Sheep and cattle are able to consume 99% of all existing plants. However, our systematic agriculture must be based primarily on serious scientific and large-scale experiments, critically adopting only what is valuable and necessary.

We have attempted to compile a list of crops, based on our own and worldwide research, but also subjected to scrutiny with respect to Soviet large-scale and scientific planning. The list must not be considered as final or unalterable. Indeed, after a few years, it may already be necessary to make serious changes to it.

The following plants can be mentioned with respect to new crops already introduced or subject to immediate introduction into cultivation in the U.S.S.R.:

Grain crops: Sorghum, soybeans, Mediterranean oats (*Avena byzantina*), early-maturing (tepary) beans [*Phaseolus acutifolius*], chick-peas [*Cicer arietinum*], French fodder lentils [*Lens culinaris*], white-seeded Persian lentils [*Lens orientalis*], which are also drought-tolerant, and lupines.

Fiber crops: Cotton for non-irrigated regions, dogbane [*Apocynum* spp.], hemp mallow [*Hibiscus cannabinus* L.], flowering maple [*Abutilon*], ramie [*Boehmeria*], teasel [*Dipsacus*], and New Zealand flax [*Phormium tenax*].

Rubber plants: 'Tau-saghyz' [*Scorzonera tau-saghus*] and guayule [*Parthenium argentatum*].

Gutta-percha plants: *Eucommia ulmoides.*

Forage crops: Sudan grass [*Sorghum sudanense*], crested wheatgrass (*Agropyrum desertorum*), alfalfa [*Medicago sativa*], yellow-flowered sweet clover [*Melilotus officinalis*], sorghum for green manure, American wheat-grasses (*Agropyrum* spp.), hairy vetch [*Vicia villosa*], fodder kale [*Brassica oleracea* var. *acephala*], fodder-melons (*Citrullus vulgaris*); for ensilage in the north: sunflowers [*Helianthus* spp.], maize, and *Amaranthus*, and in the south: velvet beans [*Mucuna pruriens* var. *utilis*] which can also be used as hay and green manure.

Tuberous plants: Jerusalem artichokes [*Helianthus tuberosus*].

Root crops: Chicory [*Cichorium intybus* var. *sativum*], sweet potatoes [*Ipomoea batatas*].

Oil plants: Castor beans [*Ricinus communis*], peanuts [*Arachis hypogaea*], safflower [*Carthamus tinctorius*], perilla [*Perilla frutescens*], and tung-oil trees [*Vernicia fordii*].

Vegetables: Rhubarb [*Rheum rhabarbarum*], which can be grown up to the Polar Circle.

Plants producing etheric oils, spices and narcotics: Tea [*Camellia sinensis*], caraway [*Carum carvi*], ommu [*Trachyspermum ammi*], fennel [*Foeniculum vulgare*], lavender [*Lavandula angustifolia*], rosemary [*Rosmarinus officinalis*],

mint [*Mentha piperita*], sage [*Salvia officinalis*], iris [*Iris*], lemon-grass [*Cymbopogon damascena*], *Eucalyptus* spp., *Citrus* spp., lemon verbena [*Lippia triphylla*], dragoon [*Dracocephalum* sp.], geranium [*Pelargonium* spp.], basil [*Ocimum basilicum*] and Damascene rose [*Rosa damascena*].
Tannin-producing plants: 'Badan' [*Bergenia crassifolia*], smoke trees [*Rhus cotinus*], tannic acacia (*Acacia dealbata* Link), *Eucalyptus* spp., and willows (*Salix triandra*, *S. cinerea*, etc.)
Insecticidal plants: Pyrethrum (*Chrysanthemum* spp.).
Fruit crops: Citrus fruits (mandarine oranges, oranges, grapefruits, etc.), figs [*Ficus carica*], pistachio [*Pistacia vera*] and persimmons [*Diospyros kaki*].
Technical woods: Cork oaks [*Quercus suber*], *Eucalyptus* spp., bamboo [*Bambusa* spp.] including grass-like species, camphor trees [*Cinnamomum camphora*].

The following crops, which are of considerable interest but which still require extensive and large-scale research, can also be mentioned:

Rubber and gutta-percha producing plants: Edison's golden rod [*Solidago edisoniana*] for the humid Subtropics, milk weeds [*Asclepias* spp.] and *Eucommia*.
Vegetables: Chayote [*Sechium edule*] and Japanese radishes [*Raphanus sativus* var. *acanthiformis* Makino].
Forage crops: Dune grass (*Elymas giganteus*) [*Leymus* spp.], which is suitable for converting parts of extensive areas of windblown sand and dunes into large-scale pastures around the lower Volga, in Kazakhstan and Inner Asia, salt-tolerant wheat-grasses [*Agropyron* spp.], which could be used after irrigation of saline deserts, biennial vetch [*Vicia picta* Fisch. & Mey], teff (*Eragrostis abyssinica*) for the humid Subtropics, *Crotolaria juncea*, for green manure in the Subtropics and velvet beans [*Mucuna pruriens* var. *utilis*].
Technical woods: Tallow trees [*Sapium sebiferum*], wax trees [*Toxicodendrum succedana*], jute [*Corchorus capsularis*] and New-Zealand flax [*Phormium tenax*].
Plants producing etheric oils: Cumin [*Cuminum cuminum*] as a spice, tuberose [*Polianthes tuberosa*], palmarosa [*Cymbopogon martini*], kumquats [*Fortunella margarita*], hyacinths [*Hyacinthus* spp.] and narcisses [*Narcissus* spp.].
Fruit crops: Peruvian cherry [*Physalis peruviana*], feijoa [*Feijoa sellowiana*], passion fruit [*Passiflora edulis*], pecans [*Carya illinoensis*, *C. ovata*], and Japanese mispel or loquat [*Eryobothrya japonica*].
Insecticidal plants: Barnyard millet (*Anabasis aphylla* L.).

The following crops can also be mentioned, although they are still subject to large-scale research and examination in comparison with corresponding crops as far as profitability, productivity and quality are concerned:

Fruit crops: Avocados [*Persea* spp.], kiwi fruit [*Actidinia chinensis*], Surinam cherry [*Eugenia uniflora*], large-fruited jujuba [*Zizyphus jujuba*], and white sapote [*Casimiroa edulis*].
Oil-producing and culinary plants: Chufa (*Cyperus esculentus*), chick-peas [*Cicer arietinum*], and Spanish dragoon [*Lallemantia iberica*].
Vegetables: Taro [*Colocasia esculenta*], and other arrowroot plants, tomatillo [*Physalis ixocarpa*], some subtropical yams [*Dioscorea* spp.] and Chinese cabbages [*Brassica pekinensis* and *B. chinensis*).
Fiber plants: Yucca [*Yucca* spp.] and jute [*Corchorus* spp.].

Rubber plants: *Cryptostegia* spp. (especially hybrids between the Madagascarian and the large-flowered species).
Plants producing etheric oils: Jasmine [*Jasminum* spp.], anisette (*Pimpinella anisetum*), wild-growing producers of etheric oils such as *Laserpitium*, *Heracleum* and thyme (*Thymus* spp.) as well as wormwoods [*Artemisia* spp.].
Forage crops: Durrah (*Pennisetum typhoideum*) for the dry southern areas, subterranean clover (*Trifolium subterraneum*) as a pasture plant, comfrey (*Symphytum asperum*), berseem clover (*Trifolium alexandrinum*) and fenugreek (*Trigonella foenum-graecum*).
Sugar-producing plants: Sugar cane (*Saccharum officinarum*) (early-maturing forms for conditions in the humid Subtropics) and American sugar maples (*Acer saccharum*).
Starch-producing plants: Seakale (*Crambe maritima*).

The plant industry under the socialistic system

The new socialistic system will introduce major changes to our common ideas about the plant industry. The development of the technical industry alongside that of agriculture and the industrialization of agriculture will change the outlook on cultivation. Already in the near future, hundreds of thousands of hectares of potatoes will be used for producing rubber, prepared from potato-alcohol. Quite likely, also maize and grain sorghum can be used in the same manner. The question has been raised as to whether the straw of our cereal grasses can be used for making paper.

Next in turn for our farm management is the widespread utilization of all kinds of plant waste for factory production of compost according to the method of A. Howard. This is considerably superior to the efficiency of manure. The production of valuable compost from straw waste as well as from the stems of cotton, sunflowers, etc., will be of great interest to us (cf. Howard and Wad, 1931) for a major expansion of vegetable crops and the laying out of new gardens within areas with a shortage of manure.

A factory for the utilization of soy meal and casein has already been built. There is a real possibility in this direction for using other beans as well. The application of 'cottonization' allows us to use hemp and hemp-mallow to a considerably broader extent. The development of major industrial centers will change the geography of our crops.

Technical perfection, the field we have now entered, enables us to utilize our old crops in a much more extensive manner. The revolution of agrotechnology, reaching all the way to using aeroplanes for seeding rice and forage grasses, will change every approach to these crops. In our opinion, the very laborious old type of cultivating rice can become mechanized by using aeroplanes for seeding and combines for harvesting. Rice can easily become a 'new' crop under the conditions of mechanized agriculture. Exchanging animals for tractors and the use of combines allows us to proceed into drier areas and the regions of the semi-deserts in spite of the well-known risk of using drought-stricken areas. The application of flax- and cotton-harvesting machines will radically change the labor cost that is estimated for the application of these kinds of crops.

Under the conditions of the new system, the old crops actually become, in essence, 'new' crops, forcing us to approach them differently.

The new crops and state-controlled cultivation of seeds

Experiments in the past demonstrated that single expeditions looking for new species – however extensively they were conducted – did not bring any serious results if they were not associated with a firm, state-controlled organization for cultivating and multiplying the seeds brought back. We have already known yellow-flowered lucerne [*Medicago sativa* var. *falcata*] for 40 years; it is listed in all the manuals as a very valuable, drought-tolerant herb. However, in the whole country, the amount of seeds of this beautiful forage plant amounts still only to a few poods (1 pood = 16.38 kilo). The difficulties when multiplying yellow alfalfa are related to the common sterility of this plant. In this sense, the yellow one differs sharply from the ordinary, blue-flowered alfalfa (*Medicago sativa* ssp. *sativa*). This one is represented by our common weedy herb, which grows in enormous quantities for thousands of square kilometers all over our country. However, we are forced to write for seeds of this herb from abroad at a high price, due to the lack of a seed-growing establishment where these seeds could be multiplied. We are still forced to order vegetable seeds from abroad as well, although there are, in fact, full opportunities for obtaining all the kinds necessary in our own country. So far, we have to calculate with a shortage of the most common forage plants.

Very valuable forage plants are tested and selected at our experimental establishments; however, so far, these have not been equipped with experimental plots large enough, since there is no powerful and well-organized seed-growing organization that could quickly master all what is valuable and bring it to good use.

An immediate development of a powerful, state-controlled organization for seed-cultivation appears to be a directly urgent and basic measure without which all the problems concerning the new crops become null and void.

The establishment of large, seed-growing state-farms that are well organized and well equipped according to the last word in technology, where work could be done with specific crops and specific strains – this is one of the primary problems facing us, without the solution of which neither plant breeding nor new crops will bring any serious results. A strong organization of seed cultivation is a guarantee for an organized and systematic exchange of crops and strains for better ones and for the expansion of the area under cultivation.

When, on behalf of the *Peasant Gazette* of the U.S.S.R. National Commissariate of Agriculture, we were organizing divisions for the introduction of new crops, wide-ranging problems had to be solved. These included, e.g. 1. all available assistance for carrying out the measures for introducing new crops and the expansion of the areas seeded with these crops; 2. a struggle for the maximum development of the reproduction in order to have seed resources for the new crops and for the introduction of new crops; 3. assistance with the creation of state-controlled and public funds; 4. all kinds of help with and development of matters dealing with the new crops; 5. assistance with new research and experimental organizations working on the new crops; 6. attracting in any manner possible collective and state-farm workers for perfecting the technology pertaining to the handling of the new crops, but also for involving them in the experimental work necessary for the production of new crops; 7. organization of a large-scale campaign for discovering new crop plants from the associations of the wild-growing flora; 8. assistance with inventive and rational activities for the introduction of new crops and for their utilization in the production of the new crops; 9. attracting the great masses for work with new crops; 10. every kind of assistance with the development of agriculture in the North and the advancement of the crops toward the North; 11. drawing the attention of the public to the problems concerning new crops for the dry and the humid subtropical areas; 12. utilization of wastelands and deserts; and 13. all kinds of help with matters concerning the development of crops that can free our industry from imported goods and that can increase our export.

This enumeration describes to a certain extent the immediate problems encountered in the struggle for new crops.

A 'research-oriented tourism' is unfolding in our Soviet country. Tens of thousands of tourists, at present uniting proletarian tourism with the Central Soviet society, offer their services, their assistance and their enthusiasm for the sake of science. As we have seen above, the mountain areas of Inner Asia, all of the Caucasus and Altai, as well as Tien-Shan, are of the greatest interest with respect to discoveries of new, previously unknown plants. The mountains and the foothill areas of the Soviet country are of maximum interest to us.

Mountain tourism could be associated with problems to be investigated. Our research organizations ought to utilize the large collective of tourists, who are ready to offer their services for the execution of problems to be studied. For this, it is necessary to immediately start preparing special instructions to help with guidance and to tell them where to search, what to look for and how to search. It is necessary to issue special instructions when looking for rubber-, tannin-, or oil-producing plants. It is also necessary to utilize the tremendous energy, enthusiasm and purposefulness, which all can be led in the correct direction.

Twenty-three years ago, a little American book, by Harwood, appeared in a beautiful translation/revision by K.A. Timiryazev, i.e. *The Earth Renewed.* [Two books, *New Creations in Plant Life* (1905) and *The New Earth* (1906) were published by W.S. Harwood, the former being the main basis of the Russian translation.] In this book, it is described how the Europeans arriving in North America altered the face of the world, how they introduced new European crops, how the expeditions of the U.S. Department of Agriculture searched all over the world and collected new and valuable plants, how the 'American Michurin', Luther Burbank, improved the strains of fruit trees and how the powerful rural economy of the United States was developed on the basis of Science.

Twenty years ago, when it was fitting to study this book, all that was written in it seemed to us as remote dreams, difficult to make real.

After reading this book, V.I. Lenin commissioned N.I. Gorbunov to take steps in order to develop the means for improving the plant industry of the Soviet Union. According to the instructions issued by V.I. Lenin, the first state-controlled seed farm was developed in the U.S.S.R. in order to rapidly develop improved strains. The All-Union Institute of Applied Botany and New Crops (presently the Institute of Plant Industry) was created according to his instructions and has, during the last couple of years, made great efforts for mobilizing the plant resources of the world.

However, we have only just begun on the second five-year plan but the ideals of the American renovation of the Earth have not yet been satisfied by us. The renovation of the Soviet world is drawn up by us in immeasurably brighter colors. We want and shall make great changes on a different scale.

There is no task more fascinating or more alluring than the creation of such a renovation of the Earth. New crops are only one of the components for the rebuilding of the Soviet world.

BIBLIOGRAPHY

Howard, A. and Wad, V.D. (1931). *The waste products of agriculture: their utilization as humus*. Oxford University Press.

Kern, E.E. (1926). *Inozemnye drevesnye porody, ikh lesovodstvennye osobennosti i lesokhozyaystvennoe znachenie [Foreign species of trees, their sylvicultural peculiarities and their importance for forestry]*. Leningrad.

Kotaro, Sirai (1929). *Research on the introduction of plants*. Tokyo (in Japanese).

Vavilov, N.I. (1926). Tsentry proiskhozhdenya kul'turnkykh rasteniy [Centers of origin of cultivated plants]. *Tr. po prikl. botan. i selek. [Papers on applied botany and plant breeding]*, **16** (2).

Vavilov, N.I. (1931a). Dikie rodichi i plodovykh derev'ev aziatskoy chasti SSSR i Kavkaz i problema proizkhozhdeniya plodovykh derev'ev [Wild relatives of fruit trees in the Asiatic parts of the U.S.S.R. and the Caucasus and the problem of the origin of the fruit trees]. *Tr. po prikl. botan., genet. i selek. [Papers on applied botany, genetics and plant breeding]*, **26**, (3).

Vavilov, N.I. (1931b). Linneevskiy vid kak sistema [The Linnaean species as a system]. *Tr. po prikl. botan., genet. i selek.*, **26** (3).

Vavilov, N.I. (1931c). Meksika i Tsentralnaya Amerika kak osnovnoy tsentr proizkhozhdeniya kul'turnykh rasteniy Novogo Sveta [Mexico and Central America as the main center of origin of cultivated plants of the New World]. *Tr. po prikl. botan., genet. i selek.*, **26** (3).

Vavilov, N.I. (1931d). Mirovoye resursy zasykhoustoychnykh rasteniy [World-wide resources of drought-tolerant plants]. *Za Novoye Volokno*, No. 12.

Vavilov, N.I. (1931e). Problemy rastitel'nogo kauchuka v Severnoy Amerike [Problems concerning vegetable rubber in North America]. *Tr. po prikl. botan., genet. i selek.*, **26** (3).

Vavilov, N.I. (1932). *Problema severnogo zemledeliya* [Problems concerning the agriculture in the North]. Leningrad.

The present state of worldwide agriculture and agricultural sciences

Impressions from a journey Through North America and Western Europe

ON BEHALF of the Narkomzem [National Commissariate of Agriculture] of the Soviet Union, the author was able to participate in the Soviet delegation to the International Conference on Problems of the Rural Economy, held at Ithaca, [NY] in the U.S.A. Almost simultaneously, there was an International Congress of Horticulture in London, which we were also able to attend. In addition, we took part in a later Pan-American Congress in Washington, where research workers gathered, concerning the rural economy of both North and South America. Following these meetings, the author had an opportunity to visit the subtropical and tropical parts of North America. We were able to visit North and South Carolina and Louisiana, to survey Florida and Arizona in great detail and to make a thorough study of California. From the southwestern states, the author traveled on into Mexico, crossing the country twice in different directions. In addition to Mexico, he was able to visit Guatemala and tropical Honduras and, on the return trip through the United States, to get acquainted with the most important scientific, agricultural and biological establishments.

The main task of the journey was first and foremost to study subtropical and tropical crops of basic interest to ourselves, e.g. crops of subtropical fruits, rubber plants, cotton, rice, and so on. In addition, the task was to get acquainted with everything new that had occurred during the past years with respect to the organization of scientific research concerning the rural economy.

In this report we will pass over the agronomical details, which were of special interest to the author in his capacity as an agronomist and botanist, and dwell mainly on the processes that concern agriculture in North America and the world in general, while selecting what was of most interest for our own crops.

Subtropical and tropical regions from a point of view interesting to the agriculture of the U.S.S.R.

The Subtropics and Tropics are of exclusive interest for the plant industry since an enormous amount of cultivated plants, grown in our country and in Europe, originally came from the mountainous areas of these zones. The native lands of a great many of our cultivated plants, even those such as rye and wheat, are found

First published in *Zvezda* [*The Star*], no. 4, pp. 119–132, 1932.

in the mountains of the Tropics and Subtropics; hence their very great practical interest in the sense of providing material for the selection of new strains. It could be said that all the basic, original varietal material of the majority of the most interesting cultivated plants on Earth is locked up within the mountains of the Tropics and the Subtropics.

North America, with its subtropical and tropical mountain areas, is of interest to us mainly because of such crops as maize, cotton and so on. In particular, all the cotton presently cultivated is based on samples of Mexican cotton. All the maize, in which we now have an enormous interest and which within the next couple of years will occupy no less than 15 million hectares in the U.S.S.R., had its origin in Mexico. Among the foreign rubber-producing plants, the Mexican guayule is of special interest to us, since it can be grown in our country in Azerbaidzhan and Turkmenistan. In addition, Mexico and Central America are the native lands of many vegetables and fruits; some of these are of immense interest for our southern regions.

EXTENSION OF THE CULTIVATED AREAS INTO THE TROPICS AND SUBTROPICS

It is necessary to add to what was stated above that the destiny of worldwide agriculture is to a considerable extent at present determined by events taking place in the Subtropics and Tropics. During the last couple of decades great changes have occurred with respect to the expansion of cultivated areas into the Subtropics and Tropics. Not long ago, before the Panama Canal was opened, farmers were frightened of the tropics with its diseases, yellow fever, overpowering tangles of vegetation and impassable thickets of trees. The primitive farmers avoided the lowland areas of the Tropics. The ancient civilizations of North and South America – the pre-Incas, the Incas, the Mayas, and so on – developed, as we now know very well, mainly in the mountain areas of the Tropics. So far, enormous expanses of tropical South America have never been seen by Man. The Tropics became available for utilization by farmers only after the opening of the Panama Canal, at which time people first became truly able to master this region.

Major events are presently taking place in the Tropics. We are, however, still at the starting point; the events are only beginning to unfold: immense plantations of tropical crops are being laid out, such as for bananas, coffee trees, rubber trees, sugar cane, cacao and coconuts. Lots of capital is directed, together with all kinds of modern technical power, into the depths of the Tropics.

In Honduras, we were able to make acquaintance with one of the worldwide industrial organizations for utilizing the Tropics, i.e. the 'United Fruits Co.', and to visit their plantations. This company is actually a 'state' of its own: it owns 73 steamships with a total capacity of 400 000 tons, its own railroads extending over 2000 kilometers, a 1000 km long tramway line, about 6000 automobiles and several hundred steam engines. This enterprise has about 200 000 acres under bananas alone (i.e. ca. 80 000 hectares), and in addition, it has subcontracted the plantations of various proprietors. In other words, this is an enormous organization in the service of which a number of experimental establishments exist, including a first-class experimental station at Tela in Honduras, which is directed by one of the most authoritative specialists on tropical plants, Dr. Popenoe.

We are only just beginning to utilize the Tropics, but already now they are becoming a serious competitor to the temperate regions as far as the production of sugar and vegetable fats is concerned.

With respect to sugar, the monopoly is unquestionably held by Cuba and Java, the production of which already considerably exceeds that of Europe, which is based on sugar beets. Coconut oil, which was unknown on the European market only 15 years ago, is presently in second place after cottonseed oil and amounts to an annual production of 5 billion pounds. It is used mainly for the production of margarine. The production of cocoa, coffee, sugar cane, quinine, rubber and coconuts in the Tropics has increased three to ten times during the last couple of decades. The advancement of such crops takes place on one continent after another. From the original native land of the rubber tree, *Hevea*, i.e. South America, its cultivation has now spread to tropical Asia, where enormous industrial-scale plantations are laid out. At present, the main producer of rubber is no longer South America, as it was not long ago, but the Malayan islands.

The temperate countries still hold the monopoly on production only as far as vegetable proteins (wheat and beans) are concerned.

The development of science and technology during the past years has directly opened up new resources in tropical and subtropical areas. In short, by means of examples of various crops of special interest for our Soviet Union, we shall acquaint ourselves with the subject concerning the movements that are taking place directly under our own eyes.

THE EXPANSION OF VEGETABLE CROPS

In reference to facts gathered from the plant industry in the United States, our attention was drawn primarily to the extraordinary increase in the acreage of vegetables grown. During the past nine to ten years the area under vegetables in the U.S.A. has literally doubled. The area cultivated with vegetables now occupies more than one million hectares in the United States. In 1929, tomatoes alone occupied ca. 150000 ha and carrots just above 40000 ha. In 1929, the general value of vegetable crops in the U.S.A. was determined to be $750000. Market garden crops have developed into field crops, and, in essence, the term market garden has, under the present conditions in the U.S., even lost its original meaning.

The enlarged production of vegetables has definitely changed the diet of the U.S.A. and reduced the consumption of cereals and fats.

The increase in the production of vegetables, the extraordinary mechanization and the use of all kinds of machinery such as seeders, harvesters and cultivators, are of exceptional interest to us. Already preliminary tests in the U.S.S.R. with some of these new American machines fully demonstrate their suitability for us.

In parallel with the development of vegetable crops, there is an extensive development of the canning industry as well, which uses many vegetable crops for preserves, and also an extensive development of transportation of the vegetables with use of special refrigerated carriages and various simplified adaptations for the transportation of fresh vegetables over great distances.

Under these conditions of reformation of the rural economy, the dispersal of vegetable crops into the areas of a country that are regular year round providers of fresh vegetables to major industrial centers is of special interest to us. Americans enjoy fresh vegetables almost all year round. During the winter, squashes come from northern Mexico, where they can be grown all the time. If you ask a New Yorker where he gets his squashes, he will inform you of the complicated geography of this crop. Every month squashes come from different areas: Colorado, California, Mexico, and so on.

THE DEVELOPMENT OF FRUIT CROPS

Although crops of ordinary fruit trees have not undergone any changes during the last couple of years in America, there has been an enormous increase, by as much as two or three times, in the area cultivated with citrus and other subtropical fruits. Thus, e.g., the amount of oranges grown in the U.S. in 1920 was estimated to be 68 000 car-loads, while in 1928 it amounted to 102 000 car-loads, although this figure does not include grapefruit, those peculiar and large citrus fruits that are presently grown in large quantities in Florida, Texas, Arizona and especially in California. Neither are limes, kumquats or other such citrus fruits included. Oranges and grapefruits are presently widely available in the United States.

With respect to the cultivation of citrus fruits, the United States can be a good example for our subtropical areas, especially those along the coasts of the Black Sea, where we are able to grow them. In spite of the comparatively mild climate in the American Subtropics, the economically-minded Americans take measures against frost. In northern Florida, California and Texas, special metal burners [smoke or smudge pots], used for heating the plants in case of frost, can be seen in almost every orange grove. Approximately 12 hours before the threat of frost, the farmers receive telegraphic information about it. Weather forecasting in America is extremely well equipped and, as a rule, the farmers always make good use of meteorological stations and the bulletins distributed according to their needs. If they receive information about a threat of frost, they ignite the little burners in the orange groves for a few hours. A special kind of mineral oil is used for the heating. This type of burner is not very complicated and could easily be produced by our own factories.

Important progress has been made in the United States while trying out frost-hardy varieties of citrus fruits. The work of finding new varieties of fruit trees has been extremely widely pursued in America. The Americans have brought back a large amount of species and varieties from the native lands of citrus fruits, i.e. China and, in part, India. At experimental stations in Florida and California, it is possible to see a very variable assortment of citrus trees. By hybridizing some of the cultivated varieties with different wild or cultivated ones brought back from Asia, new valuable and frost-hardy varieties have been obtained that, at present, are about to be taken into cultivation. Thus, by hybridization, it could be possible to raise the frost-resistance of the ordinary Japanese mandarine oranges that are grown along the shores of the Black Sea in our own country. No doubt, the utilization of such new varieties in our country could open up an opportunity for expanding the cultivation of mandarine oranges along the Black Sea coast.

Very much has been accomplished within the field of the mechanization of the fruit growing industry with respect to the care of the orchards, the fight against pests and the mechanization of sorting, packing and transporting the fruits.

THE HUMID AND THE DRY SUBTROPICS

The Subtropics of North America remind one in general of our own subtropical areas. Just as in the Soviet Union, it is also possible to distinguish between dry and humid subtropical areas in the U.S.A. To the latter belong Florida and southern Louisiana, which correspond to our Black Sea coasts. The dry Subtropics in America are found in California and southern Arizona, and to a certain degree they correspond to our subtropical areas in Azerbaidzhan and the southern parts of Turkmenistan.

In general, the amplitude of the climatic range is greater in the American Subtropics than in ours and the actual territory of their Subtropics by far exceeds that of the Soviet one. The subtropical area of California, with its variable climate and its mountain relief, is of the greatest interest to us. The amount of precipitation there varies from 200 to 1200 mm annually and some areas of California and southern Arizona remind one very strongly of our own subtropical areas of Azerbaidzhan.

The conditions found in Florida are extremely peculiar. We were able to visit Florida after a trip the same year to Murmansk [on the Kola Peninsula]. To our surprise when traveling through Florida we often saw a landscape that reminded us of Karelia: low-growing pines, endless bogs and trees covered with Spanish moss, an angiosperm that at a quick glance looks like a lichen. The subsoil horizon in some parts of Florida consists of dense limestone, barely covered with sandy soil. A very large portion of Florida is swampy. The presence of reflecting ponds among the pines enforces the superficial similarity with Karelia even more. With respect to its climate, Florida is almost tropical. The amount of precipitation is, on average, no less than 1500 mm a year. In general, the conditions in Florida are quite peculiar so that the results of experiments transferred directly from Florida into our humid subtropics must be judged with great care.

As stated above, the subtropical area of California is of the greatest interest to us due to the diversity of its climate and soil conditions as well as its crops. The number of cultivated plants, which are of variable technical importance, amounts to about 160 kinds. We know of no other country that has such a variety of cultivated plants. Different crops are brought there from all over the world and out of these, those most valuable to the plants industry have been selected.

As is well known, the date palm is a unisexual plant that needs artificial pollination to fertilize its flowers. One can still see idealized pictures of the artifical pollination of date palms in the old bas-reliefs of Babylonia. Thanks to work done by American scientists, it has been discovered that by selection of the appropriate pollen, it is possible to regulate the quality of the taste, the shape of the fruit and – most important – the rapidity with which the fruits ripen. At present, the cultivation of date palms occupies tens of thousands of acres and, in the near future, the date palm will indeed become one of the most important cultivated plants in the southern areas of California and Arizona.

Many Mediterranean plants have found a 'second homeland' in California. It is rare to see such beautiful olives in Greece and Spain or such enormous fig plantations as in California. Vineyards cover great expanses in central and northern California. With respect to research, it is in any case possible to see in California how much can be accomplished from the point of view of investigations concerning the cultivation of fruits in subtropical areas. The American standard varieties of fruits, which are selected on the basis of the enormous worldwide diversity thereof and propagated by experimental stations in California, are unquestionably of great interest for our areas in Transcaucasia, Turkmenistan and the Crimea.

THE CULTIVATION OF RUBBER PLANTS

Until recently, the cultivation of rubber producing plants was concentrated exclusively in the tropical areas. The main species of rubber trees, which are of enormous importance to the industry, i.e. *Hevea* and *Castilla*, are tropical trees that cannot grow even in the Subtropics. The development during the past decades in the tropical areas of Asia and the complete dependency of the worldwide rubber market on England and Holland – which hold the monopoly on the management of rubber plantations – has lately compelled the Americans to develop the cultivation of rubber-producing plants within the borders of the United States itself. We were able to get acquainted with the present state of the cultivation of rubber plants in North America, especially within the subtropical areas.

Of the numerous wild plants studied, the Americans selected the Mexican plant called guayule [*Parthenium argentatum* A. Gray]. This is a low-growing perennial shrub, producing a considerable amount of rubber when four to five years old. We were able to visit the native land of the guayule, the mountain areas of northern Mexico and southwestern Texas, where wild thickets of this plant are concentrated. Guayule grows at an altitude of 1600–2400 m and occupies mainly stony slopes among the mountains. The amount of wild guayule is somewhat higher in northern Mexico so that during the last decade, factories were built there to utilize the guayule. At full capacity, these factories could produce ca. 15 million pounds of rubber per annum.

The instability of Mexican politics has compelled the American capitalists to take over the cultivation of guayule and during the last couple of years they have placed large plantations of it in California, especially near Salinas. This year, they occupied an area of about 6500 acres. Wild guayule produces ca. 10% rubber of a fairly low quality, which in the form of 'black rubber' can be utilized for making tires and other articles.

The entire cultivation of this new rubber plant is in the hands of a single large American company, the 'Intercontinental Rubber Co.' All the work of this company is conducted under the most stringent secrecy. The research station in Salinas is presently closed to visitors.

The first factory for processing guayule was opened in California on February 6, 1931. No doubt the technical problems of introducing guayule into cultivation can be considered solved. In this respect, the experiment in California is quite instructive for us. Our own experiments during the last couple of years show that guayule can apparently be successfully grown in our country in Azerbaidzhan and Turkmenistan.

Now we know how the cultivation of guayule has been consolidated in the United States: the value of rubber on the world market has fallen considerably lately; at the time of the war [WWI] the price of rubber amounted to $1.5/lb, but during the last months it has gone down to 7 cents per pound. The cultivation of guayule under the conditions of expensive labor in California is without doubt disadvantageous at such a low price. Most of all, the quantity of guayule rubber is not very large. Rubber made from guayule contains only 20% resin. The minimum price that would sustain the cultivation of guayule is, according to information communicated to us, not less than 25 cents/lb. The factories using wild guayule in Mexico were already closed down last year. Furthermore, the utilization of the wild thickets of castilloa – i.e. *Castilla elastica* Sessé, a woody plant producing rubber in tropical Mexico, – is at present abandoned due to its unprofitability.

It is impossible not to mention the most recent research in the field of the cultivation of rubber-producing plants in the Subtropics, i.e. the experiments conducted by Edison. While attempting to save the United States from the necessity of importing rubber and while considering the possibility that the United States could become isolated in times of war, Edison started to look for plants with a high content of rubber. As a result of his investigations, he was able to find some species of goldenrod (*Solidago*), containing a considerable amount of gutta-percha of a high quality in the leaves. This perennial plant can give two harvests a year under the conditions in Florida: the stems grow up to 1.5 m tall and have lots of leaves, which contain a considerable amount of gutta-percha. Under similar conditions, this plant is no less interesting to us than the guayule. Edison's basic idea was to try to find such plants as could produce rubber every year and could be treated like an ordinary field crop. He solved this problem on the basis of his experiments with goldenrod. We do not yet know how successful these plants would be in our humid Subtropics. So far, Edison's work is only just beginning and the entire experimental plantation of goldenrod amounts to only one hectare.

In general, and in spite of all the interesting results with guayule and goldenrods in the United States, these crops have so far little value for us. To a considerable extent, the problem of introducing rubber-producing plants into our southern areas remains to be solved by ourselves while hunting for new plants rich in rubber within the associations of our own wild flora.

The low interest in the cultivation of rubber producing plants in the United States is, of course, justified by the fact that the problem is not particularly relevant there. Within the tropical areas close by, in Central and especially in South America, there are enormous expanses that could be better utilized for plantations of rubber trees. One of the foremost specialists on the Tropics, Dr. Sapper, states that the humid Tropics suitable for agriculture occupy easily one third of the world's land area. In other words, an entire third of our world could be utilized for agriculture, including to a great extent the cultivation of rubber plants. It is not by chance that a close friend of Edison, Henry Ford, who is well acquainted with the work of Edison, recently placed a colossal plantation of rubber trees in Brazil.

THE CULTIVATION OF RICE

Although we do not see many changes within the field of rubber plant cultivation in North and South America, we see facts of primary importance with respect to the cultivation of rice, indicating a new era for that plant industry.

The cultivation of rice, which was until recently very laborious, can be said to be specific of the overpopulated tropical and subtropical countries with cheap manual labor. So far, the cultivation of rice is concentrated mainly in China, Taiwan, Japan and India, i.e. the most overpopulated countries, where there is a low regard for the labor force.

Until recently, the cultivation of rice in the United States was concentrated mainly in Louisiana. The high price of rice at the time of the [First] World War made it feasible to transfer this crop to northern California. The promotion of the cultivation of rice in an area with expensive manual labor led unintentionally to an attempt at the maximum mechanization of its production. At present, this mechanization has reached an apogee when using airplanes to sow the rice. The speeding up of the use of airplanes was in part due to a peculiar event: in 1929, the seeding of rice in northern California was ruined by migratory birds; it appeared necessary either to cover the fields or to reseed them. Under the conditions of northern California, a delay in sowing is always risky since rice there is already at the limit of where it can be grown and each day of delay means a risk that it will not ripen. The use of seeders met with great difficulties because of the soggy condition of the fields in the spring. In northern California, the large amount of precipitation and the surplus of water at the time of sowing cause great difficulties. During the attempts to get out of the critically difficult situation, the idea of using airplanes was hatched, the more so since airplanes had already been used in the fight against insect pests. A corresponding change was made in the planes to make them able to disseminate seeds and then the fields could be reseeded by means of the airplanes.

The front of an airplane was filled with grain; it could hold about 700 lbs. Usually commercial postal planes are used for the operation and they fly at an altitude of about 20 m when sowing. As this experiment showed, the seeding was fairly even. The first attempts in 1929 gave definitely positive results. In 1930, the experiment was repeated on a larger scale and again with positive results. In particular, the use of airplanes solved a number of very difficult problems connected with the cultivation of rice, especially the fight against weeds. The possibility of sowing rice from an airplane directly into water makes it possible to save the crop from a large part of the weeds typical of it, since rice endures submersion more easily than the majority of weeds. Therefore, sowing rice in water prevents the sprouting of the major part of the weedy plants. Thus, as sown in practice, the seeding of rice from an airplane is extremely neat and the crop does not require weeding. Most of all, the use of airplanes reduces the cost of seeding to a considerable extent. The difficulties connected with irrigated crops needing furrowing could also be eliminated by the use of airplanes, which while sowing are not bothered by the furrows, which present many difficulties for the passage of seeders on the ground.

In the farmers' opinion, the problems of using airplanes for seeding seems to have been solved and this year [1932] a number of airplanes have already been contracted for mass-seeding of rice in northern California.

All the remaining operations connected with the cultivation of rice can be accomplished by means of machines that are commonly used when harvesting grain. Minor alterations in the construction of the harvesters have to a considerable extent solved the problems of mechanizing the rice harvest. Even combines are already used in areas where the harvest of rice can be accomplished in dry weather.

In other words, the very laborious cultivation of this grain that is typical of the overpopulated Asian countries can at present be completely mechanized. The harvest in California is no worse than that of Japan and even the price of rice is lower than that in Japan. From this, it can be understood what kind of changes take place at present within worldwide agriculture with respect to this crop and its expansion. In spite of the fact that manual labor in Japan is ten times cheaper than in California, America is, due to its mechanization, already able to compete with Japan with respect to the cost of producing crops of rice.

It is not difficult for us to understand the great importance of these facts as far as the expansion of harvests of rice within the Far East, the Caucasus and Kazakhstan are concerned.

THE CULTIVATION OF COTTON

The United States is a classic cotton-cultivating country. Until recently, this kind of crop was concentrated mainly in the southeast, in the Carolinas, Georgia and eastern Texas. This area is characterized by a high amount of precipitation. In general, the conditions for cotton cultivation in the United States are clearly different from those in our Turkestani areas. However, during the last couple of decades great changes have taken place within the cotton industry in the United

States. Due to the effect of the extraordinary development of harmful insects, especially the boll weevil, which has ruined whole areas of cotton, this crop has begun to be moved into drier and climatically more severe areas similar to those on our steppes in the northern Caucasus and on the Tamanskiy peninsula. Thanks to the severe winters, there are no boll weevils and the cotton is successful in areas characterized by a shortage of manual labor and a short growing season. At the same time, the level relief of western Texas allows the use of tractors and other machinery for the work.

The mechanization of cotton cultivation started in connection with the growing of cotton in the areas of western Texas by using harvesting machines while the harvesting in the original areas was, as a rule, still done by hand. The construction of cotton picking machines is still going on, but now the problem of creating cotton pickers with a great capacity can already be considered solved. We were able to see a new type of construction called a 'sled', which can already now be considered satisfactory and suitable for our new cotton-growing areas in northern Caucasus and the Ukraine, where the greatest difficulty is the lack of manual labor.

In any case, the problem of mechanizing the cotton harvest has entered a new stage and will, no doubt, be completely solved within a short time, especially in areas that produce cotton with a small number of bolls, allowing a mechanized harvest comparatively easily. After our own expansion of cotton crops into new, non-irrigated areas of the northern Caucasus and the Ukraine, the application of such machines will play a decisive role for the further expansion of this crop. Mechanized cultivation of cotton definitely opens up new horizons for this crop, allowing a considerable increase in its area and a transfer into new areas, while the production cost of cotton will be considerably reduced.

THE CULTIVATION OF SUGAR CANE

Let us briefly consider still another subtropical crop, which is not of direct interest to us but which indirectly affects our own cultivation of sugar beets.

The ordinary kinds of sugar cane cannot grow in our country and could, in any case, not be cultivated for industrial purposes.

In the United States, sugar cane is grown mainly in the southern states, i.e. Louisiana and Florida and, on a much smaller scale, in Texas. Cuba, Java and, in part, India are the main areas in the world for the cultivation of sugar cane. Cuba alone produced four million tons of sugar in 1930. Java produced approximately the same amount.

Sugar cane furnishes a considerably larger quantity of sugar per unit area than sugar beets. The cost of producing sugar from canes is about three times lower than that for beets. The application of hybridization by Dutch scientists in Java has been of decisive importance for the expansion of sugar cane cultivation. As a result of crosses between wild but stable species of sugar cane and old Chinese, low-yielding strains, it was possible to almost treble the yield per area unit. At present, sugar cane has become a deadly threat to the European sugar beet which is of immense economic importance for such countries as Germany, France and

Czechoslovakia, not to speak of our own country. In the near future, the actual cultivation of sugar beets in western Europe may be saved only by regulating the price, but even in that case, the opportunities for the cultivation of sugar cane have far from been eliminated.

We were able to observe the operation of new harvesting machinery in Louisiana. The waste from the production of sugar cane is now largely used for so-called 'Cellotex'. Boards about 1 $m^2 \times 2$ cm are prepared from the cellular tissue and lignine left after the separation of the sugar and these are now a valuable building material, more solid than wood and insulating against heat up to 35% better than wood. At present, the use of cellotex has acquired a wide distribution in the United States, especially for simple buildings, where it essentially curtails the need for fuel. At the same time, it reduces the cost of sugar production.

At the last International Sugar Conference in Brussels it was decided to limit the export of sugar to European countries by 15%. This is no doubt only a first step in the further development of the competition between the tropical and the temperate zones with respect to the cultivation of sugar.

NEW CROPS

Finally, it is necessary to mention the widespread introduction of new plants into cultivation, which is based on experiments in the United States during the past decades. At present, about 10 million acres are already occupied by crops of African sorghum, used as a grain, green manure and, in part, for the preparation of syrup. This 'work-horse of the plant kingdom' is indispensable for the dry, southern areas. Of course, we ourselves should also use the experiences of the United States for the expansion of our own acreage.

In the United States, about one million acres are used for the cultivation of sweet potatoes, which in the southern states replaces ordinary potatoes. The latter do not do well there due to the development of diseases. Unquestionably, the experiences of the United States can be used by us for our southern areas as well. In Louisiana, we were able to see new factories for producing starch from sweet potatoes.

OPPORTUNITIES FOR GLOBAL AGRICULTURE

Exceptional opportunities have at present opened up for the utilization of tropical and subtropical areas. It is possible to predict future events when considering the enormous area of the humid Tropics, which cover one third of the world's landmass. The opportunities for the Tropics with their fertility, abundant moisture and lack of winters are endless.

However, opportunities are far from lacking in the temperate zones either. When a soil map of the world is drawn up, it can be seen what enormous areas of valuable soil, suitable for agriculture, still remain unused by Man. Only about 7% of all the world's landmass is used for agriculture. It is only now, when we know the soil map of the world in general terms and have started to get relatively well acquainted with the resources for agriculture that we can see the unlimited perspectives that exist for the utilization of the natural, productive capacity of the world. If mankind should increase by three to four times, there will still be abundant resources for its nourishment. It has definitely become obvious that only a new organization of human society, i.e. only a socialistic society overcoming the barriers put up by the capitalistic system, can lead to an effective utilization of the productive capacity of our world.

The global crisis of the rural economy

Never before in the present time has there been such a deep-rooted opposition to the worldwide development of agriculture associated with capitalism.

Above, and in detail, we have considered what colossal changes for the better and what colossal opportunities there are that until recently were not expected with respect to agriculture. We have seen how endless perspectives have been opened up by science and technology for Man to control nature, and what grandiose opportunities are concealed in the world. During this period of immense opportunities, of which Man was ignorant in the past, an unprecedented and worldwide crisis is unfolding right under our eyes, enveloping both the rural economy and industry. No doubt, there is not a transient cause for the development of this crisis but it has deep roots, entangled in the barriers of capitalism. At the same time as three quarters of the world's inhabitants do not have adequate provisions for bread and at the same time as three quarters of these inhabitants live in poverty, the capitalists in North America and a number of colonial countries do not know what to do with the abundance of their products. The problem of over-production includes, in many countries, even the intelligentia. They keep themselves busy by attending international conferences. One of these international conferences discussed seriously the questions of whether to stop the production or to limit the area cultivated or whether to stop the developing mechanization of the production, which seems only to aggravate the competition. At the same time, the extraordinary curtailment of the buying power of the populations obstructs more and more the mechanization of production and reduces the funds needed for the perfection thereof. The culmination of mechanization, such as we could see with respect to the cultivation of rice, is a result of the search for ways to lower the production costs. Mechanization of production throws out more and more goods on the market at the same time as it reduces the amount of manual labor needed. In that manner, the development of a crisis is accelerated.

As far as the rural economy of capitalistic countries is concerned, a directly unprecedented and serious crisis is unfolding right under our eyes and millions of hectares of cultivated land are abandoned. Agriculture is becoming unprofitable.

The author was able to participate in an excursion, arranged by the International Conference at Ithaca concerning rural economy, in order to acquaint himself with deserted farms on hundreds of thousands of acres in the state of New York. These abandoned farms with their ruins of houses and overgrown paths and roads, situated practically next to a railroad, present a horrible picture. Lately, from 200 to 300 000 acres have been abandoned each year. Cornell University devoted a number of monographs to the problem of the deserted land. The main reason for abandoning the farms is the unprofitability and the impossibility of competing with other, more fertile areas.

A similar, but even clearer picture of this can be seen in Guatemala and Honduras: recently, banana plantations have been forced to shut down. Inexpensive rubber has forced the utilization of the castilloa forests in southern Mexico to be halted. The appearance of a pest, a *Physarium*, on bananas is presently considered a blessing, since it curtails production and reduces competition.

In California, orchards can be seen where fruits are left hanging on the trees, since the people do not know what to do with the crop. When we traveled past one of the new orange groves, a fellow agronomist said to us: 'It is fortunate that this grove is so young, since what could we do with its crop today?'

How profound the crisis actually is, can be judged by an address by the United States Department of Agriculture to the people, suggesting that the area cultivated be reduced during the next couple of years. At the same time as the Soviet Union is increasing its cultivated land at an unprecendented rate, the United States is forced to reduce its crops.

The mechanization of agriculture, still far from slowing down in the United States, makes the operation of small farms disadvantageous. Under these conditions, mechanization of farms of 70 hectares (the size of the average American farm) becomes unprofitable, since they do not offer any possibility for using tractors. A very famous economist in the United States has suggested that the number of farms be reduced. Dr. Baker, one of the foremost among economists, considers it necessary to reduce the number of farms in the United States from six to three million in order to get out of the impending difficulties in the USA.

During our journey, we concentrated our attention mainly on the problems concerning technology, since we did not have any opportunity for a detailed analysis of the worldwide crisis, except for some obvious facts: there are millions of unemployed workers in the United States, some five million thereof in Germany and more millions in England. There is an unmanageable drop in the prices at the same time as there is both a lowering of buying power and an overproduction. This picture could, in 1930–31, be seen from the Tropics to the limit of agriculture anywhere.

The growing conflicts between the development of agriculture and industry all over the world are unfolding against a background of major political events taking place. Almost every month the governments of the republics in Central and South America are overthrown. One of the new presidents (of Brazil) stated to a correspondent of a New York newspaper that he considered himself a 'bird of passage', referring to his shortlivedness. Major events are also unfolding right under our eyes in India . . . and everything that happens across our borders is of exceptional concern to what occurs in our own country.

Improvements in the organization of scientific research within the field of agriculture

We shall briefly consider what has occurred lately concerning the organization of research work within the field of agriculture.

The primary and basic fact, when considering the development of the agricultural sciences in the United States of America and in Western Europe, is the enormous growth of research. Already, within such a short span of time as five years (I was last in western Europe in 1927 and in America nine years ago), major events have occurred within the development of scientific work.

While summing up the impressions from my acquaintance with the organization of scientific work in various countries, three kinds of directions can in general be basically outlined:

First, we shall outline the definitely American type of organization of research all over the country, but built up according to local principles as far as distribution of the work on the different states is concerned. The agricultural research stations within this system are associated with the different states and their universities. The character of the work at the experimental stations is universal, involving various departments ranging from plant breeding to technology and economics. There is no narrow specialization with respect to different crops. As a rule, the stations work with many crops. At the same time, there exists both in the United States of America and in Canada a well-planned research organization at the Department of Agriculture in the form of different bureaux. In the past, we, too, had – to a certain extent – such a system.

The second type of scientific organization is a branch type, i.e. a development of institutes with different service branches and a definite kind of product. Some countries have this type of research institution exclusively: e.g. in essence, there is in Italy no general institution but only specialized institutes such as for tobacco, sugar beets, maize, root vegetables, etc. Until recently, Italy did not have a single research establishment of a complex type; the first attempt in this direction was an experimental station at Bari, but it encompassed only a few departments of the agronomic sciences.

The third type of research establishment, characteristic of the capitalistic system, consists of various independent and fairly large institutions, fortuitously springing up thanks to donors and owing to the skill of the different research workers for attracting money. There are very many such establishments in the United States and England, where it is possible to encounter institutions that literally do not know what to do with millions of dollars. These institutions are often devoted to a particular interest with respect to the concentration of this work in specific directions. The number of such institutions has increased considerably during the last couple of years, especially in the United States. They are extremely variable and not easy to classify.

Let us mention some examples of such institutions: In California, we were able to visit a research institute for food products under the direction of Alsberg, a chemist. This is a small institute with a first-class library, concentrating its work on a study of global economy and the geography of food products. A small group of scientists, especially technologists and economists, work there with small-scale technical instruments. This little institute publishes a couple of volumes annually devoted to the economy and geography of wheats and concerning plant and animal fats. There is a beautiful statistics bureau at this institute. It does not belong to any particular system but is affiliated with Stanford University.

Among such private institutions, there is also a new one for plant physiology (the Boyce–Thompson Institute), built up on funds from a millionaire, Colonel Thompson. The main capital of this institute amounts to $10 million (i.e. ca. 20 million gold-rubles). The annual budget of this establishment, working only with plant physiology, is about $7–800 000. The institute is equipped like a factory, according to the last word in science. There are greenhouses that can be regulated according to their need for light, color of glass, temperature and humidity. Every worker, even the beginners, has at his disposal everything he needs. In other words, this is an ideal set-up, such as most research workers can only dream of. Nevertheless, this establishment, which has already existed for five years, has so far not surpassed some of the smaller establishments, clearly demonstrating that the dollar far from solves all and that there can be a poverty of ideas even under first-class conditions. Neither in western Europe nor in America did we encounter such a combination with respect to the development of the theoretical sciences in public institutes and such a widespread development of branch institutes, associated with the production, such as we, ourselves, have recently started to build up.

NEW SCIENTIFIC ESTABLISHMENTS

With respect to how quickly research establishments concerned with agriculture spring up, we can, for instance, refer to France, Germany and the United States and especially to the colonial countries. France, which is most deficient in the development of its scientific work as far as agriculture is concerned, has finally during the last couple of decades started to catch up with other countries. Shortly after the war [WWI], France established a new agronomical institute (L'Institut de Recherches Agronomiques) at Versailles. It corresponds to our own Institute of Applied Botany. Using France as an example, it can be seen how strangely the development of the agronomical sciences occurs in the capitalistic countries. Only in 1919, a really major governmental research establishment was finally organized according to a definite system. Time does, of course, not stand still, and different commercial firms, needing the services of science, had themselves already started the establishment of experimental institutions. At present, France experiences a sharp conflict, a peculiar fight between the businessman and the State concerning the organization of scientific work. Thus, e.g., all plant-breeding work, practical as well as theoretical, is still concentrated in the hands of the single monopolistic establishment of Vilmorin and Andrieux, which has already existed for 200 years. The establishment of the new governmental organization of plant-breeding work corresponds, of course, badly with the interests of the Vilmorin firm. The same can be seen with respect to agro-technology: the Truffeau Institute always took a rather malevolent position against the state-operated agro-technical institute.

In spite of its great economic difficulties, Germany has developed its scientific work broadly. Recently, a number of extremely interesting new research institutes have been established. We were able to acquaint ourselves thoroughly with an institute (the Kaiser Wilhelm Institut für Pflanzenforschung) that is directed by the famous geneticist and plant breeder, Dr. Baur. In spite of the fact that this institution has existed for only three years, it has already produced valuable practical results thanks to the excellent training of its scientists, the first-class equipment, the large number of greenhouses and the first-class chemical laboratory. The results of the chemistry-assisted plant breeding are interesting. Thanks to the use of chemical color reactions, they have succeeded in selecting strains of lupines without alkaloids and sweet clover races that have a low content of coumarine. They have also found strains of tobacco with a low nicotine content. The scale of the work produced by the Germans, the industrialization of the plant-breeding work and the exceptional results of this industrialization, allowing the time required for the work to be shortened, are astonishing.

UNIVERSITIES AND RESEARCH

The tendency that has been demonstrated in our country with respect to scientific work, in particular the detachment of research from teaching, is starting to develop more and more in western Europe and in America as well. The speed with which scientific work develops, especially within the field of applied sciences, tends to increasingly isolate scientific work from teaching. Even in the United States, where agricultural experimental work is associated with universities, we could see an increasing isolation of a large number of persons devoted to pure research work. Particularly characteristic is the development lately of a large number of institutes and laboratories for pure research work such as those in the United States, England and Germany.

NEW EQUIPMENT FOR LABORATORIES

The outfitting of laboratories with new equipment seems to be a new factor in the development of research during the last couple of decades, especially in the United States. The Parnell Act, doubling the budget for agricultural research funds, was used mainly for new equipment for the experimental stations. Being familiar with many of these stations 10 years ago, we could hardly recognize them in 1930. At many modest state universities, it is now possible to see first-class chemical laboratories. The set-up in which the scientific agronomical work is done is astonishing even where there are small funds, not to speak of that in the case of large centers.

There is much that is new and interesting in the realm of equipment that we, ourselves, will no doubt be able to use in the near future. In Chicago, we were able to visit a new botanical institute, costing $200 000. This is not a large sum in relation to American conditions. We believe, nevertheless, that this is one of the very best institutions. It is associated with a university and seems, in essence, to be a laboratory for providing occupations for doctoral candidates and for individual scientific work by professors and their staff.

In this institute there are mainly departments able to work according to the highest contemporary standards within botany: departments for the regulation of temperature, light and humidity and such, for application of X-rays and radiation, first-class greenhouses for rearing the plants and excellent biochemical, chemical and microbiological laboratories.

However, the most essential thing is the mobility of the equipment. The Americans understand that the old type of laboratory, where everything was firmly affixed, is now unsuitable. When constructing the Chicago botanical institute, the builder took into account the interests such as conveyed to him by at least three generations of scientists, while considering that, with the arrival of new people and new generations, it should be possible to serve new subjects and new interests as well. All the equipment appears easy to move; the different chambers can be enlarged if it is necessary in order to expand the work; even the water pipes can be easily displaced and large instruments can be switched from one place to another, since they are placed on wheels.

THEORETICAL WORK

The following is a characteristic trait of our time with respect to the development of research work, i.e. the great attention paid to theoretical work. During the last couple of years there has been a serious change in attitude toward theoretical work, mainly within the field of the applied sciences: first-class schools have sprung up and, in the case of a number of departments, the leading role belongs at present to the United States. Particularly, major changes have occurred as far as genetics is concerned. Thus, last year a new genetics laboratory was established at Pasadena near Los Angeles, where the school of Morgan works. Much theoretical work is also performed at the Carnegie Institute and at the genetical laboratory of Emerson at Ithaca.

THE DEVELOPMENT OF EXPERIMENTAL INSTITUTIONS WITHIN THE COLONIAL COUNTRIES

The development of agriculture in the Tropics and the Subtropics compels us to pay attention to the establishment of research institutions in the colonial countries. During the last couple of decades, we have seen a large number of new institutions and experimental stations spring up in different tropical countries. Some of these are first-class establishments. Suffice it to mention the institute on the island of Trinidad and a number of research institutions on the island of Java, the scientific work of which even reaches far beyond its borders. The dispersal of agricultural knowledge is a characteristic trait of our time. The number of scientific experimental institutions for agriculture has at present been determined to be ca. 4000 all over the world.

THE INCREASING NUMBER OF SCIENTIFIC PUBLICATIONS

The enormous increase in the number of research institutes and teams of scientists is naturally reflected in the number and contents of periodical publications and books. Textbooks become out-of-date even before they appear. In order not to drop behind in science, it is necessary to constantly keep on learning.

In Germany alone, the number of agricultural journals has doubled during the last ten years. The number of books published on rural economy in Germany is exceptionally high and in this respect there is even a 'scientific hypertrophy'. German producers and publishers of books consider, apparently, matters dealing with books to be profitable and overcharge not only the Germans themselves but particularly readers beyond the borders of Germany. Monographs, reference books and collective handbooks are published in enormous quantities. Considering the fact that many of the Soviet research workers publish their works in German editions and thereby support the publishing business in Germany, the tendency of the Germans to monopolize the book trade is a well-established threat.

In connection with the intensification of publishing activities, bibliographies are beginning to play an enormous role. It is necessary to mention the exceptionally fine production of bibliographic works at the library of the United States Department of Agriculture in Washington. This is, in the fullest sense of the word, a center of bibliography and at the same time the best library in the world with respect to agronomy. In addition, it should be mentioned that all this library work and all the bibliographical work is performed exclusively by women!

The interest in Soviet science beyond its borders

In conclusion, we shall touch upon the interest shown in Soviet sciences, especially in America. At every university we were requested to lecture on the five-year plan and the reformation of our rural economy. With respect to the dead-end at which the world of agriculture finds itself and which is beginning to be felt in the United States as well, the interest in how the new country of the Soviet Union solves the problems of agiculture is unintentionally growing. There were seminars sometimes twice a night, with Soviet lectures during the International Conference on Rural Economy at Ithaca. The enormous improvements concerning the expansion of the area cultivated, the mechanization of agriculture and the development of research work that are taking place in the Soviet Union are considered astonishing even in comparison with American conditions. The advantages of large state-run farms and the collectivization of agriculture are particularly evident to the agriculture in the United States.

The International Congress of Soil Scientists in Leningrad and Moscow was of great importance for getting acquainted with our country. This congress was very successful and a large number of foreign scientists took part in it and particularly in the long excursions arranged by the congress. At the same time as the public press in America, which is completely in the hands of capitalists, were full of insinuations with respect to our lectures, there appeared increasingly frequently valuable and serious statements in scientific periodicals and journals that illustrated the successes of the Soviet Union. Statements by Dr. Marbut, the most famous soil scientist in the United States, who attended our congress of soil scientists, were recently published in the United States. In his paper, Marbut offers a prognosis for the major events that can be expected within the Soviet country and for the increase in production that – in his opinion – is entirely feasible on the basis of all he saw in the Soviet Union.

The interest in America for the U.S.S.R. is so great that – in spite of the obstacles put in the way by the government of the United States for the departure of American scientists visiting us and in spite of the withdrawal of salaries of people going to the Soviet Union – many come to visit us, as is well known.

'You are making a colossal experiment of global importance', one of the most famous scientists in Florida said to us, 'and we are delighted that this experiment is succeeding. Most of all, we wish that it shall be successful and we are willing to help the Soviet country and the Soviet students as much as possible'. This is a typical example of the attitude that meets Soviet students when visiting the United States.

The level of Soviet science allows us to hold a position on a par with that exhibited at international congresses. It is enough to state that Soviet scientific handbooks are translated in large quantities abroad into English and German. The teaching of plant physiology is done almost exclusively according to the Soviet textbooks by N.A. Maximov. The book by K.D. Glinka, *Soil Types*, seems to be the basic handbook for soil management in the United States.

It is necessary for us to take this immense interest in Soviet science and the Soviet country into consideration. We must strive toward an internationalization of the Soviet sciences and the publication of our work in English, German and other languages. According to the example of Japan, we should, evidently, organize information about Soviet sciences and our achievements.

Sympathy for the Soviet Union can be easily created among scientists all over the world. Enlisting them can be of a great practical importance.

The plant resources of the world and their utilization for socialistic agriculture

THE FLORA OF THE WORLD is at present estimated to include approximately 160 000 species of higher plants. However, this number is considerably lower than the actual one. Whole continents, such as South America and southern Africa, as well as a large part of southern Asia, have only been cursorily visited by botanists and enormous territories have never been investigated. Each year tens, even hundreds of new species are found within the borders of the Soviet Union, especially in Inner Asia and Transcaucasia, in spite of the comparatively good knowledge of that flora. The most competent botanists in western Europe, America and the Soviet Union, with whom we have discussed the question of the number of species and their distribution, agree that *we do*, indeed, *not know more than about half of the number of existing species.*

In spite of all that is known about the utilization of natural plant resources, phyto-geography still appears to be only a science in progress, since we still do not even know the regions where there is a maximum concentration of a specific variation. The first attempt in this direction during the last couple of years was made this year, according to our predecessor at the Institute of Plant Industry, Professor E.V. Vul'f [Wulff], who produced an interesting map of the distribution of the number of plant species in the world.

In spite of the fact that our knowledge is still inexact, there are nevertheless already facts of great importance that indicate in general terms the areas of maximum concentration of specific and varietal diversity. The distribution over the Earth's surface of a variety of species has established the probabilities by which we are guided when searching for new varieties and new species for the purpose of utilizing them for socialistic agriculture. Primarily, the humid Subtropics and Tropics belong to the areas that are exceptionally rich in species but also the subtropical and tropical mountain areas belong there. In this respect, particularly the region of southeastern Asia including India, Indo-China and Malaya can be distinguished by the richest of floras, containing about one third of all the specific diversity known in the world. An exceptional wealth of species is also concentrated in Brazil, along the eastern slopes of the Cordilleras, in Central America and in southern Mexico. At the same time, the enormous

First printed in: *Nauchniy Leningrad k XVII s'vezdu VKP (b)*, [*Scientific Leningrad to the XVIIth Congress of the Communist Party, sect. b*], Leningrad, pp. 197–201, 1934.

northern expanses of North America, Europe and Asia are comparatively poor with respect to the composition of species. This is many times lower than in the tropical and subtropical countries, in particular as far as the number per unit area is concerned. While the 'pygmy' republic of Costa Rica in Central America has more than 6000 species of flowering plants, including no less than 1000 species of orchids, all of Canada and Alaska, with a territory hundreds of times larger, have a similar number of species. However, the flora of Canada is much better known than that of Costa Rica, and the actual number of species in Costa Rica may be twice as high. The wealth of the flora of Peru, Bolivia and Colombia is striking. Within the limits of the Soviet Union, there is an especially high number of plants in Transcaucasia and Inner Asia as well as in the countries adjacent to them, Persia and Turkey. There is a large number of species also in the countries situated along the coasts of the Mediterranean Sea. The recent paper by A.A. Grossheim, a detailed study of the geography of the flora of the Caucasus, is interesting since it establishes for the first time relatively exactly the areas richest in plants in Transcaucasia.

Our own investigations during the past decades were directed primarily toward a study of the varietal composition of the most important cultivated plants and the maximum utilization in this respect of all the assortment thereof available in the world. A detailed study of the diversity of strains of cultivated plants by means of exact and modern methods of plant management as far as morphology, physiology, cytology, biochemistry and technology are concerned, has enabled us to work out a geographical theory concerning the origin of cultivated plants.

The investigations carried out by myself and my colleagues according to a definite plan, the organization of a large number of expeditions (I, myself, have been able to visit 52 countries during these years), but also the systematic investigations assisted by new methods of differential taxonomy such as have been worked out here, revealed the varieties in the world that are of the greatest interest to us as field and vegetable crops. Just as when studying the wild flora, a definite localization in the world of intraspecific variation could be established. The areas of a striking varietal diversity, thus established, were in many cases not expected by the science of the past. Thus, e.g., an exceptional wealth of wheat has been revealed in the small country of Abyssinia within its insignificant, cultivated area. An amazing wealth of varietal diversity of the most important cereals in the world were discovered there. An enormous number of new varieties of wheat and other crops are concealed not only in Abyssinia but also in Turkey and northwestern India. Areas with a maximum concentration of primary intraspecific and specific variations of cultivated plants have been established.

At present, we can distinguish eight main centers of varietal diversity with respect to cultivated plants all over the world, in other words, eight centers of origin of the most important plant crops. The majority of cultivated plants came from Asia. Within Asia, we can distinguish four main areas: 1. eastern China, 2. Hindustan and Indo-China, 3. Inner Asia, including the mountains of the Soviet Inner Asia, Afghanistan and northwestern India, and 4. Asia Minor, encompassing Turkey, Transcaucasia and Persia, the native lands of European fruit trees and grapevines.

In addition, we can distinguish a Mediterranean area, the native land of olive trees, carob trees and many forage plants. In Africa, the mountainous Abyssinia appears to be a collective center giving rise to a number of cultivated plants such as hard wheat and barley. In the New World, comparatively small areas, no larger than 1/40th of the total area of North and South America, appear to be original foci with respect to the development of cultivated plants. At the same time, they are distinguished by an exceptional wealth of specific and varietal potentials. All the wealth in the world as far as the genes of maize, potatoes, and American as well as the so-called Egyptian cotton are concerned is concentrated in that area. Two centers can be distinguished in America: the south Mexican one, adjoined by the northern portion of Central America, and the Peruvian–Bolivian one, including also a portion of Ecuador. The enormous territory of North and South America turned out not to be of major importance for the most-cultivated plants on these two continents.

The eight centers mentioned support a colossal treasure of various species and varieties of cultivated plants. Our attention was directed mainly to these areas. Of course, when initially organizing the expeditions we made a number of mistakes in this direction, since the concentration of the specific and varietal potentials in the world were not then known to science; they had yet to be firmly established.

Our exploratory work was pursued with a particular objective in mind: to utilize the plant resources of the world maximally for the purpose of plant breeding based on a global gene-bank.

The success of plant breeding work is determined first and foremost by the original varietal material. The successes of Michurin, Burbank and other well known plant breeders was determined to a great extent by their incorporation of original material in their work with hybridization.

As a result of the systematic and persistent work of Soviet teams, we have succeeded in opening up a treasure chest of important cultivated plants that were previously not known to science, such as with respect to wheat, maize, barley, potatoes, rye, flax, beans, melons and vegetable crops. An astonishing wealth has been revealed in the case of wild fruits, which are found in large amounts in the forests of Transcaucasia and Inner Asia. A number of new species and valuable varieties are found there that can be used directly for introduction into cultivation.

During the last couple of years, Soviet expeditions have worked their way along the entire spine of the Cordilleras, starting in California and ending in southern Chile. New species of both cultivated and wild potatoes were discovered there, which had been unknown to science until then but which were utilized locally by the Indian populations. These discoveries literally meant a revolution within potato breeding. After the preliminary publications about this, Germany, the U.S.A. and Sweden sent out their own expeditions in our footsteps for the purpose of bringing back new material for practical plant breeding.

The plant resources of the world seem endless. Even after centuries, botanists have enough work left to study the treasures distributed all over the world. There is an immense wealth especially in the subtropical and tropical countries, in the mountain areas of Southern Asia and those of the Andes in Central and South America. These have been badly studied but the penetration thereof is hampered by the barriers put up by imperialistic politics.

The socialistic reformation of agriculture and of all our rural economy has opened up unlimited horizons for creative, investigatory work. The utilization of the world's resources can be accomplished only according to a definite plan and only by organized teams that have a definite purpose in mind. The Soviet system has provided such a purpose.

The Soviet plant industry is now able to report to the 17th Congress of the Communist Party, Section b, on the excellent progress made in the utilization of the world's resources. A new treasure chest, containing the most important cultivated plants not previously known to science, has been opened. During the last couple of years, the plant industry has become a 'new science' since more of even the most important cultivated plants have been discovered during the last six or seven years than during the 200 years of previous botanical work.

New and very interesting rubber-producing plants, such as tau-saghys [*Scorzonera tau-saghus* Lippsch. and Boiss.], kok-saghys [*Taraxacum bicorne* Dahlst., syn. *T. kok-saghus* Rodin] and krym-saghys [*T. hybernum* Stev.] have been discovered and brought into cultivation, where they can be grown in a temperate climate. The problems of Soviet gutta-percha have been solved. New fiber plants have been taken into cultivation. We still have not enumerated them all, but already now the Soviet plant industry, working on a socialistic basis, is aware of the fact that it has stepped onto the correct path. The immense organization into which the plant breeding and agricultural matters of the U.S.S.R. have grown during the last three years opens up new opportunities for an unlimited utilization of the world's plant resources. The continuously growing problems of the socialistic economy, which are increasing every year, the capacity of Soviet industry and the strength and the organization of socialistic agriculture put before the scientists more and more extensive tasks. In this respect, there is a deepening difference between our two worlds: one is the capitalistic one, slowing down from its apogee and sinking into a dead-end, not knowing what to do with itself. The work of the scientists in that world has become absurd and unnecessary. The other world is the one approaching the acme of socialism. There is limitless and fascinating creative work to perform and the farther we push on, the less limited the horizon becomes.

BIBLIOGRAPHY

Wulff (Vul'f), E.V. (1937). Opyt deleniya zemnogo shara na rastitel'nye oblasti na osnove kolichestvennogo maspredelenyia vidov [An attempt to classify the world into plant regions on the basis of the quantitative distribution of plants]. *Tr. po prikl. botan., genet., i selek.* [*Papers on applied botany, genetics and plant breeding*], Ser. 1 (2).

Plant resources of the world and the mastering thereof

BOTANISTS ESTIMATE the total number of flowering plants in the world to ca. 160 000 species. The distribution of these species is rather peculiar. The immense expanses of North America, Siberia and Central Asia have an extraordinarily poor composition of species; on the other hand, some areas on the surface of our globe have an extremely rich diversity of species. One can distinguish, in particular, southeastern Asia including India, Indo-China and the Malayan islands: no less than 60 000 species can be found there. The humid tropical areas, Brazil and Central America are also very rich in a variety of species. The small republic of Costa Rica alone includes in its specific composition a greater diversity than all of Canada, U.S.A. and Alaska together. Countries rich in species diversity are also distributed along the coasts of the Mediterranean. Within the borders of the Soviet Union itself, there is a special wealth of species in Transcaucasia and the mountains of Inner Asia. While the entire array of wild species in the U.S.S.R. has been determined to 17–18 000 species, Caucasus alone has some 6000. A similar quantity is typical also of Inner Asia, where the especially rich flora of mountainous Tadzhikistan can be singled out.

The search for valuable species to be introduced into cultivation was originally started in these areas, which are very rich in their species composition. Such a localization of specific diversity is the combined result of the entire evolution of the plant kingdom and the geological changes that have taken place on the surface of the Earth. The glacial age during the last epoch played an enormous role in the extermination of plants in the northern areas. At present, the variability of the species is to a great extent determined by the conditions of climate, soil and relief.

The humid Tropics and Subtropics are especially rich in species.

As never before, investigators attempting to master the plant resources will be able to establish the present geographical localization of a wealth of plants, primarily on the basis of these important facts.

With respect to cultivated plants, we are interested not only in the species but also in the varietal diversity within the distribution areas of the species. Wheat consists of many thousands of strains, differing in inherited characteristics. Thousands of varieties of rice and barley can also be distinguished. In order to be able to control the resources of the varieties of cultivated plants, it is necessary

First printed in: *Nauka i Zhizn'* [*Science and Life*], no. 3, 15–18, 1935 not only to know the geography of the species but also to go further, in order to clear up the basic diversity of the strains and to establish the areas where the original varietal diversity of the species is concentrated.

For the improvement of our cultivated crops (wheat and other cereals), as well as those of fruits and vegetables, it is necessary to know where to find the basic, original potential of the diversity of varieties and species. In order to reveal these localities, it is necessary to study the botanical, intraspecific composition of the species, their diversity of varieties and the exact areas where a given crop originated, preferably by using specific examples.

Considering the immense variety of strains, the knowledge of botanists about the many interesting regions in the world was poor so that, according to the Soviet doctrines, which state the problem of mastering the strains of the plant resources, it was necessary to invent a method for establishing the areas of origin, where cultivated plants were formed.

For this purpose, we proceeded logically by studying one cultivated species after another as far as the diversity of the varieties is concerned. These investigations led us to understand the species as a complicated and active system, associated in its historical development with a special environment or territory and in its diversity subject to definite regularities. With respect to their diversity, genetically closely-related species appear more similar since they are subjected to the law of homologous series within the hereditary variability such as already established by myself in 1920. On the basis of this law, it was possible to explain theoretically the intraspecific composition and the absence in the hands of many European plant breeders of missing links. We proceeded logically toward an extended exploratory work, i.e. to the organization of expeditions in order to search for the wealth of varieties necessary for us, primarily that of the most important cultivated plants.

Theoretical botanical investigations have, in general terms, elucidated the basic localities of potentials with respect to the major, cultivated plants. These happen to be located within the ancient agricultural countries of southern Asia, mountainous Africa, the Mediterranean area, and South and Central America including Mexico.

During the last couple of years, Soviet expeditions have been directed into one after another of these areas. The penetration of many of these countries was far from easy: months passed just waiting for the necessary entry visas. The very appearance of Soviet research workers in the colonial countries frightened the imperialistic governments. A number of the participants in these expeditions (e.g. S.M. Bukharov and P.M. Zhukovskiy) contracted tropical diseases, which were hard to cure, and they were incapacitated for a long time.

During the last ten years, the Soviet expeditions have, as never before, investigated three quarters of the world and traveled thousands of kilometers within the borders of Afghanistan, Abyssinia, Mexico, Guatemala, Colombia, Peru, Bolivia, Chile, Brazil, India, Formosa, Korea and Japan. All the countries, situated along the shores of the Mediterranean have been studied. The length and breadth of Mexico have been penetrated, including the badly studied Yucatan. An enormous amount of material has been collected from Central

America. Caravans of Soviet expeditions have travelled along the Cordilleras from California down to southern Chile.

An immense material has been systematically collected (which is almost exhaustive with respect to the major crops) and studies of plant resources have been conducted. The areas of the initial formation of the main cultivated plants, which are of interest to the Soviet Union, have been established with an exceptional accuracy.

Decades of studies have, thus, allowed us to establish eight centers (see the map, Fig. 1):

The Chinese center of origin of cultivated plants comprises the mountainous areas of central and western China and the areas adjacent to them. This center has given rise to ca. 140 cultivated plants. This is the native land of millet [*Panicum* and *Setaria* spp.], mustard [*Sinapis* spp.], soybeans [*Glycine* spp.] and many peculiar vegetables. A very large number of fruit-bearing plants belong to China. It is the native land of tea, many citrus fruits, the camphor tree [*Cinnamomum camphora* L.], the tung-oil tree [*Vernicia fordii* (Hemsl.) Airy-Shaw] and very valuable fiber plants such as ramie [*Boehmeria nivea* (L.) Gaud.] and Chinese jute [*Abutilon theophrasti* Medic.].

The Hindustani center of origin seems to be second in importance. It includes Burma and Assam but excludes northwestern India, Punjab and the north-western frontier provinces. This is the center of rice, characterized in India by an astonishing variety of wild and cultivated forms. It is also the native land of many leguminous plants and of special tropical fruit crops, particularly citrus and mango [*Mangifera* spp.], as well as of sugar cane and of different fiber plants. Limes, oranges and some kinds of mandarines also come from here.

Side by side with the Hindustani center, we can distinguish an Indo-Malayan one, comprising Indo-China, all of the Malayan archipelago and the large islands of Java, Borneo, Sumatra and the Philippines. This subsidiary to the Indian center is rich in fruit crops, including crops of worldwide importance such as bananas and some kinds of citrus fruits.

The third, or Inner-Asiatic, center of origin of cultivated plants comprises northwestern India (Punjab, the northwestern frontier provinces and Kashmir), all of Afghanistan and the Soviet Tadzhikistan and Uzbekistan as well as western Tien-Shan. This center yields considerably to the last-mentioned ones with respect to the number of species; it is, nevertheless, of immense importance to us, since it is the native land of the soft wheats. Only there an enormous potential of varieties of soft wheat, the most important cereal in the world, is found. It is the native land of all the major leguminous plants, such as peas [*Pisum* spp.], lentils [*Lens culinaris* Medic.], chickling vetch or grass peas [*Lathyrus sativus* L.], fava beans [*Vicia faba* L.] and chick-peas [*Cicer arietinum* L.], representing an exceptional wealth of genes. Most probably cotton – [i.e. here *Gossypium herbaceum* L.] – was first introduced into cultivation here.

The fourth center, a west-Asiatic one, includes interior Asia Minor, Trans-caucasia, Iran (Persia) and the mountains of Turkmenistan. This center is remarkable mainly for its exceptional wealth of such species of cultivated wheats that – as recently established by investigators – are resistant to diseases.

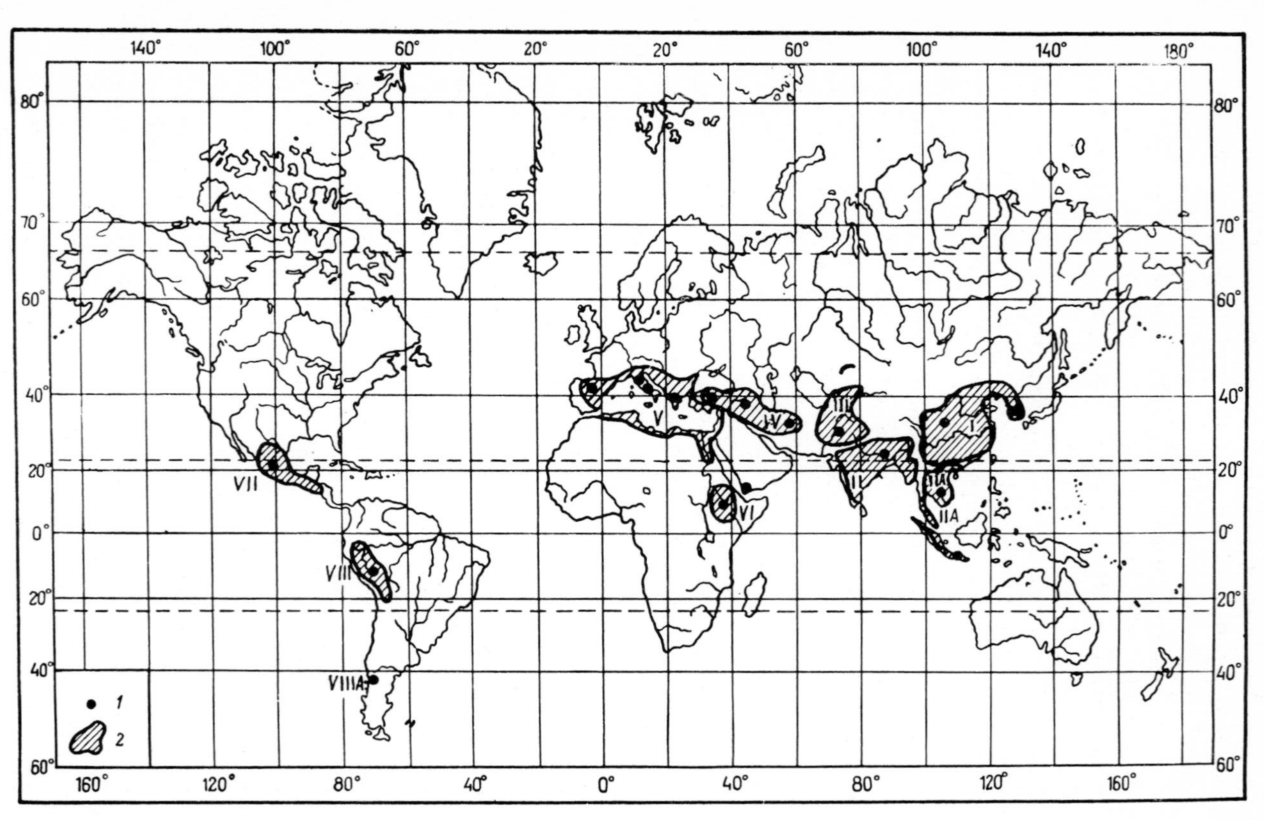

Fig. 1. Universal centers of origin (I–VIII) of cultivated plants. 1. Main foci of origin of cultivated plants; 2. centers of type formation of the most important cultivated plants.

Dozens of botanical species of wheat are endemic (i.e. occur only in a given area) in Asia Minor. Within the borders of the U.S.S.R. the diversity of wheat is especially great in Armenia, which exceeds all the rest of the areas with respect to the number of botanical varieties of wheat. A large number of wild wheats have also been found here. As far as the diversity of wheat species is concerned, this center can be distinguished from all the others in the world.

Asia Minor and the Soviet Transcaucasia constitute the main native area of rye. In contrast to the uniform type in Europe, rye appears here in an amazing variety of forms and species.

A potential for the European fruit growing industry is concentrated in Asia Minor including Soviet Transcaucasia. It is the native land of grapevines, pears, cherry plums, cherries, pomegranates, Greek walnuts, quince, almonds and figs.

All the wealth of flax in the world comes from Turkey, Iran and Soviet Inner Asia. The most valuable forage plants, alfalfa [*Medicago* spp.], Persian clover [*Trifolium resupinatum* L.] and vetch [*Vicia sativa* L.], etc., originate from Asia.

The fifth center, the Mediterranean one, is the characteristic native land of the olive tree [*Olea europaea* L.] and the carob tree [*Ceratonia siliqua* L.]. A large number of vegetable crops, including beets, originates from there. With respect to vegetable crops, this center is, besides the one in China, the most important one in the world. Many old forage plants also come originally from the Mediterranean center. Many of the plants cultivated around the Mediterranean, such as linseed, wheat, beans and chick-peas, are characterized by large seeds in contrast to the small-seeded forms found in Inner Asia, where the main center thereof is located.

Expeditions sent out by us in 1927 to Abyssinia, Egypt and Somalia revealed the peculiarity of Ethiopia and its cultivated flora and the unquestionable presence there of an autonomous main center of cultivated plants. Abyssinia must be put in first place with respect to the number of botanical varieties of wheat (i.e. those belonging to the group of hard wheats). It is also the center of formation as far as barley is concerned. Nowhere else is there such a variation of its forms and genes as in Abyssinia. This is also the native land of the peculiar cereal called teff [*Eragrostis tef* (Zuccagni) Trotter] and that of the oil-producing nug or ramtil [*Guizotia abyssinica* (L.f.) Cassini]. Flax [*Linum usitatissimum* L.] is represented there by a special form, used not for its fibers or oil but exclusively as a cereal from which a flour is prepared. In other words, in Abyssinia flax is a cereal in its own right.

Within the limits of the Americas we can distinguish the seventh or Central American center, which includes southern Mexico as well, and the eighth one or the Andean center, the territory of which covers Peru, Bolivia and Ecuador.

As demonstrated by means of the differential phyto-geographical method, the basic center of maize [*Zea mays* L.] and the native land of the American species of beans [*Phaseolus* spp.], squashes [*Cucurbita* spp.] and peppers [*Capsicum* spp.] are no doubt found within the borders of Mexico and Central America. Cacao [*Theobroma cacao* L.] comes from there and the native land of the sweet potato [*Ipomoea batatas* (L.) Lam.] is, indeed, also there.

The American cotton, the upland type on which all the cotton industry in the world is based, comes from southern Mexico. Maize (*Zea mays* L.) plays the same role there as wheat in the Old World; without it there would not have been any Mayan civilization (the Mayas were a language group among the Central American native peoples; they created a major civilization, the flowering period of which lasted from the fourth to the sixth centuries).

Within the limits of the eighth, or the Andean, center the Soviet expeditions discovered resources of cultivated plants, in essence definitely centralized, to which belong dozens of new cultivated as well as closely related wild potatoes, which were previously not known to science but utilized by the Indian tribes.

The lofty mountains of Peru, Bolivia and Ecuador are full of endemics, starting with the peculiar species of potatoes and ending with the similarly peculiar tuberous plants that are typical of this part of the world alone, e.g. oca [*Oxalis tuberosa* Molina], anu [*Tropaeolum tuberosum* Ruiz. et Pav.] and ulluco [*Ullucus tuberosus* Lozano]. Peculiar species of orach [*Atriplex* spp.] are grown here in large quantities.

The plants cultivated and the animals used in Peru and Bolivia – llamas and alpacas – are endemic and concentrated mainly in the so-called 'puna', i.e. a kind of steppe on the high plateaux at 3800–4300 m elevation. The crops there are not irrigated. It is still possible to witness the transition of wild plants into cultivated ones in that area. There is little doubt that both agriculture and husbandry in South America started just on the punas. The localization of the endemic cultivated plants and animals in the past, just as in the present, is strikingly distinct and well delimited in this center.

Besides the basic Peruvian center, it is necessary to add the small island of Chiloë, situated along the coast of southern Chile, whence the first Europeans adopted the common potato from the Indians, i.e. the kind presently known to everybody and well suited for the conditions of the long northern days.

The most significant majority of cultivated plants comes, thus, originally from the Old World. Out of the 640 important cultivated plants known to us, more than 500, i.e. five sixths of all cultivated plants in the world, originated in the Old World. The specific and varietal wealth occurs mainly in the mountain areas of southern Asia.

As a result of the work conducted by Soviet scientists during the last decades, an immense potential of species and varieties of cultivated plants and their closely related wild species has been revealed, of which neither botanists nor agronomists of the past had any knowledge. It is again necessary to work over the species and their composition with respect to the majority of our cultivated plants. Best of all, there is at present a colossal amount of material growing on experimental fields in the Soviet Union which has, in part, already been taken into cultivation on collective farms. New horizons are at present opening up for practical plant breeders. The enormous material of new species, including a multitude of new species of such plants as potatoes, offers an opportunity for a radical alteration of our cultivated plants while creating new strains that are both more frost-hardy and resistant to diseases. The method of vernalization, invented during the last couple of years by Academician T.D. Lysenko, offers an opportunity for utilizing southern strains even more efficiently, while intentionally abbreviating the vegetative period. It also opens up new possibilities with respect to our knowledge of the developmental stages of the plants when selecting pairs for hybridization.

We still have an enormous amount of work ahead of us. The area of southeastern Asia, which is extremely rich in diversity of species and of great practical importance, in particular for our Subtropics, is the one least touched upon by Soviet scientists because of the well-known difficulties. Varietal resources of citrus fruits, tea and tung-oil trees belong here. Our subtropical plant industry along the shores of the Black Sea can be built up on the basis of these. Finally, since the investigations – because of their urgency – have been rushed, they remain of an orienting nature: in essence, a global reconnaissance work has been executed, primarily revealing the approximate localization in the world of 'plant gold'. However, more importantly, a collection has been made of an immense new material of varieties that was previously not known to science but is at present at the disposal of Soviet plant breeders.

During the next couple of years, the mobilization of the plant resources will no doubt be definitely extended and every attention will be paid to mastering them for practical plant breeding and for introducing all the best thereof to the state-run and collective farms.

The phyto-geographical basis for plant breeding

Studies of the original material used for plant breeding

Local varieties and their importance

THE PRESENT VARIETAL composition of cultivated plants, grown in different parts of the Soviet Union, represented until recently the result of strains brought from different areas and countries and to a considerable extent associated with nomadic peoples and colonizers. Basically, the assortment of the plant crops reflects all the past history of our country, but in essence it corresponds to the management of the small individual farms in the recent past. It is possible to trace the dispersal of different groups of plants and strains from western Europe, the United States, Asia Minor, Mongolia or Iran. The introduction of new strains into our country during the time before the revolution had a chance-like character. Beginning during the eighteenth century, different amateurs and societies ordered without any system new foreign strains from abroad, often including very valuable ones. However, for the immense territory of our Union and for the lack, in the past, of a planned, governmental system for the introduction of new strains, the assortment ordered was usually only slowly distributed, lost its identity or disappeared altogether.

It could be said that seed-production enterprises did not exist in our country before the October revolution. Spontaneity and chance seemed to be the characteristic traits within the field of seed production of the past. We have only just started a systematic investment in varieties that meet the demands of the socialistic extension of mechanized agriculture.

At the same time, there can be no doubt that the material of strains sown for decades or even centuries and worked on in our country was subject to a natural selection or even a conscious or artificial selection and that among the local assortment ecologically and especially suitable types have been produced.

The closeness of the Soviet Union to the main centers of origin of a number of cultivated plants offered an opportunity for the selection of exceptionally valuable material for a number of crops. Some plants such as, e.g. long-staple flax, wheat, rye, clover, timothy and so on, are represented in our country by extremely valuable local strains. For instance, as demonstrated by our research

First published in *Teoreticheskie osnovy selektsii* [*Theoretical Bases for Plant Breeding*], vol. 1, Moscow–Leningrad, 1935.

with respect to flax, the northern (or mountain) populations in the European parts of the Union come in first place even against a background of a worldwide assortment. The local Inner-Asiatic and Transcaucasian material of strains, e.g. of wheat, barley, forage grasses, grapevines and fruit trees, consists of original and well adapted forms, the replacement of which should be made with great care and with consideration to all that is valuable within the local material.

The concept of 'local strain' is in practice very relative; usually both old strains, which have been subjected to natural selection during the course of decades or even centuries, and odd strains, often adopted comparatively recently but deprived of their identity or having lost their original name and pedigree, are intended. Very many so-called 'local strains' belong just to the latter category.

The presence or absence of a diversity of hereditary forms is a very important feature as far as the evaluation of local material used for plant breeding is concerned. In this respect, various plants and strains appear to be sharply different under different conditions. In some conditions the local assortments are represented by complex populations, consisting of many forms that differ with respect to both morphological and physiological characteristics; in others, the local strains consist of uniform populations more or less equivalent physiologically as well as morphologically. Of course, in the case when the old, local material (such as very often exists both in Transcaucasia and Inner Asia) is represented by its own complex of populations, it already possesses possibilities for initial plant breeding. For instance, crops of millet in the southeastern European parts of the Union are usually represented by complex populations with dozens of botanical varieties. Often the so-called local strains are comparatively homogenous and represent only themselves by their reproduction of odd but homogenous strains from abroad or from the different areas from where they were adopted some years earlier, although they have now lost their names.

Naturally, when starting the breeding of a plant, *it is first of all necessary to utilize the local material as much as possible* and to select from it the most productive and valuable forms. In the recent past the success of plant breeding in our country was based mainly on the selection of local populations of the most valuable kinds. At our plant breeding stations, our best strains of winter and spring wheat, rye, barley and flax appeared during the past years to be the result mainly of a selection among local strains. At the plant-breeding stations being established at present, our attention is turned mainly to investigations and utilization of the material of local strains when working with old crops.

When starting practical plant breeding it is first and foremost necessary to know the local assortment as well as possible. It should serve as the initial material for further improvement of the strains. It is necessary to have special nurseries at the plant-breeding stations where all the valuable local material can be preserved.

The importance of material from abroad or from different areas

Finally, it is an entirely different matter when it is a question of new crops or definitely new areas where there were no crops before. In that case, attention must be directed toward a search for especially interesting and valuable original material.

It is possible to state that all the progress in Canada and the United States within the field of the extremely widespread crops of wheat, barley, oats, rye and forage crops as well as fruit-producing crops are based on the skillful introduction of assortments from our own country, from India and from western Europe. The winter wheat grown in the dry areas of the United States is represented by strains brought from our own southern areas. The most remarkable progress has been made during the last couple of years in the U.S.A., Canada and Argentine by means of hybridization between very distant races, acquired from Europe, India and China. The successes of the plant breeders and fruit growers Michurin, Burbank and Hansen, were to a great extent based on a wide-scale application of hybridization between original material of strains from different countries. The history of plant breeding definitely demonstrates that the major improvements during the last few decades are associated mainly with a wide application of original material. This is especially evident with respect to such countries as Canada, U.S.A., Australia, Argentina and South Africa, which on the whole have introduced material from other countries. The same is also obvious in such countries that themselves harbor valuable local materials. Thus, e.g., Sweden has had considerable success with its local strains of wheat, when hybridized with the English so-called 'squarehead' strain. The same is valid also for France. In the last couple of years, Germany has started a wide-scale application of original material from all kinds of countries, while sending special expeditions in search of such material to Asia Minor, South America and India.

It is difficult to imagine our existence without such crops as sunflowers, maize, potatoes, tobacco or upland-cotton, which were introduced not so long ago from America. *In the case of new crops, exceptional attention must be paid to a systematic acquisition of varietal material*, since the very success of these crops is definitely associated with the strains selected. The fate of new crops is often determined primarily by the suitability of the strains.

The new conditions in our country with an expansion of mechanized agriculture have produced new demands on the strains with respect to quality of the product, non-shedding of the seeds, stiffness of the straws and suitability for mechanical harvesting. We have overestimated our local assortment and must start a decisive improvement of them.

Most of all, however valuable our local strains may be with respect to ecological suitability, they are still far from ideal. The spring strains of wheat along the lower Volga, in Ukraine and in western Siberia, which have been used for centuries, all suffer from drought. The strains of spring wheat often perish during wet winters even in areas where they are usually grown.

The necessity for a radical improvement of the strains in correspondence with the conditions of the dry continental climate in our country but also in agreement with the new demands posed by socialistic agriculture, means a wide-scale application of new, original material of first-class importance.

Introductions for our subtropical regions, where every attention must be paid to a skilful application of new crops and new assortments from other lands during the next couple of years, are of an especially great importance.

The theory behind introductions

The research conducted at the All-Union Institute of Plant Industry during the last couple of years has revealed a number of regularities concerning the geographical distribution in the world of plant resources, determining to a considerable degree in which direction it is necessary to search for new plants, new species and new varieties.

Botanical studies of the world are still far from being complete. Botanists actually know only one half of all the flowering plants existing in nature. The immense continents of South America and Africa, as well as countries such as Indo-China and Asia Minor are extremely poorly known. There are still large areas where botanists have never put down a foot. Even within our own Transcaucasian and Inner-Asiatic republics, hundreds of new species will, in the opinion of the most competent botanists, be discovered within the near future.

All the data, even if incomplete, that we presently dispose over as a result of our studies of the vegetation of the world, reveal a fact of utmost importance: there is a geographical localization of the processes forming the species. The geography of the plants definitely demonstrates that during the present geological epoch *the diversity of species is unevenly distributed in the world.* A number of areas can be distinguished that differ considerably in the diversity of their species. The floras of southeastern China, Indo-China, India, the Malayan archipelago, southwestern Asia, tropical Africa, the Cape Province, Abyssinia, Central America, South America, Mexico and the countries around the Mediterranean, as well as Asia Minor, can be distinguished by their unusual concentration of specific diversity. On the other hand, the northerly countries, Siberia, all of central and northern Europe and North America, are characterized by a poor composition of species.

Central Asia is astonishingly poor in species. The number of species in our own country increases definitely in the direction toward Crimea, Transcaucasia, and the mountain areas of Inner Asia, Altai and Tien-Shan. The territory of the Soviet Union, the Caucasus and the mountains and foothills of Inner Asia are astonishingly different with respect to their wealth of species, especially when the number of species per surface area is counted. The concentration of specific diversity is ten times greater in these areas than within central Europe and even more so in comparison with the northern areas of Europe.

In some localities in the world the concentration of specific diversity is surprisingly sharply expressed. Thus, e.g., the 'dwarf' republics of Costa Rica and El Salvador, corresponding in area to about 1/100th of the United States, do not yield in the number of species to that in all of North America, i.e. the United States, Canada and Alaska all together.

The interesting work by Professor A.A. Grossheim, *Analysis of the Flora of the Caucasus*, is the first one to establish Transcaucasus as the area richest of all in number of species.

When searching for new species for field crops, both within our own country and beyond its borders, it is necessary to keep in mind the localization of the specific diversity and to direct the attention especially to the corresponding areas. Thus, e.g. the remarkable rubber-producing 'tau-saghys' [*Scorzonera tau-saghys*] is found only in the Karatau mountain range in Kazakhstan, which has a wealth of endemic species, and nowhere else, in spite of special searches for it over many years all over Kazakhstan and Inner Asia. The other remarkable rubber-producers, i.e. 'kok-saghys' [*Taraxacum bicorne* Dahlst.], is found at the border with China, in Tien-Shan, which is also rich in the number of species, and 'krym-saghys' [*T. hybernum* Stevens], is located in Crimea but comes originally from the Mediterranean countries, which have an extremely rich flora.

Our research during the past years has revealed with great accuracy the areas or – as we prefer to call them – the *centers*, where the species of the most important cultivated plants were formed. Our main attention was, naturally, drawn to investigations primarily of the resources of varieties belonging to the most important cultivated plants of the greatest interest to our own country.

During the last decades the All-Soviet Institute of Plant Industry has, according to a definite plan, conducted extensive taxonomic–geographical investigations of a large number of cultivated plants. These studies concerned primarily field, vegetable and fruit crops, but also their wild relatives. Every attention was paid especially to the *intraspecific composition of different plants* and the botanical study of the varietal composition of the different Linnaean species.

The study of hundreds of cultivated plants, conducted by large teams, has led us mainly to an understanding of the Linnaean species, including those of cultivated plants, as a definitely complicated system. In our opinion, *a species is represented by a more or less isolated and complicated subspecific morphological system, associated during its genesis with a definite environment and distribution area.*

The actual studies of several hundred species revealed a scarcity of monotypic species, i.e. species represented by a single, distinct race, thus a single botanical form. Most species can be said to consist of a more or less high number of hereditary forms (jordanons or genotypes).

Detailed studies of variation within the distribution area of a species clearly revealed a regularity from the very start, inferring in particular a *parallelism*, displayed as hereditary variation of closely related species and genera and similarities between series of hereditary forms, which represent closely related species and genera. We have called this basic regularity 'the law of homologous series of hereditary variability'. The establishment of this law served as a stimulus for all our work when searching for varieties, since it revealed a large number of missing links in the systems known until then that concern the botanical varieties and forms of species of cultivated plants and their wild relatives. Theoretically, these links should have existed in the past or could still exist in the present. Naturally, the question arose about *where and in which region it would be necessary to look for these missing links*. Logically, we approached the old problem about the origin of cultivated plants but now as new and distinctly concrete missions. The essence of the solution of this problem meant an actual mastering of the original varietal potential of genes of cultivated plants.

It became eminently clear that botanists knew very little about cultivated plants. By widening the knowledge of the hereditary variability of the species, increasingly new opportunities for type-formation were discovered. *The problem concerning the worldwide resources of plant varieties arose in all its immensity.*

During the last ten years the Institute of Plant Industry has directed a large number of expeditions according to a definite plan both to areas within the Soviet Union and to those beyond its borders.

The very establishment of these basic areas of type-formation, or the geographical centers of origin of cultivated plants, was carried out by means of the *differential phyto-geographical method* invented by us. It consists of the following points:

1 A strict differentiation of the plants studied into Linnaean species and genetic groups by means of morphological–taxonomical, hybridological, cytological and immunological analyses.
2 If possible, establishment of the distribution areas of these species in the recent past, when communications were more difficult than at present.
3 A detailed determination of the composition of the botanical varieties and races of each species or of the general system of hereditary variability within the limits of the different species.
4 Revelation of the dispersal of the hereditary variability of the forms of a given species within areas or countries and an establishment of the geographical center of accumulation of the basic variability. Primary centers of origin are, as a rule, characterized by the presence of many endemic, variable traits. Under such conditions, where the endemism of a given group has an ancient origin (paleo-endemism), it can comprise not only characteristics of varieties and species but even of entire genera of cultivated plants. In reality this occurs frequently.
5 For a very exact definition of the centers of origin and the initial type-formation, additional establishment of the geographical centers of variability of closely related wild as well as cultivated plants is necessary.
6 The primary centers often include a large number of genetically dominant characteristics. As demonstrated by the direct study of the geography of cultivated plants, mainly recessive types are singled out and formed as a result of inbreeding or mutations toward the periphery of the main, ancient area of distribution or due to isolation of cultivated plants (on islands or in mountains).
7 Finally, appended to the differential phyto-geographical method, the data provided by archeologists, historians and linguists can serve a purpose. They are, on the whole, usually too general for our purpose of practical plant breeding and need correction or a more exact definition of the species and varieties.

When using this method, it is necessary to distinguish between *primary and secondary centers of type-formation*. Cases are known when the present maximum of variation of types can be the result of the similarity between species or their hybridization with each other. Thus, e.g. thanks to its mountainous condition, Spain has a general geographical position but also thanks to its history, this country has an exceptionally large number of varieties and species of wheat. The total number of types means very little, however, since – as shown by a direct analysis – the number of varieties within the limits of the different species is very small in Spain in comparison with that which we can find in actual centers of initial type-formation within this country. The large variation of wheats, turning up in Spain, can be associated with the many species brought there from other different centers.

It is natural that the application of this complicated method requires enormous, cooperative team work and a knowledge of all the world, but as a result of this application we are actually able to control the initial potential of the specific variation with respect to the morphological and physiological characteristics necessary for plant breeding. We are also able to approach in earnest an understanding of the dynamics of the evolutionary process.

The genetic investigations have demonstrated that under the external uniformity a great genetic potential for variation is often hidden. Thus, e.g., when crossing the comparatively uniform Afghani mountain peas with distant, recessive peas that are cultivated in Europe, it turned out that under the external uniformity a large number of genes were hidden, which after hybridization with the distant European peas displayed the entire range of variability of the recessive forms (Govorov, 1928).

Natural mutations and hybridization within the secondary centers can favor the development of new types, which often represent a greater interest for practical plant breeding. For instance, during our investigations of flax, it turned out that at the periphery of the distribution area, types were isolated that were especially important with respect to the length of the fibers, the height of the plants, the characteristics of the branching and the quality of the fibers. Frost-hardy kinds of wheat are essentially not associated with the primary center but with the periphery, at the limit at which the crops can be grown in the high mountains or at high latitude; in some cases it is, apparently, linked to the recessive characteristics of a number of varieties such as demonstrated by genetic investigations. Extremely interesting new recessive types of waxy maize [*Zea mays* L. convar. *ceratina* Kulesch.] and waxy beans have developed in China after the transfer of these American plants from the New to the Old World.

Of course, *the utilization of all the progress of plant breeding that is presently being made in the world and of all the new hybrid forms as original material is of enormous importance.* Acquiring material from the primary main areas and opening up an immense potential of new and valuable genes there, makes it simultaneously possible to utilize the material in every respect from the periphery as well and, in particular, also that which is the result of the most recent plant breeding in various countries.

The excellent work conducted during the last couple of years in the Soviet country by the teams at the Institute of Plant Industry concerning a systematic study of the worldwide varietal resources of plants with respect to the major, cultivated plants and the colossal new collections of material, available there, have radically changed our ideas about the varietal and specific composition of cultivated plants. Even in the case of such crops as wheat, potatoes, maize, beans, rye and flax, studied for decades by plant breeders, an enormous potential of varieties has turned up within the primary areas of type-formation in the ancient agricultural countries and in the mountain areas of Central and South America, southern Asia and Abyssinia. As a result of the Soviet investigations almost twice as many new species as previously known and a multitude of new varieties have been discovered in the case of the major cultivated plants. As far as some plants are concerned, e.g. the potato, species and varieties have been found that literally revolutionize our ideas about the foreign material for plant breeding (Bukasov, 1933).

With respect to wheat, 75% more new botanical varieties and 50% more new species have been discovered. An exceptional wealth of genes of wheat and barley turned up in the small country of Abyssinia with its insignificant area occupied by agriculture. An amazing diversity of the most important cereals in the world is concentrated in that area. An enormous number of new varieties of wheat and other crops were also discovered in Afghanistan, Turkey and northwestern India.

The areas of maximum concentration of primary interspecific and specific diversity of cultivated plants have been determined. As demonstrated by the investigations, a considerable number of cultivated plant species do not appear outside the limits of the ancient primary centers. Dozens and even hundreds of cultivated plants are still typical only of the areas where they first arose and where they may still remain, untouched by Europeans. This is particularly striking in Central and South America, where the primary centers of type-formation of cultivated plants have an extremely limited localization. This pertains also to southern Asia. In part, our own Transcaucasia and the areas adjacent to it in Iran and northwestern Turkey are some of the most interesting areas of primary type- and species-formation of wheat and rye but especially as far as fruit trees are concerned.

There, the species-building processes are especially evident in the case of such plants as wheat, alfalfa, peas, almonds and pomegranates. Apparently, it is still possible to follow the isolation of species *in statu nascendi* [being formed] and of the major genetic groups of these plants.

It is possible to establish the basic areas with great accuracy and to determine the primary specific potential even of such plants as wheat, barley, maize and cotton, which have long since been widely distributed all over the world.

The differentiating studies of cultivated plants and their closely related relatives, by means of all the biological methods presently available, demonstrated to us for the first time how to establish in detail the areas where the initial specific potentials are located.

With respect to several hundred plants, all belonging to some important crop (with the exception of ornamental and garden plants), the basic primary areas or – as we prefer to call them – *the centers of origin* of the specific and varietal potential can be determined with a comparatively great accuracy. Closely related, original and wild relatives of these species are often found in the same areas. However, for some plants such as maize, the original wild relatives have not yet been discovered.

Our initial aspirations were directed mainly toward the study of difficult objects such as wheat, rye, barley, maize and cotton which are at present widely grown all over the world and have already long since dispersed from the primary centers, where they were initially taken into cultivation. For such plants, the total areas of the species provide only superficial ideas, if attention is not paid to their formation of varieties and forms. In order to solve the problem of original areas, an application of the differential method is required, such as is already used for discovering the multitude of new varieties and new characteristics and for revealing new species of wheat, many of which turned out to have an amazingly limited localization and were first discovered by the Soviet expeditions in Abyssinia, Armenia, Georgia and Turkey.

As far as the introduction of new objectives into the study is concerned, it became increasingly evident that *there is a coincidence between the areas of primary type-formation of many species and even genera.* In a number of cases, literally dozens of species could be referred to one particular area. *The geographical studies led to the establishment of entirely independent floras of cultivated plants, that are specific for different areas.*

World centers of origin of the most important cultivated plants

When summing up the work done by the Soviet teams of plant breeders and the expeditions sent out over many years to Asia, Africa, southern Europe and Central and South America and covering up to 60 countries, and when reviewing the results of the detailed comparative studies of the colossal amount of new material of varietal and specific diversity collected, we have been able to establish *eight independent* centers in the world, where the most important cultivated plants originated. The work in this direction is not yet completed: we still know southeastern Asia very inexactly and still a number of expeditions are necessary for China, Indo-China and India in order to be able to define more exactly the centers of original type-formation of their cultivated plants and to be able to master the new material. However, we can speak with considerably greater accuracy than dreamed of ten years ago about the eight ancient and basic centers of agriculture in the world, i.e. more accurately about the eight independent areas where the plants were initially taken into cultivation. In our previous papers we delimited the centers of agriculture established according to some basic indicator-crops. There are not enough data for an exhaustive approach. In the present list we have tried as far as possible to provide a complete review of the crops typical of the different areas. We have made serious changes in and additions to our previous ideas, which were initially expressed in the book about *Centers of Origin of Cultivated Plants* in 1926. [cf. pp. 22–135]. Most of the expeditions and most of the work, while studying the varietal resources of the world, were conducted during the period from 1923 to 1933.

Let us now turn to the survey over the centers of origin (see map, Fig. 1):

I THE CHINESE CENTER OF ORIGIN OF CULTIVATED PLANTS

The mountains of central and western China and the areas adjacent to them appear as the first and largest independent center of agriculture in the world and as an origin of many cultivated plants.

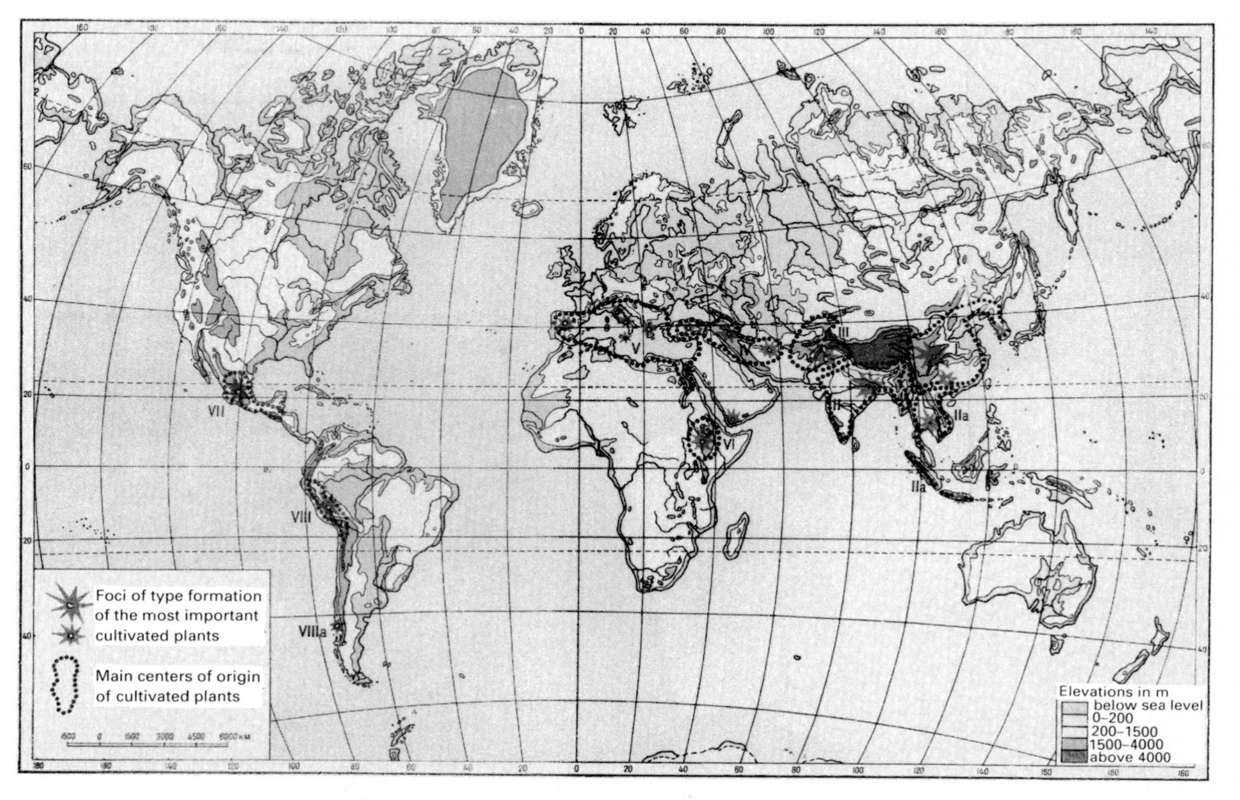

Fig. 1. Universal centers of the cultivated plants.

Below we furnish a list of the most important endemic crops (excluding ornamentals) typical of this center. The list provided is composed mainly on the basis of my own personal studies and journeys and that of the differential studies of many crops conducted during the past ten years by a scientific team at the All-Union Institute of Plant Industry (cf. *Trudy po prikladnoy botanike, selektsii i genetika* [*Papers on applied botany, plant breeding and genetics*], vols. 11–27 and later, but also the supplements to these between 1923–1934). For the tropical crops of India and China and the islands we have used sources in the literature (e.g. Watt, 1873, Ochse, Wilson, Tanaka and Akemine) and various monographs about different genera.

CEREAL GRASSES AND OTHER GRASSES

1 *Panicum miliaceum* L. – broom corn millet.
2 *P. italicum* L. – foxtail millet
3 *P. frumentaceum* Fr. & Sav. – Japanese millet.
4 *Andropogon sorghum* Brot. – sorghum.
5 *Avena nuda* L. – naked-seeded, large-seeded oats.
6 A group of naked-seeded and awnless barleys (*Hordeum hexastichum* L.).
7 A group of waxy maize, *Zea mays* L. (secondary center; typical, recessive forms).
8 *Fagopyrum esculentum* Moench – buckwheat.
9 *F. tataricum* Gaertn. – Tatar buckwheat.

LEGUMINOUS PLANTS

10 *Glycine hispida* Maxim. – soybeans.
11 *Phaseolus angularis* Wight – adzuki bean.
12 *Ph. vulgaris* L. – kidney beans (recessive forms; secondary center).
13 *Vigna sinensis* Endl. ssp. *sesquipedalis* Piper – asparagus or long beans (secondary center).
14 *Stizolobium hassjo* Piper and Tracy – Yokohama beans (E. Asia and Japan).

BAMBOOS

15 *Phyllostachys puberula* Munro, *Ph. quilioi* A. & C. Riv., *Ph. bambusoides* Sieb. & Zucc., *Ph. edulis*, A & C. Rivera, *Ph. nigra* Munro, var. *honoris* Makino, *Ph. reticulata* C. Koch, *Ph. mitis* A. & C. Riv. (this species is especially often used for food), and other species.
16 *Arundinaria simonii* (Carr.) A. & C. Riv., *A. nitida* Fr. Mitf. and other species.
17 *Bambusa mitis* Poir., *B. vulgaris* Schrad., *B. multiplex* (Lour.) Räusch, *B. spinosa* Roxb., *B. senamensis* Franch. & Sav., and other species.

ROOT VEGETABLES, TUBERS AND CORMS AS WELL AS AQUATIC PLANTS

18 *Dioscorea batatas* Decne., *D. japonia* Thunb. – Chinese yams.
19 *Stachys sieboldii* Miq. – Chinese artichoke.
20 *Raphanus sativus* L. var. *raphanistroides* (Makino) Sinsk. – radish (wild and cultivated in an enormous variety of forms).
21 *Brassica rapa* L. var. *rapifera* Metzg. – turnip (a special geographical group of east-Asiatic turnips; secondary center; Japan and the humid parts of eastern Asia).

22 *Brassica napiformis* Bailey – Napa cabbage, Chinese cabbage.
23 *Wasabia japonica* Matsum. – Japanese horse-radish.
24 *Arctium lappa* L. – burdock (Japanese gobo).
25 *Amorphophallus konjak* C. Kock – konjac (mainly in Japan).
26 *Petasites japonicus* Mig. – sweet coltsfoot (mainly in Japan).
27 *Adenophora latifolia* Fisch., *A. verticillata* Fisch. (Japan).
28 *Eleocharis tuberosa* Schult. (*Scirpus tuberosus* Roxb.) – water chestnut.
29 *Nelumbo nucifera* Gaertn. – lotus (most likely also in India).
30 *Sagittaria sagittifolia* L. var. *sinensis* Makino – arrowleaf.
31 *Zizania latifolia* Turcz. (mainly the stems and leaf sheaths, diseased by a smut, are used for food).
32 *Ipomoea aquatica* Forssk. – Chinese yung-tsai, swamp morning-glory.
33 *Trapa bicornis* Osb. – horn nut, *T. bispinosa* Roxb. – Singhara nut.
34 *Colocasia antiquorum* Schott. – taro (wild and cultivated in a great varietal diversity; belongs possibly also to India and the Sunda Islands).
35 *Lilium tigrinum* Ker., *L. maximowiczii* Regel and other species with edible corms.
36 *Elatostema umbellatum* Blume. var. *convolucratum* Makino.

VEGETABLES

<table>
<tr><td>37 Brassica chinensis L. – bok choi</td><td rowspan="5">Japan
and
Eastern China</td></tr>
<tr><td>38 B. pekinensis Rupr. – Napa cabbage</td></tr>
<tr><td>39 B. alboglabra Bailey – Chinese kale</td></tr>
<tr><td>40 B. nipposinica Bailey</td></tr>
<tr><td>41 B. narinosa Bailey</td></tr>
</table>

42 *B. juncea* (L.) Czern. var. *japonica* (Thunb.) Bailey – salad cabbage (secondary center)
43 *Peucedanum japonicum* Thunb.
44 *Aralia cordata* Thunb. – udo.
45 *Rheum palmatum* L. – Chinese medicinal rhubarb.
46 *Allium chinense* Don. (*Allium odorum* L.) – Chinese perennial onion tsyu-tsai.
47 *A. fistulosum* L. – Japanese leek.
48 *A. macrostemon* Bge. – syao-suan (N. China).
49 *A. pekinense* Prokh. in Korean and Japan (known only as cultivated, closely related to *A. sativum* L).
50 *Lactuca* sp. – asparagus lettuce.
51 *Solanum melongena* L. – eggplant (a special small-fruited group).
52 *Cucumis chinensis* Pang. – Chinese cucumber.
53 *C. sativus* L. – cucumber (a group of large-fruited cucumbers).
54 *Luffa cylindrica* M. Roem. – luffa or sponge gourd (used as a vegetable and for its fibers).
55 *Cucurbita moschata* Duch. var. *toonasa* Makino (var. *japonica* Zhit.) – small horned melon (secondary center).
56 *Actinostema paniculatum* Maxim.
57 *Chrysanthemum coronarium* L. – chrysanthemum (the leaves are eaten) and *Chr. morifolium* Ram, 'mum' (the petals are eaten).
58 *Perilla ocymoides* L., *P. arguta* Benth. – miso.
59 *Asparagus lucidus* Lindl. – clubshaped asparagus (Japan).
60 *Basella cordifolia* Lam. – Malabar spinach, maybe introduced from India.

FRUIT-PRODUCING PLANTS:

A. GROUP BELONGING TO THE TEMPERATE ZONE:

61 *Pyrus serotina* Rehd. – Chinese or sand pear.
62 *P. ussuriensis* Maxim. – Ussurian pear.
63 *Malus asiatica* Nakai – Chinese apple.
64 *Prunus persica* L., and *P. davidiana* Franch. – peach (the latter is wild)
65 *P. armeniaca* L. – apricot.
66 *P. mume* Sieb. & Zucc. – Japanese apricot.
67 *P. salicina* Lindl. – Japanese plum.
68 *P. simonii* Carr. – apricot-plum, Chinese plum.
69 *P. tomentosa* Thunb. – Chinese cherry.
70 *P. pseudocerasus* Lindl. – Chinese ying-tao.
71 *P. pauciflora* Bge.
72 *Crataegus pinnatifida* Bge. – hawthorn.
73 *Chaenomeles lagenaria* Koidz. – Chinese quince.
74 *Ch. chinensis* Koehne, *Ch. japonica* Lindl. – Chinese and Japanese quinces.
75 *Eleagnus multiflora* Thunb., var. *hortensis* Maxim., *E. umbellata* Thunb., *E. pungens* Thunb. and other species.
76 *Zizyphus vulgaris* Lam. – jujuba.
77 *Hovenia dulcis* Thunb. – Japanese raisin tree (the sweet stem of the fruit is eaten).
78 *Gingko biloba* L. – maidenhair tree.
79 *Juglans sinensis* Dode, *J. sieboldiana* Maxim. and other species of walnuts.
80 *Carya cathayensis* Serg. (mainly wild) – pecan nuts.
81 *Corylus heterophylla* Fisch., *C. ferox* Wall., *C. colurna* L. – (the latter two also grow wild in Nepal, Sikkim and Kamaon), and other species of hazelnuts.
82 *Castanea crenata* Sieb. & Zucc. – Japanese chestnut (Japan and Korea).
83 *Torreya grandis* Fort. – seeds used as nuts.
84 *Pinus koraiensis* Sieb. & Zucc. – pine nuts (Korea, Japan and Manchuria).

B. GROUP BELONGING TO THE SUBTROPICAL AND TROPICAL ZONES:

85 *Citrus junos* (Sieb.) Tan.
86 *C. ichangensis* Swingle – ichang-papeda.
87 *C. sinensis* Osb. – oranges (important secondary center), this species is extremely rich in forms.
88 *C. nobilis* Lour (most likely a secondary center), *C. ponki* Tan., *C. tarbiferox* Tan. (endemic Chinese species), *C. erythrosa* Tan. (endemic), *C. kinokuni* Tan. (endemic), *C. amblycarpa* (Hassk.) Ochse – mandarines
89 *Fortunella margarita* Swingle – oval kumquat (a typical endemic species in China, just like other species of kumquat).
90 *F. japonica* Swingle – round kumquat, *F. crassifolia* Swingle.
91 *Poncirus trifoliata* Raf. – trifoliate orange.
92 *Diospyros kaki* L., *D. sinensis* Bl. – persimmons.
93 *D. lotus* L. – date plum (wild and used for grafting).
94 *Eriobotrya japonica* Lindl. – loquat.
95 *Clausena lansium* Skeels. – wampee.
96 *Myrica rubra* Sieb. & Zucc. – yang-mei.
97 *Litchi chinensis* Sonn. – litchi.
98 *Nephelium longanum* Cambess. – pulesan.
99 *Rhodomyrtus tomentosa* Wight.

SUGAR-PRODUCING PLANTS:

100 *Saccharum sinense* Roxb. (endemic group of forms of sugar cane).

PLANTS PRODUCING OILS, ETHERIC OILS, RESINS AND TANNINS:

101 *Perilla ocymoides* L. – tsu zu.
102 *Raphanus sativus* L. var. *oleifera* Metzg. – oil-radish.
103 *Aleurites fordii* Hemsl., *A. montana* Wils., and *A. cordata* R. Br. – tung-oil trees (the latter mainly in Japan).
104 *Camellia sasanqua* Thunb., *C. japonica* L. – oil-producing camellias.
105 *Melia azedarach* L. – China-berry.
106 *Sesamum indicum* L. (special endemic group of low-growing forms; secondary center) – sesame.

SPICE PLANTS:

107 *Zanthoxylum bungei* Planch
108 *Z. piperitum* DC., *Z. planispinum* Sieb. & Zucc.
109 *Fagara schinifolia* (Sieb. & Zucc.) Engl. (var. *macrocarpa* Loes.).
110 *Cinnamomum cassia* L. – Chinese cinnamon.
111 *Illicium anisatum* (L.) Gaertn. (*I. verum* Hook. f.) – star anise (including Japan).
112 *Camellia sinensis* (L.) Ktze, (*Thea sinensis* L.) – tea.

TECHNICAL AND MEDICAL PLANTS:

113 *Sapindus mukurossi* Gaertn. – Chinese soapberry (used for soap).
114 *Eucommia ulmoides* Oliv. – Chinese gutta-percha tree (used mainly in the wild condition).
115 *Sapium sebiferum* Roxb., (*Stillingea sebifera* Michx.) – Chinese tallow-tree.
116 *Rhus vernicifera* Stokes and other species – laquer trees.
117 *Rh. succedanea* L. – tallow sumach.
118 *Broussonetia papyrifera* Vent. – paper mulberry, *B. kasinoki* Sieb. and other species (used for paper).
119 *Morus alba* L., white mulberry; also the species *M. bombycis* Koiz. and *M. multicaulis* Perr. (*M. alba* L.).
120 *Cinnamomum camphora* Nees. & Eberm. – camphor tree.
121 *Papaver somniferum* L. – opium poppy (exceptionally variable form).
122 *Panax ginseng* C.A.M. – ginseng.
123 *Aconitum wilsonii* Hort. – monkshood
124 *Smilax china* L. – China root.

FIBER PLANTS:

125 *Boehmeria nivea* Gaud., *B. tenacissima* Gaud. – ramie or Chinese silk fiber.
126 *Cannabis sativa* L. – large-seeded hemp.
127 *Abutilon avicennae* Gaertn. – Chinese jute.
128 *Trachycarpus excelsus* Makino – windmill or spinning palm.
129 *Themeda triandra* Forsk. var. *japonica* Makino.
130 *Metroxylon sagu* Rottb. (used as roof covering and as food in the Sunda Islands).

DYE PLANTS:

131 *Polygonum tinctorium* Lour. – Chinese indigo.
132 *Strobilanthes flaccidifolium* Nees – Assam indigo (used for a blue color).
133 *Rubia cordifolia* L. – Indian madder.
134 *Lithospermum erythrorrhizon* Sieb. & Zucc. – pucoon.

PLANTS FOR VARIOUS USES:

135 *Astragalus sinicus* L. – Chinese astragal (used as green manure).
136 *Cycas revoluta* Thunb. – sago palm (Japan).

The Chinese center is characterized by an exceptionally large number of cultivated plants, including representatives of the temperate, the subtropical and the tropical areas.

Three species of millet, buckwheat and soybeans and a number of leguminous plants are *the most important endemics of the temperate zone*. A number of endemic tuberous and rhizocarpous aquatic plants are specific to China. *China has to be put in first place with respect to fruit-producing species (Pyrus, Malus, Prunus). Many citrus fruits originated in China.* The cultivated flora of China is extremely peculiar and as far as its composition is concerned is sharply different from other primary centers of agriculture in the world. The vegetation as well as the animals and the food of the Chinese are extremely odd with respect to its composition. Shoots of many bamboos, a multitude of cultivated aquatic plants, including the grass *Zizania latifolia*, which is cultivated for the sake of its leafy sheaths affected by a smut, edible burdock, peculiar Chinese cabbages, gigantic radishes weighing up to 15–16 kilos, a multitude of dishes made of soybeans which replace meat and even a tofu cheese – these are some examples of the general composition of the vegetable food of the Chinese. *China can be distinguished from the other centers by its wealth of endemic species and its great specific and generic potentials of cultivated plants.* As a rule, these species are represented by an enormous number of botanical varieties and an endless multitude of hereditary forms. The diversity of soybeans, persimmons and citrus fruits can be described as literally thousands of easily distinguishable hereditary forms.

To this should be added that the multiformity of the plants in the temperate and subtropical zones is concentrated mainly within the eastern and the central parts of China.

II THE INDIAN CENTER OF ORIGIN OF CULTIVATED PLANTS

The Indian or, rather, Indo-Chinese center of origin of cultivated plants is second in importance and geographic size. It includes Burma and Assam but excludes northwestern India, i.e. Punjab and the northwestern border republics.

GRAIN CROPS:

1 *Oryza sativa* L. – rice (cultivated in the form of an enormous number of varieties but also wild growing).

2 *Andropogon sorghum* Brot. – sorghum (an especially large geographical group).
3 *Eleusine coracana* (L.) Gaertn. – finger millet (one of its centers); *E. indica* (L.) Gaertn. – goosegrass (wild, weedy, young stems used as food).
4 *Paspalum scrobiculatum* L. – kodo millet (both cultivated and wild).
5 *Cicer arietinum* L. – chick-peas.
6 *Cajanus indicus* Spreng. – pigeon peas.
7 *Phaseolus aconitifolius* Jacq. – mat bean (wild and cultivated).
8 *Ph. mungo* L. – gram.
9 *Ph. aureus* (Roxb.) Piper – mung bean.
10 *Ph. calcaratus* Roxb. – rice bean (cultivated and wild).
11 *Dolichos biflorus* L. – horse bean (cultivated and wild).
12 *D. lablab* L. – hyacinth bean.
13 *Vigna sinensis* Endl. – blackeye bean.
14 *Trigonella foenum-graecum* L. – fenugreek (independent center).
15 *Canavalia gladiata* (Jacq.) DC. – sword bean (wild and cultivated).
16 *Pachyrhizus angulatus* Rich. – yam bean or jicama (possibly also belonging to Indo-China).
17 *Psophocarpus tetragonolobus* (Stickm.) DC. – winged bean or asparagus bean.
18 *Cyamopsis psoralioides* DC. – cluster bean.

OTHER GRAINS:

19 *Amaranthus frumentaceus* Roxb. and *A. speciosus* Sims, *A. anardana* Wallich – amaranths.

VEGETABLES:

20 *Amaranthus blitum* var. *oleraceus* Watt. (wild and cultivated).
21 *A. gangeticus* L. and *A. tricolor* L. – Indian amaranths.
22 *Solanum melongena* L. – eggplant.
23 *Carum roxburghianum* Benth. & Hook., *C. copticum* Roxb. – ammi (one of its centers).
24 *Momordica charantia* L. – bitter gourd.
25 *Cucumis sativa* L. – cucumber; here also the closely related *C.hardwickii* Royle [a synonym of *C. sativa*].
26 *Raphanus caudatus* L. – oil radish (the pods are used as food).
27 *Lagenaria vulgaris* Sér. – bottle gourd.
28 *Luffa acutangula* Roxb. – dishcloth gourd.
29 *Trichosanthes anguina* L. – snake gourd (and other species).
30 *Basella rubra* L. – malabar nightshade.
31 *Pluchea indica* (L.) Less. (wild and cultivated).
32 *Anethum sowa* Roxb. (*Pseudanum graveolus* Wats.) – dill (also medicinal).
33 *Lactuca indica* L. – Indian lettuce.

ROOT VEGETABLES AND TUBERS:

34 *Colocasia antiquorum* Schott. – taro (one of its centers).
35 *Alocasia macrorrhiza* Schott. (*Arum macrorrhizum* L.) – giant taro (especially in Ceylon and on other islands; wild and cultivated).
36 *Dioscorea alata* L., *D. aculeata* L. – water yam or winged yam (wild and cultivated); perhaps also in Malaya.
37 *Curcuma zedoaria* Rosc. – zedoary, *C. longa* L. – turmeric and other species (used both as starch and for medicinal purposes).

38 *Amorphophallus campanulatus* (Roxb.) Blume – (wild and cultivated).
39 *Raphanus indica* Sinsk. – Indian radish.

FRUIT CROPS:

40 *Mangifera indica* L. – mango (wild and cultivated).
41 *Citrus sinensis* (L.) Osb. – orange (wild and cultivated).
42 *C. poonensis* Tan. – ponkan (?)
43 *C. nobilis* Lour. – King orange.
44 *C. limonia* Osb. – Canton lemon (cultivated and wild), *C. limon* Burm. f. – lemon.
45 *C. medica* L. – citron or cedrat (wild and cultivated).
46 *C. aurantium* L. – Seville orange (cultivated and wild).
47 *C. aurantifolia* (L.) Swingle – sour lime (mainly on the islands).
48 *Terminalia bellerica* Roxb. – myrobalan (cultivated and wild).
49 *Phoenix silvestris* Roxb. – date palm (wild and cultivated).
50 *Garcinia indica* Choisy – ratambi (cultivated and wild).
51 *Mimusops elengi* L. – Spanish cherry (wild and cultivated).
52 *Feronia elephantum* Correa (cultivated and wild) – elephant or wood apple.
53 *Eugenia jambolana* Lem. – Java plum, *E. jambos* L. – rose apple, wild and cultivated; one of its centers.
54 *Artocarpus integer* (Thunb.) Merr. (*A. integrifolia* L.) – champedac (wild and cultivated).
55 *Aegle marmelos* Correa (also used for resin, medicines and as a dye).
56 *Averrhoa bilimbi* L. – bilimbi (cultivated and wild).
57 *A. carambola* L. – carambole (Ceylon and the Moluccan Islands).
58 *Carissa carandas* L. – karanda (most often used wild, but also cultivated).
59 *Phyllanthus emblica* L. and other species – emblic or myrobalan (wild and cultivated).
60 *Murraya exotica* L., *M. koenigii* Sér. (wild and cultivated).
61 *Morinda citrifolia* L. – Indian mulberry (cultivated and wild).
62 *Mimusops hexandra* Roxb. (wild and cultivated).
63 *Tamarindus indica* L. – tamarind (possibly an introduction from Africa).

SUGAR PRODUCERS:

64 *Saccharum officinarum* L. – sugar cane.
65 *Arenga saccharifera* Labill – sugar palm (distributed also over the Malayan archipelago).

OIL-PRODUCING PLANTS:

66 *Cocos nucifera* L. – cocoa palm (one of its centers; mainly in southern India).
67 *Sesamum indicum* L. – sesame (main center of cultivated plant).
68 *Carthamus tinctorius* L. – safflower (one of its centers).
69 *Brassica juncea* (L.) Czern. & Coss. – Indian (Chinese) mustard (maybe a secondary center).
70 *B. glauca* Wittm. – yellow sarson or Indian colza.
71 *B. nigra* (L.) Koch – black or true mustard (a special geographical group, secondary).

FIBER PLANTS:

72 *Gossypium arboreum* L. – cotton tree.
73 *G. nanking* Meyen and *G. obtusifolium* Roxb. – tree cotton.
74 *Corchorus capsularis* L. – white jute, *C. olitorius* Roxb. – tussa jute.
75 *Crotolaria juncea* L. – Bombay or sun-hemp.
76 *Sesbania aculeata* Pers. – canicha.
77 *Hibiscus cannabicus* L. – kenaf.
78 *H. sabdariffa* L. – Jamaica sorrel (flower buds eaten).
79 *Bombax malabaricum* DC. – (cultivated and wild).
80 *Sida rhombifolia* L. – Cuba jute.
81 *Abroma angusta* L. f. – perennial Indian hemp.
82 *Sansevieria zeylanica* Willd.

SPICES AND STIMULANTS:

83 *Cannabis indica* Lam. – hemp (used as hashish or marijuana).
84 *Piper nigrum* L. – black pepper (wild and cultivated).
85 *P. betle* L. – betel pepper, *P. longum* L. – long pepper and other species.
86 *Elettaria cardamomum* Maton & White, *E. major* Smith – cardamoms.
87 *Areca catechu* L. – areca palm (indeed a secondary center; mainly in Ceylon).
88 *Alpinia galanga* Willd. and other species – galangals.
89 *Kaempferia galanga* L. – bataka.
90 *Curcuma mangga* Val. & v. Zijp. – *C. purpurascens* Bl., *C. xanthorriza* Roxb.
91 *Cuminum cyminum* L. – cumin.

PLANTS PRODUCING ETHERIC OILS, RESINS AND TANNINS:

92 *Acacia arabica* Willd. – wattle.
93 *A. catechu* (L.) Willd., (also a dye plant).
94 *A. farnesiana* Willd. – sweet acacia.
95 *Cymbopogon martini* (Roxb.) Stapf. – palmarosa-oil herb, *C. nardus* (L.) Rendle – citronella – [oil-herb used in insecticidal candles, DL].
96 *Pogostemon heyneanus* Benth. – patchouli, *P. patchouli* Hook. f. – puncha pat.
97 *Santalum album* L. – sandalwood.
98 *Jasminum grandiflorum* L. – Arabian jasmine.

DYE PLANTS:

99 *Indigofera tinctoria* L. (may be introduced from China) – true indigo.
100 *Morinda citrifolia* L. – Indian mulberry.
101 *Rubia tinctorum* L. – madder.
102 *Lawsonia alba* Lam. – henna (wild and cultivated; in cultivation already in ancient Egypt).
103 *Oldenlandia umbellata* L.
104 *Caesalpinia sappan* L. – sappan.
105 *Terminalia catappa* L. – Indian almond (cultivated and wild).
106 *T. chebula* Retz. and other species.

109 *Croton tiglium* L.
110 *Strychnos nux-vomica* L. – strychnine nut.
111 *Taraktogenos kurzii* King – chaulmoogra.
112 *Hydrocarpus anthelminticus* Pierr. – lucrabao.
113 *Oroxylum indicum* (L.) Vent.

PLANTS OF VARIOUS USES:

114 *Bambusa tulda* Roxb. – bamboo (cultivated and wild).
115 *Cedrela toona* L. – cedro (cultivated and wild; used for various purposes).
116 *Borassus flabellifer* L. – toddy palm.
117 *Ficus elastica* Roxb. – rubber plant.

India definitely appears to be the native land of rice, sugar cane, a large number of leguminous plants, many tropical fruit trees, including mango, and a multitude of citrus fruits (e.g. lemons, oranges and some species of mandarines).

Although India yields to China with respect to the number of species, when considering the presence there of rice – cultivated during the last thousand years mainly in China but introduced there from India – the position of tropical India becomes more important as far as worldwide agriculture is concerned. The fact that India is the native land of rice is indicated by the presence there of a number of wild species of rice and the discovery there of ordinary rice in a wild state and as a weed with all the attributes of wild grasses: e.g. elaiosomes, causing the seeds to drop when ripe. Also known there are forms intermediate between wild and cultivated rice. The varietal composition of cultivated rice in India is the richest in the world. The assortment in India is distinguished, in particular, by characteristics of coarse-grained primitive forms. In contrast to China and other secondary centers of crops, Indian rice is characterized by the presence of dominant genes.

Just as in the case of China, the use of a very large number of wild plants for various purposes, including a number of edible plants, is typical of India (Drury, 1873).

II. a. THE INDO-MALAYAN CENTER OF ORIGIN OF CULTIVATED PLANTS

In addition to the Indian center we distinguish an Indo-Malayan center, comprising all the Malayan archipelago, the major islands such as Java, Borneo and Sumatra, as well as the Philippines, and Indo-China.

CEREAL GRASSES:

1 *Coix lacryma* L. – coix (affiliated mainly with the islands).

BEANS:

2 *Mucuna utilis* Wall & Wight (Sunda Islands) – velvet beans.

BAMBOOS:

3 *Dendrocalamus asper* (Schult. f.) Backer (Sunda Islands).

4 *Gigantochloa apus* (Roem. & Schult.) Kurz., *G. verticillata* (Willd.) Munro, *G. ater* Kurz. (especially often cultivated on the Sunda Islands).

ROOT VEGETABLES AND TUBERS:

5 *Dioscorea alata* L. – (on the islands), *D. hispida* Dennst. – yam and winged or water yam respectively.
6 *D. pentaphylla* L. (wild and cultivated).
7 *D. bulbifera* L. – potato yam.
8 *Coleus tuberosus* (Bl.) Benth. – hausa potato (leaves, stems and tubers used for food).
9 *Phytolacca esculenta* Van Houtte – yama gobo (possibly even in India).
10 *Tacca pinnatifida* Forst. – East-Indian arrowroot (especially widely distributed on Fiji, Samoa and the New Hebrides Islands).
11 *Zingiber officinale* Rosc. – ginger, *Z. mioga* Rosc., *Z. zerumbet* (L.) Rosc. (the second species also in Japan; also used as a spice).
12 *Benincasa hispida* Cogn. – white or wax gourd.
13 *Sauropus androgynus* (L.) Merr. – katuk (Sunda Islands).
14 *Abelmoschus manihot* (L.) Medic. – muskmallow

FRUIT CROPS:

15 *Citrus microcarpa* Bge., *C. mitis* Bl. (the first species mainly on Java).
16 *Citrus grandis* Osb. (*C. maxima* Merr.) – pomelo or shaddock (mainly found on the islands).
17 *C. hystrix* DC. – papeda (on the islands).
18 *Nephelium mutabile* Bl. – pulassan (on the islands; wild and cultivated).
19 *Canarium pimela* Koenig, *C. album* Roensch – Chinese white and black olives, respectively.
20 *Areca catechu* L. – areca palm (Sunda Islands, Moluccan Islands; wild and cultivated).
21 *Erioglossum rubiginosum* (Roxb.) Brandes, – klaju *E. edule* Bl.
22 *Antidesma delicatulum* Hutch., *A. hainanensis* Merr., *A. bunias* (L.) Spr. – Chinese laurel.
23 *Musa cavendishii* Lamb., *M. paradisiaca* L., *M. sapientium* L. – bananas (the last species mainly on the islands).
24 *Garcinia mangostana* L. – mangosteen, *G. dulcis* (Roxb.) Kurz. (wild and cultivated).
25 *Artocarpus communis* Forst. – breadfruit tree, *A. champeden* (Lour.) Spreng. (Malayan archipelago).
26 *A. integer* (Thunb.) Merr. – jack fruit.
27 *Durio zibethinus* Murr. – durian (Malayan archipelago).
28 *Lansium domesticum* Corr. – duku (Malayan archipelago).
29 *Bouea macrophylla* Griff. – gandaria (Malayan archipelago).
30 *Mangifera caesia* Jack, *M. foetida* Lour., *M. odorata* Griff. – mangoes (Malayan archipelago).
31 *Baccaurea racemosa* (Bl.) Muell. – menteng (Java, the Philippines).
32 *Flacourtia rukam* Zoll. & Mor. – rukam (wild and cultivated on Java).
33 *Pangium edule* Reinw. – pangi (wild and cultivated on the Sunda Islands).
34 *Pithecolobium lobatum* Benth. (Sunda Islands).
35 *Cynometra cauliflora* L. (on the islands).
36 *Sandoricum koetjape* (Burm. f.) Merr. – false mangistan (wild and cultivated).

37 *Eugenia aquea* Burm. f, *E. cumini* (L.) Merr., *E. javanica* Lam., *E. malaccensis* L. – water rose apples
38 *Nephelium lappaceum* L. – Zanzibar litchi.
39 *Salacca edulis* Reinw. – salak palm.
40 *Rubus rosaefolius* Smith (introduced into cultivation on Java; also wild in the Philippines; extends to southern Japan).

OIL-PRODUCING PLANTS:

41 *Aleurites moluccana* (L.) Willd. (on the Sunda Islands). – candlenut tree.
42 *Cananga odorata* (Lam.) Hook f. & Thoms. – ylang-ylang (actually from Malaya).
43 *Vetiveria zizanioides* Stapf. – vetti-ver (actually from Malaya).
44 *Cocos nucifera* L. – coconut palm (actually its main center).

SUGAR-PRODUCING PLANTS:

45 *Saccharum officinarum* L. – sugar cane (one of its centers).
46 *Arenga saccharifera* Labill. – sugar palm (one of its centers, may be the main one).

SPICES:

47 *Elettaria cardamomum* Maton & White – cardamom (wild and cultivated on the Sunda Islands); also *Amomum krevanh* Pierre (wild and cultivated).
48 *Kaempferia galanga* L., *K. pandurata* Roxb., *K. rotundata* L. (on the Sunda Islands).
49 *Caryophyllus aromaticus* L. (*Eugenia caryophyllata* Thunb.) – clove tree (the Moluccans and other islands).
50 *Myristica fragrans* Houtt. – nutmeg (wild and cultivated; Moluccan and other islands).
51 *Piper nigrum* L. – black pepper.

TEXTILE PLANTS:

52 *Musa textilis* Nee – Manila hemp (the Philippine Islands).
53 *Metroxylon sagu* Rottb. – sago palm (used for covering roofs and as an edible plant on the Sunda Islands).

DYE PLANTS:

54 *Curcuma longa* L. – turmeric (one of its centers).

GUTTA-PERCHA-PRODUCING PLANTS:

55 *Palaquium gutta* Bursck. – gutta-percha tree.

Unfortunately, the region of southeastern Asia, belonging entirely to the Tropics with its rich wild and cultivated floras, is still very inexactly known and, indeed, needs additional and serious attention in the future and possibly changes in the composition of the species listed.

This center, which is affiliated with the Indian one, is especially rich in fruit crops, many of worldwide importance, such as bananas and some citrus fruits. In the composition of the wild flora there is a large number of useful plants, which have been especially well studied by the Dutch scientists on Java (Heyne, 1927; Ochse, 1933b).

III THE INNER-ASIATIC CENTER OF ORIGIN OF CULTIVATED PLANTS

The third center of development of cultivated plants occupies a considerably smaller territory. We shall call it the Inner-Asiatic one. It comprises northwestern India (Punjab, the northwestern border republics and Kashmir), all of Afghanistan, the Soviet Tadzhikistan and Uzbekistan as well as western Tien-Shan.

GRAIN CROPS:

1 *Triticum vulgare* Vill. – soft wheat (according to the latest statements by N.I. Vavilov, this is a secondary but not a primary center of the soft wheats – Remark by E.N. Sinskaya).
2 *T. compactum* Host. – club wheat.
3 *T. sphaerococcum* Perc. – shot wheat.
4 *Secale cereale* L. – rye (secondary center).
5 *Pisum sativum* L. – peas.
6 *Lens esculenta* Moench. – lentils.
7 *Vicia faba* L. – fava beans.
8 *Lathyrus sativus* L. – chickling vetch.
9 *Cicer arietinum* L. – chick-peas.
10 *Phaseolus aureus* (Roxb.) Piper – mung bean.
11 *Ph. mungo* L. (*Ph. radiatus* Roxb.) – black gram (secondary center).
12 *Brassica campestris* ssp. *oleifera* (Metzg.) Sinsk. – rape seed (one of its centers).
13 *B. juncea* (L.) Czern. – Indian or Chinese mustard.
14 *Eruca sativa* Lam. – garden rocket (from weeds; one of its centers).
15 *Lepidium sativum* L. – garden cress or pepper grass (secondary center).
16 *Linum usitatissimum* L. – flax (for linseed?); (one of its centers).
17 *Sesamum indicum* L. – sesame (one of its centers).
18 *Coriandrum sativum* L. – coriander (one of its centers).
19 *Carum copticum* Benth. & Hook. (*Ammi copticum* L.) – omum.
20 *Carthamus tinctorius* L. – safflower (one of its centers).
21 *Cannabis indica* L. – hemp (marijuana).

FIBER PLANTS:

22 *Gossypium herbaceum* L. – levant cotton.

VEGETABLES AND MELONS (GOURDS, ETC.):

23 *Cucumis melo* L. – melon (secondary center).
24 *Lagenaria vulgaris* Sér. – bottle gourd (secondary center).
25 *Daucus carota* L. – carrots (main center of the Asiatic kinds).
26 *Brassica campestris* L. ssp. *rapifera* – turnips (main center of the Asiatic turnips).
27 *Raphanus sativus* L. – radishes (one of its centers).
28 *Allium cepa* L. (sensu lato) – onion (cultivated); in a wild state, here, some closely related species: *A. pskemense* Fedtsch. and *A. vavilovii* Vved.
29 *A. sativum* L. – garlic (wild and cultivated), *A. longicuspis* E. Regel.
30 *Spinacia oleracea* L. – spinach, in a wild state the closely related *S. tetrandra* Stev.
31 *Portulacca oleracea* L. – purslane (one of its centers).

SPICE PLANTS:

32 *Ocimum basilicum* L. – basil.

FRUIT CROPS:

33 *Pistacia vera* L. – pistachio nut (one of its centers).
34 *Prunus armeniaca* L. – apricot (one of its centers).
35 *Pyrus communis* L. – pear, in a wild condition also *P. heterophylla* Reg. & Schmalh., *P. korshinskyi* Litw., *P. vavilovii* M. Pop. and *P. bucharica* Litw.
36 *Amygdalus communis* L. [*Prunus amygdalus* Batch.] – almond tree (wild and cultivated; one of its centers, very variable with respect to the intraspecific composition); in a wild state here are also found *A. bucharica* Korsh. and *A. spinosissima* Bge.
37 *Eleagnus angustifolia* L. – Russian olive (one of its centers).
38 *Zizyphys sativa* Gaertn. – jujuba (wild and cultivated; one of its centers)
39 *Vitis vinifera* L. – grapevine (wild and cultivated; in addition, the wild grapevine: *V. vinifera spontanea* M. Pop, here in a table quality, i.e. very close to the cultivated one. Such groups are so far only found in Tadzhikistan. There is no *V. silvestris* [*V. vinifera* L. ssp. *sylvestris* (Gmel.) Regel.], which gave rise to the wine grapes (Popov, 1929).
40 *Juglans regia* L. – walnuts (cultivated and wild; one of its centers).
41 *Corylus colurna* L. – Turkish hazelnuts (mainly wild in Afghanistan).
42 *Malus pumila* Mill. – paradise apple (wild, in many varieties especially in western Tien-Shan; also cultivated; one of its centers).

Concerning the number of species, this center yields considerably to the previous two. Nevertheless, it is of enormous importance to us since *this is the homeland of club and shot wheats and the native land of grain beans such as peas, lentils, chickling vetch, fava or horse beans and chick-peas, which represent an exceptional wealth of genes*. Many oil-producing plants also came from here and the Asiatic cotton was indeed taken into cultivation just here.

The particular ecological conditions and the particular wild and cultivated floras speak in favor of the necessity for uniting Punjab, Kashmir and the eastern portion of the Soviet Inner Asia into one center. In spite of the barriers of Himalaya and Hindukush, it is appropriate to unite an important part of Inner Asia and northwestern India into one center.

IV THE CENTER OF ORIGIN OF CULTIVATED PLANTS IN ASIA MINOR

The fourth center corresponds to Asia Minor in a wide sense, including Asia Minor in the strict sense, all of Transcaucasia, Iran and Turkmenistan.

GRAIN CROPS:

1 *Triticum monococcum* L. – einkorn (the cultivated 14 chromosome type), here widely distributed in its wild form; *T. thaoudar* Reut., *T. aegilopoides* Bal.
2 *T. durum* Desf. ssp. *expansum* Vav. – hard wheat (28 chromosomes).
3 *T. turgidum* L. ssp. *mediterraneum* Flaksb. – poulard wheat (28 chromosomes)
4 *T. vulgare* Vill. – an endemic, awnless group of soft wheats (the 42 chromosome

type; one of its centers). [Later N.I. Vavilov considered this to be the primary center. – Remark by E.N. Sinskaya]

5 *T. orientale* Perc. – Khorassan wheat.

6 *T. persicum* Vav. – Persian wheat (28 chromosomes; Armenia and Georgia).

7 *T. timopheevi* Zhuk. – Sanduri wheat or Timopheevi's wheat (28 chromosomes)

8 *T. macha* Dekapr. & Menabde – masha wheat (42 chromosomes).

9 *T. vavilovii* Jakubz. (*T. vulgare compositum* Tum.) – a peculiar branched wheat found by Prof. M.G. Tumanyan around Lake Van in Turkish Armenia (42 chromosomes); there a large number of wild types of *T. dicoccoides* Koern. are found and also a large number of species of *Aegilops*.

10 An endemic group of cultivated distichous barley (*Hordeum distichum* [*H. vulgare* L. convar. *distichon* (L.) Alef], var. *medicum, nigricans, nutans* and so on).

11 *Secale cereale* L. – rye; here also the wild species *S. montanum* Guss., *S. ancestrale* Zhuk., *S. vavilovii* Grossh., *S. fragile* L., *S. villosum* L. (*Haynaldia villosa* Schur.).

12 *Avena byzantina* C. Koch – Mediterranean oats.

13 *A. sativa* L. – a large number of field weeds (especially in Transcaucasia) in the form of endemic varieties.

14 *Cicer arietinum* ssp. *pisiforme* G. Pop – chick-peas (secondary center).

15 *Lens esculenta* Moench. – lentils (a large endemic group of varieties); also found here are wild *L. lenticula* (Schreb.) Alef., *L. nigricans* (M.B.) Godr., *L. kotschyana* (Boiss.) Alef. and *L. orientalis* (Boiss.) Hand.-Mazz.

16 *Vicia ervilia* Willd. – erse or French lentils – a special endemic group.

17 *Pisum sativum* L. – peas (an important, endemic group; secondary center); here are also found wild peas, e.g. *P. elatius* Stev., *P. humile* Boiss.

18 *Lupinus pilosus* L. [*L. digitatus* Forssk.], *L. angustifolius* L. – blue lupines, and *L. albus* L. – white lupines (wild and cultivated in Asia Minor).

FORAGE PLANTS:

19 *Medicago sativa* L. s. lat. – alfalfa.

20 *Trifolium resupinatum* L. – Persian clover.

21 *Trigonella foenum-graecum* L. – fenugreek (both in a cultivated and a wild condition; secondary center).

22 *Onobrychis altissima* Grossh. and *O. transcaucasia* Grossh. (two Transcaucasian species – cock's head or sainfoin; wild and cultivated).

23 *Lathyrus cicera* L. – (one of its centers).

24 *Vicia sativa* L. – common vetch (a large, endemic group of varieties in Asia Minor; the main center of development of this species).

25 *V. villosa* Roth. – hairy or winter vetch, var. *perennis* Tum. (wild and introduced into cultivation).

26 *V. pannonica* Crantz – Hungarian vetch (in cultivation and as a weed).

OIL-PRODUCING PLANTS:

27 *Sesamum indicum* L. ssp. *bicarpellatum* Hillt. – sesame (a special geographical group).

28 *Linum usitatissimum* L. – flax, linseed. Many endemic varieties; in Asia Minor there is a special endemic group, gr. *prostrata* Vav.).

29 *Brassica campestris* L. ssp. *oleifera* (Metzg.) Sinsk. – rape seed (one of its centers).

30 *B. nigra* (L.) Koch var. *pseudocampestris* Sinsk. and var. *orientalis* Sinsk. – black or true mustard (one of its centers).

31 *B. juncea* Czern. var. *sareptana* Sinsk. – Indian or Chinese mustard (secondary center).
32 *Camelina sativa* L. – cameline (cultivated and wild).
33 *Eruca sativa* Lam. var. *orientalis* Sinsk. – garden rocket.
34 *Cephalaria syriaca* Schrad. – 'belimir'.
35 *Ricinus persicus* G. Popova – castor bean.

PLANTS PRODUCING ETHERIC OILS, ALKALOIDS AND TANNINS:

36 *Pimpinella anisum* L. – anise.
37 *P. anisetum* Boiss. – anisette.
38 *Coriandrum sativum* L. – coriander (one of its centers).
39 *Lallemantia iberica* L. – (a field weed but also cultivated).
40 *Papaver somniferum* L. – poppy (a large endemic group with a high content of morphine).
41 *Rosa centifolia* L. – rose or cabbage rose.
42 *Rhus coriaria* L. – sumach.

MELONS, ETC:

43 *Cucumis melo* L. – melons; also wild forms there: *C. agrestis* Pang. [ssp. *agrestis*] and *C. microcarpus* (Alef.) Pang. [*C. melo* ssp. *dudaim* (L.) Greb.]. (All the main varieties of melons found in the world are accumulated in Asia Minor.)
44 *C. flexuosus* L. – tarra.
45 *C. sativus* L. ssp. *antasiaticus* Gabalev – Anatolian cucumber (special geographic race).
46 *Cucurbita pepo* L. – gourds or pumpkins (the largest varietal diversity is concentrated in Asia Minor); the problem of *C. pepo* needs further study.

VEGETABLES:

47 *Lepidium sativum* L. – cress (secondary center).
48 *Brassica campestris* L. ssp. *rapifera* (Metzg.) Sinsk. – turnips (secondary center).
49 *Beta vulgaris* L. – beets (in Asia Minor there is a special geographic group of forms; secondary center).
50 *Daucus carota* L. – carrots (a specially important variety of cultivated forms in Anatolia).
51 *Brassica oleracea* L. – cabbages (many endemic forms in Anatolia).
52 *Eruca sativa* L. – garden rocket (from a weed; leaves are also used).
53 *Allium cepa* L. – onion (secondary center).
54 *A. porrum* L. – leek, has many closely related wild species, e.g. *A. ampeloprasum* L. – roundheaded garlic, etc.
55 *Petroselinum hortense* Hoffm. – parsley (secondary center).
56 *Lactuca sativa* L. – lettuce (wild and cultivated).
57 *Portulacca oleracea* L. – purslane (used as a weed and in cultivation).

FRUIT CROPS:

58 *Ficus carica* L. – figs.
59 *Punica granatum* L. – pomegranate.
60 *Malus pumila* Mill. – paradise apple (one of its centers).
61 *Pyrus communis* L. – pear, and other species such as *P. salicifolia* Pall., *P. eleagrifolia* Pall., *P. syriaca* Boiss. and *P. nivalis* Jacq. – snow pear.

62 *Cydonia oblonga* Mill. – quince.
63 *Prunus divaricata* Ledeb. – cherry plum.
64 *P. cerasus* L. – sour cherry (the center may be placed in Asia Minor, which must be checked). Cherries in the wild state are not known from either Caucasus or Inner Asia.
65 *Cerasus avium* (L.) Moench. – bird cherry.
66 *Amygdalus communis* L. [*Prunus dulcis* (Mill.) D.A. Webb], *A. fenzliana* (Fritsch.) Lipsky, *A. bucharica* Korsh., *A. scoparia* Spach, *A. spinosissima* Bge. – almond trees (mainly in a wild state).
67 *Laurocerasus officinalis* Roem. – laurel.
68 *Mespilus germanica* L. – medlar.
69 *Juglans regia* L. – walnuts (one of the centers).
70 *Corylus avellana* L. – hazelnuts (cultivated and wild); also some wild species such as *C. maxima* Mill. – filbert nuts, *C. pontica* Koch., *C. colurna* L. – Turkish hazelnuts, and *C. colchisa* Alb.
71 *Castanea sativa* Mill., (*C. vesca* Gaertn.) – European chestnuts (one of its centers).
72 *Zizyphus sativa* Gaertn. – jujuba (secondary center).
73 *Vitis vinifera* L. – grapevine (in an enormous diversity of forms, cultivated and wild).
74 *Berberis vulgaris* L. – European barberry.
75 *Prunus armeniaca* L. – apricots (one of its centers; actually a secondary one).
76 *P. padus* L. – choke cherry.
77 *Pistacia vera* L. – pistachio (one of its centers).
78 *Eleagnus angustifolia* L. – Russian olive.
79 *Diospyrus lotus* L. – date plum.
80 *Cornus mas.* L. – Cornelian cherry.
81 *Crataegus azarolus* L. – mispel.

SPICES AND DYE PLANTS:

82 *Crocus sativus* L. – saffron (belongs possibly to the Mediterranean center – Greece, Italy).
83 *Rubia tinctorum* L. – madder (wild and cultivated).

This center is remarkable primarily because of its exceptional wealth of cultivated species of wheat, which has been demonstrated by the most recent research. *Nine botanical species of wheat are endemic in Asia Minor.*

Within the borders of the U.S.S.R. the variety of wheat is especially great in Armenia; with respect to the number of botanical varieties, it comes before all other areas and territories. M.G. Tumanyan has counted more than 200 varieties there out of a total of ca. 650. Wild wheats occur there in a great diversity, both mono- and distichous types. Concerning the number of species and ecotypes, this center can be distinguished from all others in the world.

Asia Minor and the Soviet Transcaucasus is the native land of rye, which is represented there by an amazing diversity of forms in contrast to the uniform rye of Europe. Both black-spiked and red-spiked rye is found there and a multitude of peculiar types, including some new species of wild rye (*Secale vavilovii* Grossh. and *S. ancestrale* Zhuk.).

An important potential for the European fruit industry is concentrated in Asia Minor:

it is the native land of grapevines, pears, cherry plums, bird cherries, pomegranates, walnuts, quince, almonds and figs. No doubt, the first gardens were established in Asia Minor. All the evolutionary stages of fruit cultivation can still be found in Georgia and Armenia all the way from wild forests, consisting almost entirely of fruit trees, through transitory stages up to the present form of horticulture, where the best of the forms are grafted on less valuable forms of the fruit trees. It can be seen how, when clearing the forests for arable land, the farmer leaves the best kinds of wild apples or wild pears standing in the fields. The most recent data have demonstrated that all the viniculture in the world and all the basic assortment of grapes originated in Asia Minor, where it is still possible to find grapes in a wild state, but fully suitable for cultivation.

With regard to a number of species such as *Medicago*, *Pyrus* and *Amygdalis* and, to a certain extent, also the wheats, a powerful speciation process has unfolded here and is apparently still going on. Thus, the development of natural polyploidization among the wheats has been established here just as among many species within the wild flora, especially in the alpine and subalpine belts.

All the world's wealth of melons and gourds comes from Turkey, Iran and the Soviet Inner Asia, and is still far from exhausted by the present plant breeding. *The most important forage plants also come from Asia Minor*: alfalfa, Persian clover or shabdar and sainfoil (*Onobrychis* spp.), fenugreek (*Trigonella foenum-graecum*) and vetch (*Vicia sativa*).

V THE MEDITERRANEAN CENTER OF ORIGIN OF CULTIVATED PLANTS

The fifth or Mediterranean center is characterized by a special composition of cultivated plants, which is, however, of a more limited importance than that of the previous center.

GRAIN CROPS:

1. *Triticum durum* Desf. ssp. *expansum* Vav. including two geographical sections: sect. *mediterraneum* Vav. and sect. *africanum* Vav. – hard wheats.
2. *T. dicoccum* Schrank – emmer (one of its centers).
3. *T. polonicum* L. – Polish wheat (one of its centers), narrowly localized; wild emmer (*T. dicoccoides* (Koern) Aarons) is also found in considerable quantities.
4. *T. spelta* L. – spelt (in the mountains of southern Germany and Asturia, Spain; possibly a secondary center).
5. *Avena byzantina* C. Koch – Mediterranean oats.
6. *A. brevis* Roth and *A. strigosa* Schreb. – sand oats (the Pyrenées).
7. *Hordeum sativum* Jess. – an endemic group of large-grained barley (secondary center).
8. *Phalaris canariensis* L. – true canary grass (the western Mediterranean area).
9. *Ervum monanthos* Desf. – single-flowered erse or vetch.
10. *Lens esculenta* Moench. ssp. *macrosperma* Bar. – large-seeded lentils.
11. *Vicia ervilia* Willd. – French erse (eastern Mediterranean area, on Cyprus and Crete).
12. *Lathyrus sativus* L. var. *macrospermus* Zalk. – large-seeded chickling vetch.
13. *Pisum sativum* L. – peas, a large-seeded group.

14 *Vicia faba* L. – fava or horse bean, large-seeded type.
15 *Lupinus albus* L. – white lupine; *L. termis* Forssk. – Sicilian or Egyptian lupine; *L. angustifolius* L. – blue lupine and *L. luteus* L. – yellow lupine.
16 *Cicer arietinum* L. – chick peas, a large-seeded group.

FORAGE PLANTS:

17 *Hedysarum coronarium* L. – sulla or cock's head (the southern part of the Apennine peninsula and Sicily).
18 *Trifolium alexandrinum* Jusl. – berseem clover (Syria and Egypt).
19 *T. repens* L. var. *giganteum* Lag.-Foss. – white clover (Lombardy).
20 *T. incarnatum* L. – crimson clover (cultivated and wild; Sardinia, the Baleares, Algeria and other areas).
21 *Ulex europaeus* L. – gorse (Portugal).
22 *Vicia sativa* L. – vetch (wild and already cultivated at the time of the Romans; one of its centers).
23 *Lathyrus gorgonii* Parl. – fodder peas (Syria).
24 *L. ochrus* DC. (wild in Italy and Spain but also cultivated).
25 *L. cicera* L. (cultivated and wild).
26 *Ornithopus sativus* Brot. – serradela (wild in Portugal, Spain and Algeria but also cultivated).
27 *Spergula arvensis* L. – corn spurry (the distribution area of the wild corn spurry is wider).

OIL-PRODUCING PLANTS AND SPICES:

28 *Linum usitatissimum* L. ssp. *mediterraneum* Vav. – flax or linseed (large- seeded); also found here are large amounts of wild *L. angustifolium* Huds. – pale flax.
29 *Sinapis alba* L. – white mustard (wild, weedy and cultivated; one of its centers).
30 *Brassica napus* L. ssp. *oleifera* Metzg. – rape seed.
31 *B. nigra* (L.) Koch – black mustard (main center).
32 *B. campestris* L. ssp. *oleifera* (Metzg.) Sinsk. – colza or turnip seed.
33 *Eruca sativa* Lam. – garden rocket (main center).
34 *Argania sideroxylon* Roem. & Schult. (Morocco).

FRUIT CROPS:

35 *Olea europaea* L. – olive tree.
36 *Ceratonia siliqua* L. – carob tree or St. John's bread tree.

VEGETABLES:

37 *Beta vulgaris* L. – beets (in very many forms; also wild *B. maritima* L.).
38 *Brassica oleracea* L. – cabbage (in very many forms but also wild; in addition, the closely related wild species *B. balearica* Pers., *B. insularis* Moris. and *B. cretica* Lam.).
39 *Petroselinum sativum* L. – parsley (cultivated and wild).
40 *Cynara scolymus* L. – artichoke (wild and cultivated); also the closely related wild *C. cardunculus* L. – cardoon.
41 *Brassica rapa* L. subvar. *rapifera* Metzg. – turnip (main center of origin of the European kinds).
42 *B. napus* L. var. *rapifera* Metzg. – swedes.
43 *Portulacca oleracea* L. – purslane (both as a weed and in cultivation; its distribution area extends to Asia Minor).

44 *Allium cepa* L. – onion, large form (secondary center).
45 *A. sativum* L. – garlic, large form (secondary center).
46 *A. porrum* L. – leek.
47 *A. kurrat* Schweinf. – Egyptian kurrat.
48 *Satureja hortensis* L. – savory.
49 *Lactuca sativa* L. – lettuce (the area also covers Asia Minor).
50 *Asparagus officinalis* L. – asparagus (also used in its wild form).
51 *Crambe maritima* L. – sea-kale.
52 *Apium graveolens* L. – celery (cultivated and wild).
53 *Cichorium endivia* L. – endive; its wild relative *C. pumilum* Jacq. is also distributed here.
54 *C. intybus* L. – chicory (wild, weedy and cultivated; area is very wide).
55 *Anthriscus cerefolium* Hoffm. – garden chervil (the distribution area of the wild chervil reaches all the way into Asia Minor).
56 *Lepidium sativum* L. – cress (cultivated and wild, secondary center).
57 *Pastinaca sativa* L. – parsnip.
58 *Tragopogon porrifolius* L. – oyster plant (wild and cultivated).
59 *Scorzonera hispanica* L. – black salsify (wild and cultivated; in Spain and on Sicily a wild salsify, *S. deliciosa* Guss., is also used).
60 *Scolymus hispanicus* L. – Spanish oysterplant or golden thistle (wild and cultivated).
61 *Smyrnium olusatrum* L. – horse parsley (cultivated and wild).
62 *Anethum graveolens* L. – dill (cultivated and wild; also used as a spice).
63 *Rheum officinale* Baill. – Chinese rhubarb.
64 *Ruta graveolens* L. – common rue (also used as a spice).
65 *Rumex acetosa* L. and other species – sorrels.
66 *Blitum rubrum* Rchb., *B. virgatum* L. and *B. capitatum* L. – blites.

SPICES AND PLANTS PRODUCING ETHERIC OILS:

67 *Nigella sativa* L. – black cumin.
68 *Carum carvi* L. – caraway.
69 *Cuminum cyminum* L. – cumin (possibly a secondary center).
70 *Pimpinella anisum* L. – anise (cultivated and wild).
71 *Foeniculum vulgare* Mill. – fennel.
72 *Thymus vulgaris* L. – thyme.
73 *Hyssopus officinalis* L. – hyssop.
74 *Lavandula vera* DC. – lavender.
75 *Mentha piperita* L. – peppermint.
76 *Rosmarinus officinalis* L. – rosemary.
77 *Salvia officinalis* L. – sage.
78 *Iris pallida* Lam. – iris (Italy).
79 *Rosa damascena* Mill. – Damascene rose.
80 *Laurus nobilis* L. – laurel.
81 *Humulus lupulus* L. – hops (mainly wild in the Mediterranean countries; its distribution area reaches far north, where it was actually taken into cultivation).

PLANTS PRODUCING DYES AND TANNINS:

82 *Rubia tinctorum* L. – madder (wild and cultivated).
83 *Rhus coriaria* L. – sumach (wild and cultivated; Italy and Spain).

PLANTS USED FOR VARIOUS PURPOSES:

84 *Cyperus esculentus* L. – earth-almonds or Zulu nuts (apparently Egypt).

This is unquestionably the native land of olive and carob trees. A large number of vegetables, including beets, originally came from here. Together with China this is the most important center of vegetables in the world. Many old forage plants also came originally from the Mediterranean center. It is interesting that almost every civilization here brought its own forage plants into cultivation: Egypt and Syria – Egyptian or berseem clover; the Apennine peninsula – sulla (*Hedysarum coronarium*) and red clover; and the Iberian peninsula – the single-flowered erse (*Ervum monanthos*). Fodder peas (*Lathyrus gorgonii)* initially came from Syria and gorse from Portugal. Many of the most important cultivated plants, such as wheat or fodder beans, bear witness with respect to their varieties of the presence in this area of secondary centers and of the great role played by Man for having already improved the plants in the distant past. *Many of cultivated plants in the Mediterranean center, such as flax or linseed, barley, leguminous plants and chick-peas, differ by large seeds and large fruits from the small-seeded kinds in Inner Asia, where their main center is located.* In the latter center, mainly dominant genes of these plants are present. As far as the crops of the Mediterranean area are concerned, it is possible to witness the great role played by Man in the selection of the best cultivated forms.

VI THE ABYSSINIAN CENTER OF ORIGIN OF CULTIVATED PLANTS

The Abyssinian center is the sixth one. The expedition made by us in 1927 to Abyssinia, Egypt and Somalia and the later, comparative studies of the large material collected have demonstrated the independence of Abyssinia with respect to its cultivated flora and the definite presence there (and in the mountains of Egypt) of an independent main center of cultivated plants.

GRAIN CROPS:

1 *Triticum durum* Desf. ssp. *abyssinicum* Vav. – Abyssinian hard wheat (in an amazing wealth of forms).
2 *T. turgidum* L. ssp. *abyssinicum* Stol. – cone wheat.
3 *T. dicoccum* Schrank – emmer.
4 *T. polonicum* L. gr. *abyssinicum* Vav. – Polish wheat.
5 *Hordeum sativum* Jess. – barley (exceptionally variable forms).
6 *Andropogon sorghum* Brot. – grain sorghum.
7 *Eragrostis abyssinica* Link. – teff.
8 *Eleusine coracana* (L.) Gaertn. (*E. tocussa* Fresenius) – African or finger millet.
9 *Pennisetum spicatum* L. – pearl millet (in semi-desert areas).
10 *Cicer arietinum* L. – chick peas (one of its centers).
11 *Lens esculenta* Moench. – lentils (one of its centers); as demonstrated by the most recent genetic studies of lentils (by Barulina, 1930) the Abyssinian lentils are partly sterile when hybridized with Afghani or Inner-Asiatic forms, which indicates their independence.

12 *Pisum sativum* L. – peas (one of its centers).
13 *Vicia faba* L. – fava or horsebeans (actually a secondary center).
14 *Trigonella foenum-graecum* L. – fenugreek.
15 *Lathyrus sativus* L. – chickling vetch (one of its centers).
16 *Vigna sinensis* Endl. var. *sinensis* (Stickm.) Pip. and var. *catjang* (Walp.) Piper – blackeye peas.
17 *Dolichos lablab* L. – hyacinth bean (late-ripening forms with variegated flowers).
18 *Lupinus termis* Forssk. – Sicilian or Egyptian lupine (in the sands of northern Abyssinia).
19 *Linum usitatissimum* L. – flax or linseed (grown as a cereal plant and also for its seeds; one of its centers).

OIL-PRODUCING PLANTS:

20 *Guizotia abyssinica* Cass. – niger seeds or ramtil.
21 *Carthamus tinctorius* L. – safflower.
22 *Sesamum indicum* L. – sesame (main center).
23 *Ricinus communis* L. – castor oil plant (one of its centers).
24 *Lepidium sativum* L. – cress or peppergrass (one of its centers, here a very variable form).

SPICES AND STIMULATING PLANTS:

25 *Coriandrum sativum* L. – coriander (one of its centers), a specific group of forms. E.A. Stoletova distinguishes it as a special subspecies.
26 *Nigella sativa* L. – black cumin (one of its centers).
27 *Carum copticum* Benth. & Hook. – ommu (one of its centers).
28 *Rhamnus prinoides* l'Hér. – (replaces hops for beer-making).
29 *Catha edulis* Forsk. (*Celastrus edulis* Vahl) – khat (wild and cultivated).
30 *Coffea arabica* L. – coffee trees.

VEGETABLES:

31 *Brassica carinata* Al. Braun – Abyssinian mustard; also an oil-producer.
32 *Allium* sp. (*A. ascalonicum* L.) – shallots.
33 *Musa ensete* J.F. Gmel. – Abyssinian banana.
34 *Hibiscus esculentus* L. – bamia, (mostly wild; in cultivation spread by Arabs).

PLANTS USED FOR VARIOUS PURPOSES:

35 *Euphorbia candelabrum* Trem. – 'milk-tea' (wild, planted instead of fences).
36 *Hagenia abyssinica* Willd. – 'kosso' (used only in a wild state).
37 *Commiphora abyssinica* (Berg) Engler – myrrh (used only in a wild state); produces resin.
38 *Indigofera argentea* L. – indigo (almost only used wild but cultivated in Egypt and Arabia).

In spite of the limited area used for agriculture, an amazing variety of forms has been found here. For instance, in Abyssinia not more than half a million hectares are occupied by wheat which, in relation to its general area in the world (determined to be ca. 160 million ha), constitutes a very small portion.

With respect to the number of botanical varieties of wheat, Abyssinia must be put in first place. Thereby the wheats of Abyssinia must in essence be distinguished botanically into special species as confirmed by genetical and physiological data. *The center of development of barley is also here. Nowhere else except in Abyssinia is there such a variety of forms and genes of barley. A number of cultivated plants are associated exclusively with Abyssinia such as, e.g. the cereal grass called 'teff' (Eragrostis abyssinica) and the oil-plant called 'ramtil'* (*Guizotia abyssinica*). Flax is represented here by a peculiar form that is used not for fiber or oil but exclusively as a cereal, from which a flour is made.

The number of plants typical of Abyssinia is not very high; Abyssinia knew practically nothing about growing fruits or vegetables until the Europeans arrived. This is mainly a realm of field crops, although in an amazing diversity of forms that thereby exist in conditions that are comparatively uniform ecologically, since the crops here are concentrated mainly in the mountains at altitudes from 1500 to 2500 m.

We have now reviewed all the centers in the Old World. As demonstrated by the expeditions sent out by the Institute of Plant Industry, starting in 1925 and ending in 1933, the New World is characterized by main centers of agriculture that are amazingly sharply localized. These belong to Central America including southern Mexico and to the central portion of the Andes in South America, comprising Ecuador, Peru and Bolivia. There we distinguish the independent seventh or Central American center, which also includes southern Mexico, and the eighth or Andean [South American–Russian editor's remark] center covering the territories of Peru, Bolivia and Ecuador.

The main wealth of cultivated plants in the New World is concentrated within the limited territory embracing the southern portions of Mexico, Guatemala, Honduras and, in part, Costa Rica.

VII THE SOUTH MEXICAN AND CENTRAL AMERICAN CENTER OF ORIGIN OF CULTIVATED PLANTS (ALSO INCLUDES THE ANTILLES)

GRAIN CROPS:

1 *Zea mays* L. – maize.
2 *Phaseolus vulgaris* L. – common or kidney beans.
3 *Ph. multiflorus* Willd. – runner beans.
4 *Ph. lunatus* L. gr. *microspermus* – lima beans.
5 *Ph. acutifolius* A. Gray var. *latifolius* Freeman – tepary beans.
6 *Canavalia ensiformis* (L.) DC. – jack beans.
7 *Chenopodium nuttalliae* Saff., – quinoa, and *Ch. ambrosioides* L. – Mexican tea.
8 *Amaranthus paniculatus* L. var. *leucocarpa* Saff. – purple amaranth.

GOURDS:

9 *Curcurbita ficifolia* Bouché (*C. melanosperma* Al. Braun) – figleaved gourd.
10 *Cucurbita moschata* Duch. – squash.
11 *C. mixta* Pang. – pumpkin.
12 *Sechium edule* Swartz – chayote.

13 *Polakowskia tacaco* Pittier (in Costa Rica) – tacacco.
14 *Sicana odorifera* Naud. – cassabanana.

ROOT VEGETABLES AND TUBERS:

15 *Pachyrhizus tuberosus* Spr. (*Cacara edulis* Kuntze) – yam bean.
16 *Ipomoea batatas* Poiret – sweet potatoes.
17 *Maranta arundinacea* L. – arrowroot (the Antilles).

SPICES:

18 *Capsicum annuum* L. – bell pepper.
19 *C. frutescens* L. – tabasco pepper.

FIBER PLANTS:

20 *Gossypium hirsutum* L. – upland cotton.
21 *G. purpurascens* Poir. – Bourbon cotton.
22 *Agavea sisalana* Perrine – sisal hemp.

FRUIT CROPS:

23 *Opuntia* spp. – tuna cacti (a number of species).
24 *Annona cherimola* Mill. (may be a secondary center) – cherimoya; *A. reticulata* L. – custard apple; *A. squamosa* L. – sweet sop; *A. purpurea* Moc. & Sessé; *A. cinerea* Dun.; *A. diversifolia* Saff.–Alamo; *A. glabra* L. – pond-apple.
25 *Sapota achras* Mill. and *S. sapotilla* Coville – sapodillas.
26 *Casimiroa edulis* La Llave – white sapote.
27 *Calocarpus mammosum* (L.) Pierre – mamey sapote or marmelade plum.
28 *C. viride* Pittier (*Achradelpha viridis* Cook) – green sapote.
29 *Lucuma salicifolia* H.B.K. – yellow sapote.
30 *Carica papaya* L. – papaya or melon tree.
31 *Persea schiedeana* Nees and *P. americana* Mill. (*P. gratissima* Gaertn.) – avocados.
32 *Psidium guajava* L. – guava.
33 *P. friedrichsthalianum* (Berg.) Niedenzu – Costa Rica guava; and *P. sartorianum* (Berg.) Niedenzu.
34 *Spondias mombin* L. – mombin; *S. purpurea* L. – red mombin or Mexican plum.
35 *Crataegus mexicana* Moq. & Sessé; – *C. stipulosa* Steud. – texocote.
36 *Diospyros ebenaster* Retz. – wild persimmon.
37 *Chrysophyllum cainito* L. – star apple or cainito (especially on the Antilles, Jamaica and in Panama, wild and cultivated).
38 *Anacardium occidentale* L. – cashew tree (Antilles and Panama).
39 *Prunus serotina* Ehrh. (*P. capollin* Zucc.) – black cherry.

VARIOUS CROPS:

40 *Agave atrovirens* Karw. – pulque or saguey agave.
41 *Cereus* spp. – cacti for pillows [so-called "pillow" cacti; DL.].
42 *Nopalea coccinellifera* Salm-Dyck – nopal or cochineal cactus.
43 *Tigridia pavonia* Ker-Gawl – cacomite (mainly in the wild state).
44 *Physalis aequata* Jacq. – tomatillo.
45 *Lycopersicum cerasiforme* Dun. – cherry tomatoes.
46 *Salvia chia* Fern. – chia (an oil plant).

47 *Theobroma cacao* L. – cocoa trees.
48 *Bixa orellana* L. – achiote (dye and fiber plant).
49 *Nicotiana rustica* L. – Aztec tobacco.

As demonstrated by means of the phyto-geographical differentiating method, *this is unquestionably the main center of maize and its closely related wild species, teosinte, and the native land of the main American species of beans, gourds, peppers and many tropical fruits. The cultivation of cocoa arose here.* It is also the native land of the sweet potato. *The American upland cotton, on which all the cotton industries in the world are based, came from southern Mexico.* Maize plays the same role here as wheat does in the centers of the Old World; without it there would have been no Mayan civilization. The fairly limited area of southern Mexico and Central America is full of cultivated endemic plants, sharply differing in this respect from the large continent of North America, where all the agriculture in the past and present was based on introduced crops.

Our investigations in 1932 and 1933 in South America made it necessary to make essential changes with respect to our old ideas about the centers of agriculture, which were established in American literature. The high mountains of Peru, Bolivia and, in part, Ecuador are no doubt the most remarkable and are unquestionably an independent center of origin of original crops. A pre-Incan, megalithic civilization was concentrated in that area. This territory, which is insignificant in area in comparison with all of South America, is distinguished by an amazing accumulation of cultivated plants and endemic animals.

VIII THE SOUTH AMERICAN (PERUVIAN–ECUADORIAN–BOLIVIAN) CENTER OF ORIGIN OF CULTIVATED PLANTS

Main endemic species at high altitudes (on the puna and in the sierras)

TUBERS:

1 *Solanum andigenum* Juz. & Buk. – Andean potatoes, very broadly distributed from Bolivia to Central America ($2n = 48$ chromosomes).
2 Other cultivated, endemic species of potatoes:
S. cuencanum Juz. & Buk. – Ecuador (24 chromosomes)
S. kesselbrenneri Juz. & Buk. – Ecuador (24 chromosomes)
S. ajanhuiri Juz. & Buk. – Bolivia (24 chromosomes)
S. pauciflorum Juz. & Buk. – Bolivia (24 chromosomes)
S. stenotomum Juz. & Buk. – Bolivia, Peru, Ecuador (24 chromosomes)
S. goniocalyx Juz. & Buk. – Peru (24 chromosomes)
S. rybinii Juz. & Buk. – Colombia (24 chromosomes)
S. boyacense Juz. & Buk. – Colombia (24 chromosomes)
S. juzepczukii Buk. – Peru and Bolivia (36 chromosomes)
S. tenuifilamentum Juz. & Buk. – Peru and Bolivia (36 chromosomes)
S. mamilliferum Juz & Buk. – Peru (36 chromosomes)
S. chocclo Juz & Buk. – Peru, Bolivia, Ecuador (36 chromosomes)
S. riobambense Juz. & Buk. – Ecuador (36 chromosomes)
S. curtilobum Juz. & Buk. – Peru and Bolivia (60 chromosomes)

3 *Oxalis tuberosa* Molina – oca; *O. crenata* Jack.
4 *Tropaeolum tuberosum* Ruiz & Pavon. – magua.
5 *Ullucus tuberosus* Lozano – ulluco.

GRAIN CROPS:

6 *Lupinus mutabilis* Sweet. – Bolivian lupine or tarwi.
7 *Chenopodium quinoa* Willd. – quinoa.
8 *Ch. canihua* O.F. Cook – cañahua (among the most cultivated species on the local puna in Bolivia and Peru).
9 *Amaranthus caudatus* L. – amaranth ('love-lies-bleeding').
10 *Lepidium meyenii* Walp. – maca.

Endemic plants of the irrigated coastal areas of Peru and the non-irrigated subtropical and tropical areas of Ecuador, Peru and Bolivia (ecological epithets: costa, ceja, yunga and montana).

GRAIN CROPS:

11 *Zea mays* L. gr. *amylacea* – soft corn, starchy corn, a large-grained form (secondary center of the variety).
12 *Phaseolus lunatus* L. gr. *macrospermus* – lima beans (secondary center).
13 *Phaseolus vulgaris* (secondary center). – common or kidney beans.

TUBERS AND ROOT VEGETABLES:

14 *Solanum phureja* Juz. & Buk. – potatoes (Bolivia, 24 chromosomes).
15 *Polymnia sonchifolia* Poepp. & Endl. – llacon.
16 *Xanthosoma sagittifolium* (L.) Schott. (*X. edule* Meyer) – yautia.
17 *Canna edulis* Ker-Gawl. – achira or edible canna.
18 *Arracacia xanthorrhiza* Bancroft (*A. esculenta* DC.) – arracachaca.

VEGETABLES:

19 *Solanum muricatum* Ait. – 'melon pear'.
20 *Lycopersicum esculentum* Mill. var. *succenturiatum* Pasq. – bush tomatoes; *L. peruvianum* (L.) Mill. – tomatoes.
21 *Cyclanthera pedata* Schrad. – pepino de comer; *C. brachybotrys* (Poepp. & Endl.) Cogn.
22 *Cyphomandra betacea* Sendtn. – tree tomatoes.
23 *Physalis peruviana* L. – Cape gooseberry.

SQUASHES:

24 *Cucurbita maxima* Duch. – winter squash or pumpkin (actually originated on the eastern slopes of the Cordilleras).

SPICES AND STIMULANTS:

25 *Capsicum frutescens* L. var. *baccatum* L. – cayenne, and *C. pubescens* Ruiz & Pav. – chili manzana peppers.
26 *Tagetes minuta* L. – wild marigold.
27 *Erythroxylon coca* Lam. – coca bush.
28 *Bixa orellana* L. – annatto.

FIBER PLANTS:

29 *Gossypium barbadense* L. (*G. peruvianum* Cav.) – sea-island or Egyptian cotton.
30 *Fourcroya cubensis* Vent.

FRUIT CROPS:

31 *Passiflora ligularis* Juss. – sweet granadilla; *P. quadrangularis* L. – giant granadilla.
32 *Carica candamarcensis* Hook.; *C. chrysopetala* Heilborn; *C. pentagona* Heilborn (3 Ecuadorian species); *C. pubescens* (A.DC.) Solms-Laub.; *C. candicans* A. Gray (2 Peruvian species) – papayas.
33 *Lucuma obovata* H.B.K. – egg-fruits.
34 *Psidium guajava* L. – guava.
35 *Annona cherimola* Mill. – cherimoya (wild and cultivated).
36 *Inga feuillei* DC. – pacay.
37 *Bunchosia armeniaca* DC. – bunchosia.
38 *Matisia cordata* Humb. & Bonpl. – chupa-chupa or sapote.
39 *Caryocar amygdaliferum* Cav. – almandra or achotillo.
40 *Guilielma speciosa* Mart. – peach palm.
41 *Malphigia glabra* L. – Barbados cherry.
42 *Solanum quitoense* Lam. – naranjilla.
43 *Prunus capuli* Cav. (*P. capollin* Zucc.) – black cherry, differs from the Central American group (mainly used wild).

MEDICINAL PLANTS:

44 *Cinchona calisaya* Wedd. and *C. succirubra* Pav. – quinine tree (mainly wild).
45 *Nicotiana tabacum* L. – tabac or tobacco.

According to the most recent genetic investigations (by Kostov *et al.*) it is possible that tobacco arose from a cross between the wild species *N. silvestris* and *N. rusbii* or species closely related to them growing in the Andes. By a subsequent doubling of the chromosomes, this became a fertile interspecific hybrid.

In the eighth center, the Soviet expeditions discovered enormous, essentially untouched reserves of cultivated plants in the form of dozens of new species, not known to science, especially of cultivated potatoes and wild species closely related to them and used by the Indian tribes. *The high mountains of Peru, Bolivia and Ecuador are full of endemic plants starting with peculiar species of potatoes and ending with odd, tuberous plants such as oca (Oxalis tuberosa), magua (Tropaeolum tuberosum) and ulluco* (*Ullucus tuberosus*), which are typical only of this part of the world. Peculiar species of *Chenopodium* are cultivated here in large quantities. The endemic cultivated plants and the odd animals (llamas and alpacas) of Peru and Bolivia are concentrated mainly on the so-called puna, a kind of steppe at altitudes between 3500 and 4500 m. The crops are not irrigated in that area. It is still possible to see the transition from wild to cultivated plants. There can be no doubt whatever about the fact that South American agriculture, just like husbandry, began just on the puna. The localization of the endemic species in the past as well as at the present is amazingly distinct and delimited here.

As far as the coastal area of Peru is concerned, it became occupied mainly during the time of the Incan civilization, i.e. at a later stage, and is sharply different from the high altitude areas. It consists mainly of desert and cultivation was feasible there only by means of irrigation. Although the agriculture of the Incas was remarkable, it was no doubt secondary, just like the Egyptian one. Both the Egyptian and the Incan civilizations were based on artificial irrigation. Until the appearance of farmers, there were no original plants in the coastal area; Peru had neither maize nor cotton. The majority of the plants in the irrigated areas of Peru were adopted from Central America and, in part, also from the eastern slopes of the Cordillera, where it is still possible to encounter peculiar endemic plants such as the quinine tree (*Cinchona*) or the coca bush (*Erythroxylon*).

In addition to the basic Peruvian center, it is necessary to add a small area, i.e. that of the island of Chiloë, which is situated along the coast of southern Chile. The first Europeans acquired the ordinary potato (*Solanum tuberosum*) from the Indians on that island, as well as *S. andigenum*, characterized by 48 chromosomes and morphologically close to the former. These were especially suitable for European conditions, thanks to their adaptation to a long day. In spite of being of great interest to plant breeders, the majority of the species and types of potatoes from Peru, Bolivia and Ecuador need a short day for a normal development and do not set tubers under the conditions of the long, bright days during the European summer months.

VIIIa THE CHILOAN CENTER

1 *Solanum tuberosum* L. – common potatoes (48 chromosomes).
2 *Madia sativa* Molina – madder.
3 *Bromus mango* Desv. – mangochil grass.
4 *Fragaria chiloënsis* Duch. – strawberry (in a wild state).

The enormous area of Brazil with its rich flora, determined by a botanist (Hoehne) to have ca. 40 000 species, has presented the world with only an insignificant number of cultivated species, of which the most important seems to be manioc, peanuts and pineapples. Thereby, it should be observed that these species are typical not of the humid tropical forests but mainly of the dry, semidesert areas of Brazil. The rubber trees, the native land of which seems to be the Amazon river basin, still exist there in a wild state but were taken into cultivation during the last couple of decades, mainly in southern Asia to where the Dutch and the English brought them.

VIIIb. THE BRAZILIAN–PARAGUAYAN CENTER

ENDEMIC PLANTS:

1 *Manihot utilissima* Pohl. – manioc (on light, sandy soils).
2 *Arachis hypogaea* L. – peanuts (on light soils).

3 *Phaseolus caracalla* L. – snail flower or bertoni bean.
4 *Theobroma cacao* L. – cocoa tree (secondary center), *Th. grandiflora* (Willd.) K. Schum. (*Guazuma grandiflora* G. Don) (the Amazon basin).
5 *Hevea brasiliensis* Müll. – rubber tree (the Amazon basin).
6 *Ilex paraguayensis* A. St. Hil. – maté or paraguayan tea.

FRUIT CROPS:

7 *Eugenia uniflora* L. – Surinam cherry; *E. uvalna* Cambess. – uvalha; *E. dombeyi* (Spr.) Skeels – grumichama; *E. tomentosa* Cambess – cabelloda (all cultivated).
8 *Myricaria jaboticaba* Berg. – jaboticaba, *M. cauliflora* (Mart) Berg – jaboticaba or Brazilian grapetree.
9 *Ananas comosus* (L.) Merr. – pineapple or ananas (on dry soils).
10 *Bertholetia excelsa* Humb. & Bonpl. – Brazil nut (in a wild condition).
11 *Anacardium occidentale* L. – cashew tree (also on the Antilles).
12 *Feijoa selloviana* Berg. – feijoa or pineapple guava.
13 *Passiflora edulis* Sims. – passion fruit or purple granadilla.

Somewhat later, although still during the pre-Columbian time, northern Indian tribes took Jerusalem artichokes [*Helianthus tuberosus* L.] and sunflowers [*H. annuus* L.] into cultivation within the territory of the present U.S.A., where these plants can still be found in the wild state.

Such a localization of the basic specific and varietal potentials of generally important and useful plants has been established by detailed phyto-geographical studies during the last couple of years. As far as a number of tropical plants within the limits of adjacent centers are concerned, they need, indeed, considerably more precise definition in the future. The centers of type-formation of the plants that are most important for the U.S.S.R. have been fairly exactly defined. All these centers are without doubt autonomous and have developed independently, as clearly demonstrated by the composition of the genera, species and varieties of cultivated plants. The peculiar agro-technology, tools and domesticated animals typical of these centers also tell us about this (Vavilov, 1934).

Essentially, it can be stated that the eight basic centers are separated by deserts or mountain ranges. The Chinese center is isolated by the enormous Inner-Asiatic deserts and the mountainous semi-deserts of Central Asia. The center in Asia Minor is isolated by the Inner-Asiatic deserts of eastern Afghanistan and western and central Iran (the Bakviy and Seistan deserts). The Inner-Asiatic center is separated from India itself by the Tar desert. The Mediterranean center is surrounded by deserts toward the south and east. Abyssinia is also surrounded by a circle of deserts. The lifeless Atakama desert borders on the high mountain areas of Peru and Bolivia, where the agricultural crops of South America were first developed. The Mexican desert-like foothills extend toward the north of southern Mexico. In other words, there is a regularity with respect to the geography of the primary centers: a presence of isolating factors, promoting an independent development of the flora and of the human settlements. During the interaction between them, independent agricultural crops were developed. The deserts were, for a long time, enormous obstacles for the primitive people, keeping them apart from each other.

The enumerations above list the most important crops in the world. Outside the centers mentioned, comparatively few plants were taken into cultivation. These were mainly forage plants, vegetables and medicinal plants, and this was done during a relatively recent time. Among the old crops of great importance that were taken into cultivation outside these centers, we can mention the date palms [*Phoenix dactylifera* L.] (actually within the oases of Mesopotamia and northwestern India and, possibly independently, in Africa) and the water melons [*Citrullus lanatus* (Thunb.)], found wild in the deserts and semi-deserts of southern Africa.

On the basis of the enormous material of seeds and plants brought back by the expeditions of the Institute of Plant Industry (up to 300 000 samples), which has been thoroughly studied by growing them in the field, *differentiating maps were drawn of the geographical localization of the crops most interesting to the U.S.S.R.* and establishing where the maximum basic diversity is concentrated. Such maps exist for wheat, oats, barley, rye, maize, millet and flax and for leguminous plants such as peas, lentils, common beans, fava beans, chick-peas, chickling vetch and gram or mung beans as well as vegetables such as carrots, tomatoes and even root fruits and potatoes (cf. Vavilov, 1926 and papers on various crops published in *Trudy po prikladnoy botanike, genetike and selektsii* [*Papers on applied botany, genetics and plant breeding*], 1923–1934).

As seen above, the primary areas of speciation of the most important cultivated plants are, on the whole, extremely narrowly localized. According to an approximate estimate (after subtraction of deserts and steep mountains inside the centers) they occupy ca. 1/40th of all the land areas. It can be seen from the above lists that the overwhelming majority of cultivated plants originated in the Old World; out of the 640 important cultivated plants (mentioned in the above lists) more than 500, i.e. 5/6th of all cultivated plants in the world, belong to some part of the Old World. The New World has given rise to approximately 100 such plants (counting all the newly established species of potatoes as a single species). *The main bulk of cultivated plants developed in southern Asia*, (i.e. more than 400 species), *which means almost two-thirds of all cultivated varieties in the world.* When proceeding from an exact number of species, the figures must be increased accordingly but, in general, the correlation is valid. The main potential of specific and varietal diversity of cultivated plants is found in the southern mountainous and tropical Asia. Africa is of lesser importance but gave rise to about 50 kinds of crops. Australia did not have any cultivated plants until recently; it is just during the nineteenth century that such Australian plants as the eucalyptus trees, acacias and casuarinas began to be selected from the composition of its wild flora. India and China, but also Inner Asia and the Mediterranean countries, are particularly rich in generic, specific and varietal potentials, furnishing almost one half of all cultivated plants.

If you take into consideration not only what has been brought into cultivation but also the utilization of wild plants, the material basis on which the great masses of farming developed can be clearly perceived, i.e. what still exists within the territories discussed.

The actual localization of cultivated plant material proved to be more distinct and more delimited than was thought during the time of De Candolle (De Candolle, 1883), for whom it was a matter of entire continents. Direct studies of the plant resources by means of the differential method has allowed us to determine exactly the main areas within the continents where cultivated plants originated.

We have presented here only summary data concerning the majority of the most important species within the limits of the large continents, although a further geographic differentiation can be made within the borders of the areas in question. For many species of importance as crops for the U.S.S.R., the actual location can be established with accuracy to within 100 kilometers, which can be considered exceptional for botanical studies.

As a result, an enormous specific and varietal potential of cultivated plants and their closely related wild species has been revealed, of which neither botanists nor agronomists had any idea in the past. The method of differentiating phyto-geographical studies of plants has fully justified itself. *In the case of the majority of the most important cultivated plants, it has become necessary to rework our ideas about the species and their composition.*

The enormous endemic material of field and vegetable crops hidden within the centers should now be widely utilized for plant breeding work within the U.S.S.R.

As can be seen, the zone of the original development of the major cultivated plants lies basically within a belt between 20° and 45° North, where the major mountain massifs of Himalaya, Hindukush, Inner Asia, the Balkans and the Appennines are concentrated. In the Old World, this belt runs latitudinally but in the New World it extends longitudinally, in the same direction as the mountain ridges.

From the species listed above that are specific to the different centers it can be seen that with respect to a number of crops a fairly important fact can be detected, i.e. that these species originated in a few areas only, each of which can be sub-divided into foci that are typical of different Linnaean species and often clearly distinguished by the physiological characteristics that are reflected or by the number of chromosomes. This can be particularly clearly seen as far as wheat, potatoes, oats, cotton and fruit trees are concerned. The old allegations by De Candolle that the native land of wheat should be Mesopotamia or the assumption by the famous Austrian phyto-geographer, Solms-Laubach, that Central Asia should be the homeland of wheat are without foundation, as shown by the data of our phyto-geographical investigations. When using wheat and oats as examples, a complicated picture of the distributions of the basic potential develops. At the same time as the potential of species grouped around the soft wheats (with 42 chromosomes) is centered between Hindukush and western Himalaya and in Transcaucasia, the species with 28 chromosomes are, on the other hand, concentrated either within Abyssinia, where they constitute a specific potential, or in Inner Asia (Transcaucasia, Turkey and northwestern Iran), which is the realm of endemic species of wheat and the genera closely related to it. The established facts lead to control over the original material for plant breeding. In turn, they pose problems concerning the original material of the most important crops. Among the new potentials that are established, a multitude of characteristics valuable for agriculture, such as resistance to diseases and the presence in nature of useful, awnless hard wheats in Abyssinia, has been revealed. In the high mountains of Peru and Bolivia, there are potatoes that possess the ability to tolerate low temperatures down to $-8°$ C. These facts proved to be of such importance that, in the footsteps of the Soviet expeditions, other special expeditions were sent out from Washington, Sweden and Germany to Central and South America. Later, German expeditions were directed to India for a special purpose, i.e. to 'collect material of soft wheat within the general geographical center of its genes'. This was the official title of the expedition according to a personal letter from Dr. Rudorf.

While, it can be said of the most important field crops such as wheat, barley, oats, flax and leguminous plants that the investigations of the worldwide varietal resources are basically completed, an enormous amount of work still remains to be done with respect to the crops of vegetables and fruit trees. It is still possible to expect major discoveries in the case of fruits in China, Asia Minor and Iran. There is enormous interest in detailed studies of northwestern India where the European field crops originated that are most important at present.

Primary and secondary crops

Our investigations led to the distinction of all the cultivated herbs into two groups. A primary group of basically ancient cultivated plants is known only in cultivation or in a wild state. These we call *primary crops*. To this group belong such plants as wheat, barley, maize, soybeans, flax and cotton. The second group consists of so-called *secondary crops*. To these belong all the plants brought into cultivation from weeds that infested the basic, primary crops.

It has been definitely revealed that cultivated rye originated from weedy rye, which infested winter wheat and winter barley in southwestern Asia and Transcaucasia, where rye is still concentrated in a considerable multiformity as a weed. In Afghanistan, rye is indeed a bothersome weed, and does not yield to wild oats with respect to the harm caused, in particular when present there as the basic type of rye, which is inclined to shedding its seeds. During the northward transfer of crops of winter wheat and barley from southwestern Asia, i.e. from the basic center of type-formation of soft wheat into Europe and Siberia, rye started to outcompete wheat and barley. Due to the hereditary frost-hardiness of the rye and its lesser demands as regards soil conditions, it supplanted wheat and it developed into a crop of its own also thanks to the will of Man, as can still be seen in various places among the mountain areas of southwestern Asia, or – within the U.S.S.R. – in the northern Caucasus.

The dominance of rye, as in the northerly parts of European U.S.S.R. and northern Europe, is to a considerable extent the result of natural selection. On the borderline between the 'struggle for dominance' between winter rye and wheat, the farmers frequently sow a mixture of rye and wheat. In Russia this is called 'surshy', and promises a more successful harvest than that of pure wheat, which can succumb to hard winters. The entire process has been observed in minute detail (Vavilov, 1917, 1926; Vavilov and Bukinich, 1929). In a way, the wheat helped transfer rye into cultivation; the genesis of cultivated rye cannot be comprehended without understanding that of the crops of wheat. However, what is most essential from the point of view of plant breeding is the presence of a fine potential of genes among the weedy types of rye in southwestern Asia.

During the search for new genes and valuable new characteristics of rye it is, thus, necessary to turn to the weeds infesting wheat.

The same kind of facts, although even more complicated, could be established in the case of oats. Thereby, it was remarkable that different species of oats, differing in chromosome numbers and typical of different centers, were originally associated with different geographical groups of distichous wheats (emmer). During the northward dispersal of the ancient crops of emmer, weedy oats, which the emmer brought along, outcompeted the emmer and became an independent crop. In the course of the search for new types and new genes of oats, the plant breeder must turn his attention to the centers of ancient crops of emmer, which appear to be preserves of a great and original diversity of genes of cultivated oats.

Similar facts were also revealed in the case of a number of other plants such as, e.g. garden rocket [*Eruca sativa*], false flax [*Camelina sativa*], mustards [*Sinapis* spp.], winter cress [*Barbarea vulgaris*], rapeseed [*Brassica rapa* f. *rapifera*] and other cruciferous plants, coriander [*Coriandrum sativum*] and a number of species of South American potatoes in Peru and Bolivia as well as tomatoes of Peru. Investigations of weedy plants from this point of view can open up new opportunities for plant breeding.

Regularities of the geographical distribution of the varietal diversity of cultivated plants

The accumulation of a large number of facts concerning the distribution of the varietal diversity within the primary centers and with respect to the dispersal from these centers has revealed a number of quite essential laws that are associated with the search for the valuable material necessary.

We have already indicated that extremely interesting and peculiar recessives can frequently be found at the periphery of centers and within isolated areas, such as mountains and islands, as a result of inbreeding or mutation. There is a large number of such cases. Some interesting geographical regularities were discovered in this way. Thus, e.g. China is characterized by the presence of peculiar forms of many secondary crops, brought there from the original centers. A great diversity of naked barley (naked grains are a typically recessive characteristic), naked-grained millet and large- and naked-grained oats (*Avena nuda*) is found there. As already mentioned above, typical recessive forms of waxy or Chinese corn [*Zea mays* convar. *ceratina*], hyacinth beans [*Dolichos lablab*] and snap [or sugar] beans [a form of *Phaseolus vulgaris*] are isolated there. The latter is characterized by the absence of the parchment-like layer in the wall of the pods, thanks to which the pods are completely edible. It is possible that such a distinction of recessive forms is also associated with the intensive selection made long ago by primitive Chinese 'plant breeders'.

The peculiar non-ligulate forms of rye, soft wheat and shot as well as club wheats are found only in Pamir, i.e. in isolated areas of Badakhshan and Shugnan, and within the borders of the Soviet Tadzhikistan as well as the Afghani Tadzhikistan. Non-ligulate hard wheat also occurs on the island of Cyprus, i.e. in corresponding isolation from its place of origin.

Geographical regularities can be observed when studying cultivated plants in the direction from Himalaya toward the Mediterranean Sea. The junction between the mountains of Himalaya and Hindukush seems to be a general preserve of an amazing wealth of primitive, mainly dominant forms of peas, chickling vetch, chick-peas and beans. These are distinguished by small seeds and small pods. The pods of the chickling vetch in Pamir, Badakhshan and Chitral tend to break open before they are completely ripe, i.e. they are characterized by a property typical of wild plants. In spite of the enormous material collected within this area (hundreds of samples of chick-peas, for instance) we did not find a single plant among them with white flowers or white seeds. The chickling vetch of Pamir and northwestern India is characterized by blue, frequently dark-blue, flowers and dark and small, mottled seeds. When advancing westward, e.g. into Pamir, we can already see an increase in varieties and an appearance of recessive forms with white seeds and rose-colored flowers. The countries around the Mediterranean are almost exclusively characterized by big-seeded, white-flowered and white-seeded, highly cultivated types.

The same is valid for lentils. In Chitral and on the border to India and Afghanistan, these are represented by types with small black seeds that are very different from the large and disk-shaped seeds of the Mediterranean forms. The seeds of the beans in Sicily and Spain are seven to eight times larger than those of the beans in the areas around Kabul (Afghanistan) and Badakhshan. The same pertains to chick-peas as well. The Mediterranean countries are characterized by large-seeded flax and large-grained types of wheat and barley. In Algeria we observed a type of onions where the bulb alone weighed up to two kilos.

In other words, in the direction from Himalaya toward the Mediterranean there is, in the case of a number of species, an increase in the number of recessive characteristics and an appearance of larger seeds and fruits, a fact that is of quite essential importance for the orientation of a search for the most valuable forms.

As far as the distribution of resistance to diseases is concerned, definite regularities can also be observed. Some geographical groups are, on the whole, characterized by resistance to specialized species of smut, rust, mildew and other parasitic fungi, and also to bacterial diseases. Thus, e.g. the American group of grapevines lack resistance to phylloxera and mildew. The east-Asiatic species of apples, pears and chestnuts differ sharply with respect to their behavior toward diseases in comparison with our west-Asiatic and European cultivated species and varieties. The sesames of Abyssinia, southwestern Asia, India and Japan react very differently to various bacterial diseases. The hard wheats of the Mediterranean countries differ sharply from the comparatively susceptible soft wheats of Abyssinia and parts of Egypt in their expression of immunity to brown and yellow rust. The distichous forms (emmers) of various geographical regions also differ in their resistance to yellow, brown and stem rusts, and also in their relation to mildews, ranging from practically complete resistance to a significant susceptibility.

An extremely interesting regularity was observed during the study of the varietal material brought in from the mountains of Arabia, i.e. Yemen ('Arabia felix'). In this country, which is adjacent to the Arabian deserts with their typical climatic conditions, extremely fast-ripening types of all the herbaceous crops grown there in the high mountains have developed. From there we have the fastest-ripening types of wheat in the world, which outdistance all other varieties when grown in our southern regions. The same is characteristic of barley, lentils and fenugreek [*Trigonella foenum-graecum*]. The blue-flowered alfalfa from Arabia can also be said to grow and complete its cycle of development very quickly.

Different geographical groups are characterized by biological specialities. Thus, e.g. chickling vetch [*Lathyrus sativus*] of Afghanistan, Pamir and Tadzhikistan is characterized by a tendency to self-fertilization while, on the other hand, the European–Mediterranean forms have a tendency to cross-fertilization. The boreal types of forage grasses have a tendency toward apogamy, which is of decisive importance for specific methods of plant breeding.

When starting to work with different plants, the plant breeder must study the phyto-geographical peculiarities of the species with which he will deal. That kind of knowledge can help him in his practical work and enable him to look for the original material most valuable for his purposes.

Finding valuable forms far from the primary centers

During the search, one sometimes runs into the fact that exceptionally valuable types can be discovered far from the primary centers. Thus, e.g. the well-known brand of oranges, 'Washington Navel', was found in Brazil, although the main native land of citrus fruits is in southeastern Asia. At present, enormous plantations in the U.S.A. are occupied by this type which apparently mutated or came from fortuitous seedlings from Brazil. The famous 'Jaffa oranges' in all likelihood came from a bud mutation found in Palestine. In essence, the highly interesting recessive forms of cultivated plants within the secondary center in China also belong to this category.

The results of a conscious selection over the last few decades or even centuries are undoubtedly of great interest. The utilization of that kind of selected material can facilitate plant breeding work to a considerable extent.

Original material for forage plants

Only two centers are characterized by endemic, cultivated forage plants, i.e. those in Asia Minor and in the Mediterranean region. These gave rise to a number of valuable crops such as alfalfa, species of clover, chickling vetch [*Lathyrus sativus*], sulla [*Hedysarum coronarium*], serradellas [*Ornithopus* spp.] and vetches [*Vicia* spp.]. These centers were in the distant past associated with the taming of the main domesticated animals and the development of husbandry (Vavilov, 1934).

The introduction into cultivation of herbaceous forage plants occurred during a comparatively late period. The accultivation of forage grasses required several centuries. With respect to the main cultivated plants we turn, for instance when searching for new types, to the centers where they were first introduced into cultivation. As far as forage plants are concerned, we can find an enormous supply of species and varieties within the associations of wild plants. Thereby, in essence, the stage of selection of species, not to mention that of varieties, is not yet past. This is the stage that plant breeders of the majority of forage grasses and leguminous plants have just entered upon.

General experiments during the last couple of decades have revealed the exceptional value of wild European and Siberian floras with respect to utilization of original material for taking new forage plants into cultivation. Hence, it is interesting that the American flora, which is fairly well studied and repeatedly used for introducing new plants into cultivation (cf. Malte and Kirk in Canada and Piper and Hitchcock *et al.* in the U.S.A.), has with few exceptions not revealed any species able to compete with European forage plants. Only one Canadian species, *Agropyrum tenerum* Vasey, attracts some attention, but this one occupies only a limited area in North America. In the U.S.S.R. it is of interest only for a few areas in western Siberia. There is a large number of grasses and leguminous plants within our own Union, which deserves serious attention as original material to be introduced into cultivation. A Swedish expedition under the direction of Turesson, sent to western Siberia and Altai in 1928, discovered a fact of exceptional interest. It turned out that some forage plants in the foothills of Altai and in western Siberia were distinguished by very rapid development and a large vegetative mass, which was associated with a chromosome number doubled in comparison with that of the corresponding species in northern Europe. According to the opinion of the Swedish plant breeders, some west-Siberian species are of great interest for plant breeding and surpass as forage plants those taken from the local strains found in northern Europe.

The theory of climatic analogy when dealing with introduction

When selecting species and varieties for the U.S.S.R. it is, of course, natural to consider the climatic growing conditions and, if possible, to bring in strains from areas more or less similar to those in our country. Knowledge of the climate in the countries from where the original plant-breeding material is taken is of primary importance.

It must, however, be definitely stated that the question concerning climatic analogy cannot be solved simply in the manner used during the recent past. The study of the dispersal of plants clearly bears witness to how complicated this process is.

Some crops and strains are amazingly cosmopolitan. Many forage plants and vegetable crops belong to this category. The Danes export seeds of vegetables to many countries all over the world. In many of our own areas, which are quite different from the conditions in Denmark, Danish vegetable seeds are the very best, according to data on the strains from the State Experimental Stations. They can be successfully grown in the U.S.S.R. as far north as the Polar Circle and all over the Soviet territory. Southern Sweden exports seeds of forage grasses all over the world as well. The Swedish strain of oats, 'Victory', has really 'conquered' the world; it succeeds well in the U.S.S.R. in western Siberia as well as in Ukraine but also in Western Europe under different climatic and soil conditions. The 'Petkuss' rye, brought in from northern Germany, can be equally successfully grown in the U.S.S.R. as in Germany. Many herbaceous ornamental plants are also distinguished by an amazing universality. Petunias, fuschia, marigolds, lobelias, resedas, snapdragons, nasturtiums, dahlias, asters and stocks [*Matthiola*] and many others can, when seedlings are used, be grown as far north as the Polar Sea, although they originate from the Tropics and the Subtropics. In other words, some plants and strains behave in cultivation as true *cosmopolitans*.

On the other hand, the majority of foreign strains of spring and winter wheat, as well as of barley, are, as is evident, of little interest to us even when transferred into very similar conditions. The best standards in the world, occupying enormous areas in the U.S.A. and Canada, such as, e.g. the 'Marquise' wheat, did not succeed in any important areas suitable for their cultivation but were surpassed by our local strains. Many strains belonging to this group appear to be *specialized*. However, there are also interesting exceptions with respect to these crops.

Thus, e.g. the 'Lyutestsens' 062 wheat from Saratov proved to be unexpectedly suitable for the Primorskaya District of the Far East under climatic conditions quite dissimilar to those of Saratov. The Australian strain of spring wheat, 'Aurora', grows successfully in Sweden and Finland and in our country in the northern European parts of the U.S.S.R. A number of Argentinian strains of wheat, which grow in that country under conditions that could be called subtropical, where there are no winters and where the amount of precipitation is 1000 mm or less, can, as shown by direct experimentation, to be successfully grown in the U.S.S.R. under conditions such as those in the Leningrad district or those of the Northern Territory. The barley collected by us from the mountains of Abyssinia proved to be excellent under the conditions around Leningrad in spite of the essential difference in climate: the long summer days in the Russian north in contrast to the short days in Abyssinia. Without further breeding, the peas from Abyssinia turned out to be some of the best under the conditions of Leningrad. On the other hand, the Abyssinian spring wheat from the same locality succeeded badly around Leningrad and in general in all the European parts of the U.S.S.R.

The blue-flowered and the white-flowered alfalfa of southern provenance can survive the winters along the Polar Circle at Hibiny [on the Kola Peninsula]. The awnless broomgrass, collected around Voronesh, is not destroyed by the frost in the Far North (according to experiments conducted by Z. Zherebina).

There are, of course, no fully analogous climatic or soil conditions. Among meteorological factors in general, temperature and precipitation are taken into consideration in the case of introduced plants. However, the length of the day, which definitely varies with latitude, is of essential importance. Physiological experiments made by Allard and Garner and other scientists, but also the geographical experiments conducted by ourselves, have demonstrated the great importance of photoperiodism for vegetation. Beets and radishes from southwestern Asia turn from biennials into annuals in the Soviet north. Among the root crops, Afghani radishes are converted in the north into an annual oil-producing plant lacking thick roots (according to experiments by E. Sinskaya).

Direct experiments have demonstrated the necessity for great prudence when applying the theory of climatic analogy to introduced materials. The facts mentioned above can be supplemented by observations made on Peruvian and Bolivian potatoes. In spite of what could be expected, some species and varieties of potatoes from these equatorial countries grow very well along the Polar Circle, where they not only produce tubers but even set fruits.

Contrary to expectations, the fruit and seed production of a number of South American strains of potatoes succeeds especially well in the Far North.

In general, considerable constraint is demanded when drawing parallels between vertical belts in the mountains and latitudinal zones. The difference in photoperiodism demonstrates the anomaly of the elementary scheme that draws parallels between vertical belts and latitudinal zones.

While being guided by the data on climate and soil conditions during the selection of species and strains, it is at the same time necessary not to overestimate the importance of climatic analogy. Thus, e.g. a suggestion based on the climatic data of Inner Asia and made by such a great authority on cotton cultivation as G.S. Zaitsev, i.e. that cotton would not succeed in the northern Caucasus or the southern Ukraine, has not been corroborated. Under the conditions of the light, sandy soils of the northern Caucasus and the southern Ukraine, cotton, when densely sown, reaches a stage of full development and produces ripe capsules. This could not be deduced from the conclusions drawn on the basis of data from Inner Asia, where cotton crops are irrigated and the plants are set far apart.

Therefore, it can be stated with confidence that direct research is necessary for establishing the suitability of new crops under various kinds of conditions.

Organization of quarantine for introduced material

The development of extensive introduction of new plants and strains makes it necessary to simultaneously build up a quarantine system for avoiding the importation of new parasites and new pests together with the new plants and strains. The organization of a quarantine inspection constitutes an indispensable part of the introduction of plants. Each parcel of seeds must be subjected to fumigation and be treated with fungicides and insecticides. In the case of doubtful material, it must be sent to special quarantine nurseries for study. The organization of special greenhouses for quarantine purposes is also necessary.

This is why the importation of plants must be centralized and strictly controlled.

Problems concerning new crops and plant breeding

While the material resulting from thousands of years of experimentation by farmers of the Old and New Worlds can be utilized with respect to cereal plants used mainly for food, when it comes to crops of technical plants in the wide sense as well as crops of seed- and fruit-producing plants, the utilization of the resources offered by the wild floras of the world and of the U.S.S.R. are of great importance for plant breeding.

Hunger and the constant search for food forced primitive people to select what was valuable from the point of view of nutrition and suitable for cultivation in the temperate and subtropical areas. An enormous amount of work has no doubt been done in this direction during the course of the millenia by ignorant 'plant breeders'. One can only be amazed by the diversity of species and strains of wheat, barley, maize, millet and leguminous plants that were already known by primitive civilizations. To find competitors to the cereal grasses known to us at present will not be easy: to replace wheat, rice and maize by other crops is very difficult. In any case, botanical research made during the last 50 years all over the globe has not produced any serious promises in this direction. It is only possible to alter the existing brands of wheat, to replace one kind by another and to create new species by hybridization and to widen the horizons in this direction. But the search for new genera of cereals does not promise any great success.

However, it is possible to find quite a few new types among the old, edible plants, as demonstrated by the discovery during the last couple of years of the non-alkaloid types of lupines.

As we have seen, almost boundless opportunities exist with respect to forage plants, where literally hundreds of species deserve to be examined by plant breeders for use as green mass but also as a source of grains. Recent work done at the Italian and Portuguese research stations has demonstrated that much can still be achieved in the case of leguminous forage plants. The Institute of Plant Industry has succeeded in finding a lupine containing 21% fat and 30% albumin, i.e. a lupine of a value equal to that of the soybean but in contrast to the soybean suitable for sandy soil conditions.

Great opportunities also exist when searching for new technical plants to be taken into cultivation. In this respect, the farmers of the past achieved comparatively little. Even as far as fiber plants are concerned, which provide raw material for the barest necessities, the example of the Indian hemp or dogbane [*Apocynum cannabinum*], taken into cultivation during the last couple of years, has shown that much more can still be done.

It is interesting to mention that the great difficulties encountered when introducing Indian hemp into cultivation were overcome thanks to the selection of special ecological types that proved to be particularly suitable for cultivation.

The field of medicinal plants and that of plants producing etheric oils have hardly been touched upon.

Many opportunities also exist for taking into cultivation such new rubber-producing plants that are suitable for being grown in temperate and subtropical zones. The 'tau-saghyz' [*Scorzonera tau-saghus*], the 'kok-saghyz' [*Taraxacum bicorne*] and the 'krym-saghyz' [*T. hybernum*] are shining examples, as is Edison's goldenrod [*Solidago edisoniana*]. The introduction of these plants into cultivation brought up the problem about strains again, since, as shown by the investigations, different types of 'tau-saghyz' and goldenrod (which has several races) differ sharply with respect to the yield of rubber. *The problem concerning new crops cannot be distinguished from that concerning strains.*

The area of dye- and paper-producing plants has barely been touched upon either.

The wild types of almonds, cherry plums [*Prunus cerasifera*] and walnuts are of exceptional value among wild fruit trees, since they yield the best strains thereof that are presently cultivated. Such types are found in the mountain areas of Inner Asia and Transcaucasus, not to mention the wealth, quite untouched upon, of resources in eastern Asia where there are literally hundreds of species of wild fruit trees.

The vernalization method and its importance for the utilization of the plant resources of the world

The method of vernalization, invented by T.D. Lysenko, has opened up extensive opportunities for the utilization of the worldwide assortment of herbaceous crops. All our old and new strains, as well as all the worldwide assortment, should from now on be tried out with vernalization since, as demonstrated experimentally during the last couple of years, this can lead to amazing results and can literally transform strains, converting them from being unfit for a given area under ordinary conditions into productive and high-quality kinds. Thus, e.g. strains of winter barley, which will not develop properly under ordinary conditions around Leningrad, since they do not set ears when seeded in the spring and succumb to the winter when sown in the fall, did after vernalization not only set ears under those conditions but some of them even surpassed in yield the best strains of spring barley that were bred for the north.

Some of the vernalized strains of flax produced an excess of tall stems.

Without question, we must make an overnight revision of all the assortment of plants in the world, including also all our basic assortment of bred and local varieties, with respect to their reaction to vernalization. The method of vernalization is a powerful tool for plant breeding of many grass crops, enabling us to grow southern, subtropical strains in our north, where they would not develop under ordinary conditions.

During the last couple of years, research concerning the application of vernalization at Hibiny on the Polar Circle and in Leningrad, using an enormous amount of material of different crops, has demonstrated the exceptional possibilities for utilizing the assortment of plants from all over the world for the purpose of plant breeding by means of vernalization. For the first time, a collection of living plants of all kinds of barley from all over the world could be represented for us under the conditions around Leningrad, i.e. strains which would not be able to set ears there under normal conditions.

Conclusions

We have only just started a systematic study of the worldwide assortment of plant resources but have already revealed enormous reserves not touched upon and unknown to scientific plant breeding in the past. The enormous potential of species and varieties opened up will be subjected to intensive investigation by means of all the newest methods. The tasks for the immediate future consist of classification of the enormous diversity of strains of the most important crops, not only on the basis of their botanical and agronomical characteristics but also with respect to the utilization of methods pertaining to physiology, biochemistry and technology. The dogmas of biochemical and physiological taxonomy of cultivated plants have to be revised in the near future. The collosal potential revealed within the centers of the main type-formation and speciation of cultivated plants must be studied not only by botanical taxonomists but also by physiologists, biochemists and pathologists. An immense field of very fascinating and urgent work has opened up for genetical efforts directed toward the selection of the most rational combination of gene pairs.

BIBLIOGRAPHY

Barulina, E.I. (1930). Chechevitsa SSSR i drugikh stran [The lentils of U.S.S.R. and other countries]. *Tr. po prikl. botan., genet. i selek.* [*Papers on applied botany, genetics and plant breeding*], Suppl. no. **40**.

Bukasov, S.M. (1930). Vozdelyvayemye rasteniya Meksiki, Gvatemaly i Kolumbii [Plants grown in Mexico, Guatemala and Colombia]. *Tr. po prikl. botan., genet. i selek*, Suppl. no. **47**.

Bukasov, S.M. (1933). *Revolyutsiya v selektsii kartofelya* [*Revolution in the plant breeding of the potato*]. Leningrad.

De Candolle, A. (1883). *L'origine des plantes cultivées*, [*The origin of cultivated plants*]. Paris.

Drury, (1873). The useful plants of India. In Watt, *Dictionary of economic plants*, 2nd edn. London.

Govorov, L.I. (1928). Gorokh Afghanistana [The peas of Afghanistan]. *Tr. po prikl. botan., genet. i selek.*, **19**, (2).

Heyne, K. (1927). *De nuttige planten van Nederlandsch Indië* [The useful plants of Dutch East India], 2nd edn. Buitensorg, Java.

Ochse, J.J. (1933a). *Fruits of the Dutch East Indies.* Buitensorg, Java.

Ochse, J.J. (1933b). *Vegetables of the East Indies.* Buitensorg, Java.

Popov, M.G. (1929). Dikie plodovye derev'ya i kustarniki Sredney Azii [Wild fruit trees and bushes of Inner Asia]. *Tr. po prikl. botan., genet. i selek.*, **22** (3).

Schiemann, E. (1932). Entstehung der Kulturpflanzen [Origin of cultivated plants]. *Handbook d Vererbungswiseensch.*, edn. **15**.

Trudy po prikladnoy botanike, genetike i selektsii [*Papers on applied botany, genetics and plant breeding*], 1923–1934 and supplements to these.

Vavilov, N.I. (1917). O proizkhozhdenii kul'turnoy rzhi [On the origin of cultivated rye]. *Tr. Byuro po prikl. botan.* [*Papers from the Bureau of Applied Botany*], **10** (7–10).

Vavilov, N.I. (1922). The law of homogenous series in variation. *J. Genet.*, **12** (1).

Vavilov, N.I. (1926). Tsentry proiskhozhdeniya kul'turnikh rasteniy [Centers of origin of cultivated plants]. *Tr. po prikl. botan. i selek* [Papers on applied botany and plant breeding], **16** (2).

Vavilov, N.I. (1927a). Geograficheskie zakonomernosti i rasdelenii genov kul'turnykh rasteniy [Geographical regularities and the dispersal of genes of cultivated plants]. *Tr. po prikl. botan., genet. i selek.*, **17** (3).

Vavilov, N.I. (1927b). Mirovye tsentry sortovikh bogatstv (genov) kul'turnykh rasteniy [Main centers of the wealth of strains (genes) of cultivated plants]. *Izd. Gos. In-ta opyt Agronomii* [*Publication of the National Institute of Experimental Agronomy*], **5**.

Vavilov, N.I. (1928). Die geographische Genzentren der Kulturpflanzen [The geographical gene-centers of cultivated plants]. In *Verhandl. des V. Internat. Kongr. für Vererbungswiss., Berlin 1927*, [*Acta of the Vth International Congress of Genetics*, Berlin 1927]. Leipzig.

Vavilov, N.I. (1929a). Problema proiskhozhdeniya kul'turnykh rasteniy v sovremennom ponimanii [The Problem of the origin of cultivated plants from the present point of view]. In *Dostizheniya i perspektivy v oblasti prikladnoy botaniki, genetiki i selektsii* [*Progress and perspectives within the fields of applied botany, genetics and plant breeding*]. Leningrad.

Vavilov, N.I. (1929b). Geograficheskaya lokalizatsiya genov pshenitsy na zemnom share [Geographical localization of the genes of wheat in the world]. *Dokl. AN SSSR*, [*Lectures at the Academy of the Sciences*], ser. A. no. **11**.

Vavilov, N.I. (1931a). Dikie rodichi plodovykh derev'ev aziatskoy chasti SSSR i Kavkaza i problema proiskhozhdeniya plodovykh derev'ev [Wild relatives of fruit trees in the Asiatic parts of the U.S.S.R. and the Caucasus and the problem of the origin of the fruit trees]. *Tr. po prikl. botan., genet. i selek.*, **26** (3).

Vavilov, N.I. (1931b). *Linneevskiy vid kak sistema* [The Linnaean species as a system]. Moscow, Leningrad.

Vavilov, N.I. (1931c). Meksika i Tsentral'naya Amerika kak osnovoy tsentr proiskhozhdeniya kul'turnykh rasteniy Novogo Sveta [Mexico and Central America as a basic center of origin of cultivated plants in the New World]. *Tr. po prikl. botan., genet. i selek.*, **26** (3).

Vavilov, N.I. (1931d). Rol' Tsentral'noy Azii v proiskhozhdenii kul'turnykh rasteniy [The role of Central Asia for the origin of cultivated plants]. *Tr. po prikl. botan., genet. i selek.*, **26** (3).

Vavilov, N.I. (1931e). The problem of the origin of the world's agriculture in the light of the latest investigations. In *Science at the crossroads*. London.

Vavilov, N.I. (1932). The process of evolution in cultivated plants. In *Proc. of the VIth Internat. Congr. of Genetics, New York* 1932, Vol. **1**.

Vavilov, N.I. (1934). Mirovye ochagi rastenievodstva i zhivotnovodstva [World centers of plant industry and husbandry]. *Tr. 2-y Bsesoyoz. konf. po evolyutsii domashnikh zhivotnykh pri Akademii Nauk SSSR* [*Papers from the 2nd All-Soviet conference concerning the evolution of domesticated animals at the Academy of the Sciences of the U.S.S.R.*]. Leningrad.

Vavilov, N.I. and Bukinin, D.D. (1929). Zemledel'cheskiy Afghanistan [The Agricultural Afghanistan]. *Tr. po prikl. botan., genet. i selek.*, Suppl. **33**.

Vavilov, N.I. *et al.* (1931). Pshenitsy Abisinii i ikh polozhenie v obschey sisteme pshenits [The wheats of Abysinnia and their position within the general system of the wheats]. *Tr. po prikl. botan., genet. i selek.*, Suppl. **51**.

Watt (1873). *Dictionary of economic plants*, 2nd edn. London.

Zhukovskiy, N.M. *et al.* (1933). *Zemledel'cheskaya Turtsiya* [Agricultural Turkey]. Leningrad.

Asia – the source of species

THE GENERAL NUMBER of species of flowering plants that are known to botanists all over the world amounts to about 160 000. These species are not evenly distributed over the surface of the globe. Extensive areas of North America, Siberia and the central parts of Asia are characterized by extreme poverty with respect to the number of flowering plant species. On the other hand, some areas of the world have an incredibly rich diversity of plant resources. Southeastern Asia, embracing India, Indo-China and the Malayan archipelago, constitutes one of these areas. No less than 6000 species of different flowering plants have already been described from that part of the world.

As far as cultivated plants are concerned, botanists and agronomists are interested not only in the number of species but also in the diversity of the varieties within the limits of each of these species. There are thousands of different varieties of wheat, cultivated rice and barley. In order to understand the varietal resources of cultivated plants, it is necessary not only to study the geography of the entire complex of a species but also to know the primary region, where they are distributed, and the exact location of the original source of the varietal diversity.

During the last ten years, Soviet scientists have organized a series of expeditions in order to collect varietal resources of the most important cereal crops. About three fourths of the world has been studied, including Afghanistan, Asia Minor, Persia, India, Java, Formosa, Korea, Mongolia, Syria and Palestine, as well as some countries in South and Central America, Africa and Europe.

These expeditions were undertaken with a definite plan in mind, which was based on a theory worked out concerning the origin of cultivated plants. According to this plan, Soviet scientists started to collect worldwide resources of varieties, seeds and plants mainly within those areas where there is the greatest diversity of species and varieties of cultivated plants within the primary areas of their origin. These ten years of investigation have created opportunities for us to establish exactly described primary areas of origin of cultivated plants.

First published in English in the journal *Asia*, Febr. 1937, pp. 113–114; reprinted in Russian in *Rastit. Resursy* [*Plant resources*], vol. 2 (4), pp. 577–580, 1966.[1]

[1] Since the journal *Asia* was not available to the present translator, the Russian text has been translated anew into English. – D.L.

The majority of the world's cultivated plants have their origin in Asia. Out of 640 major cultivated plants, about 500 originated in southern Asia. We have established five important areas of origin of cultivated plants within Asia alone.

The first of these has been named 'the Chinese center' and includes the mountain areas of central and western China and the plains adjacent to them. This center gave rise to no less than 140 different cultivated plants. This is the native land of different species of millet, mustard, peculiar leguminous plants and many unusual vegetables. An enormous number of fruit-producing plants originated in China; no other area is as rich in wild fruit trees as China. It is the native land of various trees, tea bushes and trees producing camphor and tung oil and a very valuable fiber plant, the ramie.

The second area of origin with respect to cultivated plants, is called 'the Hindustani center', which includes Burma, Assam and a large portion of India but excludes northwestern India (i.e. Punjab and the northwestern border provinces). This is the native land of rice, including a large number of wild and cultivated types thereof. It is also the homeland of many leguminous crops, various kinds of tropical fruits and species of lemons and oranges, mango, sugar cane and other plants.

'The Indo-Malayan area of origin' is the third one, and includes Indo-China, the Malayan archipelago together with the large islands such as Java, Sumatra and – according to some authors – also Borneo and the Philippines. This is by far the richest area of local (native) tropical fruit trees. Some, such as bananas and some citrus fruits, are of worldwide importance.

The fourth center is 'the Inner-Asiatic' one, including northwestern India (Punjab, the northwestern border provinces and Kashmir), all of Afghanistan and the mountainous portions of Soviet Turkestan (Uzbekistan, Tadzhikistan and a part of eastern Turkmenistan).

This area is not only rich in various local (native) species of cultivated plants but is of great importance as the native land of soft wheats. We discovered a great diversity of that cereal here. It is also the native land of leguminous crops such as ordinary (garden) peas, chick-peas, lentils, and so on, in a great diversity of forms.

The fifth area of origin in Asia is the 'southwestern-Asiatic center', which comprises Asia Minor, Transcaucasia, Iran and western Turkmenistan. This area is remarkable for the wealth of wheat species that are endemic in southwestern Asia. Within the borders of the Soviet Union, there is an especially high number of wheat species in Armenia. A great variety of wild wheats has been found within that area. There can be no doubt whatever that this is the basic native land of cultivated wheat. Asia Minor and Transcaucasia are the areas of origin of rye, which is represented there by a large number of varieties and species.

Southwestern Asia seems also to be the native land of the fruit trees, now cultivated in Europe. It is the homeland of grapevines, pears, plums, cherries, pomegranates, quince, walnuts, almonds and figs.

Asia Minor and Soviet Turkestan house a great wealth of melons. Southwestern Asia is also the native land of forage plants such as alfalfa, strawberry clover [*Trifolium fragiferum* L.] and vetches.

In addition to the five Asiatic centers of origin of cultivated plants, we have established four centers in other parts of the world: 'the Mediterranean' one; the small 'Abyssinian' one – the native land of several endemic species; the 'southern Mexican–Central American' one – the homeland of the Mayan civilization and also of maize and cotton; and 'the Peruvian–Bolivian' one – the native land of the Incan civilization. In this center many species of potatoes originated and this is where the llama and the alpaca were first domesticated.

The map included[2] shows clearly the main centers of origin of cultivated plants. All these centers are independent. Their autonomy is clearly established by the fact that in each of them there are different genera, species and varieties of cultivated plants and wild species related to them. Their independence can also be proved by a comparison between the primitive agricultural tools and the domesticated animals used in each area.

Our studies have quite definitely established the fact that Asia is not only the native land of the majority of the presently cultivated plants but also of our main domesticated animals such as cows, yaks, buffaloes, zebus, sheep, goats, horses and swine. Domestication of fowl had its beginning in southern Asia, India and Indo-China. India is the native land of peacocks, zebus and buffaloes. The basic genera of cows and other large kinds of horned cattle, the eastern (or Arabian) type of horses, sheep and goats came from the Iranian area, if Asia Minor, the mountains of the Soviet Turkestan, and Afghanistan are included under this concept.

It is interesting that the basic species of the large kinds of horned cattle originated in areas closely adjacent to each other. The homeland of the zebu is the savannahs of India. The wild buffalo still inhabits the valley of the Ganges, the native land of this animal. A great variety of wild and tamed yaks has been discovered in Himalaya, Tibet and Pamir. Farther west, on the Iranian high plateau, we find a great diversity of the ordinary European kinds of large-horned cattle on the steppes. Archeological, geographical and genetical research has demonstrated that Asia is the native land of the large kinds of horned cattle.

Different ecological conditions gave rise to correspondingly different types of the large kinds of horned cattle. In low-lying, hot and humid areas, we find buffaloes and in the cold mountain areas, yaks. On the hot and humid steppes, we find zebus but in the arid areas of the steppes in southwestern Asia, the ordinary kinds of horned cattle. During the process of evolution, these animals were separated and adjusted to the different conditions. It is still possible to trace the roots of their origin. Some of these species are closely related and are able to hybridize.

It is of interest to mention that all the main centers of origin of cultivated plants and domesticated animals are separated by deserts and/or high mountains. The southwestern-Asiatic center is isolated from the Inner Asiatic one by the deserts of western Afghanistan and eastern and central Iran. In turn, the Inner-Asiatic center is separated from the Hindustani center by the Tar desert.

Two thirds of all the diversity of cultivated plant species originated from

[2] Not included in this edition but cf., e.g., the map on p. 312. – D.L.

southwestern Asia. The greatest potential of specific and varietal diversity is found in southern Asia and the greatest potential of specific and varietal diversity of cultivated plants is accumulated in the southern, mountainous and tropical parts of Asia. India and China are especially rich in the number of endemic genera, species and varieties.

If we pay attention not only to modern cultivated plants but also to the species of wild, edible plants, it is possible to form an idea about the main sustenance basis of the enormous numbers of inhabitants in these areas.

Recently, investigations carried out at the Leningrad Institute of Plant Industry have established exactly that Asia is the basic, native land of major forage crops as well. As a result of the brilliant research by Dr. E.N. Sinskaya at the Institute, it was recently revealed that the native land of alfalfa, the most important forage crop in the world, is located in Transcaucasia and Iran. Within this area, Dr. E.N. Sinskaya observed that this plant is still found in the processes of speciation. In the mountains of Transcaucasia, it is possible to follow the evolution of the species of blue-flowered and yellow-flowered alfalfa from the original chaos of types.

Iran and Afghanistan are the native lands of the Persian or shabdar clover [*Trifolium resupinatum*]. In Siberia, especially in the mountains of Altai and in the Far East, very valuable wild grasses have been discovered, which ought to be taken into cultivation in the future.

The importance of Asia as the primary land of an enormous amount of cultivated plants and domesticated animals is evident from these well established facts. The majority of the most important cultivated plants and domesticated animals has already been used by Man for thousands upon thousands of years. The recent results of our research have revealed that, within these areas, there are still enormous resources available and suitable for improving our present material and our present strains of plants.

The area around Himalaya and the mountains of China, Indo-China and southwestern Asia possesses inexhaustible riches, which are still inadequately explored.

Many facts indicate that southern Asia may also be the basic, primary homeland of mankind. A great variety of human races of all kinds of skin colors are found there. In China and Japan, very important fossils of humans have been found. India is the homeland of some anthropoid species of apes as well. Tropical and subtropical conditions, healthy mountain climates, the opportunities for isolation of small tribes within the mountain areas and the abundance of animal and plant food were all favorable for the development of mankind. Southern Asia is a true cradle of life. In this respect, Asia occupies a unique position among the continents of the world. In the future, we can expect new discoveries there of great importance, which will throw light over the origin and evolution of the flora, the fauna and even Man himself.

Plant resources of the world and their utilization for plant breeding

THE VARIETAL COMPOSITION of field crops of pre-revolutionary Russia reflected to a great extent the historical fate of agriculture in our country and the role of the environment in which it had developed for centuries. The proximity of our country to the basic centers of origin of the most important cultivated plants, such as wheat, barley, rye, flax and clover, has opened up opportunities for natural and artificial selection of an exceptionally valuable assortment: our local strains of flax, wheat, rye and barley, as well as clover, represent an extraordinary value and do not have an equal in any other country. With respect to drought- and frost-resistance, as well as the quality of the grain, the winter wheat of our country has a worldwide reputation and has become widely distributed in the U.S.A. during the last decade.

In spite of the great importance of the plant industry in our country, the composition of the strains used was little known in the past. The breeding thereof was, to a great extent, haphazard, although extremely valuable populations were frequently present.

The extensive organization of plant breeding during the Soviet era has first and foremost presented the research workers of our country with the task of evaluating the varietal resources.

The start of research in this direction was already authorized before the Great October Revolution by the Bureau of Applied Botany for agricultural studies and by a committee from the Department of Agriculture on the initiative of P.E. Regel, who started the systematic research concerning the botanical complexes of cultivated wheat, barley and oats.

The evaluation of the wealth of varieties all over the country – a general inventory of the strains – began only during the Soviet era. The studies of the varietal composition during the Soviet era comprised all the crops ranging from cereals and leguminous crops through technical crops to those of fruits and vegetables, and all the agricultural areas. As the studies proceeded, an exceptional wealth of species and varieties was revealed. The varietal resources of our country have already turned out to be even richer than known before the revolution as far as the most important field crops are concerned. An amazing

First published in *Matematika i Estestvenie SSSR: Ocherki razvitiya matematicheskikh i estestbennykh nauk za dvatsat' let*, [*Mathematics and Natural Science in the U.S.S.R.: Essays on the development of mathematics and natural science during the past 20 years*], pp. 575–595, Moscow–Leningrad, 1938.

wealth of species and varieties of wheat, rye, fruits and alfalfa were discovered in the Caucasus and Inner Asia.

In the past, foreign material was brought into our country to a great extent by pure chance. The socialistic management imposed the task of the universal utilization of the best material from abroad that is suitable for our conditions, side by side with our own assortment.

The best Soviet and foreign strains were taken into cultivation on a broad scale and on the basis of systematic, comparative studies of field crops by official testing of the strains, which was organized in 1924 in the form of the Institute of Applied Botany and New Crops. Among these strains, especially important roles were played by the Swedish-bred oats, the West-European 'beer' barley and the American strains of maize, cotton and citrus fruits. An extensive importation from abroad that also belongs to the Soviet period is that of new crops such as Sudan grass [*Sorghum sudanense* (Piper) Stapf], American wheat grasses [*Agropyrum* spp.] and gutta-percha producing *Eucommia ulmoides* Oliver, the tung-oil tree [*Vernicia fordii* Hemsl.], geraniums [*Pelargonium* spp.], etc.

While first and foremost utilizing local varieties and the best assortment of those tested in practice by the advanced capitalistic countries, we began again to rework our theories concerning original material for plant breeding and the introduction of specific and varietal resources from the basic primary and secondary centers of type-formation of cultivated plants.

Success in plant breeding is determined to a great extent by correct selection of the original material. From where does it come? The haphazard construction of collective nurseries, with which the plant breeder had to work when approaching practical breeding, led, during the Soviet time, only to a veritable growth of problems relating to the systematic study of the original plant material of the most important cultivated plants, later on including also all the valuable store of cultivated plants in the world in the plant breeding work. The construction according to the plan of a register of all the wealth of strains in the world led, in turn, to work on the geography of cultivated plants.

In my paper on 'Understanding Soft Wheat' (Vavilov, 1923), summing up results based on a large material from all kinds of countries, a hypothesis was launched concerning the precise localization as regards the distribution of species and their formation of varieties. The principle of evolution compels us to turn our attention to the historical moments during the development of the different crops, to the relationship between cultivated plants and the original wild types and to the regularities of the dispersal of entire species and of species composed of different units.

While critically comparing all the available data about the geography of cultivated plants in my book on *Centers of Origin of Cultivated Plants* (Vavilov, 1926), I put forward the idea about the necessity for a systematic investigation of the wealth of varieties and species. On the basis of the existing but still insufficient knowledge at that time it was already possible to outline enormous resources of species and varieties that have not been touched upon in Asia Minor, the Caucasus, Soviet Inner Asia, the mountain areas of Iran, India, interior China, the Mediterranean area and Abyssinia. As far as cultivated plants of the New World were concerned, the exceptional role of the area along the

Cordilleras definitely had to be added. The importance of southern Mexico and Central America and the agricultural areas of Peru and Bolivia, as well as southern Chile, was especially evident with respect to such crops as potatoes, cotton and maize. These areas coincide at the same time with the distribution of important agricultural crops.

It is natural that the study of the variety and diversity within our own country should come first. Proceeding from one crop to another and subsequently, including all kinds of field, vegetable and horticultural crops of plants, Soviet expeditions went for the first time through all the agricultural areas of our country according to a systematic plan. Hundreds of expeditions covered enormous expanses, while discovering reservoirs of specific and varietal resources that have not yet been touched upon.

In spite of all the practical importance of this section of applied botany, which is directly related to questions concerning practical plant breeding and all kinds of plant industry, it actually turned out to be a field almost untouched upon. Soviet scientists had to assume the duty of creating a new section of botanical science, i.e. *the study of original material for plant breeding*.

The importance for the socialistic rural economy of a number of species and crops that were originally brought in from other countries, forced us to pay great attention during the past few years to exploratory work outside the borders of our own country. The year 1923 can be said to be the start of systematic expeditions from the All-Union Institute of Plant Industry (at that time the Department of Applied Botany and New Crops under the National Institute of Experimental Agronomy), searching for varietal material beyond our own borders. The first such expedition (led by V.E. Pisarev) was sent out in 1923 to the agricultural areas of Mongolia. In 1924, the expedition led by N.I. Vavilov and D.D. Bukinich went to Abyssinia, thoroughly investigating all the agricultural areas of that country (Vavilov and Bukinich, 1929). As far as field, vegetable and melon crops were concerned, varietal resources that had been hardly touched upon were discovered for the first time, including a large number of new botanical varieties. Between 1925 and 1928, expeditions led by S.M. Bukasov and S.V. Yusepchuk were directed to Mexico, Guatemala, Colombia and Bolivia, Peru and Chile to study and collect crops from the New World, especially those of potatoes, maize and cotton (Bukasov, 1930). During 1926 and 1927, an extensive expedition was organized (and led) by N.I. Vavilov to investigate cultivated plants of all the Mediterranean countries, also including the islands of Cyprus, Crete, Sicily and Sardinia. This expedition collected a large amount of material from Abyssinia and Egypt as well (Vavilov *et al.*, 1931; Orlov, 1931). V.V. Markovich made collections of cultivated plants in India and on the island of Java during 1926–1928. In 1928, the expeditions by F.D. Likhonov to Yugoslavia and by N.I. Vavilov to Bavaria and other mountain areas of southern Germany were organized. For three years, starting in 1927, P.M. Zhukovskiy did research in Turkey, subsequently including a major portion of Anatolia as well. He collected an enormous amount of material of all kinds of field and vegetable crops. The results of these expeditions are published in the form of a large paper, 'Agricultural Turkey' (Zhukovskiy, 1933).

In 1929, E.N. Sinskaya made a long expedition to Japan, collecting valuable material of crops (Sinskaya, 1930). In 1929, expeditions to western China, led by N.I. Vavilov and M.G. Popov (Vavilov, 1931b; Popov, 1931), were sent out and the same year an expedition led by Vavilov went to Japan, Korea and the island of Taiwan. In 1930, a brief expedition led by Vavilov was sent to Mexico, Guatemala and Honduras (Vavilov, 1931a). Expeditions by N.I. Vavilov to Central and South America, including Ecuador, Peru, Bolivia, Chile, Argentina, Uruguay, Brazil and the islands of Trinidad, Puerto Rico and Cuba were undertaken during 1932–1933.

In addition, an enormous amount of material of subtropical and decorative crops from Italy, Spain, Holland, Iran, Turkey, Japan, California and Florida was acquired by the Subtropical Office of the National Commissariate of Agriculture by means of specialized trips made by trained research workers.

A colossal amount of material, amounting to several tens of thousands of samples, was collected during the past few years under the Soviet regime and planted out on experimental fields as well as on state and collective farms. For this, it was necessary to multiply the acquisitions of the most valuable kinds of material for the most important vegetable and field crops while proceeding according to a systematic plan established by the National Commissariate of Agriculture and other commissions. The best standards of the greatest value for our crops have been multiplied during the Soviet era and are, in part, already included in the composition of the Soviet agricultural assortment.

Never before has our country tested such a colossal assortment from all over the world. The worldwide selection of the most important crops of interest to the Soviet Union has been subjected to strict testing.

In addition to the study of cultivated plants by specialized scientists in our country, wild forage grasses, plants producing etheric oils and tannin and those producing rubber were investigated for the first time during the last couple of years for the purpose of introducing the best thereof into cultivation. Very valuable centers of wild forage grasses were discovered within the areas of the Caucasus, Altai and the Far East. With respect to alfalfa, E.N. Sinskaya was able to establish the type-formation of the blue-flowered and the yellow-flowered types of alfalfa and the species adjacent to them within the area of Transcaucasia. For the first time, detailed research was done on the wild fruit trees of the Caucasus, Inner Asia, Kirgizia, Kazakhstan and the Far East as well as Crimea. We can also mention the surprising discovery of rubber-producing plants, 'kok-saghyz' and 'tau-saghyz' [*Taraxacum bicorne* and *Scorzonera tau-saghus*, respectively], which are presently being cultivated.

The enormous amount of material collected by the expeditions in the form of seeds and living plants was subjected to detailed research for a number of years. Monographic studies of different, major crops by means of comparative harvests in various areas revealed for the first time the varietal wealth of our own country and, to a certain extent, also that of the whole world. The results of these studies are now published in the large edition of *The Cultivated Flora of the U.S.S.R.* by the All-Union Institute of Plant Industry. At the end of 1937, five of the 20 volumes planned had appeared. With respect to those plants that are of great importance for our plant industry, *The Cultivated Flora of the U.S.S.R.*

includes practically all the specific and varietal diversity of the entire world, while taking their importance for practical plant breeding into consideration. Consequently, this publication is, in essence, the first attempt at a compilation of the cultivated flora of the world with a detailed evaluation of the intraspecific varieties. The possibility of compiling such a 'cultivated flora' on the basis of an enormous worldwide variety of plants, which have been discovered and analyzed for the first time according to a definite plan, already bears witness to the level that the scientific plant industry of our country has reached during the past two decades. With respect to many, even most important cultivated plants, this *Flora* reveals for the first time a specific and varietal diversity that was not previously known. Almost one half of the botanical species of cultivated plants and their wild relatives closely related to them, have been described for the first time and have actually been made known during the Soviet era. Indeed, we are not mistaken when we state that during the last 20 years, Soviet research has discovered a no lesser number of cultivated plants and their wild relatives than all those described during the past 200 years between the time of Linnaeus and that of our own. A number of crops such as wheat, oats, rye, potatoes, leguminous plants and melons have, in essence, been thoroughly revised taxonomically.

The basic task of a wide mobilization of specific and varietal plant resources was first and foremost a practical one, i.e. the procurement by our Soviet plant industry of the most valuable foreign strains of material, in other words, the creation of a new material basis for plant breeding in the sense of a basic 'building material' necessary for creating new and valuable strains.

In order to include the enormous diversity of new species and varieties discovered thanks to Soviet research, it was necessary to develop immediately a systematic plan for studying the foreign material and for entering it into a strictly scientific system.

The evolution of cultivated plants took place both in space and time with an important participation by Man during the selection and alteration of the plants. The extensive material collected for many plants all over the world, must be comprehended and strictly systematized, and can be understood with respect to its importance to plant breeding only in the light of the theory of evolution. The stages of evolution reveal more profound traces than it is possible to suggest by direct studies of cultivated plants. Within the areas of southwestern Asia, the Caucasus and the mountain areas of Central and South America as well, it is actually possible to expose, in detail, the evolutionary stages of many cultivated plants and to trace the relationship between the cultivated and wild plants. With respect to many plants, all the successive evolutionary stages could be revealed by the presence of the related links that still exist.

Investigations of wild, closely related species revealed essentially that among them were valuable properties such as resistance to diseases, frost and drought and other characteristics. At the very border between cultivated and wild forms there are, in many cases, no distinctions during the initial stages. This can be clearly seen, e.g., as far as the fruit trees of the Caucasus and the mountains of Inner Asia are concerned. It is still possible to find an entire range of transitional varieties of almonds, figs, pistachio nuts, walnuts and grapevines in the mountains of Kopetdag in Turkmenia. Within the borders of Soviet Azerbaidzhan, it is possible to observe all the successive evolutionary stages of pomegranates and quince from wild, small-sized, bitter and acid fruits to gigantic and sweet cultivated ones. The same is, to a great extent, valid also in the Caucasus and other areas for such cultivated plants as rye, oats and hemp.

Direct studies *in situ* of the type-forming processes led to a clearing up of the different stages of the evolutionary process. With respect to such crops as wheat, barley, maize and flax (linseed), for which the divergence between the wild and the cultivated forms is distinctly expressed, new links have already been discovered, allowing us to explain the successive stages of differentiation of the species and their relation to the wild types. In this respect, Asia Minor, including Transcaucasia, turned out to be a territory where a wealth of plants species and even genera are particularly concentrated, and to be an area that gave rise to the present European crops, where especially the initial stages of speciation of cultivated plants are distinctly revealed. In that area, the Soviet scientists discovered for the first time species of cultivated and wild wheat and a number of species, as well as a multitude of varieties, of wild and cultivated rye and oats.

All the modern methods of biology ranging from the ordinary morphological methods of taxonomy to anatomical, cytological, genetical, physiological, chemical and technical ones, were needed for the study of the large amount of material of specific and varietal resources.

In contrast to the previous fragmentary work in this direction, the Soviet research on cultivated plants is, most of all, characterized by complexity. The very concept of 'botanical species', based on these studies, allowed us to develop an understanding of the species as a complete system, which unfolded during a historical process of type-formation, is often exceptionally multiform and, during evolution, was linked both to the environment and time as well as to the effects on it exerted by Man.

The studies of the wild material collected demanded that physiological, cytological and chemical methods be worked out. Thus, new cytological methods, worked out by G.A. Levitskiy, made it possible to approach again the problem of differentiation of the species of wheat and the genera closely related to them. The method, recently worked out by I.I. Tumanov, A.A. Votchal and others for evaluating frost- and drought-resistance, made it feasible to present, in exact figures, a physiological evaluation of the specific and varietal diversity under definite conditions. The studies of the developmental stages made, by T.D. Lysenko, have made it possible to approach more profoundly varietal differences with respect to the vegetative period. The application of vernalization widened the possibilities for utilizing the worldwide collection of grasses. Even southern varieties of winter cereals can be grown normally as far north as the Polar Circle, thanks to vernalization.

The mobilization of the varietal resources stimulated the work of biochemists and revealed a multitude of new facts of great importance. Major differences with respect to the content of vitamins in various strains of fruits and vegetables were revealed in the case of wild and cultivated fruits. Important regularities were discovered concerning chemical variability and the effect of external conditions.

During evolution and dispersal from the basic centers of type-formation, the cultivated species differentiated just like their wild relatives into distinct ecological–geographical groups. In my essay on the 'Phyto-geographical basis for plant breeding' (Vavilov, 1935), I made an attempt to list the basic centers of development and concentration of a wealth of varieties of 640 cultivated species. Practical research has demonstrated that, in addition to the primary areas where links to the wild species are especially well reflected and where, frequently, a botanical diversity of types is found, the secondary areas of development of crops are of great importance. There, plants have for a long time been subjected to new environmental conditions and the effects of selection. Our studies have already compelled us to include extensive areas within the centers of origin. Further research revealed an even greater importance for the differentiation of the species into ecological–geographical groups, distinguished by a complex of morphological and physiological characteristics, particularly of the latter.

Thus, e.g., we have no doubts whatever that, in the light of a multitude of facts, Inner Asia, including Transcaucasia, is the basic territory where the type-forming process unfolded and led to the development of the present species of wheat, barley and rye. There, an enormous potential of species and even genera is concentrated, consisting of genetically distinct units and including a multitude of endemics not known anywhere else in the world. In Transcaucasia alone, all the four genomes of wheat, i.e., the four sets of chromosome numbers, which are quantitatively distinct and characteristic of all the basic diversity of wheat, have been discovered. In this area alone, the cultivated species are concentrated in an amazing variety and just there, and in adjacent Syria and Palestine, a great variety of wild wheats is also concentrated. Within this territory and areas adjacent to it, dozens of species of *Aegilops* and *Haynaldia* are accumulated, i.e., genera very closely related to wheat.

The maximum diversity of species and varieties of rye – ranging from wild types of weedy rye to cultivated crops – can be found in the Caucasus but especially in Transcaucasia and in Turkey. In Syria, Palestine and the foothills of eastern Azerbaidzhan and the adjacent areas of the mountains in Inner Asia, one can observe an enormous variety of wild types of rye that are closely related to the cultivated ones. In other words, in Asia Minor and Transcaucasia, as well as in the foothills of Inner Asia and Iran, adjacent to them, a basic material is found from which the cultivated species of wheat, barley and rye developed.

However, far from this large area, within the Mediterranean countries, in Abyssinia and even in China, important agro-ecological groups of wheat and barley have developed in the course of the millennia. Typical complexes of physiological or morphological characteristics distinguish the groups in question from the original types in southwestern Asia. Thus, e.g., eastern Asia is characterized by the presence of a peculiar awnless barley and an awnless type of rye, which are definitely not present in the southwestern areas of Asia or other parts of the world. The Chinese forms of barley and soft wheat are resistant to brown and yellow rusts. Very valuable and peculiar perennial types of wheat have developed there. The Chinese forms of wheat and barley no doubt represent a great practical interest to be utilized by us for plant breeding purposes. Within the Mediterranean area, extremely large-seeded types of these plants, which are also resistant to many fungal diseases, have been developed.

Even more striking examples can be mentioned with respect to our fiber flax, which is characterized by tall growth, single stems and a high quality of first class fibers. Such forms have, indeed, developed during the course of two millennia in the Russian Northland as a result of selection and environmental conditions. There is nothing like this in the basic centers of origin of cultivated flax, neither in the Mediterranean area nor in southwestern Asia.

Maize, which in all likelihood had its initial development in southern Mexico and Central America, where its association with wild species can be established, underwent changes when transferred into South America – into Peru – leading to the formation of extremely large-grained and late-ripening forms that are not found in Mexico or in Guatemala.

These studies actually led to a renewed working out of the geography of cultivated plants and to a hypothesis about the differentiation of cultivated plants during their evolution as well as to the localization of the various inherited characteristics, the knowledge of which is of essential importance for practical plant breeding.

The kind of discoveries that were made as a result of the new systematic approach to actual control over the plant resources for the purpose of securing valuable foreign material for the Soviet plant industry can best be seen when using the potato as an example. Soviet scientists (S.M. Bukasov and S.V. Yuzepchuk) have, during the last decade, determined 18 new species of cultivated potatoes within the limits of South America, in addition to the one already known in the past. They are still preserved in South America as primitive crops, grown in the mountains by American Indians, but unknown to the botanical science of the past. In addition, dozens of new species of wild potatoes were discovered, among which groups of species typical of Mexico turned out to be resistant to the most damaging disease of the potato, i.e., *Phytophthora*. Among the wild species at high altitudes in the mountains of Peru, types were discovered that are able to withstand frost down to $-8°$ C [ca. 16° F] without the leaves or the tubers being destroyed. At present, this new material is utilized for practical Soviet plant breeding. Strains have already been selected from it and grown on fields of collective farms. Problems of which plant breeders of the past did not even dare to think are presently not only beginning to be solved but have to a great extent already been solved, as shown by the strains resistant to *Phytophthora* and frost.

Dozens of species of cultivated and wild wheats have been discovered. For the first time, literally thousands of new botanical varieties of plants have been observed. Three fourths of the botanical varieties of wheat from all over the world that are known at present by science, have been discovered during the last 25 years by Soviet scientists. The same can be stated in the case of barley, rye, flax and leguminous crops.

Thus, major steps forward have been taken during the past few years with respect to a mobilization of specific and varietal resources of the most important crops of interest to the Soviet Union; valuable new materials have been discovered and areas where specific and varietal riches are concentrated have turned up. All this new material has been put to use for practical plant breeding.

The practical problems facing the Soviet scientists for the purpose of being able, as fast as possible, to utilize the colossal new material, have during the last couple of years consisted primarily of starting to rework the classification of cultivated plants. Now we are entering a new stage of scientific work.

In order to be able to utilize the varietal diversity, it is necessary to put it into a general botanical system. The establishment of differences in physiologically and agronomically valuable characteristics appears even more important. It is natural that at the very beginning of the investigations of the specific and varietal diversity the plant industry cannot control such characteristics as resistance to diseases, drought or frost. The systematic acquisition of a colossal new material has forced us to approach the creation of a new classification in addition to the botanical system on the basis of which, primarily, the physiological characteristics should be established. The importance of resistance to many of the diseases that affect cultivated plants makes it necessary to enter this important factor into the system of classification as well.

The scale of work and the efforts spent on it can, to some extent, be understood when considering the multiformity of the crops both with respect to the varieties composing them and the complexity of an exact evaluation of their differences in the case of resistance. A team of Soviet scientists has actually tackled all this work during the last couple of years.

A parallel and comparative study of many crops that were associated during their evolution with a single but wide territory revealed a multitude of general regularities, allowing an interpretation of the phenomena and facilitating an exact establishment of their differentiation into different groups.

The law established by myself in 1920 and called 'the law of homologous series' for inherited variation was, at the time, based mainly on morphological characteristics that serve for botanical classification, but has actually been confirmed by the enormous new material of cultivated and wild plants.

The differentiation of the different species into agro-ecological and geographical groups revealed a multitude of parallels not only as far as the morphological characteristics are concerned but – and this is particularly important – also with respect to physiologically, ecologically and agronomically valuable properties. A striking parallelism was revealed within different areas, which was especially evident in the case of species close to others or that of closely related genera, but apparent also as far as different families are concerned.

To an important extent we now know where to look for valuable characteristics, in what areas we can find the most drought-tolerant types, where especially productive and large-fruited or large-grained types are concentrated, where to find types with the most functional straw, which is resistant to attacks by different pests, where to find types with the stiffest straw, which is resistant to being felled by rain and where to look for straw that is immune to fungal diseases.

Homologous and analogical series appear to be the result, on the one hand, of a combination of origin and genetic relationship and, on the other hand, of the results of both the effects of specific environmental conditions and a specifically directed selection.

Comparative studies of agro-ecological groups of the most important annual crops of plants in the Old World conducted by us under different conditions, and a comparison of these with the conditions found at the basic localities, revealed a number of regularities as far as the differentiation of species during their successive evolution is concerned.

Let us list the main *agro-ecological groups* that, to a great extent, are general for wheat, barley, rye, oats and flax as well as leguminous crops (peas, lentils, chick-peas, vetches, beans and chickling vetch).

1 *The Syrian Group*, the main sites of which are found in the highland and foothill areas of Syria, Palestine and Transjordania. This area is characterized by mild winters and precipitation falling during the autumnal and spring months. The summers are dry.

The following special properties are characteristic for the annual crops of this area (hard wheat, barley, flax, peas, chickling vetch, chick-peas, lentils and vetches): low growth, strong slender stalks, drought tolerance and, particularly during the later stages of development, early ripening, small leaves, small seeds, small flowers and slender and strong stems. The seeds are not shed. Some leguminous plants, such as vetches and peas, appear to be typically amphicarpous (self-propagating in the soil). The developmental stages, both the early one (in the spring) and the late one (in the fall), are short. The plants are tolerant to low temperatures during the early stages of development and to drought, as well as high temperatures, especially during the late stages.

2 *The Anatolian Group*. The main sites are found in the mountain areas of interior Anatolia, which are characterized by a dry climate but have an adequate amount of precipitation during the summer. The particular characteristics of the same plants as mentioned above are: low to medium growth, comparatively strong stems and medium-sized grains, beans and spikes. The early and the late stages of development are short. The plants are drought-tolerant during the later stages of development but require heat, especially during the late stage.

3 *The Armenian–Georgian Group (the Caucasian Mountain Group)*. The main sites are located in the mountain areas of Soviet and Turkish Armenia, in Georgia and in the highlands of Karabakh and Ossetia, which are distinguished by short vegetative periods and comparatively moderate temperatures. The amount of precipitation is adequate, although there are periods of drought. Peculiar ecotypes of soft wheat have developed there under the conditions of a typical mountain-steppe. They have narrow leaves and may be the prototypes of the European Banatka and Rusak standard brands. The main sites of the peculiar species of 'Persian wheat' and forms of soft wheat close to it are found here. The large group of cultivated barley is characterized, in this area, by a candelabra-like growth with evenly tall culms and narrow leaves (gr. *colchicum*). Narrow-leaved types of vetches and peas are also concentrated here. Ecologically, this is a group that has proved to be cosmopolitan during its subsequent

dispersal, when it became widely distributed over the plains and forest-steppes of Europe. This group is, on the whole, characterized by rather small seeds, slender culms, interrupted spikes, soft and slender awns and seeds that are easy to thresh. On the whole, the group is tolerant of periodic drought and is represented both by winter, semi-winter and spring types; correspondingly, they have more-or-less long developmental stages. They need a specific amount of heat and are comparatively susceptible to the common European fungal diseases.

4 *The Azerbaidzhan–Dagestan Foothill and Basin Group*. The sites are found in basins in the mountains of Dagestan and Azerbaidzhan and in the foothill areas of the Karabakh highland. The winters are relatively mild, suitable for winter crops of wheat and barley. The crops grown are both irrigated and non-irrigated. A special group of gigantic hard and soft wheats, barley, rye, peas, as well as vetches, has developed there. Its special characteristics are: tall growth, strong leafiness, broad leaves, large spikes, large grains and large beans; the stalks are thick, the culms stiff. The entire group is characterized by long vegetative periods. Hard winter wheats are concentrated in this area, as they are nowhere else on earth. In general, this group is mesophilic in its requirements for moisture but demands a specific amount of heat during the ripening period. It is relatively resistant to the common European fungal diseases. Under favorable developmental conditions, it is exceptionally productive.

5 *The Transcaucasian Humid–Subtropical Group*. The areas of western Georgia and the coast of the Black Sea, including Abkhazia and Adzharistan, belong to this group, as do the humid areas of northwestern Anatolia, Lenkoran', the Zakatal'skiy area within the borders of Azerbaidzhan, as well as the northern provinces of Iran (the Astrabadskayan, the Mazanderanskayan and the Gilyanskayan ones). These areas are characterized by plentiful precipitation and comparatively mild winters.

Peculiar types of late-ripening soft wheat have developed in this area, such as the striking, endemic one-rowed wheat, the Georgian distichous one and 'Timofeyev's wheat', which is distinguished by an amazing resistance to fungal diseases. The main sites of very late types of flax with a candelabra-like growth-form, which are usually sown in the fall, are here. The cereals are mainly winter or semi-winter types. Hygrophily, late ripening, tall growth, leafiness and a relatively satisfactory resistance to fungal diseases are specific for this area.

6 *The Iranian–Turkestani Group* is found in the mountain and foothill areas of Inner Asia, northern Afghanistan and Iran. This group is characterized by plants of low to median height and is exceptionally susceptible to all the European fungal diseases. The group is relatively tolerant to drought and high temperatures, in particular during the later stages of development. It is very dependent on heat for normal development. The culms of the cereals are, just like the stems of leguminous plants, not very strong and are very prone to falling over; the spikes are large and the grains are difficult to thresh and do not shed easily. The awns of the wheat, barley and rye are coarse and brittle.

Special subgroups have developed in the oasis of Khiva as well as in the one at Kashgar (in Chinese Turkestan). The oasis of Khiva is distinguished by very late-maturing types of flax and peas. The wheats are represented by small-grained, winterhardy types with short straw and narrow leaves. Extremely cold-hardy types of wheat have developed in the dry conditions at Kashgar.

The majority of the crops in Iran and Afghanistan as well as in Inner Asia is irrigated. Under natural conditions (in the case of non-irrigated crops), correspondingly fairly drought-resistant types have developed. The tendency of the grains to sprout in the spike under wet weather conditions is one of the negative characteristics of this group. Because of the exceptional susceptibility to European fungi and the tendency of the grains to sprout in the spike, this group is, as such, not suitable for a large portion of the European area.

7 *The Inner-Asiatic Group of the High Mountains.* To this area belong agricultural areas, which are dispersed as far as Hindukush in Afghanistan and the areas of high mountains in Tadzhikistan, Himalaya, Badakhshan, Chitral, Ladak and Tibet. The area is characterized by a dry climate. Its crops are usually irrigated. The vegetative period is short and destructive effects of low temperature frequently appear during the summer.

The biological characteristics of this group are: early ripening, not very high demands for heat, a fast initial development of the plants in connection with irrigation, a satisfactory mass of leaves, medium tall or even tall growth (e.g. giant types of spring rye), tolerance to low temperatures, but a strong susceptibility to European fungal diseases. Peculiar recessive kinds of peas have been discovered here, as have non-ligulate types of soft, shot and club wheats as well as of rye.

8 *The Indian Group.* Its area is characterized by a climate with monsoons and typical downpours of rain during the summer months. Different agro-ecological provinces can be distinguished. The areas concentrated on alluvial soils in northwestern and northeastern India can in particular be sharply distinguished, as can the areas that belong to the Indian central plateau, represented mainly by dark-colored soils.

The special characteristics of this group are: low growth, few leaves, not very bushy growth, delicate and short awns of cereal grasses and grains usually well filled. As a rule, fast-ripening types have developed that are tolerant to dry air and comparatively resistant to low temperatures during early development but dependent on high temperatures during the late stages before ripening. The first and second stages of development are short and the seeds are small. On the whole, the annual crops grown in India are distinguished by their susceptibility to the European fungal diseases.

9 *The Arabian or Yemenite Group.* This group grows in a small center of ancient agriculture distinguished by early-ripening types of wheat, lentils, peas and alfalfas. In general, the vegetation is distinguished by low growth and small, narrow leaves. The group is, on the whole, characterized by fast development, as a result of which the majority of the types characteristic of this area can be distinguished by a tolerance to summer heat. The group, as such, is characterized by low productivity. The initial stage of development is very short and the second one is also short.

10 *The Abyssinian Group.* Its main sites belong within the high mountain

areas of Abyssinia and Egypt, at an altitude of 1800–3000 m.s.m. This area is characterized by a climate that is relatively wet during the vegetative period. Harvests usually coincide with the dry period. Spring types are normally grown and the crops are not irrigated.

The particular properties of this group are: narrow leaves and a relatively low growth of cereal grasses, leguminous plants, flax and wheat. The plants are comparatively tolerant to low temperatures, particularly at first. The vegetative periods are medium long to short. The barley and wheat of this area are distinguished by an exceptional morphological diversity in relation to the comparative uniformity of the ecology. One of the characteristic properties of the Abyssinian barley and peas is their cosmopolitan qualities: they can be successfully grown as far north as the Polar Circle but succeed well also under steppe conditions. In addition to the mesophilic subgroups of plants, sown at the beginning of the rainy period, there are also xerophilic ones, sown at the end of the wet season and subject to the effects of drought.

11 *The Chinese–Japanese Group*. The main sites belong to the damp interior areas of eastern China, Japan and Korea, which are characterized by a monsoon climate. Intensive cultivation here has produced particular types, which have no doubt been subjected to the effects of selection.

The special properties of Chinese–Japanese wheats and barleys are: low growth, stiff and thick culms and small grains. The variation in type here is extremely large and comparable with the diversity of the conditions, in relation to the vegetative periods and ranges from winter types to record-fast ripening spring types. The stages of development during the spring are: a long initial one and a short second one. The winter and semi-winter types are distinguished by small, narrow leaves and a sharp reaction to vernalization, which shortens the time required for ripening. The winterhardiness of the winter and semi-winter crops is slight, just like their tolerance to drought. This group is, on the whole, not very demanding with respect to heat for ripening. Full-spiked types of both barley and wheat predominate but thin-spiked types are also found, especially in the coastal belt.

A characteristic physiological property of the Chinese barley and wheat is the fast ripening of the grains, which apparently is associated with the small size of the grains and the absence of awns or the presence of short awns only. Both Japan and China are characterized by peculiar awnless and short-awned types of barley but also by definitely awnless wheats. Among the Chinese soft wheats, many kinds have been discovered that are immune to brown and yellow rusts.

On the whole, this group is undoubtedly secondary with respect to its origin, of which the absence in China of closely related wild forms and the limited composition of the species bear witness. Nevertheless, several thousands of crops there have developed typical complexes of characteristics, which are seen nowhere else in the world, but which are very important for plant breeding purposes, such as fast ripening, complete absence of awns, densely flowered spikes in the wheats, stiff culms and characteristics of immunity.

12 *The Mediterranean Group*. The main sites are located along the coastal areas of the Mediterranean, which are characterized by mild winters and a

concentration of precipitation mainly in the fall and spring. Islands such as Sicily and Sardinia also belong to this group. Cereals are mainly sown late in the fall in this area.

The special characteristics of this group are: relatively fast growth, large-sized seeds, large spikes, long awns (typical of wheat and barley), slender culms, which in the case of wheat are stiff as in the case of some other plants and which have thick walls and partly filled cores. The life-form is a spring or semi-winter type. The vegetative period is medium to long-lasting. The stages of development consist of an initial medium long one and a second short one. All the annual Mediterranean crops are characterized by plants developing quickly during the initial stage of the vegetative phase. The winterhardiness is only slight. The entire group is, on the whole, tolerant of atmospheric drought during the later stages of development and requires heat during the ripening period. A characteristic property of the entire group is the presence of an enormous number of types resistant to fungal diseases and to various kinds of rust and smut as well as to other parasitic infections.

In general, the group is very productive under optimum conditions of development. The plants are usually very bushy and branched. This is contrary to what is the case, e.g., in India, and especially in the case of the powerful development and dimensions of pods and seeds. Many types with light-colored seeds have developed within the Mediterranean area.

13 *The Egyptian–Cypriot Group*. This group is distinguished by fast ripening, slender and stiff culms, narrow leaves and low growth. The complex of characteristics is especially typical of the wheats and the barleys of Egypt and Cyprus. Like the previous group, this one is also characterized by an initially rapid growth.

14 *The South-European Group*. The main sites are located in southern France, Yugoslavia and northern Italy. It is, in general, distinguished by large seeds, tall growth and large fruits. These are the 'giants' among the varieties of the types discussed.

The most productive kinds of wheat on the Earth are concentrated in this area, to which also belong strains representing the English type of wheat. Among the soft wheats of Lombardy, we find the largest-grained types of soft wheat with the largest spikes; they are also distinguished by tall and stiff culms and broad leaves and ripen comparatively late.

15 *The Steppe (or Plains) Group* covers the steppes, plains and forest–steppe areas of Europe and Asia, including the area of Hungary and the plains of Poland and Romania as well as the forest–steppe areas of the European and Asiatic parts of the U.S.S.R. During the last couple of decades, the steppe types have also become widely dispersed over the prairies of Canada, the U.S.A., Argentine and South Africa.

This is the realm of the awned, soft spring wheats of the Banatka type, which are distinguished by excellent winterhardiness and glassy grains. The most winter-hardy kinds of barley are concentrated on the steppes in the foothills of the Caucasus.

Comparatively tall growth, medium leafiness and the presence of spring as well as winter types are the special characteristics of the steppe group. On the whole, the steppe types are susceptible to fungal diseases and are distinguished by a medium frost-hardiness and tolerance to atmospheric drought as well as to dry soils, but they do require warm temperatures when ripening. The vegetative period is medium–long.

16 *The West-European Group* covers the lowland areas of England, the Netherlands, Denmark, Sweden and northern Germany. Its special characteristics are: high productivity under optimum conditions of growth, hygrophily, stiff culms, broad leaves, large spikes and grains and a good response to fertilizers. This group is particularly distinguished by the results of conscientious plant breeding during the last 200 years.

The wheat, barley, oats, rye, peas and beans within this extensive area are distinguished by high productivity, large seeds and high quality. The distichous 'beer-brewing' barley is mainly concentrated here.

17 *The West-European Mountain Group* comprises the areas of the Alps, Tirol and Bavaria. The particular characteristics of this group are: relatively rapid ripening, not very large spikes and small grains, winterhardiness, frost-tolerance and little demands for heat as well as hygrophily. The developmental stages, both the first and the second ones, are long.

18 *The Boreal Group* embraces the northern areas of Sweden, Finland and those of the European and Asiatic parts of the U.S.S.R. The special characteristics of this group are fast ripening, hygrophily and soft spikes, tall growth, and culms that are not very stiff and prone to lodging. The first stage of development is short, the second one long.

Rapid ripening under conditions of fairly low temperatures and a comparatively satisfactory tolerance of frost are typical of this group. In general, this group is of medium tolerance to the European fungal diseases. The productivity is high under optimum conditions of growth and the spikes and awns are soft.

These are the main groups. For the sake of brevity, we have abstained from subdividing them. It is self-evident that the different areas are joined by transitional types, especially as a result of the changes in conditions there. However, on the whole, the areas listed are adequately presented with respect to the range of ecological types, which have developed there during a thousand-year-long period of agricultural practice in the Old World.

During the last hundred years, the crops mentioned above have been transferred to America and Australia and are grown there over extensive areas, covering several billions of hectares. However, under the new conditions there it is still possible to trace the agro-ecological groups of cultivated plants from the Old World. Thus, the Mediterranean types of hard and soft wheats introduced to Mexico, Guatemala and Colombia are only slightly different from the original Spanish types, which were characterized by slender and stiff culms and large grains. The kinds of leguminous crops introduced to the New World still preserve their basic characteristics to a great extent. This can, of course, be explained by the similarities in the growing conditions. The winter wheat of the Banatka type, introduced to America from Europe, retains features typical of our Crimean kinds and is only slightly different from the original type. The

Mediterranean oats, distinguished by their immunity to crown rust and smut, have been introduced to the southern areas of the U.S.A. and are widely distributed there, as in Chile and Argentina, while still preserving their characteristics.

However, major changes were made to the old ecotypes, when practicing plant breeding by means of hybridization during the past century, resulting in a combination of different agro-ecological groups.

Canadian plant breeders have done outstanding work by crossing European wheat with a fast-ripening Indian one with stiff culms. In that manner, a new agro-ecological group originated, combining the characteristics of two different ecotypes. The origin of the famous 'Marquise' wheat, the standard brand most widely distributed in Canada and the U.S.A., was of this type. It is generally possible to spot the characteristics typical of the Indian wheat in such a hybrid, e.g. the low growth, the stiff culms, the perfect grains of high quality which are not shed, combined with the high productivity and the excellent frost-hardiness inherited from the European steppe wheat. The numerous grains in the spike are to a great extent inherited from the Indian base of the 'Marquise' and its derivatives. Australia has adopted quite a few of the Indian and European types of wheat. The majority of the Australian hybrid brands remind one of the Indian as far as rapid ripening, low growth and stiff culms are concerned. The same applies to the wheats of South Africa.

Argentina followed another course. The condition of mechanized agriculture with harvesting being done exclusively by means of combines made it necessary to pay attention to stiff straw and non-shedding grains. At the same time, the presence there of brown, yellow and stem rusts necessitated a selection of resistant types. This problem has been brilliantly solved by crossing a Mediterranean group, initially introduced by settlers from Spain, with Chinese brands. The Argentinian wheat derived its resistance to brown and yellow rusts from the Chinese ones. The famous standard, 38 MA, and others were created in this manner. The Argentinian group of wheats is distinguished by non-shedding grains, early ripening and coarse awns, reminding one to a great extent of the typical Mediterranean types, but also by the fast ripening and high quality of the grains, derived from China.

An analogical scheme of agro-ecological differentiation also applies to other plants in the New World. Maize, leguminous plants and cotton display a number of interesting regularities during their evolution in the direction from southern Mexico toward Colombia, Ecuador, Peru and Brazil but also northward to Arizona and the central parts of U.S.A. Crops from southeastern Asia, inherited from India and China (e.g. rice and soybeans), display typical evolutionary trends due to their agro-ecological differentiation. Such problems should be worked out in the near future.

Based on an analysis of the groups, it is possible to see that the most valuable but contrasting qualities are distributed in different groups in relation to the environmental conditions and the selection that has been made.

One of the most important tasks of modern plant breeding is to combine the most valuable characteristics of different agro-ecological groups. Work is also going on in that direction.

In the light of the agro-ecological studies in the U.S.S.R. with respect to many crops, such as wheat, barley, rye, flax, leguminous crops and fruits, the Caucasus has turned out to be of exceptional importance. During the historical process of development there near the main areas of type-formation of those plants now cultivated in Europe, a large number of contrasting ecological types of botanical species of cultivated plants have also been produced. The most amazing facts for explaining the resistance of the groups to diseases have turned up only in this area. Timofeyev's wheat (*Triticum timopheevii* Zhuk.) is such a species, distinguished by its exceptional resistance to practically all the diseases of wheat. The Persian wheat (*T. persicum* (Perciv.) Vav.), which is typical of the mountain areas of Georgia, Dagestan and Armenia, possesses amazing properties such as grains that do not germinate in the spike during rainy weather, and resistance to mildew and to brown, yellow and stem rusts. In the mountains of Dagestan, a whole group of productive, endemic types of naked-grained distichous barley, lentils and other plants has been discovered. The peculiar Azerbaidzhan–Dagestani foothill group of cereal grasses and other annual crops possesses an exceptional productivity and large seeds, in this respect reaching record levels in competition with even the most productive types from southern Europe. Valuable hard winter wheats are concentrated in this area and are known nowhere else except in this area. The most winterhardy types of winter barley were discovered in the foothills of the northern Caucasus. At the same time, a group of awned, soft winter and spring wheats, formed historically in the mountain and steppe areas of Armenia and southern Georgia, most likely gave rise to the 'Banatka' and 'Poltavan' types, which presently occupy an enormous territory on the plains of the Old World and the New World as well. As far as rye is concerned, we can find all the transitional forms there, from wild to cultivated ones, with an exceptional diversity in both.

On the whole, the Caucasus represents an amazing area of development and differentiation with respect to cultivated plants. The importance of this becomes increasingly evident when applying the thorough methods of physiology, ecology and phytopathology. The enormous amplitude of conditions displayed over a relatively limited area and the vicinity to the main centers of type-formation of the most important cultivated plants are the reasons for the amazing focus there of specific wealth as far as such crops as wheat, barley, rye, flax, leguminous plants and fruits are concerned.

As a result of the great team work among Soviet scientists during the Soviet era, a colossal wealth of species and varieties has been revealed, which opens up unlimited prospects for practical plant breeding. The plant breeder now knows where to look for the necessary 'building material': rather, he actually controls it. To a great extent, we have learned of the level to which mankind has reached as a result of thousands of years of work. A clear picture of the general composition of cultivated plants in an amazing diversity of forms and physiological constitution has developed. The investigator now knows where a particularly valuable characteristic or a special gene can be found.

In turn, a problem arose concerning how to work out a theory for the selection of pairs, based on the diversity of the species and varieties discovered. By breaking down all the intraspecific variation of the major cultivated plants into agro-ecological groups, we have recently been able to approach an explanation of the most expedient combinatory method in the form of so-called cyclic hybridization and to enter upon a cycle of gradual hybridization of agro-ecological groups. This work has only just started but already a number of facts have been outlined with respect to the inheritance of quantitative and qualitative properties. Using a combination of actual agro-ecological groups, direct experiments, repeated within the varietal combination *within the limits of the species* of different plants, have as a rule revealed the dominance (although not always complete) of the maximum quantitative expressions of different characteristics such as the dimensions of fruits, seeds or leaves and the height of the plants. The recessiveness of characters, such as small seeds, small fruits and small leaves or low growth in relation to the dominant types with large seeds, large fruits and tall growth was also revealed. The characteristics of long fibers, unquestionably one of the last links in the evolution of flax, has during the past few thousands of years definitely shown the dominance of tall growth over low growth. Many hybrids occupy an intermediate position but most approach a maximum quantitative expression. We have never, in a single case among hundreds of crosses of flax or linseed, found a reverse phenomenon, i.e. dominance of low growth, small seeds or small capsules. We have found the same thing applying to wheat and the interspecific crosses of barley, peas, chickling vetch and other leguminous plants. As a rule, the offspring of hybrids lie somewhere between the original ancestral types with respect to quantitative expressions of different characteristics.

Thousands of combinations, made between various characteristics, have also revealed comparatively rare ones, such as when – during a hybridization – it is found that a combination of quantitative expressions occurs as a result of the cross and that there is an additive effect of the quantitative expressions of both parents, resulting in larger fruits and taller growth than expressed by either parent.

The hybrids between the Turkish winter flax with a candelabra-like growth type and the long-staple flax can serve as a striking example of this kind. As a result of this cross, types were obtained that displayed both a tall and bushy growth (a dominant character) and a much taller growth than that of the original long-staple flax. In the fourth and fifth generations, types appeared that surpassed by one third the original long-staple flax with respect to the length of the stems. Our observations are similar in the case of the hybrids between Chinese small-grained and small-spiked barley and the Inner-Asiatic and Turkish types. As a result, the height of the plants, and also the dimensions of the leaves, almost doubled and new, considerably more productive types than those of both original parents developed. The same was observed by T.K. Lepin in the case of crosses between different agro-ecological groups of wheat.

Thus, the possibility for a systematic approach to obtaining the combinations necessary by means of hybridization can be outlined. We actually started reworking the theory about the choice of pairs. The theory concerning the stages of development allows us to approach, considerably more so than previously, a feasible control over vegetative periods in order to select pairs for obtaining early-ripening types. Cyclical hybridization has led to the development of a number of new forms, while determining the regularities valid for this purpose. Thus, it is quite common to find the appearance of smooth-awned types of barley derived from ancestors with serrated awns or the appearance of awnless types after crosses between awned ones, due to the special selection of ecological–geographical groups. Soviet genetics has entered upon a path toward the exploitation of biological syntheses both in theory and in practice. Parallel and comparative studies of different crops and the production of parallel hybridization involving special agro-ecological groups will allow us to establish the general laws of genetics.

Thus, during the past period, we have to a great extent concluded the first stage of the investigations, i.e. the control over already existing genotypes. With respect to a number of species such as clover, alfalfa and some other wild plants, it is still proper to utilize mainly the existing forms of historical ecotypes, which so far have been inadequately involved in plant breeding.

We are now entering the stage of biological synthesis. The approach can follow two lines: an intraspecific one and one of more distant hybridization. The latter line is more difficult but also opens up greater perspectives.

Soviet science has already followed the path of overcoming the infertility of hybrids between distant parents. In principle, a possibility already exists for synthetically producing new species and genera by means of hybridization. The experimenters have learned to turn infertile hybrids into fertile types by doubling the sets of chromosomes. The appearance of polyploids and amphiploids, which occur relatively rarely in the world of wild plants by natural and artificial reproduction, has opened up new opportunities for producing species that are isolated from the original types but not hybridized.

The ideas of I.V. Michurin concerning the practical utilization of distant hybridization, which we promoted boldly in our country, and have been realized for different kinds of plants, have already led to the creation of new and valuable types. Hybrids between distant species of fruits (made by I.V. Michurin), between crops of berries (made by I.V. Michurin and M.A. Rozanova), between rye and wheat (produced at the Saratov Byelotserkovsk station), wheat-grasses and wheat (made by N.V.Tsitsin and B.A. Vakar), perennial rye and wheat (made by A.I. Derzhavin), sunflowers and Jerusalem artichokes (made by N.A. Shibrya), and between synthetically obtained cultivated tobaccos (made by D. Kostov and S.A. Egiz) as well as between soft wheats (produced by O.N. Sorokina) have been realized during recent times.

In practice, there are prospects for overcoming the difficulties, which are still rather substantial, for the completion of the work of plant breeders, whose ultimate purpose is to explore various types and to combine all the necessary valuable characteristics. In principle, the problem of producing new species experimentally can be considered solved.

This is the situation within applied botany, plant breeding and genetics within the U.S.S.R. Summing up, it can be stated that during the last 20 years a new, rational theory concerning the potential of the original species of cultivated plants has to a great extent been created. Now a stage of actual control over biological synthesis is being outlined, exploiting the theory and practice of producing new types synthetically both by means of intraspecific and distant interspecific hybridization.

The discoveries made by the Institute of Genetics within the Academy of the Sciences of the U.S.S.R. on the threshold of the end of a 20-year period, signal a new cycle of work, directed toward application of the theory of genetics as a basis for practical plant breeding.

BIBLIOGRAPHY

Bukasov, S.M. (1930). *Vosdelyvayemye rasteniya Meksiki, Gvatemaly, Kolumbii* [*The plants cultivated in Mexico, Guatemala and Colombia*]. Leningrad.

Orlov, A.A. (1931). Vazheyshie prakticheskie i botanicheskie formi yachmeney vida, *Hordeum sativum* [The most important practical and botanical forms of a barley species, *Hordeum sativum*]. *Tr. po prikl. botan., genet. i selek.* [*Papers on applied botany, genetics and plant breeding*], **26**.

Popov, M.G. (1931). Mezhdu Iranom i Mongoliey [Between Iran and Mongolia]. *Tr. po prikl. botan., genet i selek*, **26** (3).

Sinskaya, E.N. (1930). Kratkiy ocherk sel'skokhozyaystvennogo rastenobodstva v Yaponii [Short essay about the agricultural plant industry in Japan]. *Tr. po prikl. botan., genet. i selek*, **22** (2).

Vavilov, N.I. (1923). K poznanniyu myakikh pshenits [Toward understanding soft wheat]. *Trudy po prikl. botan., i selek.* [Papers on applied botany and plant breeding], **13** (1).

Vavilov, N.I. (1926). Tsentry proiskhozhdeniya kul'turnykh rasteniy [Centers of origin of cultivated plants]. *Tr. po prikl. botan. i selek.*, **16** (2).

Vavilov, N.I. (1931a). Meksika i Tsentral'naya Amerika kak osnovoy tsentr proiskhozhdeniya kul'turnykh rasteniy Novogo Sveta [Mexico and Central America as basic centers of origin of cultivated plants in the New World] *Tr. po prikl. botan., genet. i selek.*, **26** (3)

Vavilov, N.I. (1931b). Rol' Tsentral'noy Azii v proiskhozhdenii kul'turnykh rasteniy [The role of Central Asia in the origin of cultivated plants]. *Tr. po prikl. botan., genet. i selek.*, **26** (3).

Vavilov, N.I. (1935). *Botaniko-geograficheskie osnovvy selektsii rasteniy*, [The phytogeographical basis for plant breeding]. Moscow, Leningrad.

Vavilov, N.I. and Bukinich, D.D. (1929). *Zemledel'cheskaya Afghanistan* [*Agricultural Afghanistan*]. Moscow, Leningrad.

Vavilov, N.I. *et al.* (1931). Pshenitsi Abisinii i ikh polozheni v obshchey sisteme pshenits [The wheats of Abyssinia and their position within the total system of wheat]. *Tr. po. prikl. botan., genet. i selek.*, Suppl. **51**.

Zhukovskiy, P.M. (1933). *Zemledel'cheskaya Turtsiya* [*Agricultural Turkey*]. Moscow, Leningrad.

The important agricultural crops of pre-Columbian America and their mutual relationship

THE AUTONOMOUS DEVELOPMENT of the ancient civilizations in southern Mexico, Central America, Peru and Bolivia, which were very successful within the areas of fine arts, science, technology and rural economy, independently from the Old World, is an amazing geographical and historical fact the significance of which should no doubt still be contemplated within the context of world history.

All established and indisputable facts clearly demonstrate that Man arrived in the New World relatively recently and mainly by a northern route, bringing him first into North America and later on into Central and South America. New archeological and anthropological research by Hrdlicka and others in Alaska and on the Aleutian Islands has established repeated passages of people from Asia at least 2000 to 3000 years before the present. There is an interesting paper, written in short-hand by Dr. Hrdlicka at the Ethnographical Institute of the Academy of the Sciences of the U.S.S.R. and dated June 14, 1939, on the topic of the 'Anthropological and Ethnographical relationship between America and Asia'. In all likelihood, people came to the American continent in canoes made of skin (the usual means of transportation in coastal areas in the past). The link between the ancient migrants from Asia and the present type of Indians has been established. The epithet 'Indian' was first used by Columbus, who, when he had already returned to Europe, did not suspect that he had discovered a new continent but believed that the land found was only a short distance from India. As is well known, the American continent was already to a great extent settled by Indian peoples when discovered by the Europeans. The most competent and careful investigators, like Hrdlicka, consider it likely that the period of initial immigration occurred between 5000 and 15 000 years before our time. Up to that time America was unknown to Man. Both archeological and anthropological documents clearly bear witness to this, as does the absence of anthropoids or remains of fossilized, primitive peoples in America. From Asia, Man brought only the dog, which is indicated by finds of fossilized remains from pre-

Dedicated to the President of the Academy of the Sciences of the U.S.S.R., Academic V.L. Komarov.

First published in *Izd. Gos. Geogr. O-va* [*Publications of the National Department of Geography*], vol. 71 (10).

Columbian America; Man was evidently unable to bring with him any other domesticated animals or cultivated plants.

When the author of this paper traveled through North, Central and South America between 1921 and 1933, while carrying out the main task set before him of collecting varieties of the plants resources for Soviet plant breeding purposes, he had an opportunity to make acquaintance *in situ* with the material commonly used for agriculture and the plants cultivated by peoples who had settled both subcontinents before they were discovered by Europeans.

In 1921, this author began studying the agricultural crops preserved on the Indian reservations in the northern states of North America. Such crops have also been excellently described in a book by Will & Hyde in 1917. Later on, in 1932, the author was able to visit isolated Indian villages in northern Arizona together with the well-known botanist and ecologist, Dr. H. L. Schantz, at that time president of the University of Arizona.

In 1930, the author had an opportunity to become thoroughly acquainted *in situ* with the agricultural crops of Mexico, Guatemala and British Honduras, and in 1932–33, with those in Yucatan and the Central American republics and also with those of Peru, Bolivia, Chile, Ecuador, Argentina and Brazil (cf. Figs 1 and 2).

In addition to his personal observations of the relics of ancient agriculture in North and South America and his research on the endemic plants that are cultivated in the New World, the author also studied the extensive amount of material of ancient Indian crops that is collected in the museums in New York, Chicago, San Francisco, Mexico City, Lima, Rio de Janeiro, London, Paris and Berlin, where an extensive amount of material concerning the crops of ancient peoples is accumulated.

The important literature on the problem in question, which is unfortunately very scattered, was also utilized to a great extent by the author. Among these books, the old but useful review by Max Steffen (1883) and the voluminous work by Kaerger (1909) should especially be mentioned. Among the more recent works, the book by Carrier (1923), the valuable works by the Argentinian botanist Parodi (1935) and the book by the German geographer Karl Sapper (1936), as well as that by the Chilean archeologist Latchan (1936), should be cited.

As far as the agricultural crops of Peru and Bolivia are concerned, the papers by Cook (1916, 1925) are especially interesting. Both these papers are accompanied by excellent illustrations of the plants endemic in Peru, as well as of the methods used for their agro-technology. The bibliography, assembled and published in 1932, by E.E. Edwards and the library of the U.S.D.A. in Washington, concerning the agriculture of the American Indians, referring to all the important papers published up to June 1932, also deserves to be mentioned.

A large and valuable material for the study of Indian agriculture in North America is also found in different publications by the state universities, in the publications of the American Library of Congress and in American journals such as *Geographical Review*, and *The National Geographic Magazine*, but also in

Fig. 1. Stone tool [metate] for grinding flour, Yucatan. Photo: N. I. Vavilov.

publications by the Smithsonian Institute in Washington (especially those by the Bureau of American Ethnology) and, in general, in books dealing with the ancient civilizations of the New World. Among the latter, we can mention works by Bastian (1876–1884), Wissler (1922–1924), Kroeber (1923, especially vol. XIII), Spinden (1928), Merrill (1929) and the 'American Aborigines' (1933).

It should be stated that, as far as the studies of the ancient crops of the New World are concerned, the state of the published material is considerably better than with respect to that of the Old World, except for such areas as Egypt, to the agriculture of which a number of monographs have been devoted.

The present essay is intended as a review of my observations and investigations concerning the agriculture of pre-Columbian America, the endemic plants cultivated there and the mutual relationship between the centers of agriculture in North and in South America.

The enormous territory of North and South America has, with respect to the

Fig. 2. 'Foot plows' for working the fields such as widely used in the high-altitude areas of Peru and Bolivia. Photo: N. I. Vavilov.

historical development of agricultural crops, revealed a striking geographical localization during the not very distant past.

While the nomadic economy was associated with the hunting of deer and bison (in N. America), and guanaco (in S. America), even during the time, in the distant past, when the hunting and gathering of many wild plants and fishing for food was already widely practiced on both the American subcontinents, the development of permanent agricultural civilizations, which left an amazing amount of documentation about the persistent work and genius of the ancient peoples, reveals a *striking affinity first and foremost between two limited territories: on the one hand the southern, mountainous areas of Mexico and the adjacent parts of Yucatan and Central America, and, on the other hand, the areas of Peru, Bolivia and those contiguous with them in the mountains of Ecuador, but also a relatively isolated one in southern Chile.*

In his interesting review dealing with the Indians of Central and South America, Wissler distinguished the following 12 areas which, at the time when America was discovered by the Europeans, corresponded to the main kinds of food used by their inhabitants.

1 The area of North America adjacent to the Great Lakes, where the nomadic Indian peoples existed mainly by hunting deer and collecting fruits and roots of edible plants; Wissler calls this territory *the area of the deer hunters.*

2 The prairie area of North America, where the Indian peoples lived mainly by hunting bison and gathering wild plants; Wissler named it *the bison hunter's area.*

3 The northwestern and norther portion of California, where people existed mainly on fisheries (there is a large number of fishes, e.g. salmon and herrings, in the rivers and the sea) but also by hunting as well as gathering fruits, nuts and mushrooms; Wissler called it the *area of the salmon and herring fishermen.*

4 The area east of the Mississippi and south of the Great Lakes, where they lived by hunting deer, turkeys and, in part, bison, and by collecting roots and utilizing wild rice (*Zizania aquatica*) and sunflowers [*Helianthus* spp.]. According to Wissler, this is the *area of the hunters and gatherers of wild plants.*

5 The peoples of *the Californian–Texan area* with its rich vegetation existed mainly by collecting fruits and bulbs and using tortoises, snakes, fishes, insects, ants and termites as food.

6 Further, Wissler distinguished the agricultural subtropical and tropical areas embracing the mountains of Mexico and Central America, where the people lived mainly on different plants. This was *the main area of the maize growers.*

7 The peoples of the *West Indian Island area* existed mainly by fishing and gathering mussels, hunting and, in part, on agricultural crops.

8 The peoples of the enormous territory of tropical South America, concentrated mainly in the Amazon basin, lived to a great extent by cultivating manioc, (*Manibot aipi* and *M. utilissima*), ground nuts [*Arachis hypogaea* L.] and *Canna edulis.* A large quantity of fruits and plants were also used here but hunting wild animals also occurred. This is *the main area of the manioc growers.*

9 The high-altitude areas of Peru and Bolivia, with people existing mainly

on potatoes and other tubers as well as on goose-foot species such as quinoa and cañahua [*Chenopodium quinoa* and *C. canihua*], as well as the meat of domesticated llamas and alpacas. This was *the main area of the potato growers.*

10 *The island of Chiloë* in the southern part of Chile (Araucania), where they existed on crops of potatoes (*Solanum tuberosum*), the recently almost-extinct species of 'mango chil' (*Bromus mango*) and by collecting the pine nuts of araucarias.

11 *The area of Gran Chaco* and the adjacent areas of South America east of the Andes, characterized mainly by hunting of guanacos, a species closely related to the llama, and the American ostrich [*Rhea*], as well as by collecting the seeds, berries, tubers and nuts of wild plants. Extracts of the leaves of the wild Paraguayan tea (*Ilex paraguayensis*) served as a stimulating beverage. Wissler called this the *area of the guanacos.*

12 The ancient people of *Tierra del Fuego* and the adjacent Pacific coast existed by fishing, collecting algae, molluscs and sea-urchins.

Mexico and Central America were the main agricultural areas in North America; in South America, these were found mainly in Peru and Bolivia. All the statements by the conquistadors bear witness to this: these agricultural centers were already distinguished during the sixteenth century by their concentration of large settled populations and their development of important crops. Remains thereof are preserved in the form of amazing pyramids, ruins of temples and astronomical observatoria, city-like structures and a colossal amount of ceramic material. Peru is especially rich in ceramics which, up to the present, could be found in 'living museums' where a limitless amount of ceramics was preserved harmlessly for thousands of years under desert conditions; they now fill the museums of Europe. These ceramics are of special interest to us since some of the vessels depict the endemic plants and animals of Peru in the form of sketches and shapes.

The first ancient centers of agriculture in the New World, i.e. Southern Mexico and Central America, were relatively thoroughly described in the paper about 'Mexico and Central America as the Basic Center of Origin of Cultivated Plants in the New World' (Vavilov, 1931).

Additional research, conducted in 1932 both in the republics of Central America and on the Yucatan, and during a visit to the farming villages around the famous ruins of Chichen Itza, enhanced for me even more the exceptional importance and the originality of the crops concerned, and their complete independence.

The *absence of animals within the rural economy stands out* as the basic factor characterizing the ancient Mexican agricultural civilizations associated with the Mayan people and the Aztecs, Zapotecs and others that branch off closely from them; this is in contrast to what was common in the Old World. The great civilization of the Mayas knew only domesticated turkeys (*Meleagris gallopavo*) and a particular white goose (*Anas boschas*). In the past all agricultural work was done by hand until the introduction of European draught animals. On the Yucatan peninsula, in the forested areas close to Chichen-Itza, and in a number of areas in Guatemala, Honduras and southeastern Mexico, it is still possible to

Fig. 3. Chayote (*Sechium edule*), endemic in Mexico and Central America. Photo: N. I. Vavilov.

observe the primitive methods of the past: the clearing of stony soil by hand and planting by means of a wooden stick. During the dry season, a greater or lesser portion (milpas) of the woody and herbaceous vegetation is burned and cleared off and later used for agricultural crops. The clearing of fields is done in the same primitive way, by means of wooden poles. Holes are then made in the ground with a digging stick and seeds of maize, squashes, beans and other plants put into them. Grains are stored in special granaries. Fruit-bearing plants such as guava [*Psidium* spp.], papayas [*Carica papaya*, etc.] and different kinds of sapotes [*Pouteria, Casimiroa*] and plants belonging to the genus *Annona* are also grown.

The rich flora of Central America and Mexico, which in numbers considerably exceeds that common to the extensive area of the present U.S.A., Canada and Alaska (cf. the estimate of the numbers of wild plants in various parts of North America in Vavilov, 1931), allowed the primitive peoples to take the valuable original material into cultivation, on the basis of which the particular and independent agricultural civilization of the Mayas developed. Cultivated endemic plants (Fig. 3), which do not appear beyond the limits of the territory in question, are still preserved there. These are, e.g.: the maguey (*Agave atrovirens* Karw.), which is used for making the beverage called pulque, and a number of fiber-producing agaves (*A. lechugilla* Torr., etc.); endemic species of squashes (*Cucurbita mixta* Pang.) and tomatillo (*Physalis aequata* Jacq.); the root-fruit called jicama (*Pachyrhiza angulatus* Rich.) and *Cacara edulis* Kuntze among the beans; cacomite (*Tigridia pavonia* Ker-Gawl, belonging to the family Iridaceae), often grown side by side with edible bulbs in Valle de Mexico; purple amaranth (*Amaranthus cruentus* L.), an endemic Mexican cereal especially characteristic of the Mexican high plateau; chia (*Salvia chia* Fern), an oil plant; the Mexican siruela plums or mombins (*Spondias mombin* L. and *S. purpurea* L.); the Mexican hawthorn (*Crataegus mexicana* Mocq. & Sessé); different kinds of cacti (*Cereus* spp., *Opuntia* spp., and *Nopalea coccinellifera* Salm-Dyck); and peppers (*Capsicum mexicanum* Hazenbusch as well as *C. annuum* and some forms of *C. frutescens*).

There is no doubt that maize was the main staple within the territory in question right from the beginning of agriculture. In all likelihood, the actual development of agriculture, and even all the settled civilizations in southern Mexico and the adjacent parts of Central America, were associated with the presence of the original wild strains of maize which, unfortunately, no longer exist or have not yet been discovered. The fact that the original native land of maize is found here is indicated mainly by all the well-established evidence in the form of weeds among the maize, which are especially widely distributed in the foothills north of Mexico City and in Guatemala. They belong to the wild species of teosinte (*Euchlaena mexicana* Schrad.) but also to a different species, closely related to maize, *Tripsacum* [gama grass], which is present there. Although the teosinte is morphologically very different from maize (with respect to the inflorescence), it produces more-or-less fertile hybrids when crossed with the latter. As regards the structure of the chromosomes, both genera have much in common. Their isolation into different genera is conditional. As demonstrated by the recent and thorough studies by Mangelsdorf and Reeves (1938), the teosinte may, during the distant past, have arisen from a hybrid between primitive forms of maize and species of *Tripsacum*. Teosinte is distinguished from maize by the inflorescence, which breaks up when ripening, (i.e. a typically wild characteristic) but at the same time by high productivity. To the north of Mexico City, we encountered crops of corn among which weedy teosinte amounted to ca. 30% of the total number of plants. Beadle (1939) has recently stated that teosinte, just like popcorn, pushes out the endosperm from the hard seed cover when the seeds are heated and, thus, the grain becomes edible. Proceeding from such facts, Beadle was inclined to believe that, during the course of a long period of time, the Indians selected mutant forms from teosinte that were close to the popcorn type of maize.

The exceptional diversity of the biological and morphological types of Mexican maize, which distinguish it all over the world, speaks in favor of the hypothesis concerning the origin of maize as a species of its own within this area, as does the original introduction of it into cultivation here. Both indurate, evert indentate and amylaceous types and a multitude of varieties with extremely long ears can be found, which can with difficulty be referred to the ordinary European types of cultivated maize. Some of the Mexican types appear to have characteristics that are artifically united from the various groups, which had become distinguished during the distant evolution of the maize.

In Mexico, maize reaches from sea-level to an altitude of 3150 m. It is grown both in the humid, tropical areas where the annual amount of rainfall exceeds 2000 mm and in the dry areas of northern Mexico and Yucatan. Correspondingly, the types of Mexican corn have differentiated into contrasting ecotypes.

The territory discussed is also exceptionally rich in wild and cultivated species of beans, such as *Phaseolus vulgaris* L., *Ph. multiflorus* Willd. and *Ph. lunatus* L. As we have demonstrated, there is a particularly striking diversity with respect to the seeds of the ordinary beans (*Ph. vulgaris*) in Guatemala.

A large number of species and types of squashes are also concentrated in Guatemala, such as *Cucurbita ficifolia* Bouché (syn. *C. melanosperma* Al.Brown), *C. moschata* Duch. and *C. mixta* Pang.

The variation of peppers (*Capsicum annuum* L.) in Mexico and Guatemala is amazing, as is that of chayote (*Sechium edule* Schwartz) and cocoa (*Theobroma cacao* L.). It cannot be denied that the sweet potato (*Ipomoea batatas* Poiret), too, originally came from Central America. Its wide distribution in cultivation among the Indians in Central America and southern Mexico and its enormous varietal diversity both in Central America and the West Indian islands indicate this.

The territory in question is no doubt also the native land of an important group of a cultivated species of cotton. A typical representative of this is the ordinary upland cotton, *Gossypium hirsutum* L., at present widely distributed in all the cotton-growing areas of the world. The best standards in the world of American cotton (Acala, Big Boll and Durango) originated from local, Mexican strains. On the Yucatan we observed a number of forms of wild cotton, which were close to the upland type. Cook observed a large number of small, endemic species. *Gossypium palmeri* Watt. was discovered in Central America and southern Mexico but the adjacent islands appear to be the main focus where the long-staple *G. barbadense* L. and *G. purpurascens* L. were taken into cultivation.

This territory is also typical for a multitude of endemic species of fruit-bearing plants such as avocados (*Persea schiedeana* Nees and *P. americana* Mill. – syn. *P. gratissima* Gaertn. –, etc.), guavas (*Psidium guajava* L. and other species), sapotes (white ones, *Casimiroa edulis* La Llave, and yellow ones, *Lucuma salicifolia* H.B.K.), perennial species of *Annona* (*A. cherimola* Mill, *A. muricata* L., *A. squamosa* L. and *A. reticulata* L., i.e. cherimoya, soursop, sweetsop and custard apple, resp.), mamey (*Calocarpum mammosum* (L.) Pierre) and papaya or melon tree (*Carica papaya* L.).

All these plants are to an equal extent originally associated with genesis within the area discussed. In the wild forests covering the territory of Mexico, Guatemala and the other Central American republics, wild fruit trees can be seen in abundance and are represented there by an almost unlimited number of types, connecting the presently cultivated forms with the typically wild ones. Wild and cultivated kinds of avocados, guavas, annonas, sapotes, mombins and hawthorns, which are distributed in southern Mexico and Central America, are represented by a full range of transitional types. You can see the process of type-formation of the majority of the endemic species in that area taking place right under your own eyes. (Lists of the cultivated endemic plants of Mexico are found in Vavilov, 1931.)

It is also interesting that the bees of the Mayas belong to the genus *Melipona*, in contrast to the domesticated bees of the Old World, which belong to the genus *Apis*. The *Melipona* species are distinguished either by the absence of a stinger or the presence only of a blunt one in the case of the female. These bees were called 'angelitos' by the conquistadors. The characteristic of the honey produced by these bees (*Melipona domestica* and *M. fulvipes*) is that it does not crystallize when stored for a prolonged period of time. Apiculture was first introduced by the Spaniards in 1518–19 on the Yucatan peninsula (Nordenskiöld, 1929; Sapper, 1935). During the time of the conquistadors, apiculture was famous in the states of Jalisco and Campeche. Beeswax was, however, not used for candles. In Chichen-Itza and the surrounding villages and at Merida we observed, even in 1932, peculiar and complex so-called multiple hives, constructed out of tree butts. In the past the Mayas had apparently used the wild honey of bees belonging to the *Melipona* species.

In addition, as is well known, the Mayas multiplied the coccineal (*Coccus cacti*, a scale insect producing a red dye) in great amounts. Therefore, copses of the cochineal cactus, *Nopalea coccinellifera* (L.) Salm-Dyck, were grown and protected in every way possible.

As can be seen, many of the endemic plants of Mexico and Central America constitute not only species but also genera. There can be no doubt that the agricultural crops of Mexico and Central America arose independently and without any influence from the Old World. However, although some of the investigators, who have already been mentioned above, still try to see an influence from the Great Egyptian civilization on the ancient art of Mexico and Yucatan (Smith, 1917, 1933), a comparative analysis of cultivated plants and the domesticated animals of the Old World and the New World demonstrates very clearly the independence and autonomy of the Mayan civilization. Alfonse De Candolle (1883) leaned toward this opinion. This hypothesis has been completely settled by one of the best contemporary scholars acquainted with the floras of North and South America, i.e. Dr. Merrill, who has devoted a number of interesting essays to this problem (Merrill 1931, 1933, 1934, 1938).

All the Mexican agriculture was essentially non-irrigated in the past. Before the time of the conquistadors, just as in the present, a type of vegetable gardening on floating islands called 'chinampas' existed just south of Mexico City. It was of limited importance, however, and was apparently used in the past mainly for growing vegetables and decorative plants. Irrigation was developed in northern Mexico only during the nineteenth century.

In the Yucatan, it is still possible to observe the primitive types of agricultural tools such as stones for grinding flour [metates] and wooden digging sticks. The fields in Yucatan, those on the chinampas south of Mexico City and those in Guatemala around Antigua, frequently carry a mixture of different cultivated plants: beans wind themselves around the maize and in between them grow different kinds of squashes. Mixed crops appeared to predominate in ancient Mexico just as it does to a great extend in the south up to this day.

The manually worked crops of the Mayas, like those of the Aztecs and Zapotecs, naturally had to be intensely managed. The absence of farm animals forced the people to limit the size of the fields to small plots, to carefully cultivate the small areas and to use special methods when caring for the plants such as, e.g. breaking open the spadix of the maize when the ears were ripening.

This was widely practiced in ancient Mexico and apparently is still done in Guatemala in order to promote early ripening and, possibly, to fight insects that damage the ears. Growing the plants on small plots made it possible to pay full attention to every plant. Indeed, the first 'plant breeders' may have arisen in this manner.

The cultivated flora of Mexico and Central America retains, even now, an aspect of originality in spite of a considerable influx of plants from the Old World. In essence, the ancient American endemic cultivated plants still predominate there.

When proceeding from the north, from the present United States to the Mexican high-plateau, the traveller sees before him a typically treeless landscape of agaves and cacti, the typical endemic plants of pre-Columbian America. The peculiarity of the landscape is enhanced by the widely distributed endemic weeds, covering wasteland and the borders of fields and represented by plants such as species of marigolds (*Tagetes*), zinnias (*Zinnia elegans* Jacq., *Z. mexicana* Cav. and *Z. multiflora* L.), and cosmos (*Cosmos bipinnatus*, etc.) but also species of the peculiar Mexican weed, *Tithania tubaeformis* Cass., reminding one at a quick glance of wild sunflowers.

Many local strains of maize, papaya, beans, fruit trees and cotton have attained great perfection here. In Guatemala, near Antigua, we collected local flinty maize, which is striking due to the dimensions of its ears (up to 40 cm or more in length) because of which it is preferably grown there.

The Mexican strains of upland cotton have still not been surpassed by breeding. A number of annonas, especially *Annona cherimola*, are distinguished by first-class taste and large dimensions.

When summing up the endemic, cultivated plants of the New World we arrived, in contrast to Cook (1925), at the opinion that the largest number of cultivated plants in the New World, especially the important ones such as maize, upland cotton and other long-staple species of cotton, beans, cocoa, fibrous agave and other plants, are linked geographically with the original centers and their initial introduction into cultivation particularly in Mexico and Central America. The introduction into cultivation was, apparently, undertaken mainly by the Mayas.

Besides cultivating plants, the ancient peoples inhabiting Mexico and the adjacent areas of Central America utilized a multitude of wild plants for various purposes. Janovski (1936) mentioned up to 1100 species of wild plants, belonging to 444 genera and 120 families, which were used to a variable extent by the Indians on the North American continent. It should be noted that in the mountains of Mexico, different kinds of agave were used instead of papyrus for making paper (Richter, 1938). In the lowland areas of Mexico, different species of fig (*Ficus*) were employed for the same purpose.

Continuing south from Mexico and Central America, passing through the forested tropical area of Colombia and the lowlands of Ecuador and penetrating into the high-altitude areas of Peru and Bolivia, the traveler enters a 'new world'. Again, everything appears strange: both the peculiar relief of the bare mountain plateau, covered by the Peruvian needle grass, *Stipa ichu* Kunth (syn.

S. jarava), and the wild and cultivated flora, as well as the definitely strange fauna. This is the so-called puna, a high-altitude Peruvian–Bolivian 'steppe', where a pre-Incan, so-called megalithic culture of Indians was concentrated. Although only slightly different from the Mayan one, this was a decidedly original civilization.

In contrast to Mexico and Central America, an original form of animal husbandry was developed in connection with the agriculture in the mountain areas of the Andes, the realm of the llamas and alpacas, those peculiar domesticated South American animals that are genetically related, although distantly, to the camels of the Old World. Externally, the alpaca (*Lama pacos* L.) reminds us of sheep. The llama (*L. glama L.*) is even stranger. In addition to these animals, the guinea pig (*Cavia porcella* L.) and a goose (*Anas moschata* L.) are domesticated here. In the high-altitude mountains of Bolivia, around La Paz, you can still encounter the vicuña (*Lama vicugna* Mol.) and the guanaco (*L. huanachus* Mol.), which are closely related to the llama. Among these, the guanaco is closer to the llama and can produce fertile hybrids with it. Attempts to domesticate the guanaco have been made recently (Barreda, 1936). Llamas and alpacas are represented by variable races, in particular as far as the fur and the meat are concerned. Breeds of alpacas that are sharply different with respect to the kind of fur and their constitution can be distinguished. Even now, when farm animals from the Old World, especially sheep, are widely spread in high-altitude Peru and Bolivia, the llamas and alpacas constitute a significant base for the rural economy. Alpacas are used exclusively for their fur and meat. The meat of the alpaca is particularly esteemed. The llamas are also used as pack animals, being able to carry loads up to 20–30 kilos. Neither llamas nor alpacas can be used for farm work. Only the male llamas are used for transporting loads. Neither of the two animal species is used for milk, although the female llamas produce a considerable amount thereof.

The geographical factors were of great significance when determining the localization of the basic agricultural centers of the New World, both in the case of Mexico and Central America as well as Peru and Bolivia. The foothills of Mexico frequently consist of open, treeless areas, which are easily accessible to people. The comparatively dry, wooded areas of Yucatan with its sharply distinguished dry season were also relatively accessible and easily mastered by an agricultural civilization. In this respect, the high altitude areas of the Peruvian and Bolivian steppes are even more suitable (i.e. the punas) than the humid tropical forests of the eastern slopes of the Andes, which are very difficult to penetrate because of their unmanageable vegetation (Troll, 1931; this is an extended report from the XXIVth International Congress of Americanologists, 1931, and has excellent maps of the cultivated zones of Peru and Bolivia). The lowland areas of the Pacific coastal belt of South America, starting in Colombia and encompassing a considerable portion of Ecuador, are represented by a typical subtropical mass of forest, which is difficult to penetrate. The low-altitude and foothill areas from Ecuador down to Chile are characterized by lifeless desert areas.

The puna turned out to be easily accessible, especially for the primitive farmers. Original and special Andean agricultural crops were developed there in open areas on comparatively rich soils and based on the existing wild flora, which is represented to a great extent by root-vegetables and tuber-forming plants. In addition to domesticating llamas and alpacas, the ancient farmers of high-altitude Peru and Bolivia took into cultivation, for the first time, a significant number of local tuberous plants, especially a number of potato species. The expeditions to Peru and Bolivia between 1927 and 1933 by the All-Union Institute of Plant Industry, discovered a great diversity there of species of wild and cultivated potatoes not known to Science but consisting of an entire system of species distinguished by their chromosome numbers as well as by their morphological and physiological characteristics. Species of potatoes were discovered on the puna that were amazing because of their cold-hardiness, e.g. *Solanum acaule* Bitt. ($2n=48$) and *S. bukasovii* Juz. ($2n=24$).

Within the complex of the cultivated flora, a number of cultivated species of potatoes were identified by S.V. Yuzepchuk and S.M. Bukasov. The following list of endemic species of potatoes was produced by S.M. Bukasov, taking all the most recent data into consideration:

1 *Solanum phureja* Juz. and Buk. – Bolivia (24 chromosomes);
2 *S. ajanhuiri* Juz. and Buk. – Bolivia (24 chromosomes);
3 *S. macmillanii* Juz. and Buk. – Bolivia (24 chromosomes);
4 *S. goniocalyx* Juz. and Buk. – Peru (24 chromosomes);
5 *S. stenotomum* Juz. and Buk. – Bolivia,Peru,Ecuador (24 chromosomes);
6 *S. caniarense* Juz. and Buk. – mountainous Ecuador (24 chromosomes);
7 *S. kesselbrenneri* Juz. and Buk. – mountainous Ecuador (24 chromosomes);
8 *S. tenuifilamentum* Juz. and Buk. – Peru and Bolivia (36 chromosomes);
9 *S. juzepczukii* Buk. – Peru and Bolivia (36 chromosomes);
10 *S. mamilliferum* Juz. and Buk. – Peru (36 chromosomes);
11 *S. chocclo* Juz. and Buk. – Peru, Bolivia, Ecuador (36 chromosomes);
12 *S. cuencanum* Juz. and Buk. – mountainous Ecuador (36 chromosomes); and
13 *S. curtilobum* Juz. and Buk. – Peru and Bolivia (60 chromosomes).

Solanum andigenum Juz. and Buk., the Andean potato, is the one most widely spread among the cultivated potatoes in Peru and Bolivia. This species is genetically very close to the ordinary potato, *S. tuberosum* L., and has the same chromosome number ($2n=48$), but it differs by a whole complex of morphological characteristics: less leafiness, a more open growth form and a short period of flowering. It forms tubers during short days. As a species, *S. andigenum*, as is well known, surpasses by far *S. tuberosum* in its resistance to phytophthora.

In addition to potatoes, strange tuberous plants such as ulluco (*Ullucus tuberosus* Loz.), oca (*Oxalis tuberosa* Mol. and *O. crenata* Jack.) and anu (*Tropaeolum tuberosum* Ruiz and Pavon) have been taken into cultivation.

Among the cereal plants, two species of goosefoot – quinoa (*Chenopodium quinoa* Willd.) and cañahua (*Ch. pallidicaule* Aell.) – are cultivated. The latter belongs to the puna and is locally most frequently cultivated in Bolivia and Peru. Also grown are an amaranth, 'Indian or Inca wheat' (*Amaranthus caudatus* L.), maca (*Lepidium meyenii* Walp.) and a lupine (*Lupinus cunninghamii* Cook); an excellent photo of this species is included in the paper by Cook (1925).

Apparently, the main species of pepper, *Capsicum bolivianum* Hazenbusch, was also taken into cultivation here. It is distinguished by a columnar inflorescence (rather than the rotate one of the ordinary pepper), small pods and small, brown seeds.

Until the arrival of the Europeans, the agriculture of the Andes was exclusively managed by hand. In the high altitude areas of Peru, as in the mountains of Ecuador, it was not irrigated. The open character of the punas of Peru and Bolivia offered opportunities for using considerable areas for agricultural purposes which, in turn, led to the invention of the peculiar 'foot-plow' [cf. Fig. 2], still widely used by the Indians in the mountains of Peru and Bolivia. The 'foot-plow' allows pressure to be applied on it via a wooded projection. In essence, this is neither a 'plow' nor a 'foot-plow', as the Americans call it, because it is operated by a single person, but it represents a step forward in comparison with the wooden digging stick, used in Central America and Mexico. At present the 'foot-plow' is made of iron; in the past it was entirely made of wood. The soil of the puna can be said to be exceptionally fertile; toward the south it is frequently alkaline.

In special literature dealing with the ancient agriculture of Peru and Bolivia, there is considerable confusion concerning the centers of agriculture in the New World. This applies particularly to the papers by Cook (1916, 1925), with respect to his evaluation of the independence and importance of the Peruvian–Bolivian center. As a result, there is a mix-up with later archeological strata that are found in the low-altitude and mid-altitude areas of irrigated agriculture in Peru.

The phyto-geographical and climatological data demonstrate clearly the necessity for a sharp delimitation between the ancient high-altitude agriculture on the puna, which is as a rule not irrigated and which is represented by the mountain steppes of Peru and Bolivia, and the historically later, irrigated agricultural crops of the desert and semi-desert areas on the western slopes of the Andes. The ancient agriculture of the mountain steppes definitely developed independently from that in Mexico and Central America during the pre-Incan period. Archeological data indicate this as well as the definitely independent, original flora and fauna. There are still mainly Indian inhabitants here and, in essence, the character of the agriculture and the entire rural economy preserve primitive traits. When driving northward from the capital of Bolivia, La Paz (at an altitude of 3854 m) to Lake Titicaca, situated on the border between Peru and Bolivia, the traveler sees before him a very unusual picture: on the lake float canoes, braided from a bulrush, *Scirpus riparius*; on the slopes toward the lakeshore, goosefoot, i.e. quinoa and cañahua [*Chenopodium quinoa* and *Ch. pallidicaule*) are grown just like cereals. There are also peculiar species of potatoes, the tubers of which reach enormous proportions, frequently weighing up to 1 kg each. In addition to the potatoes, there are crops of strange tuberous plants: oca [*Oxalis tuberosa* Mol.], ulluco [*Ullucus tuberosus* Loz.] and anu [*Tropaeolum tuberosum* Ruiz and Pavon]. Alpacas and llamas graze on the puna. Ancient, cyclopean stone structures can also be seen there.

Not only are the cultivated flora and the domesticated animals of the high-altitude areas of Peru and Bolivia strange, but also the methods of using them are different. Some species of potatoes are eaten raw. At high altitude a so-called 'chuño' [potato starch] is prepared from the potatoes: when harvested, the ripe tubers are allowed to freeze during the cold nights, wrung out by trampling them by foot and after a considerable time washed out for the separation of a dissolved, white substance which is then dried and used for food. This kind of product is distinguished by a particular durability and can be stored for several years.

In contrast to what is done in Mexico and Central America, the farmers of Peru and Bolivia apply manure in great amounts; the excrements of the llamas and the alpacas are used for this purpose. Apiculture was not practiced here in the past.

As can be seen, the world of cultivated plants and domesticated animals in Peru and Bolivia, just as in the adjacent areas of the Andes, differs sharply from that of Mexico and Central America. When comparing them, there is no doubt that they represent two different and important agricultural civilizations, which definitely developed independently of each other and on the basis of different species and genera of animals and plants. Entire complexes of agricultural crops bear witness to this fact.

It is necessary to distinguish the later, irrigated crops of the low-altitude and foothill areas of the dry subtropical and tropical Peru and Bolivia, associated with irrigation and canals dug during past centuries in the coastal desert near Lima itself, and the comparatively more humid areas adjacent to them with a milder climate and a subtropical vegetation (e.g. the area around Cuzco) from the agricultural crops on the puna with their tuberous plants and goosefoot cereals.

The coastal zone of Peru and Chile is distinguished by an extremely dry climate. In a number of areas the inhabitants do not see any rain for years on end; the poor vegetation of the stony hillocks (lomas) surrounding Lima can exist only because of the wet fogs and the dew. Agriculture is possible only thanks to irrigation.

The expansion of the high-altitude ancient agriculture in the region of the punas of Peru and Bolivia and their adjacent areas, in combination with the increase in population during the early development of the Indian civilizations, made it necessary to apply artificial irrigation and to move the farmers into the desert and semi-desert areas. The development of irrigation is concentrated particularly along the multitude of small rivers running parallel from the Andes down toward the Pacific Ocean and making a network of canals feasible.

The transfer into the low-altitude zones with a mild tropical climate widened the opportunities for agriculture to a considerable extent. Accordingly, a tropical and subtropical vegetation was directed into the low-lying areas: late-ripening strains of maize, the Peruvian tree-cotton (*Gossypium arboreum* L.) and tropical fruit trees like guava [*Psidium*], sapotes (*Pouteria* and *Casimiroa*) and papayas [*Carica papaya*]. Some of these crops were selected from the complex of the wild flora of the adjacent humid tropical and subtropical zones of Peru; others were apparently adopted from the Mayas in the north.

The period of irrigation in ancient Peru, associated with the construction of canals, roads and grandiose buildings, actually depended on the introduction of a number of more widely distributed plants from Central America and Mexico. From Quito, the present capital of Ecuador, to southern Chile (a distance of 5000 km) there was already a network of roads before the arrival of the conquistadors. Thus, the introduction from the north of such crops as maize, beans, cotton and different fruits was possible. The introduction as such was not automatic since it was conditional with respect to the low-lying areas, which are situated along the equator and distinguished by short days and the presence of irrigation and intensive cultivation, making it necessary to select original ecotypes and local strains as well.

When adopting the crops from the north, the farmers of Peru and Bolivia also applied many of their own original methods of using the plants. While maize in Central America and Mexico is used mainly in the form of 'tortillas' (baked, thin flat-cakes), in Peru and Bolivia it is preferably used in the form of a porridge, made from the grains. Some strains of maize in Peru, especially those known under the names of 'Cuzco' (which is also the name of the old mountain capital of Peru), produce amazingly large grains, surpassing by three or four times those of ordinary maize. Thus, Peru and Bolivia are distinguished by exceptional farinaceous and starchy strains of maize.

The Peruvian maize is represented by a great diversity of strains as far as its colors are concerned. It is used both in the form of entire ears or as boiled grains, corresponding to the selection of differently colored sets of ears. In the market places of La Paz, Arequipa and Cuzco one can frequently see how the Indian merchants and the farmers display a whole range of ears on their counters, according to color, just as it pleases them. The diversity of the Peruvian corn with respect to color has long attracted attention to this aspect.

Based on the diversity of the color genes, some American authors (e.g. Emerson, Campton and Collins) overestimated the importance of Peru as far as the origin of maize plants was concerned and considered Peru the true, native land of maize. In our opinion, this is incorrect. Comparative studies of maize in Central America, Mexico and Peru and Bolivia demonstrate a considerably more limited amplitude of the varieties of this plant in Peru. The Peruvian strains differ among themselves according to inadequately known taxonomic characteristics, colors and some other characteristics associated with large grains (e.g. a weak development of the ears and the ability of the grains to germinate). There are no such contrasts in the structure and shapes that are characteristic of the maize in Mexico and Central America. Nor are there any of the closely related wild plants such as teosinte, and Peru actually seems to be on the periphery of the distribution area of *Tripsacum*.

We do not see any data that could indicate decisively that the Mexican maize could have been adopted from Peru. All the data indicate rather the opposite, if you distinguish strictly the actually endemic crops of high-altitude Peru, where corn is absent, from those of the Peruvian low-altitude agricultural areas, which were subjected to extraneous influences.

It is very strange how the American scientists (particularly Cook), who repeatedly traveled to Peru and Boliva and studied Mexico and Central America, mixed up the historical strata of the agricultural civilizations. Usually they combined into a single unit the ancient, high-altitude, unquestionably independent and distinctly situated Peruvian–Bolivian crops with the irrigated crops of the low-altitude zone of Peru. The latter were a product of later influences and made use of cultivated plants of Central America together with wild flora from the humid, tropical areas on the eastern slopes of the Andes in Peru and Bolivia.

Of course, the very ancient, irrigated crops of the foothills, lowlands and subtropical and tropical areas of Peru and Bolivia are also of great interest since they reveal the high level of technology necessary for the construction of the mighty terraces and the intense type of cultivation that already existed before the arrival of the Europeans.

Below, we shall furnish some lists of plants which, in all likelihood, were taken into cultivation on irrigated soils from the complex of local floras or were adopted from the adequately moist areas of the eastern slopes of the Andes. With respect to a number of the plants listed, it is not impossible that they originated from introduced plants that were already cultivated in other areas.

ROOT VEGETABLES

Polymnia sonchifolia Poepp. and Endl. (syn. *P. edulis* Wedd.), llacon (fam. Compositae) – found in the Andes both in a wild and a cultivated state.

Xanthosoma sagittifolium Schott., malagna – possibly introduced to Peru from Central America.

Canna edulis Ker-Gawl., canna – other edible species of canna are found in cultivation on the Antilles islands.

Arracacia xanthorrhiza Bancr. (syn. *A. esculenta* DC.), arracacha or Peruvian carrot – distributed especially among the Chibcha people on the Bogota Plateau. In Colombia it reaches up to 2850 m. elevation; rare in Peru and Bolivia; a staple crop on the Bogota Plateau.

VEGETABLES:

Solanum muricatum Ait, pepino.

Lycospermum esculentum Mill. var. *succenturiatum* Pasq. and *L. peruvianum* (L.) Mill., tomatoes.

Cyclanthera pedata (L.) Schrad. and *C. brachybotrys* (Poepp. and Endl.) Cogn., caigua and achocha.

Cyphomandra betacea (Cav.) Sendt., tree tomato.

Physalis peruviana L., Peruvian cherry – mainly distributed as a weed on irrigated soils.

SQUASH:

Cucurbita maxima Duch, Hubbard squash – actually taken into cultivation from the eastern slopes of the Andes, and *C. andreana* Naud. – distributed in Uruguay and Argentina (Parodi, 1935, has a fine picture of *C. andreana*). Apparently the kind of *C. maxima*, which in America is called Hubbard squash, was taken into cultivation here from a basic, wild type.

SPICES AND STIMULANTS:

Capsicum peruvianum Hazenbusch, *C. pubescens* Ruiz and Pav., peppers – mainly used in a wild condition.
Tagetes minuta L., wild marigold.
Erythroxylon coca Lam., coca bush.

FRUIT CROPS:

Passiflora ligularis Juss. and *P. quadrangularis* L., passion fruits or granadilla and giant granadilla, respectively.
Carica pubescens (A. DC) Solms-Laubach and *C. candicans* A. Gray, papayas.
Lucuma obovata H.B.K., yellow sapote.
Psidium guajava L., guava – wild and cultivated as well as independently taken into cultivation in Mexico.
Annona cherimola Mill., cherimoya – also independently taken into cultivation in Mexico.
Inga feuillei DC., pacay.
Matisia cordata Humb. and Bonpl., chupa chupa.
Caryocar amygdaliferum Mutis, achotillo.
Guilielma speciosa Mart., peach palm.
Malphigia glabra L., huesito or Barbados cherry.
Solanum quitoense Lam., naranjilla or lulo.
Prunus capuli Cav. (syn. *P. capolina* Zucc.), capulin.

NARCOTIC PLANTS:

Nicotiana tabacum L., tobacco – according to the latest genetic research (by D. Kostov, S.A. Egiz and others) it is correct that this is the ordinary tobacco, which originated from a hybrid between the wild species *N. silvestris* and *N. rusbii*, both of which inhabit the south-American Andes, or some species close to the latter, followed by a doubling of the chromosome number (amphidiploidy), making this interspecific hybrid a fertile species.

Apparently also cotton, *Gossypium barbadense* L., was adopted from Central America in addition to maize, lima beans and ordinary beans. The cotton species is now represented by an ecological group of its own with large capsules, *G. peruvianum* Cav.

As far as the number of fruit crops such as guavas, sapotes, annonas and passion fruits are concerned, they are distributed in the wild form in the humid tropical and subtropical areas of both southern Mexico and Central America and the Pacific, coastal portion of South America. They were, therefore, undoubtedly taken into cultivation in different places.

Thus, within the borders of Peru and Bolivia we distinguish first of all a main center of high-altitude agriculture, i.e. on the puna and the adjacent sierras (mountains) and the cejas (the forested edges thereof) which have an original flora and fauna, and secondly, an additional center of vegetable gardening that developed both on the basis of crops imported from the north and the use of the wild flora of adjacent humid areas. The crops of the coastal deserts and the dry areas of the foothills and the mountains adjacent to them must not be considered as a basic but as an additional or secondary center. Among the endemic floras of the irrigated mountain areas, the coca bush (*Erythroxylon coca*), which was taken into cultivation in Bolivia and Peru, should particularly be mentioned. Its leaves were, and still are, used by the Indians as a stimulating product (the leaves are masticated). Such original species as papaya (*Carica pubescens* and *C. candicans*) are used in Peru.

When Cook, in his paper about 'Peru as a Center of Domestication' (1925), dwelt on the high level of agro-technology among the pre-Columbian civilizations of Peru, especially the terraced crops, he mixed up as a single endemic unit the crops produced both at the high-altitude and in the low-altitude vegetable growing areas and presented a long list (of more than 70 species) of plants which, in his opinion, were taken into cultivation within Peru. In his list many plants are incorrectly included, such as pineapple, ground nuts [*Arachis hypogaea*], bottle gourds (*Lagenaria*), beans and maize; the same plants also occurred under several different names. This list appears to be quite incorrect and not edited from a botanical point of view; the same species are repeated several times. There are also a number of definite mistakes with respect to the names. Thus, e.g. *Carica papaya* is a species endemic to Central America and Mexico; other species of *Carica* are typical of Peru (see above). In short, Cook arrived at a mistaken conclusion concerning the definitely exceptional role of Peru as a basic center of origin of cultivated plants in the New World with which he also associated the origin of maize and beans.

Already, during pre-Columbian times, the great agricultural civilization of the Mayas and the pre-Incan and Incan ones in the Andes of South America could not but affect the people settled over the extensive territory of the two continents. Maize was transplanted from its main area of origin into more northerly areas such as Arizona and the areas of the present states of Montana, North and South Dakota as well as into New Mexico, correspondingly changing into new forms and ecotypes due to the effects of natural and artificial selection. The primitive 'plant breeders' selected from it the fast-ripening strains, which have a vegetative period sharply different from the late-ripening Mexican ones, especially those grown at low elevation in tropical Central America and Mexico. On the basis of subsequent changes (mutations and hybridizations), a typical Arizonan ecological–geographical group of maize formed with long coleoptiles and distinguished by an exceptional drought-tolerance. Successively, types of dent corn, pod corn, flint corn, popcorn and sweet corn were developed by the Indian peoples settled within the area of the present U.S.A. On the Indian reservations in Arizona, far from the present-day civilization, corn grown according to a primitive agrotechnology can still be observed. Necessity forced the farmers in these dry regions to work out particular methods for intensive care of the maize plants: putting the seeds into deep furrows, repeatedly pushing up the soil around them or putting screens of stones at the side of the plants in order to protect them from the destructive effects of winds and drought.

When studying the crops of the Navajo tribe in northern Arizona in 1932, we collected innumerable types of flint corn, differing by the colors of its grains, ranging from dark blue over red, pink and mottled to yellow and white.

Apparently the primitive 'plant breeders' used to pay attention – they still do – to the colors of the ears.

Especially drought-resistant types of tepary beans (*Phaseolus acutifolius* A. Gray) were apparently originally selected from the wild flora within the borders of Arizona and Texas. Special, fast-ripening forms of squash and ordinary beans were also produced there. The Indians in North Dakota and other states also took particular types of squash into cultivation such as the 'Hubbard' squash, which belongs to *Cucurbita maxima* Duch., as well as the pumpkin, *C. pepo* L. The wild sunflower (*Helianthus annuus* L.), still an abundant weed on wasteland and along the roadsides of the U.S.A. and parts of southern Canada, was brought into cultivation and subjected to selection by some of the northern tribes of Indians together with the utilization of the so-called Jerusalem artichoke (*H. tuberosus* L.). It seems that these plants were taken into cultivation somewhere on the prairies along the border between Canada and the U.S.A.

Agriculture was only an additional occupation of the northern Indian tribes, as shown by a study of the materials used for the crops; their main occupation was hunting bison and deer. Fishery and the collection of shellfish played a great role along the coasts of both the Pacific and the Atlantic but also around the Great Lakes and along the great rivers. The Indians also collected large amounts of wild plants such as wild rice (*Zizania aquatica* L.), as they still do. Here and there new species of wild plants were incidentally taken into cultivation such as, e.g., the squashes. On the whole, however, only an insignificant number of plants were accultivated by the Indians within the territory of the northern U.S.A. and southern Canada.

It is still possible to clearly see the exceptional importance of southern Mexico and Central America as a basic center of origin and introduction into cultivation of the endemic plants grown in North America. Essentially, the peoples of the extensive continent mainly achieved a selection and distinction of the corresponding ecological types of maize, beans, squashes and cotton. Even during modern times, within the era of the Europeans, a number of nut-producing plants (e.g. black walnuts and pecans) and a large number of decorative plants, endemic in North America such as sweet acacia, *Acacia farnesiana* (L.) Willd., the tulip tree, *Liriodendron tulipifera* L., American magnolias and the honey locusts, *Gleditsia triacanthas* L., as well as American ash trees, *Fraxinus* spp., have been taken into cultivation and are now widely used for artificial plantations of forests and shelter belts all over the world, including the U.S.S.R.

Within South America, the agricultural crops were, in the past, associated mainly with the narrow belt of the Andes, but already during pre-Columbian time, they spread to southern Chile and gave rise to the peculiar but fairly limited Araucarian civilization concentrated in southern Chile and on the island of Chiloë. Under the conditions of long daylight at 38–40° S. latitude and a mild climate with a significant amount of rain, our ordinary potato (*Solanum tuberosum* L.) was taken into cultivation from among the species of wild types. Subsequently, it spread from there into the European countries. In contrast to *S. andigenum* Juz. and Buk., which is a typical short-day plant, *S. tuberosum* forms tubers in the normal manner during long daylight. In Araucaria, such oil-producing plants as the madia or Chilean tarweed (*Madia sativa* Mol.) and the peculiar mango chil (*Bromus mango* E. Desv., which is no longer grown) were also taken into cultivation. The distinctive conifer, *Araucaria imbricata* Pav. was and is still of great nutritional benefit [in the form of pine nuts; D.L.].

The civilization of the Chibcha tribe on the Bogota plateau in Colombia is represented by one of the smallest agricultural centers, situated apart from the Andean range. The language of the Chibchas is clearly different from that of the peoples inhabiting Central America, Peru and Bolivia.

The crops of the Chibchas consisted in particular of plants adopted from Peru and from Central America. From there came species such as the Andean potato, *Solanum andigenum* Juz. and Buk.; at the same time a number of potato species, e.g. *S. rybinii* Juz. and Buk. and *S. boyacense* Juz. and Buk., were independently taken into cultivation here from the wild flora. Both these species are characterized by $2n = 24$ chromosomes instead of $2n = 48$, which is typical of the ordinary potatoes (*S. tuberosum* and *S. andigenum*). The arracacha (*Arracacia xanthorrhiza* Bancr.) is especially widely grown here. It is a tuberous plant, possibly independently taken into cultivation here.

Apparently, a special kind of pepper, *Capsicum columbianum* Hazenbusch, was also brought into cultivation here. It differs from the Mexican pepper by ripening later but also by the structure of the seeds (the presence of a thin rim) and by being difficult to cross both with ordinary cultivated peppers and other kinds thereof. The coca bush was also introduced from Peru and Bolivia and from Central America came maize, represented by flinty types.

While reviewing the data concerning the civilization of the Chibchas in Colombia, Kroeber (1923) considered it an intermediate link between the Central and the South American ones. 'It does not', Kroeber (1923, p. 281) writes, 'have a single element of any exclusive or characteristic significance'.

The Indian tribes spreading out along the valleys of the great rivers (Orinoco and the Amazon as well as Paraña) into the heart of the South American continent found among the rich flora of Brazil enough edible plants, roots and fruits. While mainly occupied by hunting, they also gathered seeds, fruits, roots and tubers of the wild plants and took, when settled, some plants thereof into cultivation. Those most important were two kinds of manioc: the non-poisonous *Manihot aipi* Pohl. and the poisonous one, *M. utilissima* Pohl.; the latter requires time-consuming boiling of the roots. Manioc is now one of the most important cultivated crops in tropical South America and replaces potatoes there. The groundnuts [peanuts], *Arachis hypogaea* L., and the pineapple, *Ananas comosus* (L.) Merr., were also taken into cultivation there.

In essence, these are all the pre-Columbian crops of Brazil and the interior, tropical South America. Later on, during the European era, the rubber tree, *Hevea brasiliensis* Muell., gained great importance. Plantations thereof have mainly been located outside America, in tropical Asia, where they were grown from seeds gathered in the Amazon basin. During the last couple of decades, the Paraguayan tea, the maté, *Ilex paraguayensis* A. St. Hil., has been taken into cultivation as have been a small number of fruit- and nut-producing plants such as the pitanga, *Eugenia uniflora* L., the uvalha, *Eu. uvalha* Cambers, the grumichama, *Eu. dombei* (Spreng.) Skeels, and the cabellioda, *Eu. tomentosa* Cambers, the jaboticabas, *Myricaria jaboticaba* Berg and *M. cauliflora* Berg, the feijoa, *Feijoa selloviana* Humb. and Bonpl., the passion fruits, *Passiflora edulis* Sims, and the brazil nut, *Bertholletia excelsa* H. & B. (still growing mainly in the Amazon basin).

It should be mentioned that the main endemic crops of tropical South America, manioc, groundnuts and pineapple, are in essence plants of the arid or relatively dry, open tropical and subtropical zones. The areas where these plants are distributed were more accessible to the primitive people than the humid tropical forests with their overwhelming vegetation.

The Amazon basin has an exceptionally rich tropical flora, characterized by hundreds of species of palms and thousands of fruit- and nut-producing species. It still remains in a primitive condition although in the 'Sea of the Amazon' it is even possible to encounter military posts. Even the brazil nut tree, *Bertholletia excelsa*, which is unique in the plant kingdom because of its enormous dimensions and its gigantic fruits, is used mainly in the wild state here. In 1933, we were able to study the heaps of wild fruits brought in from the neighboring forests to the markets of Para-Belem in the estuary of the Amazon. As a rule, these fruits surpass our ordinary European crops of fruit in taste and are all edible. Hundreds of species of plants, including the native species of cacao, *Theobroma grandiflora* (Willd.) Schum. (syn. *Guazuma grandiflora* G. Don), and other species of which the pulp, or the mass in which the seeds are buried, can be eaten, are to a variable extent used by the Indians and the black settlers along the Amazon river.

Before the Europeans arrived, there were no native cultivated plants in Patagonia, Argentina or Uruguay. A number of species from the humid tropical slopes of the Andes, and mainly species of the quinine tree (*Cinchona succirubra* Pav. and *C. calisaya* Wedd., the native land of which is in Peru, Bolivia and Ecuador) became objects of cultivation only during the second half of the nineteenth century, both in South America itself and in tropical Asia. Seeds of the quinine trees in South America initially gave rise to the Dutch monopoly of this crop in Java.

The above sketch is a picture of the agricultural civilizations of North and South America before the arrival of the Europeans such as drawn by us as a result of a synthesis of the extensive material obtained according to instructions. It is still possible to trace in detail the original centers of the agricultural civilizations and to establish the native lands of the most important cultivated plants in South and North America. As can be seen from this essay, it is still possible to study the primitive civilizations of the Indians *in situ* within the area of the Andes in Bolivia, Peru and Ecuador, in Guatemala, Honduras and on Yucatan. The Europeans still enjoy the results of the efforts of the primitive farmers of North and South America.

It is now possible to trace the geographical evolution of the most important cultivated plants within the limits of the New World and also to begin to master the varietal and specific resources of North and South America, of interest for our own crops. As a result of these discoveries, we have an enormous, original varietal material on hand of the crops most important to us, such as potatoes, maize, cotton, beans, and so on.

Concerning what Mexico and Central America have presented to the world, we can particularly mention maize, at present covering more than 90 million hectares in all kinds of countries, and cotton (mainly the Mexican upland and the Central American long-staple species), now cultivated over more than 40 million hectares of land. The Mexican upland cotton ('Acala') has served as the original material for a number of the most important strains bred within the U.S.S.R.

The Andean center of agriculture also turned out to be of considerable importance. The crops of potatoes, which began to be cultivated in the Andes, presently fill more than 20 million hectares all over the world. The annual harvest of potatoes exceeds in value that of all the stores of gold and silver taken from Peru and Bolivia by the conquistadors, who, in addition to taking the bejewelled metals, also destroyed the major part of the native populations themselves. According to estimates by Americans, 4/7th of the value of the farm products in the U.S.A. at its present worth, consists of crops adopted by the Europeans from the Indians (Edwards, 1933).

The discoveries by the Soviet expeditions with respect to new species of wild and cultivated potatoes are presently used for practical plant breeding both in our own country and beyond its borders. Swedes, Germans, Englishmen and Americans followed in the footsteps of the Soviet expeditions (Bukasov, 1930, 1933a, 1933b; Bukasov and Sharina, 1938).

Apparently the primitive Indians already used as crops all the important edible plants from the complex of the wild flora in the New World. Unlimited horizons are, however, still open with respect to opportunities for utilizing the technical crops in a wide sense. The flora of Brazil alone is estimated at approximately 40 000 species by the German botanist Hoehne. To master this flora, botanists need the assistance of biochemists and technologists.

The scientists of our time must again turn to the rich flora of the New World. There can be no doubt that North and South America, including the ancient centers of agriculture in Central America and southern Mexico, will in the future present many more agricultural crops to the world.

BIBLIOGRAPHY

Barreda, C.A. (1936). La domestication de la vicuna. *Rev. zootech*, Buenos Aires, **7** (236).

Bastian, A. (1876–1884). *Die Kulturländer der Alten America*, [*The civilized countries of ancient America*]. I–III. Berlin.

Beadle, G.W. (1939). Teosinte and the origin of maize. *J. Hered.*, **30** (6).

Bukasov, S.M. (1930). Vozdel'vayemye rasteniya Meksika, Gvatemala i Kolumbia [The plants grown in Mexico, Guatemala and Colombia]. *Tr. po prikl. botan., genet., i selek.* [*Papers on applied botany, genetics and plant breeding*], Suppl. **47**.

Bukasov, S.M. (1933a). Kartofeli Yuzhnoy Amerike i ikh selektsionnoe znachenie [The potatoes of South America and their importance for plant breeding]. *Tr. po prikl. botan., genet. i selek.*, Suppl. **58**.

Bukasov, S.M. (1933b). *Revolyutsiya v selektsii kartofelya* [*Revolution in the breeding of potatoes*]. Leningrad.

Bukasov, S.M. and Sharina, N.E. (1938). *Istoriya kartofela* [*The history of the potato*]. Moscow.

Carrier, Lyman (1923). *The Beginnings of Agriculture in America*. New York.

Cook, O.F. (1916). Staircase farms of the ancients. *Nat. Geogr. Mag.*, **May**.

Cook, O.F. (1925). Peru as a center of domestication. *J. Hered.*, **16** (2).

De Candolle, A. (1883). *L'origine des plantes cultivées* [The origin of cultivated plants]. Paris.

Janovksi, E. (1936). Food plants of the North American Indians. *U.S. Dept. Agric., Misc. Publ.*, **237**.

Kaerger, K. (1909). *Landwirtschaft und Kolonization im Spanischen Amerika*, [*Agriculture and colonization in Spanish America*], Vol. 2. Leipzig.

Kroeber, A.L. (1923). *Anthropology*. New York.

Latchan, R.E. (1936). *La agricultura precolumbiana en Chile y los paises vecinos*, [*Precolumbian agriculture in Chile and adjacent countries*]. Santiago.

Mangelsdorf, P.C. and Reeves, R.G. (1938). The origin of maize. *Proc. Nat. Acad. U.S.A.*, **24** (8).

Merrill, E.D. (1931). The phytogeography of cultivated plants in relation to assumed pre-Columbian Eurasian-American contacts. *Am. Anthropol.*, **33** (3).

Merrill, E.D. (1933). Crops and civilizations. *J. Am. Mus. Nat. Hist.*, **33** (3).

Merrill, E.D. (1934). The problem of economic plants in relation to man in pre-Columbian America. *Proc. 5th Pacific Sci. Congr.*

Merrill, E.D. (1938). Domesticated plants in relation to the diffusion of culture. *Bot. Rev.*, **4** (1).

Nordenskiöld, E. (1929). L'apiculture indienne [Indian apiculture]. *J. Soc. Am. de Paris, Novelle Ser.*, **21**, (1).

Parodi, L.R. (1935). Relaciones de la agricultura prehispanica con la agricultura Argentina actual [Relations between the pre-hispanic agriculture and the present Argentinian agriculture]. *Anal. Acad. Nac. Agr. Vet. de Buenos Aires*, **1**.

Richter, Oswald (1938). Untersuchungen an Papiere aztekischer Völker aus Kolumbischer und vor-Kolumbischer Zeit und über chinenziche, türkische, buddistische, soghquische und andere Papiere aus der Turfanfunden, [Investigations of the papers of the Aztec people from Columbian and pre-Columbian times as well of the Chinese, Turkish, Buddhist, Soghish and other papers from the Turfan collection]. *Faserforschung*, **13** (2).

Sapper, Karl (1935). Bienenhaltung und Bienenzucht in Mittelamerika und Mexico [Beekeeping and apiculture in Central America and Mexico]. *Ibero-Am. Arch. Berlin*, **9** (3).

Sapper, Karl (1936). *Geographie und Geschichte der indianischen Landwirtschaft*, [*Geography and history of the Indian agriculture*]. Hamburg.

Smith, G.E. (1917). The origin of the pre-Columbian civilization in America. *Science*, **15** (1158), March 9th.

Smith, G.E. (1933). The influence of ancient Egyptian civilization in the East and in America. In *The making of man*.

Spinden, H.J. (1928). Ancient civilization of Mexico and Central America. *Am. Mus. Nat. Hist.*, New York.

Steffen, Max (1883). *Die Landwirtschaft bei dem Altamerikanischen Kulturvölkern* [*The landscape of the ancient American civilized peoples*]. Leipzig.

The American Aborigines, their Origin and antiquity (1933). In A collection of papers by ten authors, assembled and edited by D. Jennes. *Publ. 5th Pacific Sci. Congr.*, Toronto.

Troll, C. (1931). Die geographischen Grundlagen der Andiden Kulturen und des Incareiches [The geographical basis for the Andean civilizations and that of the Inca realm]. *Ibero-Am. Arch., Berlin*, **5** (3).

Vavilov, N.I. (1931). Meksika i Tsentral'naya Amerika kak osnovoy tsentr proizkhozhdeniya kul'turnykh rasteniy Novogo Sveta [Mexico and Central America as the basic center of origin of cultivated plants of the New World]. *Tr. po prikl. botan., genet. i selek.*, **26** (3).

Verrill, Hyatt (1929). *Old civilizations of the New World*. Indianapolis.

Will, George F. and Hyde, E. (1917). *Corn among the Indians of the Upper Missouri*. St. Louis.

Wissler, Clark (1922). *The American Indian*, 2nd edn. New York.

Wissler, Clark (1926). *The relation of nature to man in aboriginal America*. New York.

Reply to the article by G. N. Shlykov, 'Formal genetics and consistent Darwinism'

THE LATEST ARTICLE by G.N. Shlykov, 'Formal Genetics and Consistent Darwinism', published in the journal *The Soviet Subtropics*, nos. 8–9, 1938, is unparalleled as far as misinformation is concerned.

At the start of the article, this author states that the geographical theory about the origin of cultivated plants, in short, 'the theory about centers of origin', which was launched by myself and on the basis of which the Institute of Plant Industry, exceptional with respect to the range of its work, was set up, has been a 'grandiose fiasco'.

During the work of searching for species and varieties of cultivated plants to improve our own strains by providing them with drought-tolerance and disease resistance, I based my work on Darwin's theory of evolution. According to this theory, the development of organisms and alterations to them proceeded both in space and time from special natural-historical areas with a concentration of original genera and species, just as phytogeographic data indicate. The very epithet, 'center of origin of species', is taken from Darwin, who considered the localization of centers of origin of species to be very important (cf. the 1937 translated edition of his vol. XII).

The concept of the 'origin of cultivated plants' was coined by me in 1926 in the book *Centers of Origin of Cultivated Plants*. In 1926, I received the Lenin prize, just established at that time, for this scientific paper. This book and the development of the ideas in it met with exceptional approval in agronomical and botanical literature all over the world, the more so since in this book an enormous amount of material that had already been obtained by the work of Soviet scientists, was concentrated in a new form.

In agreement with the ideas held by myself and my colleagues and also with those of scientists from other countries, extensive research and expeditions were undertaken, which far from resulted in any 'fiasco', but led to discoveries of a new specific and varietal wealth, of which Science in the past was unaware.

As far as the important cultivated plants of interest to the Soviet Union are concerned, what is accepted within botany with respect to these geographical centers of origin, which have so extensively engrossed us, is not some kind of a 'dot on the map' but major areas, displaying a multitude of new species.

Therefore, Soviet expeditions (led by Dr. S.M. Bukasov and Dr. S.V.

First printed in *Sov. Subtropiki* [*The Soviet Subtropics*], no. 6, pp. 54–6, 1939.

Yuzepchuk) were sent out on the basis of my theory to the mountain areas of South America and Mexico, which proved to have dozens of new species of cultivated and wild potatoes. Some of these possess characteristics that are very valuable to us (such as cold-hardiness, resistance to diseases and a high productivity of protein and starch). They are widely used at present for Soviet plant breeding.

New species of wheat have been discovered in Transcaucasia, Asia Minor and Abyssinia. The explorative work in the areas of the initial development of species and genera of cultivated plants led to the discovery of enormous varietal resources. Many practical and valuable characteristics of the species were revealed within areas such as India, Abyssinia, Afghanistan, Iran, western China, Mexico, Peru and Bolivia.

According to the suggestions by the VOKS [All-Union Society for Cultural Relations with Foreign Countries], we reported on these achievements not only in different countries (France, England, Denmark, Sweden and the U.S.A.) but also at international congresses. The Institute of Plant Industry enjoys wide acclamation not only within the U.S.S.R. but also far beyond its borders. The Departments of Agriculture in the U.S.A., England, France and other countries have repeatedly sent their students to learn from us.

Plant breeding all over the world is done on the basis of our results. One expedition after the other was dispatched from the U.S.A., England, Sweden and Germany to centers of origin of cultivated plants established by us. In all handbooks on plant breeding abroad, the discoveries made by Soviet scientists are considered basic within this field.

When directing our searches, we considered it absolutely certain that finding strains ready and entirely suitable for being widely taken into cultivation within the U.S.S.R was unlikely. The actual position of the ancient centers is at low latitudes and often in mountain areas with short daylengths and specific conditions different from those of our own. Therefore, all the exploratory work embraced the 'building material' we were able to collect, i.e. species and varieties with necessary but different characteristics that could then be used for hybridization by Soviet plant breeders.

If original material should somehow be required for plant breeding in order to take this or that species into cultivation, it is impossible to ignore the concept of the geography of the species or the fact that varietal and specific diversity actually exist and that these are basic facts of evolution.

Thus, the claims made by G.N. Shlykov concerning the 'grandiose fiasco' of my theory and the exploratory work performed by the Institute of Plant Industry are nothing but falsehoods.

New material of species and varieties in the form of thousands of samples are sent out annually from the Institute of Plant Industry to plant breeding stations.

At a conference of the Stakhanovites ['shock workers', i.e. workers of exceptional achievement in their fields] at the Kremlin, Academician T.D. Lysenko evaluated the work with the introduction of plants as 'enormously useful' (cf. the report of the conference published by the State Publishing House of Agricultural Literature, Journals and Posters).

In 1927, I published a paper about 'Geographical Regularities of the Distribution of Genes of Cultivated Plants', in *Trudy po prikladnoy botanike* [*Papers on Applied Botany*]. On the basis of my research, I explained the accuracy necessary in order to verify the disappearance of a number of morphological characteristics of so-called dominant character in the direction away from the geographical centers of origin of the species out toward the periphery of the distribution areas, where the plants are usually represented by a concentration of recessive properties.

Thus, e.g. within the initial distribution areas of the species of rye – i.e. in Asia Minor and Transcaucasia – we encounter a great variety of color forms among the weedy and even the cultivated types of rye, such as red or black spikes or velvety ones. At the same time, at the periphery – in northern Europe and the European parts of the Soviet Union – we meet predominantly with white-spiked types.

Abyssinia – an ancient center of agriculture – is characterized by the presence of peculiar colorations in the case of wheat, e.g. such as violet grains. Barley there is often represented by black-spiked types and distichous varieties, which are genetically dominant.

Ancestral types of rye, wheat, barley, flax and rice differ in basic characteristics such as brittle spikes and seeds that are shed when ripe, all of which favor self-dispersal. As can be seen by means of hybridization, and as genetically expressed, these characteristics are dominant, whereas the non-brittle, cultivated and later-developed types are recessives. On the Mediterranean islands and in Pamir, there are interesting kinds of wheat and rye with leaves of a simplified structure. These, too, are recessive.

Within the secondary area of distribution in Asia, maize – introduced there from America during the last century – has developed a peculiar, so-called waxy type, unknown in its native land. Genetically this, too, appears to be a recessive and not a dominant characteristic.

When studying western China (Chinese Turkestan), we observed an amazing recessive form of flax with white flowers, white seeds and narrow petals, rather reminding us of the flowers of a *Stellaria* than of those of ordinary blue-flowered flax. We also found many other interesting recessive types there. Apparently, these are genetically linked to the results of isolation, inbreeding and recessive mutations. In their native areas, such characteristics are ordinarily only rarely expressed. They also vanish in the case of hybridization with the original common types. As demonstrated by our research, some physiological properties appear to be recessive as well.

There are also exceptions, such as we were able to demonstrate in the case of the awnless wheat and awnless barley of Eastern Asia. Quantitative properties (e.g. dimensions of grains or fruits) reveal themselves in a different manner: increased dimensions of fruits and seeds are often observed toward the periphery of the distribution areas in conformity with the environmental conditions.

We continue to work with the study of the evolution of plants. On the basis of the colossal amount of material gathered, a number of the principles are becoming more complicated but also more accurately defined. Basically, the principle of the geographical centers holds firm and can be confirmed by the practical exploratory work. As demonstrated, within the ancient initial areas we have first and foremost not only a presence of several dominant characteristics but also entire botanical species, which, during later evolution, have not yet reached the periphery of the original center.

Thus, we discovered a number of species of cultivated wheat in Transcaucasia displaying interesting and practical properties such as resistance to diseases, but which have not yet spread across the border to the Caucasus. In the Cordilleras there are dozens of species of cultivated potatoes, which do not reach beyond South America but are still preserved within the main center of origin.

In my paper, I mentioned that the nature of the dominant and the recessive characteristics is far from being explained and that Bateson's hypothesis, explaining the dominance of forms as the presence of genes but the recessiveness as an absence of the corresponding genes, appears to be wrong. I indicated, rather, the great agricultural value of some recessive genes.

Being of a different opinion, which is based on numerous facts, I wanted to demonstrate clearly that recessive characteristics are not always inferior to the dominant ones. Under certain conditions they can be very valuable for agriculture and during our investigatory work we paid great attention to this fact.

Concerning animals and Man himself, I pointed out that during the evolution of its laws, the recessive characteristics of plants are, to a certain degree, also applicable to humans and that the historical process of isolating recessives was positive from the point of view of the main historical civilizations, since it gave rise to the great northern civilizations in addition to the very important ancient ones.

My basic opinion is that secondary, peripheral centers, where we can still encounter an exceptional variety of genes, are of no less importance than the ancient agricultural centers.

When I speak of original centers of agriculture, it is necessary to keep in mind that they can still be relatively primitive, even though they belong to civilizations that are thousands of years old. Mexico, Peru, Bolivia and southern Asia as well as Asia Minor are represented by countries with very old agricultural crops, in one way intensely different from ours and not mechanized, but still producing exceptional results based on creative and selective work. It is enough to mention that the maize with the largest grains was produced in Peru by the ancient Indians; these strains have not been surpassed by any others created by modern plant breeders.

The conclusion drawn from all this by G.N. Shlykov makes me question his fairness. When studying countries with old civilizations, I tried to respect the people who settled the ancient agricultural territories and to learn from them.

At the end of his paper, G.N. Shlykov, by sheer habit tries to accuse me of anti-Michurinism. It is intolerable that a co-worker at the Institute of Plant Industry does not know that just this institution was the first one to publish the entire paper by I.V. Michurin in 1923 after convincing him to review his work in order to make it widely available. Already, in 1920, I considered him of great importance and I agreed with the Soviet public opinion of him, which I had long held at the time when this paper appeared, and I knew of the importance of his work for plant breeding. This happened when I.V. Michurin modestly conducted his little-known work not at Michurinsk but in Kozlov. In 1923, I especially ordered a number of workers to go to Ivan Vladimirovich to study his achievements. The question arises: is it necessary for G.N. Shlykov to give false evidence about all this, to view it through a mirror of distortion?

I.V. Michurin himself even sent his own best students to study under me (P.N. Yakovlev, A.V. Petrov and F.K. Teterev). Continuing his work, we at the Institute of Plant Industry have extensively applied hybridization and worked toward overcoming the infertility of hybrids, etc.

BIBLIOGRAPHY

Darwin, C. (1937). *The Origin of Species*. Moscow, Leningrad.

Kol', A.K. (1936). Rekonstruktsiya rastenievodstva SSSR [Reforming the plant industry in the U.S.S.R.]. *Sots. rekonstruktsiya sel. Khoz-va* [*Socialistic reformation of agriculture*], No. 10.

Michurin, I.V. (1924). *Itogi deyatel'nosti v oblasti gibridizatsii plodovykh*, [*Results of activities within the field of hybridization of fruit trees*]. Foreword by N.I. Vavilov. Ed. by V.V. Pashkevich. Moscow.

Shlykov, G.N. (1936a). *Introdyktsiya rasteniy* [*Introduction of plants*]. Moscow.

Shlykov, G.N. (1936b). *Introdyktsiya i genetika*, [Introduction to plants and genetics]. *Sots. rekonstruktsiya sel. khoz-va*, [*Socialistic reformation of agriculture*], No. 9.

Vavilov, N.I. (1919). *Immunitet rasteniy k infektsionnym zabolevaniyam*, [*The resistance of plants to infectious diseases*]. Moscow.

Vavilov, N.I. (1935). *Nauchnye osnovy selektsii pshenitsy*, [*Scientific basis for breeding wheat*] Moscow, Leningrad.

The theory of the origin of cultivated plants after Darwin

THE STUDY OF the races of domesticated animals and cultivated plants played a great role for the concept held by Charles Darwin concerning evolution. A synthesis of the data on the variability in a domesticated condition filled the first chapter of his *On the Origin of Species* and two volumes of his *Variation of Animals and Plants under Domestication*, which appeared in 1868.

During Darwin's remarkable activities in the Tropics, botanical and zoological objects were studied alternately. Being mainly a zoologist during the early years of his activities and while devoting large monographic studies to the systematics of the invertebrates, he became more of a botanist during the second half of his life, as shown by the large monographs devoted to flower biology. As is well known, the cycle of works by Darwin was crowned by his theory on the origin of Man. A characteristic trait of the great naturalists of the eighteenth and nineteenth centuries was to be encyclopedists and to pave new roads for Science, as a result of which the historical condition of the development of the natural sciences in which they found themselves during the time of the industrial revolution also had a strong effect on agriculture. In Darwin's case, this was especially extensively and deeply expressed. Such a range was necessary for him to develop a multi-faceted, well-formed and profound theory of evolution: on the one hand, he was familiar with animals, on the other with botanical objects. Plants have in many respects a great advantage over animals for the study of the evolutionary process, especially for experimental investigations. Breeding work was widely practiced in England in the case of both animals and plants.

While approaching the studies of variation and evolution of cultivated plants in a phyto-geographical setting, Darwin relied first and foremost on the monumental work of the Genevan botanist, Alfonse De Candolle, *Géographie Botanique Raisonnée*, which appeared in two volumes in 1855 and where considerable space was devoted to the origin of cultivated plants. In addition to this, Darwin zealously studied the particular breeding work in England and the works by the foremost German and French investigators of cultivated plants.

The libraries of the learned frequently reflect the historical paths of their personal creative work.

Lecture presented at the Darwinian session of the Academy of the Sciences on November 28, 1939. First printed in *Nauka* [*Science*], no. 2, pp. 55–75, 1940.

When working during 1913 and 1914 in the personal library of Darwin, which was preserved in its entirety after his death and at that time was located in the Botany School of the University of Cambridge, where Darwin's son, Francis Darwin, was professor of plant physiology, and when looking over the large and beautifully selected collection of books – which was exceptionally large for that time – I had the opportunity to see how thoroughly Darwin studied the works of his forerunners with respect to the history of crops and the breeding of plants. Books by the best English, French and German plant breeders and experts on cultivated plants such as Shireff, Le Coutère, Metzger and Loiseleur-Deslongchamps were annotated with remarks by Darwin himself (cf. the interesting *Catalogue of the Library of Charles Darwin*, compiled by H.W. Rutherford in 1908, and now in the Botany School, Cambridge). In the text and at the ends of books, Darwin noted down facts and ideas that were especially important to him. On the basis of his library it is, to a certain extent, possible to follow the course of the creativity of this great scientist and the enormously laborious work preceding his general conclusions.

In contrast to De Candolle, who was mainly interested in establishing the true native lands of cultivated plants on the basic of taxonomic–geographical, historical and linguistic data but also on the relationship between cultivated and wild plants, Darwin was first and foremost interested in the evolution of the species, the appearance of subsequent changes to which the species were subjected when taken into cultivation, the amplitude of the variability of the organisms under the influence of the conditions of cultivation and the role of selection. Unlike De Candolle, every plant interested Darwin, not only because of itself but as an integral unit for explaining evolution. Different species were, in essence, considered by Darwin as illustrations of the basic idea he wanted to argue. In contrast to the detailed codex of De Candolle, the book by Darwin on *Variation of Plants and Animals under Domestication* is of characteristic significance as a persistent idea toward a revelation of the dynamics of the evolutionary process. Although the books by De Candolle, whom Darwin considered a most competent judge of facts and with whom he had an extensive correspondence, were still of great interest for his actual subject; the more concise chapters on cultivated plants were distinguished by Darwin as the richest source of ideas. The great scientist can be perceived in his general conclusions and, in each category of facts, the searching mind and his passionate involvement in the arguments about his evolutionary concept are evident.

Darwin was most interested in the problem of the monophyletic and polyphyletic origin of cultivated plants and domesticated animals. Could not the multiform types of cultivated plants and the large amplitude of the variations be explained as participation of the same species in their development, in a role of hybridization? Considering the role of the latter phenomenon, Darwin most definitely leaned toward the hypothesis that the decisive role for evolution was to be found in the intraspecific, hereditary variability in combination with selection, both natural and artificial. He was especially interested in the range of variation of those species during the development of which hybridization of distant taxa had taken place. He produced remarkable examples of pumpkins, which were already well-studied at his time by the French scientist Naudin, and which showed an astonishing quantitative range of inherited variation within the limits of various kinds of pumpkins, expressed as fruits a 1000 times larger in size. This is still one of the most extreme examples of intraspecific quantitative amplitude.

Darwin studied maize, vegetables, fruit trees, potatoes, berries and decorative plants from every point of view and was interested in the multiformity revealed by the species and in the quantitative and qualitative differences as displayed by the hereditary variability of the species. He constantly collected information about a number of strains of various species, citing figures amazing for his time: the types of roses then already numbered several thousands. He studied especially thoroughly the evolution of the dimensions of gooseberry fruits [*Ribes uva-ursi* L.], demonstrating how cultivated strains had reached a weight of 53 g each by means of selection from wild gooseberries with a mean weight of 0.5 g per fruit.

In short, the section loaded with facts about the origin of cultivated plants demonstrated that Darwin diligently collected facts from the most different sources, which reveals his amazing erudition. In a humorous manner, he called himself 'a millionaire of facts'.

One of the ideas running through his *Variation of Animals and Plants under Domestication*' appears to be that the variation in type can be expressed particularly in such characteristics and properties that are of interest to Man and, thus, reveal in which direction the selection is going. The synthesis of the facts clearly demonstrates both the enormous hereditary variation of the species and the actual effects of the selection as well as the possibilities for a progressive advance toward the improvement of both genera and species. With a clarity amazing for his time, Darwin demonstrated that races of animals and strains of plants can be created on the basis of a union of contrasting heredity and as a result of selection. The great and immortal merit of Darwin is the fact that he drew attention to the dialectics of interrelated variation, heredity and selection, thereby opening up opportunities for progressive breeding. That is why the works of Darwin became the foundation for the theory and practice of selection, which is basic to successful breeding work. Never, before Darwin, had the idea of variation and the enormous role of selection been advanced with such clarity, definition and substantiation.

How has the theory about the origin of cultivated plants fared after Darwin and to where can it be further directed?

In 1883, the year after Darwin passed away, the book by Alphonse De Candolle, *L'Origine des Plantes Cultivées*, appeared. This was also a work of fundamental importance for its time. In this book, De Candolle revised and expanded the chapters on cultivated plants, published earlier in *Phytogéographie Raisonnée*, accepting the ideas of Darwin concerning the role of natural and artificial selection. In essence, he did not dwell on the problem of evolution but concentrated his attention exclusively on the establishment of the native lands of cultivated plants.

'My objective', De Candolle wrote in the preface to his book, 'is mainly to retrace those original conditions and localities of every plant characteristic for it when first taken into cultivation'.

There is no doubt that De Candolle recognized, indeed, that three-quarters of the genera of cultivated plants determined by Linnaeus were either relatively incorrect or incomplete.

The classical work by De Candolle, saturated with a multitude of well-classified and actual facts, appears – in spite of its value – to be one-sided and illustrates only the original native lands of cultivated plants and the relationships between cultivated plants and their wild original or closely related species.

During the past half century since the death of Darwin and the publication of De Candolle's *Origin of Cultivated Plants*, knowledge about the various crops has expanded a lot. The problem concerning the origin of cultivated plants and domesticated animals has attracted the attention of many scientists: botanists as well as archeologists, historians, philologists and agronomists alike.

Some of the problems are linked to the history of the crops and to archeological, historical and linguistic investigations. We must mention the classical and historical work by Victor Hehn, *The Cultivated Plants and Domesticated Animals during their Dispersal from Asia to Greece and Italy but Also to the Rest of Europe*, which has appeared in numerous editions with revisions, and, as a continuation of it, the book by Bertold Laufer, *Sino-Iranica*, where data based on linguistics link the plants of China and Iran.

Other authors collected interesting factual material concerning the history of the plant crops, e.g. Gibault in his interesting treatise of the history of vegetables. The American scientist Sturtevant (1919) gathered information about the history of many cultivated plants. The valuable work by Bushan on *The Ancient History of the Cultivated Plants of the Old World Based on Prehistoric Discoveries* (1895) is also worth mentioning. A large number of scientists devoted themselves to the crops of the New World (e.g. Carrier, Cook, Merrill, Safford, Bailey, etc.). The problem of the interrelationship between wild and cultivated plants interests many botanists. In this respect, the book by Thellung (1930) is of particular interest.

The work by Academician V.L. Komarov, *The Origin of Cultivated Plants*, which appeared in two editions, is also devoted to this subject. In it, the author launched the theory about the great role of hybridization for the origin of many cultivated plants. Finally we should mention the extensive review by the German scientist, Elizabeth Schiemann, which appeared in 1932 and where a thorough attempt is made at a genetic illustration of the problems concerning the origin of some cultivated plants. Literally hundreds of valuable works have been devoted to different crops.

Numerous monographs have appeared, devoted to cereal grasses (such as papers by F. Koernicke, A. Schultz, G. Solms-Laubach, J. Percival, R. Regel, K.A. Flaksberger, A.I. Maltsev, P.M. Shukovskiy, M.G. Tumanyan, A.A. Orlov, V.F. and V.I. Antropov, A.I. Mordvinkina, N.N. Kuleshov, L.L. Dekaprelevich, etc.); on the leguminous plants (papers by L.I. Govorov, V.S. Muratova, E.I. Barykina, N.R. Ivanova, G.M. Popova, V.S. Felotov, E.E. Ditner, F.M. Balkind, etc.); on oil-producers and root crops (works by E.N.

Sinskaya, G.M. Popova, V.P. Zosimovich); on cucurbitaceous plants (by K.I. Pangolo); and on cotton (papers by G.S. Zaytsev, Kern and Harland). A number of valuable papers are devoted to the potato (by S.M. Bukasov, S.V. Yuzepchuk and R. Salaman). A large number of works concern fruit crops, berries and grapes (papers by V.V. Pashkevich, M.G. Popov, Yu. N. Voronov, G.A. Rubtsov, M.A. Rozanova, N.M. Pavlova, A.M. Negrul', P.A. Baranov, K.F. Kostina, I.N. Ryabov, N.V. Kovalev, V.A. Rybin, A.D. Strelkova, L.A. Smolyaninova, Kharyuzova and others).

A number of monographs have appeared that are devoted to cultivated plants of different countries such as Turkey (P.M. Shukovskiy), Afghanistan (N.I. Vavilov and D.D. Bukinich), Abyssinia (a number of authors), Iran (E.G. Chernyakovskiy), China (Swingle, Wilson, Wagner, etc.), Japan (E.N. Sinskaya, Tanaka), India (Watt and A. and G. Howard and Gamm), Mexico, Guatemala and Colombia (S.M. Bukasov), South Africa (N.A. Basilevskaya), the oases of the Sahara (Chevalier) and so on. These works develop the ideas of Darwin and De Candolle as applied to various plants. This was a period, when the ideas of Darwin and De Candolle were mastered and a time of growth and differentiation of the knowledge concerning cultivated plants.

As can be clearly seen from the above-mentioned reviews by Schiemann and Komarov, a great portion of these papers was written by Soviet scientists.

The great development of breeding work during the twentieth century, the flood of successful plant breeding activities and the creation of a multitude of plant-breeding establishments indicated a renewed interest in the Darwinian problems concerning the origin and evolution of cultivated plants. The question about the native land of cultivated plants, worked out by De Candolle mainly on the basis of herbarium specimens and literature data, has taken on a special meaning and realism. The necessity of mastering practical breeding of original material forces us to approach a systematic study of the intraspecific composition of the most important cultivated plants, a multifaceted study of the species and the geography of the varietal diversity and a study of the hereditary variability. It is important to apply genetical, cytological, physiological, biochemical and technical methods more intensely for such investigations.

The plant breeder, like an engineer, is interested in the structure of the original material. As shown by experiments all over the world, the success of plant breeding is to a significant extent determined by the correct selection of the original species or varieties. The widespread application of plant breeding during the past decades, using hybridization and the possibility of utilizing closely related wild species or geographically distant forms for improving cultivated plants and providing them with this or that kind of valuable properties, depended in turn on the currently heightened insight into the composition of the species of cultivated plants and the relatives closely related to them, and on the elucidation of the phylogenetic relationships.

The plant breeder is faced with problems of where to find components, which ones to use for hybridization and exactly what the valuable properties of the wild plants or their closely-allied relatives are. The question is: what is represented in the species of cultivated plants that have evolved and occupy enormous areas and thrive under different environmental conditions, and what stages of evolution have the different species passed through? Thus, there is a necessity for a wide-ranging geographical approach to the study of the evolution of the species, reaching from the original native lands, where they were first taken into cultivation, to the final link during their evolution. A logical investigator must therefore approach the problem of the origin and the evolution of cultivated plants to the fullest, while applying new and realistic methods.

There is a clear necessity for a distinction between the stages of evolution with respect both to the initial origin and introduction into cultivation of species of cultivated plants within different areas and their relation to the corresponding wild species and varieties, and to the subsequent evolution of the plants dispersed from the basic centers and subjected to changes under new conditions as well as to the further effects of natural and artificial selection.

As the basis for exploratory work in the sense of establishing areas of initial type-formation we find, in essence, the ideas developed by Darwin in Chapter XII of his *On the Origin of Species* (1859), concerning the individual geographical centers of origin.

Considering the problem of the evolution of species from a geographical point of view, Darwin unfailingly arrived at an understanding of the relationship between the cultivated species and a definite, individual area.

'He who neglects this view', Darwin wrote, 'rejects the *vera causa* of ordinary generation with subsequent migration and calls in the agency of a miracle'.

'Hence it seems to me, as it has to many other naturalists, that the view of each species having been produced in one area alone and having subsequently migrated from that area as far as its powers of migration and subsistence under past and present condition permitted, is most probable.'

And further: '... the belief that this has been the universal law, seems to me incomparably the safest'. (Darwin, 1859[1]).

Translated into the language of contemporary biogeography, Darwin's geographical idea about evolution consists of each species being located within its initial, original area and evolution being historical; therefore, the knowledge of the source of the species and its path of geographical dispersal is of decisive importance for understanding the course of its evolution, for mastering its stages and for tracing the dynamics of the evolutionary process.

When starting a wide-ranging geographical investigation of the important cultivated plants, we soon became convinced that the geographical centers of the initial origin of the important cultivated plants, such as established by De Candolle, were too general and sometimes encompassed entire continents or territories on a scale too wide for conducting a serious exploratory work. The old collective Linnaean species, which were frequently determined from herbarium specimens, proved after investigations enhanced by cytological, geneti-

[1] Vavilov quotes here a 1937 Russian translation of Darwin, but the translator has used Darwin's original 1859 text. D.L.)

cal and physiological methods to be complexes of biological, physiological and morphological characteristics and, as a rule, to have distinct distribution areas.

According to Darwin, but in contrast to De Candolle, it was necessary to pay attention to the basic areas of origin of a plant and to the study of the evolutionary stages through which it passed under the effects of cultivation and environmental conditions as well as to the effects of natural and artificial selection.

During the course of the last 15 years, Soviet botanists and agronomists have made wide-ranging expeditions to study the most important cultivated plants of interest for the Soviet Union, both within the areas where the plants are cultivated at present and those where their basic, initial type-formation and introduction into cultivation took place.

Proceeding on the basis of the geographical and evolutionary theories of Darwin, systematic studies of the most important plants were planned, including all the evolutionary stages, from the initial areas, where the linkage to the wild types can still be traced and where the phylogenetic relation between different wild species and cultivated forms can be established, via the subsequent further historical dispersal of the species up to the final links with contemporary plant breeding. This work could, of course, be done only by a team of scientists working according to a definite plan, the basis of which was the principle of evolution.

Consequently, the investigations covered the major part of the agricultural areas all over the world, while paying attention mainly to the Soviet Union itself and the countries contiguous with it. All the agricultural areas of North and South America were studied comparatively thoroughly. The Cordilleras were perused and the main agricultural mountain massif of Africa and a considerable portion of the agricultural areas of the Asian continent were also studied. Enormous collections of material were made, determined at present to be more than 200 000 samples, which are subjected to a multifaceted study of the living material, sown or planted under various conditions, while applying all modern methods available.

These studies have not only doubled our knowledge of various cultivated plants but have also led to the discovery of a large number of new species and an enormous amount of botanical varieties, both of cultivated plants and of the wild plants, closely related to them and all previously unknown to Botany and, even more so, to plant breeders.

In the case of a number of species, these investigations led to definitely new facts, radically altering our ordinary concepts about the most important cultivated plants. Thus, in addition to the single species of the potato, previously held to be adequately known, dozens of cultivated and wild species were discovered, which are true species in the full sense of that word and which had been subjected to a sharp genetical divergence in the sense of Darwin and which are now clearly isolated from each other and difficult to hybridize. They have different chromosome numbers and constitute complexes of biological and morphological characteristics.

Within Asia Minor, the Caucasus and Abyssinia, new species of cultivated and wild wheat and rye were discovered in the form of, literally, thousands of previously unknown botanical species, varieties and forms. Among these new species and varieties valuable characteristics were found, such as immunity complexes, resistance to many infectious diseases, drought-tolerance, stiff straw, high yield, and so on.

The penetration into the areas where cultivated plants initially developed, i.e. those already called 'centers of origin' by Darwin, led to the discovery of enormous specific and varietal resources, which exceeded every expectation.

The enormous diversity discovered within the limits of different species made it necessary to differentiate the concept of species. Following the principles of taxonomists dealing with cultivated plants, e.g. Koernicke or Seringe, we were compelled to go on with a differentiation of the species and at the same time we discovered new and major taxonomic units, i.e. species in the sense of Linnaeus. The usual framework of classification proved to be too narrow for the inclusion of the diversity of all the types. An attempt was made at a revision of the differentiating taxonomy.

While deepening the studies of cultivated plants and their closely allied relatives, a complicated system of hereditary forms was cleared up. It proved to represent special species of cultivated plants, widely dispersed beyond the borders of the original land. The facts concerning the enormous variability of cultivated plants, put forward by Darwin, proved to be even more important. Direct statistical analyses of the events can demonstrate what changes have taken place during the course of time.

In addition to the single species of potato [*Solanum tuberosum* L.], Soviet botanists at present distinguish 18 cultivated species, associated geographically with their origin within the Andes of South America and sharply differentiated into a complex of morphological and physiological characteristics, different chromosome numbers and incompatibility when hybridized. In addition, botanists further distinguish dozens of new, tuber-producing species of potatoes discovered within the territory of the Cordilleras and the ranges projecting out from them.

In addition to the five or six species of wheat known at the time of Darwin, contemporary botanists distinguish more than 30 species and subspecies thereof. Among the hundreds of botanical varieties of wheat discovered during the 80+ years since Darwin and De Candolle, we distinguish at present more than 1200 botanical varieties, each of which consists of a multitude of well-differentiated forms.

As is well-known, Darwin put much emphasis on an exact recording of the composition of the flora and its complex preservation, in the manner devised by the Linnaean Society for taking an inventory of the new species described. This proved to be perfectly feasible for the majority of plants. The major portion of the species of cultivated plants can therefore be almost directly linked to the original wild types. The later inclusion of agricultural areas, which have only been slightly studied both within the borders of our own country and beyond it, and the general inclusion in the final total of various cereal, vegetable and fruit crops within the sphere of investigations, led to the definite establishment of entire floras of cultivated species belonging to definite floristic areas.

I will try to furnish a brief review of the investigations, which are unfortunately still incomplete and in need of additional data, in particular concerning tropical Asia. However, they allow us to draw definite conclusions concerning entire, comparatively well localized original floras.

In 1883, De Candolle established the local origin of 247 cultivated species. A larger total of cultivated plants, not counting the decorative ones, has been carefully enumerated during the last couple of years by Prof. E.V. Wulff at the All-Union Institute of Plant Industry. He estimated it to be about 1500–1600 species, including in this number only fully-established crops. On examination, hundreds of additional 'candidates' for becoming cultivated plants were revealed, especially among the forage plants and those producing etheric oils and medical drugs.

While revising considerably the scheme of geographical centers of origin of cultivated plants, which was launched by myself in 1926 (Vavilov, 1926), I shall dwell briefly here on the conclusions to which the investigations of cultivated plants have led up to the present time.

The total area cultivated all over the world is at present estimated to be approximately 850 million hectares, which amounts to ca. 7% of the entire landmass. Out of the total number of 1500 species of edible, technical and medicinal cultivated plants, I will take about 1000 of the most important plants into consideration. These actually occupy no less than 99% of all the cultivated area. The remaining 500–600 species, together with all their varieties, occupy less than 1% of all the territory under cultivation.

Asia appears to be the area furnishing the highest number of cultivated plants. Its share of the 1000 species concerned is about 700, i.e. ca. 70% of all cultivated plants. Approximately 17% have arisen in the New World. Before the arrival of the Europeans, Australia had no cultivated plants and it is actually only during the last century that its eucalyptus and acacia trees have begun to be widely grown within the tropical and subtropical areas of the world.

The following seven main geographical centers of origin of cultivated plants can be distinguished within the borders of the continents (cf. Fig. 1):

1 *The South-Asiatic tropical center* embraces the territory of tropical India, Indo-China, southern China and the islands of southeastern Asia. This enormous territory with a rich wild flora, which is determined numerically to be approximately one fourth of the specific diversity of the worldwide flora, gave rise to about one third (ca. 33%) of the entire established number of cultivated plants. This is the native land of rice, sugar cane and a large number of tropical fruit and vegetable crops.

Within this large geographical center or area, three foci can be distinguished, which are considerably different with respect to the complex of cultivated plants characteristic of them.

(a) The Indian focus (with the richest cultivated flora);
(b) the Indo-Chinese one, including southern China; and
(c) the island focus, including the Sunda Islands, Java, Sumatra, Borneo, the

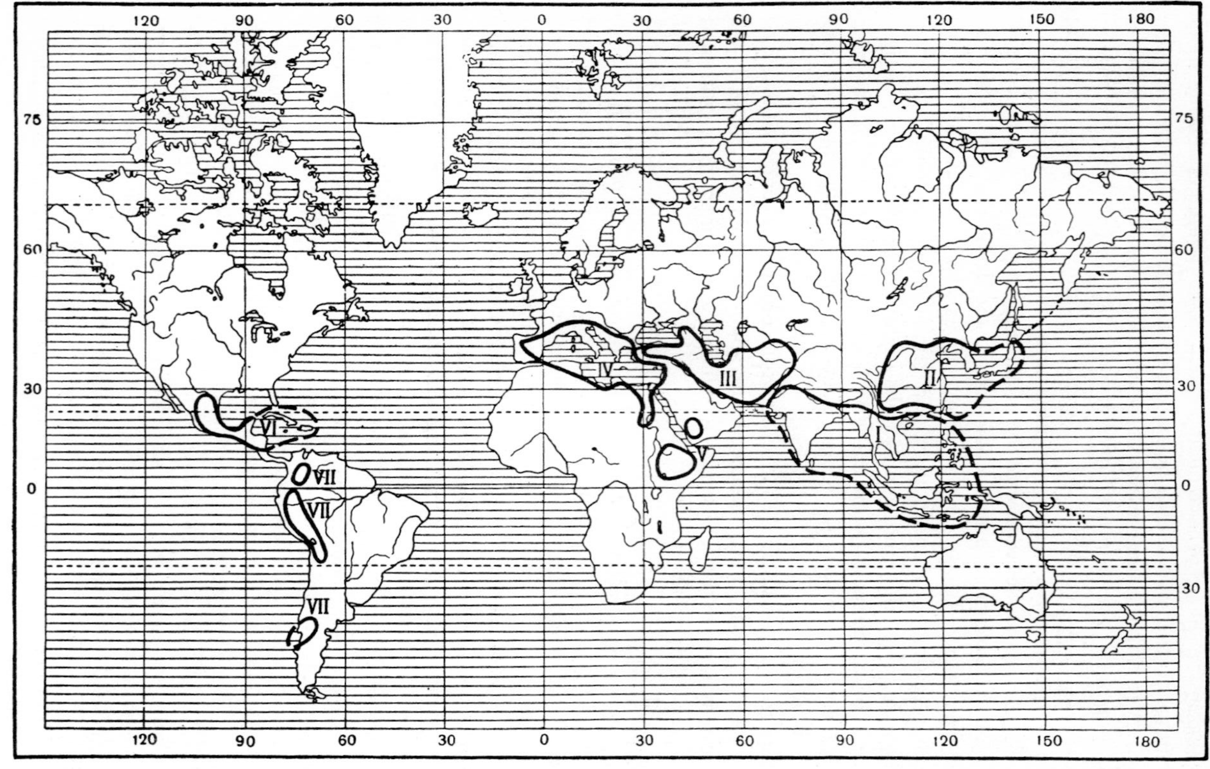

Fig. 1. Center of origin of cultivated plants. I. The tropical south-Asiatic center; II. the east-Asiatic center; III. the southwestern-Asiatic center; IV. the Mediterranean center; V. the Abyssinian center; VI. the Central American center; and VII. The Andean (South American) center.

Philippines, etc. This focus gave rise to the smallest number of cultivated plants, according to the investigations made by Dr. Merrill.

More than one fourth of the world's inhabitants (more than one half billion) still live in tropical Asia. In the past, this number was even more important.

2 *The East-Asiatic center* includes the temperate and subtropical parts of eastern China and the major portions of Taiwan, Korea and Japan. This is the native land of species such as soybeans, different species of millet and the majority of vegetable crops, as well as a very large number of fruits. As far as the composition of wild and cultivated fruits is concerned, China occupies, indeed, first place in all the world. The total number of cultivated plants, not including the decorative ones, is determined to be approximately 20% of the world's total, i.e. about 200 out of the 1000 plants discussed.

Within this center can be distinguished a main Chinese focus and a secondary, mostly Japanese one.

It is well known that about one fourth of the world's population (ca. one half billion) lives within this territory.

3 *The Southwestern-Asiatic center* covers the interior foothill area of Asia Minor (Anatolia), Iran, Afghanistan, Inner Asia and northwestern India. The latter is floristically linked to Iran (as far as cultivated plants are concerned). It also covers the Caucasus, the cultivated flora of which is genetically associated with that of Asia Minor. This center can be distinguished into the following foci:

(a) The Caucasian focus with the majority of endemic species of wheat, rye and fruits – as explained by comparative cytological and immunological investigations, this is the most important focus of origin in the world as far as wheat and rye are concerned;
(b) the Asia Minor focus, including the interior portion of Asia Minor (Anatolia), interior Syria and Palestine, Transjordania, Iran, northern Afghanistan and Inner Asia (including Chinese Turkestan); and
(c) the Northwestern-Indian one, in addition to the Punjab and the adjacent provinces of northern India also covering Belutchistan, southern Afghanistan and Kashmir.

As demonstrated by detailed research, the southwestern-Asiatic geographical center is a very important center of origin of those species that are now cultivated in Europe, both the cereals and many leguminous plants as well as almost all the European fruits, including grapes. The total complex of species of cultivated plants genetically associated with this territory is estimated to be approximately 14% of all cultivated plants in the world (ca. 140 out of the 1000 mentioned above).

Wild relatives of wheat, rye and various kinds of fruits are concentrated within the exceptional specific diversity in this area. In the case of many important cultivated plants, it is possible to trace an uninterrupted sequence back from the cultivated to the wild forms and to identify the links existing between the wild and the cultivated forms.

4 *The Mediterranean center* embraces the countries situated along the coasts of the Mediterranean Sea.

This remarkable geographical center, characterized by the great ancient civilizations of the past, gave rise to about 11% of the species of cultivated plants (ca. 110 out of the 1000 discussed). Among these we can mention olives, the St. John's bread or carob tree and a multitude of vegetable and forage plants. In addition, it is still possible to trace the close association between the origin of various crops and particular territories such as, e.g., the Iberian peninsula, the Appennines, the Balkan peninsula, Syria and Egypt. Each of these foci is typical of original species of forage plants, such as sulla (*Hedysarum coronarium*), berseem or Egyptian clover [*Trifolium alexandrinum*], gigantic white clover [*T. repens*], gorse [*Ulex europeus*], single-flowered vetch [*Vicia articulata*], French vetch [*V. narbonensis*] and a kind of chickling vetch [*Lathyrus gorgonii*]. In other words, it is still possible to differentiate this center with respect to the localization of the origin of various plants and their introduction into cultivation.

5 Within the borders of the African continent, the tiny *Abyssinian center* can be distinguished as an independent geographical area, characterized by a number of endemic species and even genera, such as the cereal grass called teff (*Eragrostis abyssinica*), the oil-producing plant called ramtil, nug or Niger seed (*Guizotia abyssinica*), a special kind of banana (*Musa ensete*) and the coffee tree [*Coffea arabica*]. In part, this is also the native land of grain-sorghum. The total number of species of cultivated plants genetically linked to Abyssinia and the mountains of Egypt adjacent to it does not exceed 40, or approximately 4% of the world's cultivated flora.

To this center belongs also the somewhat peculiar *Arabian mountainous* or *Yemenite* focus, which reflects influences from both the Abyssinian and the southwestern-Asiatic centers and is characterized by extremely fast-ripening kinds of cereals, leguminous plants and alfalfa.

In addition to endemic genera, the presence of peculiar endemic cultivated species and subspecies of wheat and barley is characteristic of Abyssinia. In spite of the absence here of wild wheat and barley, Abyssinia is distinguished by peculiar cultivated species of wheat and by an amazing wealth of endemic forms of wheat and barley, cultivated under rather primitive conditions. This can apparently be explained by the isolation of agriculture among the mountains of Abyssinia, the fairly ancient type of agriculture and the vicinity to the main centers of origin in the case of wheat and barley. As every botanist knows, the flora of all the mountain chains running along East Africa from the Cape Colony to Himalaya is characterized by distinct complexes of genera and even species. This condition is especially clearly shown by the classical investigations made by A. Engler.

Within the New World, a remarkable localization of the type-formation of important cultivated plants could be established.

6 A *Central American geographical center* can be particularly distinguished within the wide-ranging territory of North America. It embraces southern Mexico as well and can be subdivided into three foci:

(a) the mountainous South-Mexican focus;
(b) the Central American one; and
(c) the West Indian focus.

From this geographical center originate about 90 edible, technical and medicinal species of plants (out of the 1000 discussed) such as maize, upland and other American long-staple cotton species, a number of bean species and gourds, cocoa and also – in all likelihood – sweet potatoes, yams, peppers and many fruits such as guava [*Psidium guajava*], different kinds of sapotes [*Pouteria* spp., *Casimiroa* spp.] and avocados [*Persea* spp.].

7 The *Andean center* within South America covers parts of the Andes. We can distinguish three foci in that area:

(a) The Andean focus proper, including the mountain areas of Peru, Bolivia and Ecuador. This original focus seems to be the native land of many tuberous plants, primarily of a large number of cultivated potatoes such as described for the first time by Soviet scientists, but also of tuber-producing species such as oca (*Oxalis tuberosa*), ulluco (*Ullucus tuberosus*) and anu (*Tropaeolum tuberosum*). This is a mountain–steppe vegetation, thriving at high altitudes, from 3000–4500 m. The llama and the alpaca were domesticated here. The quinine tree and the coca bush came from the humid montane forest east of the Andes.

(b) The Chiloan (Araucarian) focus, situated in southern Chile and on the neighboring islands, gave rise to the species of ordinary potato, *Solanum tuberosum*. In contrast to the Peruvian, Bolivian and Ecuadorean potatoes, which usually form tubers during the short equatorial days, the ordinary potato does so in the conditions of medium–long days at a latitude of 38–40°S. in southern Chile. This species can also be successfully grown under the conditions of long daylight and could therefore be brought directly from southern Chile and used as crops in Europe. In addition to the potato, oil-producing plants such as madi [*Madia sativa*] and mango chil [*Bromus mango*], no longer cultivated, came from this area.

(c) The Bogotan focus in eastern Colombia was established by the Soviet botanists, Drs S.M. Bukasov and S.V. Yuzepchuk. The crops are raised at a high altitude (up to 2800 m elevation). Some species of potato were independently taken into cultivation here by the Chibchan people as, perhaps, was the root fruit called arracacha (*Arracacia xanthorrhiza*).

Some plants were also brought into cultivation from the wild flora found outside the centers listed. Thus, the date palm was introduced into cultivation in the oases of Arabia, southern Mesopotamia and, perhaps, in the Sahara as well. Watermelons were initially cultivated in South Africa, in areas apparently adjacent to the Kalahari desert. Manioc, pineapple and groundnuts and – during the recent European-influenced time – the rubber tree, were taken into cultivation within the interior areas of tropical South America.

During the last decade, Paraguayan tea (*Ilex paraguayensis*) has been taken into cultivation from the wild flora in northern Argentine. In North America, the Indians were already cultivating 'Jerusalem artichokes' (*Helianthus tuberosus*) before the times of Columbus.

Some plants – such as hemp, millet, apples and pears – were apparently taken into cultivation at different times and in different areas of the Old World. Because of this, the exact site where they were first cultivated is impossible to establish. The total number of plants taken into cultivation outside the main centers enumerated above is comparatively small. They amount to less than 3%

of the 1000 discussed. It is only during the past couple of decades that an extensive use of the local and wild vegetation was made in Europe and America in order to introduce into cultivation new forage plants and plants producing etheric oils, medical drugs and tannins.

As can be seen, the main geographical centers of original introduction into cultivation of the majority of the plants grown are associated not only with floristic areas, distinguished by rich floras, but also with ancient civilizations. Actually, the seven large centers distinguished correspond to the sites of ancient agricultural civilizations. The tropical south-Asiatic center is associated with the high-level Indian and Indo-Chinese ones. The most recent excavations reveal the great age of these cultures, which are contemporary with those in Asia Minor. The east-Asiatic center is associated with the ancient Chinese civilizations and the southwestern-Asiatic center with the ancient ones of Iran, Asia Minor, Syria and Palestine. The Mediterranean center was, several centuries before the present, already associated with the Etruscan, Hellenic and Egyptian civilizations, which are estimated to have existed about 6000 years ago. The comparatively primitive Abyssinian civilization has deep roots, and was contemporary with the Egyptian one or perhaps even pre-dated it. Within the New World, the Central American center is linked to the great Mayan civilization, already in existance before Columbus, which was enormously successful in science and the fine arts. The Andean center is associated with the remarkable pre-Incan and Incan civilizations (Vavilov, 1939).

Floristic analyses show that there is no direct agreement between the number of species or the wealth of the wild flora of various areas in the world and the number of plants taken into cultivation. The total number of plants in the world known to botanists is estimated to be approximately 200 000 species. The very rich flora of South America, estimated at more than 50 000 species, i.e. one fourth of the world's flora, furnished very few cultivated plants. Tropical Africa, characterized by a flora of not less than 13 000 species, has also given rise to very few currently cultivated crops. The Cape province, with a flora that is remarkable because of its richness and that is determined to contain 7000–8000 species, has only recently started to be utilized, mainly for its decorative plants. Only the *Pelargonium* species of the Cape flora have become agriculturally important as a valuable source of etheric oils. Similar statements can be made about Australia.

The qualitative composition of the flora, the presence of developed, agricultural civilizations and the correspondingly large masses of populations played an enormous role in the past for the utilization of the wild flora.

As the investigations have shown, the majority of the species of cultivated plants typical of the geographical centers in question, and associated with ancient agricultural civilizations, did not disperse beyond the borders of the initial areas of type-formation because of their remarkably low agricultural value, geographical isolation or other causes. The majority of the species are still mainly used by the people who took them into cultivation. Out of the number of cultivated plants, determined at present to be ca. 1500 species (not including decorative ones), not more than one fourth have dispersed far from their initial centers of origin or where they were originally brought into cultivation. These have been subjected to further major changes. This applies to an even greater extent to the wild species that are closely related to the cultivated ones and endemic to the area concerned.

This is especially obvious within the Soviet Union, e.g. in the Caucasus, which is exceptionally rich in cultivated and wild endemic species of wheat, rye and fruit trees, many of which have not yet spread across its border.

This fact is of primary importance for plant breeding but was not adequately studied until recently, either by botanists or by plant breeders.

Above, we have dealt with the basic areas of initial origin and introduction into cultivation of the majority of the currently cultivated plants. Now we shall proceed to the further evolution of cultivated plants.

When selecting plants from the wild flora that corresponded best to the needs, as far as fruits, grains or roots are concerned, and taking them into cultivation, the primitive farmer was by necessity also in part a plant breeder, who altered the initial population, in essence creating new strains according to his needs. He selected, e.g., cereal grasses that do not shed their seeds or have brittle spikes or plants with larger and sweeter fruits or bigger roots. The powerful effect of artificial selection already prevailed within the original centers of agriculture.

As can be seen from their geographical distribution, the primary areas are situated mainly within mountains or foothill areas with a more or less steppe-like or forest–steppe type of vegetation and, thus, easily accessible to primitive farmers.

Corresponding to the mountainous conditions, these territories are characterized by very variable macro- and micro-climates and show the effects of factors of geographical isolation. In agreement with this, there is a great diversity both with respect to the species of plants and the intraspecific variation as well as the diversity of minor hereditary units. As shown above by indisputable research, a powerful process of intraspecific type-formation already occurred within the basic territories, as far as both cultivated plants and their wild relatives are concerned.

Different species (e.g. wheat, barley, rye and flax) which have dispersed beyond the borders of their original centers and which occupy wide-ranging distribution areas and have often spread, together with Man, all over the world, can, based on Darwin's theory, be expected to display a wide range of inherited variability. The further evolution of a given species could also show how a xerophytic plant turned into a mesophyte and, later, even into a hygrophyte; how few-seeded forms became multi-seeded, how cross-pollinators became self-pollinators or how forms previously susceptible to infectious diseased developed resistance to them.

As shown by unquestionable research, the greatest alterations concern the physiological characteristics, with respect to which natural selection is especially effective. When studying, e.g., species of wheat or barley, which occupy enormous territories of agricultural areas all over the world, it can be seen how the corresponding environmental conditions, the manner of cultivation and both natural and artificial selection have resulted in independent and sharply distinct populations. Correspondingly, the environmental conditions and selection give rise to different ecological–geographical groups which, on the one hand clearly illustrate the enormous hereditary plasticity and, on the other, the powerful role of natural and artificial selection.

The theory developed during the last couple of years concerning ecological differentiation of species, and about ecotypes from which species are formed, is in essence the same as the theory developed by Darwin linking alteration of organisms to environmental effects and directed selection.

When pursuing the evolution of cultivated plants from a center toward the periphery, it is possible to discover a special regularity of the physiological and morphological structure of various species in agreement with the ecological conditions of agricultural areas all over the world. Much research has been done in this respect. Actual studies of populations, both of local and bred strains of cultivated plants, have revealed regularities, disclosing various facts which cannot be understood without ecological studies concerning the separation of the plants from their original environmental conditions.

Thus, e.g., when studying the evolution of wheat and barley, which have dispersed from Asia Minor and, together with Man, spread all over the world, we can see how a special structure has developed in China (within a secondary center of cultivation of wheat and barley) under the conditions of intensive cultivation and the effects of a monsoon climate. The heavy precipitation during the summer and the associated development of saprophytes and parasites resulted in a natural and artificial selection of forms maturing as fast as possible and with an abbreviation of the following stage of setting grains. In spite of the intensive Chinese and Japanese manner of cultivation, this has led to an extreme reduction in the dimensions of the grains. Modern wheat and barley in China set seed two to three times faster than the ordinary prairie and forest–steppe types do. In addition, these changes are correlated with a reduced size of the awns and have led as far as to the development of awnless barley and especially of awnless wheat, which is endemic to China and Japan. Such development can be traced in the case of different species and different genera as well.

In contrast, in the countries around the Mediterranean, under conditions with a typical distribution of precipitation during fall and early spring, types of cultivated plants have developed that have exceptionally large seeds or fruits. This is associated with the speeding up of the initial growth in correspondence with adequate precipitation during the early stages of development. In correlation with the large seeds, flowers, leaves and other organs are also enlarged.

In any given case, natural and artificial selection proceeds basically in a single direction. The large seeds and fruits, developed under Mediterranean conditions, seem to be due to major ecological factors that determine the development during the early stages.

In Inner Asia, where there is a dry climate, ectoparasites such as rust and mildew do not find favorable conditions for development. A corresponding selection for resistance did not occur there. As a rule, when transferring Inner-Asiatic strains to European conditions, where there is a humid climate, we find them almost without exception to be susceptible to such diseases. On the other hand, in the conditions of a Mediterranean climate, with a remarkably high humidity favoring contamination by parasitic fungi, resistance to such diseases has been achieved due to both artificial and natural selection. In that area, the plant breeder can find remarkable, original varietal material and specific types, developed during thousands of years, that are resistant to both leaf and crown rust. As far as different diseases are concerned, the optimal conditions have been achieved in different manners: under some conditions forms have developed that are resistant to one kind of disease, but under different conditions to another kind. This has been revealed in the case of a number of infectious diseases affecting cereal grasses in China, India and Abyssinia.

When comparing morphological and physiological characters of local species and strains of cultivated plants and species closely related to them with a detailed map of agricultural conditions all over the world, we begin to understand the more or less regular formation of characteristics and the distribution of the corresponding complexes of characteristics.

During its evolution, a species undergoes great changes during the dispersal from its original center of origin with respect to both genetically recessive and dominant characteristics. At the start of the investigations, which concerned mainly the morphological characteristics, I assumed (Vavilov, 1926) that recessive characters usually increased due to the dispersal and isolation of species of cultivated plants. The ordinary morphological characteristics of the wild forms of cereal grasses and other plants – e.g. brittle spikes, fruits breaking open when ripe, scattering the seeds, etc. which favor self-dispersal – usually seemed to be genetically dominant properties.

Peculiar recessive, white-flowered, white-seeded strains of flax, reminding one rather of Stellarias than of ordinary flax, as well as a multitude of other strange, recessive forms of other cultivated plants, have been discovered in isolated oases of western China (at Kashgar) surrounded by the Himalaya, Pamir, the K′un-Lun ranges and the Takla-Makan desert. In the mountains of Cis-Pamir and on the islands in the Mediterranean, peculiar recessive forms of wheat and rye with simplified (non-ligulate) leaves have developed. As a result of inbreeding (of closely related strains) of maize, a number of strange recessive forms, e.g. the so-called 'waxy corn', have developed within China.

Subsequent studies of the genetic behavior of quantitative but also of physiological characteristics have demonstrated that in the case of many of these we find rather the opposite process going on in the direction toward the periphery and during the dispersal of genetically dominant properties.

Large grains and large fruits seem rather to be dominant characteristics within the limits of the various species. The same also applies to a number of physiological characteristics. As far as the characteristics discussed are concerned, the origin of the genetically dominant ones is apparently associated with the role of selection, a fact to which Fisher (1930) drew attention.

From the examples mentioned above concerning the appearance of dominant, awnless, secondary forms of barley in China, where forms have developed in connection with natural selection far from the center of origin of the basic kind of cultivated barley, which are distinguished by small grains and decreased yield, it can be seen, when working out the evolutionary processes, how complicated the interrelationship is.

On the basis of actual data from studies of the subsequent evolution of various species of cultivated plants and their dispersal in correspondence with the environmental conditions, we have succeeded in revealing the distribution areas of forms with structures, that are definitely inherited and characterized by properties frequently valuable to plant breeders. At present, scientists know not only about the range of the variation but also the areas where one is most likely to find the different properties of the greatest interest to them.

Darwin was already very familiar with the facts concerning parallel or analogous variability of different genera and species. We have revealed a very large amount of such facts, based on detailed studies of cultivated plants and those closely related to them, and generalized these facts in the form of 'the law of homological series' in the case of hereditary variability. The essence of such a kind of parallelism and its nature can be explained on the one hand by physiological characteristics, manifested in particular in relation to closely related species and genera, and, on the other hand, by parallelism and similarity possibly being the result of effects of different conditions or different directions of the selection. In the latter case this can also be seen with respect to phylogenetically distant species, genera and even families.

The enormous plasticity of the species, noted by Darwin on the basis of his studies of cultivated plants, was later explained by a large number of facts concerning striking similarities that seem to be acquired by species developing under different conditions.

In his time, K.A. Timiryazev drew attention to the remarkable paper by N.W. Singer, which was devoted to the weeds among cultivated flax, and published in 1909 under the title of 'On the Species of False Flax and Corn Spurry in Crops of Flax'. In this outstanding work, N.W. Singer demonstrated that, under the effects of environmental conditions and cultivation, flax attracted specialized weeds such as false flax [*Camelina linicola* and *C. sativa*], corn spurry [*Spergula arvensis*] and rye grass [*Lolium* spp.] because of their genetic, biological and morphological characteristics that corresponded to those of the basic crop itself and because these weeds were hard to eliminate. This remarkable example of one species simulating another is the result of natural and artificial selection, based on the hereditary variability inherent in the species.

During the last couple of years we have succeeded in revealing a large amount of facts concerning various morphological similarities between different species of wheat, oats, and leguminous plants in such cases where the species or genera contaminate the crops of others or when they grow under similar conditions. In Abyssinia, we discovered environments that are typical of Abyssinian hard wheat, but where the crops were contaminated with soft wheat. The latter could only with difficulty be distinguished from the endemic Abyssinian hard wheat. The difference between them was revealed by their chromosome numbers, the difficulty in crossing them and in some reactions concerning their resistance to parasitic fungi. We were able to observe similar cases in Dagestan with respect to soft wheat, which is contaminated in that area by hard wheat. In western Georgia, two endemic species of wheat (*T. monococcum* and *T. dicoccum*) were discovered frequently growing together and were difficult to distinguish from one another even by specialists. We were able to separate them only by their chromosome numbers, their different resistance to rust and the difficulty of hybridizing them. In Syria, a weedy (42-chromosome) club wheat was found among the original 28-chromosome wheat; the former could only be distinguished with difficulty from the latter by the spike and the general aspect.

A few years earlier, a new species of Persian wheat (*Triticum persicum*) was described by myself. It is distinguished by 28 chromosomes and an exceptional resistance to various fungal diseases. Recently, a multitude of forms of soft wheat (*T. vulgare*) was discovered within the native land of the Persian wheat in Dagestan, Georgia and Armenia. According to its vegetative characteristics, the spike and the grains, these were difficult to distinguish from the Persian wheat, but differences were discovered with respect to immunity, the number of chromosomes and difficulties were encountered when crossing them.

In the Pri-Kama area, we discovered among the emmer [*T. dicoccum*] an interesting group of weedy oats, which, with respect to the articulation of its flowers, reminds one of the structure of emmer. When threshed it does not separate into individual spikelets as is the usual case with ordinary oats, but mimics the emmer spike and lands in the granary together with the emmer. Among the crops of large-grained poulard wheat (*T. turgidum*) in Portugal, we found peculiar forms of weedy sand oats with large, inflated spikelets similar to those of the wheat in question. This weedy form is known by us as *Avena turgida* and is sharply different from the oats ordinarily grown there (*Avena brevis*), which have contracted grains.

Among the relict forms of 'jumping flax' [?*Linum crepitans* (Boem) Denn.] with capsules that crack open when ripe, and which are still grown here and there in the Ukraine and Portugal, E.N. Sinskaya discovered a weedy false flax [*Camelina linicola*], which is also distinguished by pods that break open. In the case of ordinary flax, the capsules do not open when ripe, and the weedy false flax does not open its ripe pods either.

All these examples illustrate the amazing hereditary variability of the species of cultivated plants as well as of weedy species, and the role of an unconscious selection during their formation.

As could be shown by the above examples of rye, oats and various cruciferous plants, the very introduction into cultivation of a number of annual plants is the result of natural selection. Rye and oats are troublesome weeds among wheat, barley and emmer in the native lands of these species. During the transference of the ancient crops of wheat, barley and emmer northward or into mountainous areas with less favorable conditions, rye and oats, being less demanding plants, replaced the original ancient crops and became independent, cultivated plants. This process can still be observed in the Caucasus, Inner Asia, the Povolzhe area and the Pyrenées. In essence, the northerly countries, where enormous areas are used for rye and oats, utilized the result of such a natural selection, which converted these field weeds, which in the past accompanied wheat, barley and emmer, into independent crops.

In Dagestan, it is still possible to observe in great detail how, during the transfer from sea-level into high elevations among the mountains, one cereal is 'changed' into another, in this case wheat into rye. We can see there how, as in a natural laboratory, it is possible to study the powerful effect of natural selection during the formation of populations that correspond to certain conditions.

Geographical research, enhanced by the available methods of modern genetics, cytology and physiology, also leads to an understanding of the basic problems concerning the origin of the species and to a comprehension of the dynamics of type-formation. As is well known, Darwin, who was concerned with the concept of species while dwelling on its conditionality, produced the following hypothesis for the lack of a sharp delimitation between species and varieties: 'Varieties are incipient species'. Using plants that dispersed widely during their evolution as examples, it can be clearly seen how, due to the effects of environmental conditions and geographic isolation, different ecological–geographical populations begin to be distinguished into subspecies and even species while undergoing great morphological and physiological alterations.

Consequently, detailed studies of cultivated plants, progressing from the center of an area toward its periphery, demonstrate how – within various areas – structures sharply different from the original ones begin to form under different extreme conditions. For instance, when comparing the wheat and barley of Abyssinia and China with the original forms in Asia Minor, or the European types with those in Asia Minor, we can, in essence, talk quite correctly about a formation of new species at the periphery of the distribution area. The large-grained soft wheat of India, which was distinguished by Percival from the basic species, *Triticum sphaerococcum*, belongs to this type. The same applies to *T. spelta*, the so-called spelt wheat, which developed at the periphery of its area in the Alps, Tirolia and the Pyrenées. In spite of their secondary nature, the Abyssinian and Chinese barleys can absolutely correctly be distinguished at least as special subspecies. It is difficult to imagine more sharply contrasting characteristics than, e.g., those of wheat and barley in India and those in western Europe. According to the complex of morphological, physiological and biological characteristics, they can with full reason be distinguished as special species.

In contrast to the original cross-pollinating populations of rye in Asia Minor, those in the Far North have become self-pollinating. Under the conditions of irrigation of the crops in high-altitude Cis-Pamir, rye has turned into populations of spring-types with extremely large anthers, large grains and large spikes, correctly deserving to be called gigantic.

As demonstrated by hybridization experiments, such groups and populations, which are isolated geographically and ecologically, frequently display genetical differentiation in the form of difficulties in crossing with each other. This supports Darwin's basic hypothesis, i.e. that 'varieties are incipient species'. His concept of 'varieties' corresponds rather to 'geographical races' in the modern sense.

There can be no doubt that one of the most important pathways of differentiation is the factor of geographical isolation under new ecological conditions. This can be clearly followed within the evolution of cultivated plants and can be proven by a large amount of actual material.

As demonstrated during the past couple of years in the case of flowering plants, genetic divergence can follow different pathways. Based on indisputable research, many cultivated plants and the wild species closely related to them are represented by so-called polyploid series of chromosome numbers, resulting in multiple numerical ratios with which distinct geographical isolation is associated. Wheat, oats and cotton are polyploids and tobacco is distinguished into species differing in chromosome numbers and numerical ratios. Such multiple ratios can be the result of a direct doubling of the chromosome number of interspecific hybrids (allopolyploidy) which, as a rule, leads to the result that such hybrids become fertile. The wide natural distribution of the phenomenon of polyploidy, which is displayed to an extreme extent for at least half of the species of flowering plants, demonstrates its importance for genetic isolation. The dependency of the origin of the polyploids on physiological and chemical factors, as demonstrated during the last couple of years, and also the particular frequency of polyploid series in foothill and mountain areas, show that in part this category can be explained by the appearance of the divergence that interested Darwin so much. It may not at all be general but it is fairly widespread and it is necessary to take it into consideration when dealing with a number of important cultivated plants.

Summing up the investigations concerning the origin and evolution of cultivated plants, modern scientists, following the work of Darwin, have expanded it, which leads to an affirmation of the following steps of the evolutionary process:

1 The basic material for natural and artificial selection is the hereditary variability of the species, i.e. mutations such as minor ones on a quantitative and physiological scale but also major ones, including the remarkable appearance of polyploids among the flowering plants. After causing experimental difficulties, the mutation processes have increasingly become the cause of physiological effects.

2 Hybridization, which gives rise to a huge amount of material for selection, plays in part a great role in the formation of cultivated plants. Distant hybridization has played an important role in the origin of a number of cultivated species.

3 During subsequent evolution, the dispersal of the species, their occupation of new areas and factors of ecological and geographical isolation are of decisive importance.

4 Natural and artificial selection is a basic, decisive factor in evolution and in the formation of adaptation and the creation of species as living structures.

While balancing the total development of the theory of evolution and one of its essential divisions, the theory of selection, we can clearly see that the evolutionary principles of Darwin, i.e. the historical concept of species and the study of the hereditary variability corresponding to the environment and to selection, are decisive for the understanding of cultivated plants. These hypotheses of Darwin's penetrate all plant breeding work, starting with the original material and the establishment of the basic areas of origin of the species, and ending with the creation of new strains, which in essence are new stages in evolution. With full reason, plant breeding can be considered as evolution, directed according to the wishes of Man. We are faced with an enormous task of mastering cultivated plants and of creating forms that will satisfy the demands posed by a socialistic rural economy. This is impossible to conceive without a historical, ecological–geographical understanding of species and their evolution.

Darwin's theory of evolution is a basic and unique theory that has stood firm for more than 80 years. During their immediate work, botanists, zoologists, geneticists, plant breeders and ecologists as well as phyto-geographers are affected by this universal theory, and it is only on the basis of this that it is possible to understand the process of evolution and the functioning of organisms.

BIBLIOGRAPHY

Darwin, C. (1859). *On the Origin of Species.* London. (Translated into Russian in 1937.)

De Candolle, A. (1855). *La géographie botanique raisonnée.* Paris.

Fisher, R. (1930). *Genetical theory of natural selection.* Cambridge.

Thellung, A. (1930). *Die Erstehung der Kulturpflanzen* [The evolution of cultivated plants]. Munich.

Vavilov, N.I. (1926). Tsentry proiskhozhdeniya kul'turnykh rasteniy [Centers of origin of cultivated plants]. *Tr. po prikl. botan i selek.* [*Papers on applied botany and plant breeding*], **16** (2).

Vavilov, N.I. (1939). Velikie zemledel'cheskie kul'tury dokolumbovoy Amerike i ikh vzaimootnosheniya [The important agricultural crops of pre-Columbian America and their relationships]. *Izv. Gos. geogr. o-va* [*Publication of the State Department of Geography*], **71** (10).

Introduction of plants during the Soviet era and its results

A review of the work with introduction of plants at the All-Union Institute of Plant Breeding, 1921–1940

THE HISTORY of worldwide plant industries during the past century clearly shows the enormous and frequently decisive role played by correctly introduced new plants and new varieties. The entire production of the plant industry in countries such as Canada, Brazil, Argentina, Australia and South Africa is based on cultivated plants adopted from abroad. The worldwide production of raw vegetable material during the nineteenth and twentieth centuries was associated with the transfer of crops from one continent to another. The prosperity of Canada is based entirely on European crops and that of Brazil on the transfer of the coffee tree from Abyssinia and the sugar cane from tropical Asia. All the well-being of Argentina and Australia is built on the successful introduction of crops from the Old World. The greatest events within the worldwide plant industry are, in essence, linked to introduced materials, which, in turn, were directly associated with the agricultural activities of the peoples and with the stages of development of the human societies.

The concept of introduction can be subdivided into three parts:

1. The adoption of new species and varieties of cultivated plants from other countries.
2. The transfer of crops from one area to another within the borders of a country. The build-up of cotton and rice cultivation within new areas during the Soviet era can serve as an example.
3. Introduction of new plants from their associations within the wild floras, both native and foreign.

The closeness of the Soviet Union to the main centers of origin of the most important plants now cultivated in Europe, such as wheat, barley, rye, fruit trees and a number of vegetables, even assured – in the past – the exceptional value of our local varietal complex of cultivated plants, which is the result of long-term natural and artificial selection. Even today, our local strains of wheat, rye, barley, flax, clover and fruit trees are, when compared with foreign ones, distinguished by an exceptional cold-hardiness, drought-tolerance and a considerable productivity, such as required under our conditions. Our northern long-staple flax has no competitors, nor have our red clover, our winter wheat

Lecture presented to the Conference on Botanical Gardens at the Academy of the Sciences, January 1940. First published in *Izbrannie Trudy*, [*Selected Papers*], Vol. 5, no. 5, pp. 674–689, 1965.

of the 'Banatka' type or the hard and soft spring wheats grown on steppes and prairies. However, the necessity of developing horticultural and vegetable gardening already made it imperative in the past to turn our attention to foreign strains and crops.

The serious introduction of plants was apparently first associated with the Greek colonies on the Black Sea coast of the Caucasus and Crimea. The laurels, olive trees and groves of stone pines are the remnants of these ancient introductory activities. Watermelons, originating from deep within Africa, had already penetrated into the Povolzhe area some centuries before our time. According to an order from Czar Aleksey Mikhailovich, black soil was brought from Astrakhan for the purpose of assuring normal growth of watermelons near Moscow. Bold attempts were made to grow southern crops, even including date palms and almonds.

The birth of industrial capitalism under Peter the First [the Great] was associated with the introduction of many new crops, including potatoes. The Free Economical Society, organized in the middle of the eighteenth century, developed a considerable amount of activity writing for seeds from abroad. 'Apothecaries' gardens' were established in St. Petersburg, Moscow, Kiev, Astrakhan, Lubny, Yekaterineburg and Tobolsk, where they began to grow strange medicinal plants. As is well known, the history of botany actually began with the establishment of such 'apothecaries' gardens'. Together with this activity began a considerable development of the work with the introduction of plants (Kovalevskiy, 1929).

I shall not dwell on this early history of introduced material but shall only mention that the early introduction of crops, such as that of sunflowers, is linked to what was happening in these early gardens.

An important role for the introduction of plants was played by foreign colonists, especially the Germans, who during the second half of the eighteenth century introduced a number of valuable strains of vegetables and potatoes. Attempts were already being made during that century to introduce foreign strains of oats. It is impossible to overlook the introduction of tobacco, which had already penetrated into the Ukraine at the end of the fifteenth century in spite of the prosecution of smokers by the governmental authorities. Severe measures were taken in the war against the introduction of tobacco even before Peter the Great. During the eighteenth century, foreign dye plants such as madder [*Rubia tinctorum*], safflower [*Carthamus tinctorius*], dyer's woad [*Isatis tinctoria*] and dyer's rocket [*Reseda luteola*] came from abroad into the cultivated flora of our country. Teasel [*Dipsacus sylvestris*] was also introduced. Under Peter the Great a number of new vegetables such as asparagus [*Asparagus officinalis*], celery [*Apium graveolens*], artichokes [*Cynara scolymus*], purslane [*Portulaca oleracea*] and Jerusalem artichokes [*Helianthus tuberosus*] appeared. The cultivation of lemons and pineapples was also started in protected areas.

A great role for further introduction was played by the Nikita Botanical Garden [in the Crimea], which was created mainly during the nineteenth century and which devoted its activities to the introduction of new crops and strains from abroad. A remarkable period of well-planned introduction of valuable varieties of fruits, grapes and decorative plants is associated with the names of its first directors, Steven and Hartwig. This became very important, not only for the south coast of the Crimea but also for other areas of the European parts of our country.

The history of the introduction of subtropical crops to the Black Sea coast of the Caucasus is very instructive (cf. the excellent review by S.G. Ginkul about the 'Introduction and Naturalization of Plants within the Humid Subtropics of the U.S.S.R.', 1936). Among the measures taken in our Subtropics in the past we can mention the energetic activities at the Suchumi and Sochi stations and at the botanical garden of Batumi. Expeditions from a department of the Independent Party, led by the agronomist, I.N. Klingen, and Professor A.N. Krasnov, made systematic collections of materials for gardens and field crops in China, Japan, Ceylon and Himalaya to be used for our Subtropics. They also facilitated the establishment of the first large-scale plantations of tea, citrus fruits and bamboo, as well as other subtropical crops, on the coast of the Black Sea. In this connection, it is impossible to forget the activities of various amateurs and enthusiasts, such as V. N. Chernyavskiy, A.N. Vvdenskiy and N.N. Smetskov. The results of these introductions, according to an estimate by S.G. Ginkul, led to the fact that the number of plants introduced to the Black Sea coast of Caucasus and the belt between Sochi and Batumi could be set at 1050 species and varieties, as well as horticultural strains, in all belonging to 319 genera. The outstanding activities of horticulturalist L.I. Simirenko, concerning the introduction of fruit crops brought into our country, resulted in an exceptionally valuable collection of fruits, now widely dispersed into gardens.

The remarkable activities of the agronomist and plant breeder V.V. Talanov are associated with the Soviet era. He displayed exceptional persistency and energy when organizing the introduction of more valuable strains of field crops from America. The introduction of better strains of maize, millet, Sudan grass [*Sorghum × drummondii*] and American, non-rhizomateous wheat grasses [*Agropyrum spp.*] are connected with his name.

The activities concerning introductions affected in the past also such sections as the planting of green belts, which is important even at present. The poverty of the dendrological complex of Eastern Europe in comparison with that of the forests in North America, China and the Far East has long since turned the attention of foresters and dendrologists to the introduction of foreign species. To a great extent, the entire history of forestry, park management and the cultivation of forests and trees on the steppes – the 180th anniversary of which will be celebrated in 1943 – is associated in our country with the introduction of foreign woody species. Found primarily in gardens and parks, the foreign species – or as they are usually called, 'exotics' – underwent an initial testing period where they were spread into forests and used for plantations of green belts. In such areas as the Ukraine and the northern Caucasus, the North American sweet acacia [?*Albizzia julibrissin*], honey locust [*Gleditsia triacanthos*], Pennsylvanian green ash [*Fraxinus pennsylvanicus*] and the American maples [*Acer* spp.] have become the most widespread kinds.

The role of the Forestry Academy and particularly that of the late Professor

E.L. Wolf must be mentioned. They tested more than 3600 species and varieties of trees in and around Leningrad. We owe a great many introduced species and the cultivation of a multitude of valuable imported woody genera to them, and also to the well known firm of Regel and Kesselring.

All our culture, including that of indoor plants, mirrors in essence the results of spontaneous introductions during the past. Indian weeping fig [*Ficus benjamina*], Chinese hibiscus [*Hibiscus rosa-sinensis*], South American begonias and South African geraniums have long since become attributes of every home.

The activities of the botanical gardens concerning introduced material, taking the form mainly of an exchange of seeds of wild flora, has taken on an extremely wide-ranging character. In the interesting book by Professor G. Kraus (1894) about the history of introductions to botanical gardens, it is stated that in 1890 an estimated 25 500 plants were counted in the Petersburg Botanical Garden, i.e. considerably more than even in the Kew Gardens of London. However, this introduced material of the past is poorly represented in the plant industry except for that of the Nikita Botanical Garden which, in essence, is an agricultural establishment.

The Department of Applied Botany, under a committee of the Department of Agriculture (later changed into the Institute of Applied Botany and New Crops and finally into the Institute of Plant Industry), was – until the October Revolution – occupied mainly with research on our own local strains, although it also included the introduction of new crops into its program.

This continued until 1921. During the drastic change in matters concerning introduced plants that occurred after the October Revolution and during the change-over to a systematic introduction of material, the interest shown by V.I. Lenin in these matters played a great role. When he was ill he read the book by Harwood, *The New World*, translated by K.A. Timiryazev, in which the importance of the introduction of new plants for the development of American agriculture is described. Therefore, Vladimir Ilich [Lenin] emphasized the necessity of turning our attention to these matters also in our own country. Thus, the All-Union Institute of Applied Botany and New Crops developed its first link with the Lenin Academy of Agricultural Sciences. Radical changes took place within the Soviet plant industry with respect to the composition of the crops. There was a remarkable increase in technical crops, an expansion of agriculture toward the north and east and a development of subtropical agriculture, which in turn led to a systematic utilization of research units from all over the world. The exceptionally dry year of 1921 made us turn our attention to an extensive introduction of all the best material from countries most similar to ours with respect to climate, e.g. mainly standard strains from Canada and the U.S.A. to be used for our steppe and forest–steppe areas. The Soviet period of large-scale introduction of plants had begun.

Thanks to the direct participation from the very beginning of a scientific team, the Department of Applied Botany under the Agricultural Research Committee (i.e. the later Institute of Applied Botany and New Crops) started to introduce new standard varieties. Due to a wide-ranging enactment during the first decades of Soviet authority, a complete change took place within the field of plant industry, especially as far as the old, local brands of cotton in Inner Asia were concerned. Thus, the so-called Zavodskiy mixture of upland cotton, which had already been introduced during the 1880s, and what remained of the Asiatic cotton (*Gossypium herbaceum*) was exchanged for improved standard varieties of American cotton together with a simultaneous application of a selection of the best individual strains. The foremost Canadian brands of spring wheat were sent by the thousands of hundredweights to the Soviet Union. During this period, an extensive and systematic utilization of what was available of Swedish and German breeds of field crops was made for the belts of the U.S.S.R. that do not have any black soil. Valuable and well-bred strains of Swedish and German oats, such as 'Pobeda', 'Gold Rain' and 'Leytevitskiy' were taken into cultivation. They still occupy millions of hectares in our country. Czechoslovakia furnished valuable 'brewing' barley. Standard American brands of maize such as 'Minnesota 13', 'Minnesota 23', 'Northwestern', 'Lemming', 'Sterling' and 'Ivory King' found a second homeland in our country. A mass-introduction of the best standard strains of vegetable crops from Denmark, Germany, France and the U.S.A. also took place. Varietal material for subtropical crops was acquired in an enormous quantity, unprecedented in the past, and based on data from our research with them in the greenhouses used for introduced material. Special expeditions brought back material from China and Japan for crops of tea and citrus fruits. Other expeditions were sent to Algeria, Spain, Italy, southern France and Japan to bring back large amounts of a standard assortment of crops for our dry and humid subtropical areas. Special trips were also made to Turkey and Iran to purchase the best brands to be used in our dry Subtropics.

In addition to the plants already discussed, we should also mention the great importance of the imported foreign varieties of alfalfa, sudan grass [*Sorghum × drummondii*] and various other kinds of sorghum and beans. The altered assortment of apples, pears, plums, apricots and peaches proved exceptionally successful and was a radical example of the complex of brands presently grown in our gardens. This also led to a considerable extension of the period over which fruit can be used for the canning industry. Egyptian cotton has begun to be grown over extensive areas.

comparison with the enormous range of practical material that has been introduced during the last two decades. All the best standard varieties from western Europe, Canada, the U.S.A., Argentina and Japan have been subjected to 'governmental examination' on Soviet fields. No less than 10%, or 140 million hectares, of the total cultivated area are at present occupied by introduced species and varieties that have been propagated in our country.

Experiments with introduced industrial plants were to a great extent based on the hypothesis about climatical analogy and furnished extremely instructive and positive results. This can be seen, e.g. with respect to maize, west European oats, 'brewing' barley, fruit and vegetable crops, as well as many subtropical plants. At the same time, some mass introductions proved – frequently after great expenditure – to be less well-suited for our country than such crops of our own as winter and spring wheat, flax and clover, which already belong to our native crops. The 'champions' among the American brands had to yield to our ancient but better bred Soviet varieties of wheat, barley, rye and flax.

Direct, large-scale Soviet research showed, in complete agreement with research in the U.S.A. and Canada, the necessity for a concurrent development of plant-breeding work based on a wide-scale application of original material. Even the remarkable and very advanced brand of American upland cotton of the 'Acala' type needs further work, using individual selection. The same is apparent with respect to many vegetable crops. Just as in America, we are faced here with problems concerning plant-breeding work, assured by a systematic collection of original material, which is especially necessary for a large-scale application of hybridization. This is one of the most basic methods used for plant breeding. Using Canada, the U.S.A., Australia, Argentina and Italy as examples, the history of worldwide plant breeding clearly demonstrates the importance of correctly selected original material for successful breeding work.

This has led us to study mainly the U.S.A., which is especially well represented by its work with introduced material in order to carry out definitely successful breeding work. Systematic organization of the introduced material for successful breeding started in the U.S.A. at the end of the nineteenth century and is associated with the name of Dr. Fairchild. This was recently well documented by Fairchild himself in a semi-fictional autobiography entitled *The World Was My Garden*.

Numerous expeditions from the Department of Agriculture in the U.S.A., dispatched by the Bureau of Plant Introductions, included, by and by, all the agricultural areas of the world within their sphere of activities. Such American plant importers as Frank Meyer, Carleton, Hansen, Westover and Whitehouse have repeatedly been seen in our country. The results of their work are periodically published in the well-known *Inventories*. In 1927, work with introduced material was begun in Australia under the supervision of the energetic Dr. Wenholz. Borrowing from Soviet research, organized work with introductions – which during the last decades assumed a very wide range – was started in Germany on the initiative of the plant breeder Erwin Baur.

Unfortunately, the organization for introducing plants to the U.S.A., although technically perfectly equipped, does not attain the appropriate level from a scientific point of view. The book by Fairchild indicates this. After carefully studying it, we could not find any theory behind the introductions. In essence, all matters point to an empirical search for plant material more-or-less suitable for the U.S.A. The book by Fairchild confirms only the exceptional opportunities existing for the American work with introduced material, including special yachts equipped with the latest word in technology and constructed for the purpose of obtaining the plants to be introduced and for traveling around the world in comfort. In this sense, the work at the Arnold Arboretum near Boston really seems much more serious, e.g., as when sending out expeditions under the outstanding scientist Wilson.

Nevertheless, in activities concerning introduced material, the work carried out in the U.S.A. is very instructive, especially after a detailed acquaintance with it. All the introduction of cultivated plants and, to a considerable extent also that of wild plants, is centralized and subjected to quarantine. Within the organization of the Bureau of Plant Introductions there are splendid nurseries in different states, e.g. in Florida and California as well as in Washington. In addition, there are research stations for introduced material in Puerto Rico and around the Panama Canal. The Arnold Arboretum has a branch in the southern tropical part of Cuba at San Fuegos. The wide-ranging collections made by Scoffield, Carleton and Harland in the case of cereals, by Piper and Westover in the case of forage plants and by Swingle and Whitehouse in that of subtropical plants and fruits, as well as the collections made by Fairchild himself, have furnished an enormous amount of material for practical plant breeding. The complex of nurseries in the U.S.A. for the material collected is remarkable because of its variety and is undoubtedly of interest to us with respect to material introduced into the Soviet Union. Documentation concerning the expeditions that have been made are concentrated mainly in the archives of the Bureau of Plant Introductions in the form of unpublished but comparatively well-preserved material. During the last couple of years, the attention of the Bureau of Plant Introductions has focused especially on potatoes, decorative plants and drug plants.

When, in 1921, we were entering upon a systematic organization of the work with introduced plant material, and proceeding on the basis of the successful Soviet plant breeding, we considered mainly the necessity for making use, in general, of the research done in the capitalistic countries of western Europe and America. Not having an opportunity during the short period of time when I visited America in 1921 to acquire all the necessary varietal material, we opened a small office in New York on behalf of the National Commissariat of Agriculture in order to be able to continue my work. As a result of the activities of this office, over a period of three years, an enormous amount of seed material from seed firms and research stations was obtained. Introduction of field crops, cereals and technical crops began and was, subsequently, expanded to include vegetables, and cucurbitaceous and fruit crops. The activities of this office did not neglect any of the important plant-breeding stations in the U.S.A. or Canada, nor any of the important seed firms. Studies for this work in Washington at the Bureau of Plant Introduction, the collection of a very large material of strains from research stations and private firms, and preliminary research led us to realize the necessity for a radical change in all the work concerning the introduction of plants. It had to be based on the theory of evolution and data from phyto-geography. From this point of view the pathway proved, in essence, not to be very well paved. Only in rare cases had the material collected by the American introducers been subjected to serious taxonomic and phyto-geographical investigation. As a rule, the material collected was not scientifically identified. Only what was more-or-less suitable could be extracted from it; all the rest could, at most, be preserved without having to sow it first. The catalogues of the seed firms, even the first class ones, which offer a large number of strains, did not as a rule indicate the origin of the material, which made the utilization and evaluation of it difficult, not to speak of the confusion which usually characterized the varietal nomenclature or – to put it bluntly – the strains themselves.

The first plant-breeding stations in our country, created during the first decade of the twentieth century, started as a rule with laying out nurseries for foreign seed material ordered from the catalogues of two well-known foreign firms, the French 'Vilmorin-Andrieux' and, especially, the German firm 'Haage & Schmidt' in Erfurt. This material was used for replenishing our local strains.

In essence, the composition of the varieties grown in our country in the past, i.e. up to the time of the October Revolution, was unknown. The first systematic review attempted concerning our varietal resources was carried out by the Institute of Plant Industry only during the Soviet era. This investigation revealed an exceptional wealth of local cereal varieties, including species of wheat in Transcaucasia unknown until then. Proceeding on the basis of the theory of evolution and the origin of cultivated plants and their variability, as well as on the basis of research on foreign material, directed by Soviet scientists, it became eminently clear to us that the nurseries used for the collections were imperfect. It became obvious that, in essence, the local material of strains, especially those from areas basic to the development of forms of the most important cultivated plants, was neglected by the seed firms and the research stations. For the purpose of controlling the cultivated species and genera and their closely allied wild relatives and providing Soviet plant breeders with original material, it was evidently necessary to make systematic, phyto-geographical studies of the varietal composition of the species, starting with the original territories – i.e. within the centers of origin of cultivated plants – and working out toward the periphery of their distribution areas, including the results of modern plant breeding. Such problems could be solved only as follows, i.e. by means of special expeditions.

In addition to studying the varietal composition of the plants grown in our country, the Institute of Plant Industry started in 1923 to send out one expedition after another. Out of 180 expeditions dispatched from the Institute during the last two decades, 140 operated inside the borders of the Soviet Union, while the remainder were sent out across our borders. With respect to the crops of interest for our Union, these expeditions covered a major portion of all the agricultural areas in the world (65 countries). They collected an enormous amount of material from North and South America, all the countries situated around the Mediterranean, Abyssinia and Egypt. A considerable portion of the Asiatic continent was also studied.

The theory about introductions did actually grow to a great extent out of the facts revealed by phyto-geographical research and the investigations made by expeditions. The first draft of the theory concerning the search for introductions was presented in 1926 in my book about the *Centers of Origin of Cultivated Plants*. Now, I will only briefly allow myself to touch upon the basic problems concerning this theory, not being able to dwell adequately on the theory about introduced plants, which in my opinion is directly linked to problems of the origin and evolution of cultivated plants.

The prime theory, which is basic for all introduced material, is the theory of evolution according to Darwin, actualized by its applicability to cultivated plants. When applying it to any group of cultivated plants or to individual species, it becomes necessary to establish the primary areas of development of the plants and their origin, i.e. those areas that Darwin called 'centers of origin of species'. Such centers can be established for cultivated plants by searching for the territories where the link between the cultivated and the wild plants has still not been broken, where closely related plants still exist and where, as a rule, the original varieties of an intraspecific complex can still be found. Summing up the great team work carried out, it should be mentioned that the majority of cultivated plants have still not spread beyond the borders of their original native lands.

According to a very thorough estimate by Professor E. Wulff, the total number of cultivated plants can be determined to be 1500–1600 species, not counting the decorative herbaceous and woody ones. I speak here only of species already definitely taken into cultivation, not of prospective candidates, even if they have been tested within comparatively limited areas. The overwhelming majority of cultivated plants that have been described has, as we have been able to establish, not crossed the borders of their original centers. Only about 400 species have dispersed beyond their original geographical centers. The research, including that on both edible and technical crops, vegetables and fruits, cleared up the localization of entire cultivated floras, typical of special geographical centers. In my lecture on the origin of cultivated plants to the Conference on Darwin in the November of 1939, I dwelt at length on the complexes of these cultivated floras and their geographical affiliation (Vavilov, 1940).

At present we distinguish seven major cultivated floras belonging to seven geographical centers. These are: tropical south Asia, eastern Asia, southwestern Asia (including the Caucasus), the Mediterranean area, Abyssinia, Central America (including southern Mexico) and the Andean geographical center, covering the mountains and foothills of Ecuador, Peru and Bolivia.

Within the seven cultivated floras mentioned, which belong to the seven main geographical centers discussed, the largest number of species come from tropical Asia. This is followed by eastern Asia, with southwestern Asia in third place. Comparatively few cultivated plants originated from the Mediterranean area, Abyssinia and the New World. The richest flora is that found in South America, estimated to include about one third of all plant species in the world, but it has furnished only 8% of all cultivated plants; however, this figure does not include any decorative ones. This only shows that the large number of species within a particular flora is not of decisive importance but that the character of the flora is.

The fact is that a distinctive localization of the original specific diversity of the overwhelming majority of cultivated plants is not paradoxical. Even at present the area of the globe that is cultivated is estimated to be only about 850 million hectares, i.e. it does not occupy more than 7% of all the landmass. Thereby, the greatest expansion of the cultivated area occurred mainly during the second half of the nineteenth and during the twentieth centuries. The enormous expanses of prairies in the U.S.A., Canada, Argentina and Siberia became the subjects of cultivation only during the last two centuries. The ancient agricultures, with which are associated the introduction into cultivation of the majority of the most important plants, were situated in nearby territories, mainly in the geographical centers discussed, which are usually located in foothill and mountain areas.

The investigations of the geographical centers in question and the later studies of the material collected as field crops and grown under various conditions, revealed a multitude of new, previously unknown species. The well-executed, although modest Soviet expeditions have exceeded all expectations by establishing a multitude of new species of cultivated and wild potatoes and a lot of new species of wheat and rye, as well as many kinds of pears and cherry plums [*Prunus divaricata*]. In addition to an exceptional diversity of species of both cultivated plants and, in particular, of their closely allied wild relatives, the scientists encountered within these centers an exceptional variety of forms within the limits of the species. Such small countries as Abyssinia or the Soviet and Turkish Armenia are distinguished by an amazing wealth of wheat varieties, far exceeding that in all the rest of the world. The climate and soil conditions of the mountainous countries are distinguished by an exceptional diversity, which is related to the processes for the differentiation of forms into ecotypes that are so sharply expressed just there.

Of course, some species arose outside the centers in question but in the past these were few and they can be easily counted. A wide-ranging utilization of the world's flora has only recently begun, other than in the main areas of the major cultivated plants discussed, especially with reference to the introduction into cultivation of forage and technical plants and those producing etheric oils.

As shown by direct research, different species of cultivated plants, widely dispersed beyond the borders of their original centers, have undergone major hereditary changes. Even species not subjected to hybridization display great changes that are definitely inherited. Definite ecotypes, corresponding to the environmental conditions, have developed on the basis of major and minor mutations, including the phenomenon of polyploidy, and natural as well as artificial selection; these are frequently very different from the original types. Species widely dispersed from the initial center display, as a rule, a complicated system of ecotypes. Our experiments with introduced plants show decisively the necessity for a different approach to the species concept as such. The association of the majority of cultivated plants with mountain conditions is a decisive prerequisite for the presence of the great diversity that is already found within the initial centers.

Success of the introduced plants is frequently connected with a correct utilization of the concept of variety and that of ecotypes. 'Species' appears to be a historical and ecological concept. As a rule, a species appears as a complicated system of forms and is frequently differentiated in agreement with the diversity of the environmental conditions and far from indifferently, whatever part of the specific diversity the introduced material embraces. Entire ecological groups of various species are characterized by resistance – or, inversely, susceptibility – to diseases, by drought-tolerance or by hydrophily.

Until recently, in matters concerning introduced material (especially in the case of dendrology), the hypothesis of climatic analogy developed by Hendrich Meyer predominated, i.e. that species for introduction must be looked for from corresponding climates. This theory is no doubt of great general importance and has furnished good results in practice. Both the American and our Soviet introduction of plants were to a great extent built on this idea. Meyer's theory, based on a large number of data, has played a great role and will, no doubt, be of use also in the future. With respect to trees, our theory demands that consideration be taken also of the ecotypes and to the complexity of the species, something to which Hendrich did not pay any attention. Frequently, the limited initial area depends on the presence of some kind of barrier. However, a species can also be definitely cosmopolitan, as demonstrated, e.g., by the acacia [*Albizzia julibrissin* (Willd.) Durazz.], introduced by us from the state of Virginia. The climatological analogy as such could be exchanged for an ecological analogy, i.e. a more complex consideration of the interrelationship between the environment and the development of the plant. Ecology and climatology are not the same. Particularly large discrepancies can be observed during the application of this theory to herbaceous plants. During our experiments with important introduced plants, facts were revealed that were in sharp contrast to Hendrich Meyer's scheme. Thus, e.g., Abyssinian cultivated plants, especially barley and leguminous plants, grew better, not worse, than they did at home in their native land when transplanted northward from the equatorial conditions with short days into the boreal conditions with long days. The same applies in the case of cultivated and wild species of potatoes from Ecuador, Peru and Bolivia, which develop excellently at the Polar Circle, better even than around Moscow or in the Ukraine.

It is hard for a phyto-geographer to imagine a more amazing event than the behavior of many decorative herbaceous plants at the Polar Circle. Dozens of species, originating from Mexico, Chile, Peru and the Mediterranean area, grow as beautifully at the Polar Circle as in their homelands. The Mediterranean snapdragon [*Antirrhinum majus*] is able to complete a full cycle from seed to seed at the Polar Circle. Nasturtiums [*Tropaeolum* spp.], common weeds in the foothills of Peru, i.e. approximately on the equator, can be successfully grown in Leningrad and even at the Polar Circle. When their roots are protected during the winter, peonies [*Paeonia* spp.], originally introduced from Guatemala and Mexico, develop profusely and flower at the Polar Circle even better than in their own homelands. Barley, collected from all over the world and growing under different conditions, sets seed normally even under the conditions at Hibiny [on the Kola Peninsula] on the Polar Circle. Even Mediterranean oats, sown in their native land in the fall, can usually set seed normally and ripen at the Polar Circle, thanks to vernalization.

It is necessary to distinguish specialized plants with a limited capacity for adaptation from the cosmopolitan ones. Thus, the State Seed Control has demonstrated that, e.g., 'Lyutestsens 62' spring wheat can grow only within the forest–steppe and steppe zone of the European part of the Soviet Union and in the area around Amur, but holds first place among the crops there.

Studies of the evolution of cultivated plants on the basis of an ecological map of the world reveals a number of regularities during the formation of ecotypes and the distribution of different characteristics specific of both the different species and of closely related species, genera and families.

He who introduces plants from abroad must have a wide-ranging outlook. In addition to a basic knowledge of phytogeography, he must learn about the history of the plant industry, the ecology of cultivated plants and those closely related to them, and must know the history of plant breeding all over the world. He must adhere to a strict scheme. Odd facts are frequently linked to the history of introduced plants. Thus, e.g., the well known American brand of oranges, 'Washington Navel', was originally discovered in Brazil (in the province of Bahia), although the basic native land of oranges is in eastern Asia.

Development of taxa, including everything up to speciation, occurs not only within the initial centers but also at the periphery. Different forms and different geographical races can, after lengthy isolation under contrasting environmental conditions, become converted into species.

For the modern introduction of plants, not only are improved standard versions of great interest but also the wild or semi-wild forms, which can be objects of hybridization. During the last couple of years, genetics has extensively widened the possibilities for distant hybridization by means of artificial restoration of fertility due to the doubling of the number of chromosomes. The horizons of an introducer of plants from abroad now stretch around the entire world.

During the last 20 years of large-scale systematic introduction of plants, Soviet plant breeding has acquired a rich material of species and varieties at its disposal. Successively, the work at the Institute of Plant Industry with introduced plants has embraced field crops, technical plants, vegetables, fruits and subtropical crops. Special attention has been paid during the last couple of years to new kinds of crops, such as those of tung-oil trees [*Vernicia fordii*], jute and medicinal plants new to us, as well as to rubber-producing plants. The range of the introductions can be judged by the number of samples brought back: the collections of wheats made by the Institute of Plant Industry amounted most recently to 36 600 samples, that of maize to 10 022, of leguminous plants to 23 636, of vegetables to 17 955, of fruits and berries to 12 651 and of forage plants to 23 200; the total number of samples brought back amounts at present to ca. 250 000. All this enormous amount of material consists not just of a collection but of a 'bank' of living plants, providing the Soviet plant breeders and the Soviet plant industry with original material. In contrast to the American introduction of plants, the Institute of Plant Industry is lately occupied not only with gathering material and reproducing it, but also with fitting it, as far as possible, into a strictly scientific system. In addition, special nurseries have been built for the distribution among the research stations all over the country of imported material to be tested under conditions different from those natural to the imported plants. The stations range from the Polar Circle to the periphery of Turkmenia and the Far East. Some of them have now been transferred into the control of the republics.

An equal parceling out of the material collected is, as a rule, not justified. With respect to the conditions of our climate, it is really impossible to preserve a worldwide assortment even of wheat. Thus, e.g., during the winter of 1927–28, out of 10 000 samples of a collection of winter wheat, sown at Kuban' [in SW Siberia], 95% perished. This matter needs considerable centralization and regulation of the distribution of the nurseries as well as constant control over the purity of the collections. The latter is especially difficult in the case of cross-pollinators. For the purpose of a sensible utilization of the varietal material, problems arose not only concerning its arrangement within the ordinary botanical system, even a differentiated one, but also concerning its arrangement, if possible, within a system that takes physiological and ecological characteristics into consideration in conformity with its utilization within different areas.

The first stage of the botanical research has already, to a great extent, been outlined and formulated in the book about *The Cultivated Flora of the U.S.S.R.*, which is, in essence, the first attempt toward a general differentiation of the cultivated flora with respect to all plants of interest to the Soviet Union. During the development of the work with the introduced plants, the Institute recently concentrated its attention on revising the agronomical classification by using data concerning all the physiological and biological differences. This is an extremely tedious task, requiring the revision of almost every method used. If the variety of parasites common only to wheat is taken into consideration, this proves to be different within different areas and, therefore, the complexity of this work can be understood. Basically, and with respect to the most important cereal crops, this work is in part completed and will be published within the near future. Thereby, a more precise definition of the collections has been achieved, i.e. a reduction of the enormous diversity, which is hard to overlook, into a well-composed system, which is to a considerable extent general for all plants, which, during their evolution and their history, were subjected to the effects of different ecological conditions.

The subsequent stages, to which we have actually already turned with respect to a number of crops, consist of selection of special, original material according to the revised theory concerning the selection of pairs in conformity with various conditions for use both during intraspecific and distant hybridization. The present development of genetics, which has considerably widened the plant-breeding basis to include distant species (allowing, e.g. wheatgrasses to be crossed with wheat and sorghum with sugar cane), and physiological research have, during our time, opened up new perspectives. The enormous role played by photoperiodism, discovered by Allard and Garner and thoroughly revised by Soviet physiologists, has provided us with an extremely interesting outlook.

Geographically, the harvests obtained during the last couple of years according to a definite plan while using a special selection of varieties of different geographical provenance have revealed extremely interesting regularities. Thanks to the long daylength in the north and the low temperatures at the start of the season, vegetative development eventually proceeds extremely fast and even late-spring types of barley and oats are able to ripen under near-arctic conditions. The work done by I.A. Kostyuchenko and T. Ya. Zarubaylo demonstrates the appearance of a natural vernalization in the case of cereal ripening, which has also been confirmed by English scientists. Under the special conditions in Pamir, infertile triploid species of potatoes became fertile thanks to a doubling of the chromosome numbers.

The difficulties encountered, especially in the case of tropical crops, lie in finding protected ground for the preparation of seedlings. Thus, we have succeeded in making perennial, first class quinine trees produce annual crops. This year we laid out the first state farm plantation of quinine trees on the Black Sea coast. The effect of heteroauxin and other hormones has promoted rooting extensively, which until recently presented difficulties during the vegetative propagation of many tropical and subtropical plants.

Statistical data concerning the utilization of the most interesting areas of the world (except for those in the Soviet Union) demonstrate that material from western Europe, the Mediterranean, North and South America, eastern Asia, southwestern Asia and Abyssinia can be adequately widely used.

The worldwide collections that have been made are extensively utilized by various plant-breeding establishments and laboratories. No less than 1 million parcels of specimens have been sent out from the center in Leningrad alone during the last 10 years. Including what has been sent out from the research stations of the institute, this figure grows to 1.5 million. All this enormous amount of material serves as a basis for trying out new strains both by means of direct selection and, in part, by means of hybridization. A number of strains have already been put into production. No less than 254 varieties have been put into production by the institute itself and its research stations; one half of this consists of fruit and berry crops. In this figure, 63 crops are not included, since they were taken into cultivation after the introduction of material for the industry on the recommendation of the Institute and on the basis of research there. In addition, 52 varieties have already been introduced by other establishments, but based on collections obtained from the Institute of Plant Industry. More than 200 varieties are available from the State Seed Control, many of which appear to have exceptionally good prospects. In all, out of the worldwide collection, the introduced varieties occupy more than 2 million hectares, in addition to the large amount of introduced varieties, mentioned above, which were obtained as a result of introductions for industrial purposes.

Among the most interesting introduced plants we can mention is the new species of 'Khoranka' wheat, collected by us in Syria and Palestine. This year it occupies some 90 000 hectares. A number of strains of spring and winter wheat, resistant to brown and yellow rust, and the extremely drought-tolerant varieties of hard wheat from Tunisia, Palestine, Morocco and Syria also belong here. More than half of the varieties of barley selected and now growing in the U.S.S.R. originate from the worldwide collections of the Institute of Plant Industry. Mediterranean large-grained oats, resistant to crown and leaf rust, have been taken into cultivation in Azerbaidzhan. A large number of sorghum varieties and leguminous plants have also been taken into cultivation, as have been a number of varieties of alfalfa and sunflowers, especially for ensilage; these covered ca. 80 000 hectares in 1939.

Since I have no time to dwell at length on potatoes now, it should only be briefly mentioned that the attempts made to introduce them have exceeded all expectations. A large number of previously unknown species of cultivated potatoes, as well as wild ones were discovered among the crops of the Indians in Peru, Bolivia, Colombia and Ecuador. These are distinguished by valuable characteristics such as an exceptional frost-hardiness and resistance to diseases. This enormous amount of material has definitely opened up exceptional outlooks for practical plant breeding. On the basis of the Soviet discoveries made by Drs. S.M. Bukasov and S.V. Yuzepchuk, both Soviet and foreign plant breeding have progressed by using intervarietal and interspecific hybridization. A number of valuable strains, tolerant to low temperatures and resistant to *Phytophthora* are already being grown on the fields of cooperative farms. Strains of potatoes resistant to canker have been taken into cultivation. They are particularly necessary for us at present, because of the wide distribution of potato canker in the Ukraine.

More than 1 million grafts have been made of new strains of fruit trees, providing plants for 6000 hectares for propagation in different nurseries. Out of the 600 samples of berries studied, the Institute has selected the most valuable, under the direction of Dr. M.A. Rozanova. Of these, 47 are now state-controlled brands, which amount to 42% of all the kinds of berry crops in our country.

I also want to point out the remarkable kinds of large-fruited wild grapes with a high sugar content, which originate from Amur and which have now been taken into cultivation.

A number of varieties of castor beans [*Ricinus communis*], oil flax, ground- or peanuts [*Arachis hypogaea*], sesame plants [*Sesamum indicum*], perilla [*Perilla frutescens*], and safflower [*Carthamus tinctorium*] have also been introduced. Long-staple flax, introduced from the worldwide collection, occupied more than 60 000 hectares in 1939.

I should also mention the introduced crops of the quinine tree [*Cinchona* spp.] based on the material collected by the expeditions to Peru and Bolivia. At present, jute has also been taken into cultivation. In 1940, industrial plantations thereof covered 165 hectares.

Among the new crops, introduced from the material imported by the Institute, we can also mention the tung-oil tree [*Vernicia fordii*], a number of bamboo species and a large number of eucalyptus trees, including the so-called 'lemon-eucalypt', as well as some medical plants such as jaborandi [*Pilocarpus jaborandi*], the coca bush [*Erythroxylon coca*] and the Mexican orach [*Atriplex hortensis*].

During the early years of its activities, the Institute undertook extensive introduction of tree species, both of technical and decorative importance, and organized nurseries for trees in Pushkin, Minsk, near Moscow, around Tula and at Kamennaya Step', near Kharkov, at Otrada Kubanskaya, Pyatigorsk and on the Apsheronskiy Peninsula as well as at Suchumi, Maykop' among the mountains of Turkmenia and around Tashkent. An enormous amount of woody plants is accumulated there.

Simultaneously, the Institute worked on a review of all previous attempts at spontaneous introduction, which has a 200-year history. Research stations for dendrology, organized within the system of the Institute of Plant Industry, were managed on the basis of original material. The Tulsa forest–steppe station of the Institute, now incorporated into the Goszelenkhoz system [state agricultural farm system], has in its dendrological collection thousands of species and forms of plants, which are widely used in practice. Otrada Kubanskaya supplies hundreds of kilos of seeds annually. The station at Pyatigorsk, supported by the Institute of Plant Industry, has grown into an independent establishment, introducing a wide assortment of woody plants.

In connection with the establishment of a special forestry and forest-ameliorating institute during 1930–31, the work with woody species was to a considerable extent transferred to that establishment. However, for a short time, work within this field was done exclusively by an enthusiastic team of dendrologists at the Institute of Plant Industry, among which we should mention N.G. Vekhod, D.D. Artsybashev, A.V. Gurskiy, V.M. Bortkevich, S.G. Ginkul, H.D. Kostetskiy, H.N. Kobranov and E.E. Kern. As far as the practical results of their work is concerned, we should mention the northward extension of crops of black walnuts [*Juglans nigra*], a tree valuable from a dendrotechnical point of view. We should also mention the work by the Tulsa Station toward the discovery of the valuable tannin-producing sumach [*Rhus coriaria*]. The china berry [*Melia azeradach*] is now widely dispersed in Inner Asia thanks to the work done by the Institute.

In connection with the introduction of plants, the question concerning quarantine is of exceptional importance. Ignorance and carelessness, which occurred in the past, resulted in that together with what was imported, plants of no value and dangerous parasites were introduced, such as *Phyloxera*, *Odium* and mildew on grapevine and *Phytophthora* on potatoes. It is therefore interesting that in the native land of the potatoes, in Peru, Bolivia and Chile, there is no *Phytophthora*. Potatoes were not cultivated in Mexico in ancient times. The red aphid and the *Sphaerotec* of the gooseberry were introduced in the same manner. Hence, it is necessary to centralize matters concerning introductions as is done in many countries, starting with the U.S.A. Australia has organized a quarantine service at all ports. Quite a lot has been accomplished in our country in this respect during the last couple of decades. There are quarantine stations at Pot', Batumi, Tbilizi, Odessa, Krasnodar, Baku and Vladivostok.

We also have quarantine service in Moscow and Leningrad (at the Institute of Plant Industry). Every sample is, as a rule, subjected to quarantine and to control by phytopathologists and entomologists. The Institute is equipped with special greenhouses for the control of specially dangerous objects such as potatoes.

The object of quarantine is first and foremost to prevent diseases from being brought in but at the same time to make it possible to introduce very valuable material and to disinfect it. It is impossible to underestimate the authority of the State over the quarantine matters, especially in the fight against imported pests. However, at the same time, the entire orientation of the quarantine must be such that pests are rendered harmless but in such a way that the introduction of the plant material is not prevented. Recently, the management of the quarantine system amended it by introducing a system of internal quarantines in connection with the spread of diseases within particular areas of our country. This is, to a great extent, based on the experiments made herewith in the U.S.A.

A great deal of work has been carried out. With respect to field crops, this embraces practically every kind of crop from all over the world.

In our country, as far as such crops as wheat, barley, rye, oats, flax, leguminous plants, fruits and grapes are concerned, we now have an exceptional wealth of strains in a 'bank', which has no equal beyond our borders. Research concerning such crops as vegetables, and subtropical and textile plants continues, because some is still inadequate.

We are now faced with problems where we need assistance, especially from botanists. Just like the plant-breeding stations, the Institute of Plant Industry concentrates on work with material, the agricultural importance of which is already established, but which requires an approach that will differentiate the varieties. The task of the exploratory work with respect to finding new plants should be concentrated mainly at botanical institutes and botanical gardens. A decisive change is necessary, concerning the approach to the introduced material itself, from the general point of view, such as with regards to the differentiating comprehension thereof and the registration of the ecotypes constituting those species, which has become obligatory for agricultural plants. It is necessary for a network of botanical gardens to take a more active part in the 'green revolution'.

Problems arise when looking for new plants to be used as industrial raw materials such as textile plants, dye plants, rubber-producing plants and drug plants. In this respect, botanists should cooperate with chemists and technologists. The most-important task at present is to secure winterhardy forage plants for 20 million hectares of pasture. We still need to order seeds from abroad, but those brands are not always suitable as forage plants in our country. For those who intimately know the forage plants it is definitely evident that a solution to the forage problem is linked to a utilization of both the wild and the cultivated resources of forage plants in our own country. Here, more than elsewhere, a differentiating, ecological approach is necessary.

Next in turn is the problem concerning a more systematic investigation of the plant resources in our own country, especially the wealth of its specific diversity. Simultaneously, the country must be provided with necessary and promising original material. In addition to the flora of the U.S.S.R., those of China, Himalaya, Asia Minor, the Mediterranean area and the Cordilleras promise fair prospectives in this respect.

The chasm that exists between the botanical gardens and the plant-breeding stations and similar establishments, such as the Institute of Plant Industry, is eminently obvious to us. We cope with difficulty with the problems of improving and controlling cultivated plants, the number of which already amounts to 300 species, not counting the decorative ones. The task of studying the wild flora and its innumerable resources is a matter for competent botanists. Problems such as those with wild fruit trees and wild forage plants, where a cooperative work between botanists and agronomists is necessary, nevertheless mean a lot of very fascinating but also urgent work for each of us (although our Soviet teams of botanists and agronomists already amount to a whole army) and for every establishment, considering the size of our country and the growing needs of the industry.

BIBLIOGRAPHY

Ginkul, S.G. (1936). Introduktsiya i naturalizatsiya rasteniy po vlazhnykh sub-tropikakh S.S.S.R. [Introduction and naturalization of plants for the humid subtropics of the U.S.S.R.]. *Izv. Sukhum. Subtrop. Botan. Sada* [*Publ. from the Sukhumi Botanical Garden*], No. 1.

Kovalevskiy, I.V. (1929). Ocherk sel'skokhozyaystvennykh kul'tur i introduktsii ikh v Rossii v XVIII veke [List of agricultural crops and their introduction into Russia during the 18th century]. *Izv. Gos. In-ta opyt agronomii* [*Publ. from the State Institute of Agronomical Research*], **7** (6).

Kraus, G. (1894). *Geschichte der Pflanzeneinfürungen in die europäischen Gärten*, [History of plant introduction into European gardens]. Leipzig.

Vavilov, N.I. (1940). Uchenie o proiskhozhdenii kul'turnykh rasteniy posle Darvina [The theory of the origin of cultivated plants after Darwin]. *Sov. Nauka* [*Soviet Science*], No. 2.

New data on the cultivated flora of China and its importance for Soviet plant breeding

IN A RECENT REVIEW of the number and composition of plants cultivated in our world, the Japanese scientist Akemine mentions about 1200 species, not including the herbaceous and woody ornamentals. China has to be placed near the top among the countries where these cultivated plants originated, especially as far as her main provinces in eastern Asia are concerned. Various species of millet, soybeans and a large number of root vegetables and tuberous plants, as well as plants cultivated in water (which is characteristic of that country), are very important endemic plants, typical of China. China must definitely be put in first place with respect to the wealth of fruit species. Many kinds of citrus fruits originate from there.

The cultivated flora of China is extremely peculiar and very different in composition from that of other primary centers of agriculture in the world. The food of China, both animal and vegetable, has a very strange composition – from shoots of various bamboo species, peculiar plants such as *Zizania latifolia*, grown for the sake of its leaf sheaths infested by a smut, edible burdocks [*Arctium* spp.], unique Chinese cabbages and giant radishes weighing several kilograms to a multitude of dishes made of soybeans, replacing meat, and a cheese called 'tofu'. This is the conventional composition of the vegetable foods of the Chinese.

China is outstanding in comparison with the other centers as far as the wealth of endemic species is concerned. Many plants such as soybeans, adzuki beans [*Vigna angularis*], persimmon [*Diospyros kaki*] and citrus fruits are represented by an amazing diversity of forms. According to my own estimates, the origin of more than 200 species is linked to China. Literally thousands of species of woody and herbaceous decorative plants, which can be seen in parks all over the world but especially in the U.S.A., originate from China. The famous American scientist, Wilson, who studied the flora of China and collected plants for the Arnold Arboretum, called one of his books: *China – the Mother of Gardens*. With respect to the plant resources used for subtropical crops, China is of exceptional interest (as is Japan, which adopted its cultivated plants mainly from China) as far as such crops as tea, many citrus species, persimmons, tung-oil trees [*Vernicia*

First published in *Izv. AN SSSR* [*Publications from the Academy of the Sciences of the U.S.S.R.*], *ser. biol.*, pp. 744–747, 1958. [This paper was published posthumously after the name of Vavilov had been rehabilitated in 1957].

fordii] and species of bamboo are concerned. These are the main crops of our own subtropical agriculture along the coast of the Black Sea and at Lenkoran. They originated from China, where the basic generic and specific potential of these plants is found. In the past, just as during the last couple of years, much attention was – and still is – paid to the endemic, cultivated species of China and Japan, which were collected and acquired in large numbers for our humid subtropical areas.

The modern agriculture of China is actually not based on this amazing wealth of endemic species but on adopted crops, such as rice, wheat, barley and maize.

Out of the 86 million hectares at present used for agriculture in China, 19.5 million are used for wheat, about 6 million for barley, not less than 7 million for maize and 18.4 million for rice, which was introduced in the past from India. These crops have, until recently, not been given much attention by scientists. It was assumed that wheat imported by China from southwestern Asia and Asia Minor, i.e. the main native lands of wheat and barley, was of no particular interest. The peculiar forms of awnless barley found in China did not attract their attention. In 1936, I became convinced that this was wrong. Three years earlier, thanks to assistance from plant breeders in Nanking, Dr. Cheng *et al.*, but especially thanks to Dr. Love, the famous American plant breeder, who directed the practical plant breeding in China during the last couple of years, we had obtained an outstanding collection of wheat samples, collected by expeditions visiting all the agricultural areas of western and northern China where wheat is grown, such as Henan, Hunei, Shenxi, Shaanxi, Shandong and Hangsu, etc., also including the province of Sichuan, where wheat is comparatively sparsely grown.

The samples mentioned were sown in different areas of the U.S.S.R., from Leningrad to Inner Asia and the northern Caucasus. An epidemic of brown rust in 1936, which infested the wheat of the northern Caucasus, necessitated much research during the last couple of years, concerning the resistance of wheat to rust which revealed unexpectedly an important fact – the amazing resistance of the Chinese soft wheat to brown rust. Although resistance to this disease is a characteristic and fairly common property of hard wheats and other species of little interest as crops, the overwhelming majority of the strains of soft wheat were and are usually considered strongly susceptible to brown rust. The vast majority of the European and Soviet strains of wheat known to us are fairly susceptible to the species of rust, which are widely spread and very detrimental within our country. Hundreds of samples of the Chinese wheat, sown in different areas, were definitely distinguished by a physiological resistance to brown rust. Thus, one of the characteristics most necessary for our own Soviet plant breeding was discovered within the complex of Chinese wheat.

The overwhelming majority of Chinese wheats proved especially to be characterized by fast ripening. Some of them matured in two weeks less than our ordinary strains of wheat. In addition, the dry year of 1936 and the reduced harvests as a result of the distinct appearance of a drought plus the lack of rain revealed a normal development of the grains of the Chinese wheat under those circumstances, thanks to the extremely short time it requires for ripening. The

Chinese wheat is of exceptional interest as far as the short time required for ripening is concerned – a very important property especially during dry conditions – and opens up new opportunities for practical plant breeding.

Thorough investigation of Chinese wheat demonstrated that it possesses, in addition to this very valuable physiological characteristic, a number of morphological ones such as, e.g., lower growth than that of our ordinary strains. A most interesting fact is the presence of groups characterized by their awns. A number of Chinese forms are without even a rudiment of awns, i.e., they are entirely awnless. We know of no such forms among the European wheats. In addition, there are important groups of wheat with short awns, and slender awns as well as some with ordinary long awns, as is typical of our own common wheats. Special groups with twisted, awnlike appendages and so-called inflated forms thereof are also encountered.

While not dwelling on the morphological differences, it should nevertheless be mentioned that the grains of the Chinese wheat are distinguished by comparatively small dimensions, e.g., being about one fourth the size of those of our ordinary strains. However, although they differ in volume and roundness, they are particularly valuable for their mycological properties. The appearance of many flowers in the ears, and, consequently, many grains, is also typical of them. Therefore, they are characterized by a comparatively high yield. In the case of several forms, the number of florets amounts to six or seven at the center of the spike in comparison with the four or five that usually develop. The total number of botanical varieties in China has been estimated at several dozen.

All that has been stated led me to the conclusion that it was expedient to distinguish the Chinese groups of wheat as a particular subspecies, *Triticum vulgare* ssp. *sinicum*. The province of Henan seems to be the main area of its development. According to Chinese annals, cultivation of wheat already existed in that area 2000 years before our time; the other provinces where wheat is grown are adjacent to Henan.

The distinction of Chinese wheat as a particular subspecies also requires consideration from the points of view of botanists, geographers and plant breeders. Plant breeding all over the world has profited from the Chinese varietal material of wheat. When studying the genealogy of the best foreign varieties, it is, for instance, evident that the new Italian wheats, which are outstanding because of their resistance and which were developed by the plant breeder Strampelli, are created in part from Chinese and Japanese forms and have equals among the Chinese wheat as far as the short straw, fast ripening, the reduced size of the grains, resistance to brown rust and a particular golden coloration of the ripe ears and culms are concerned; the latter characteristic usually distinguishes the Chinese group of wheat.

Investigations of Chinese wheat have, in any case, revealed new opportunities for practical plant breeding of the most important grain in the world. On the basis of this, we have recently carried out a major task of hybridizing the Chinese wheat with our best Soviet standard varieties, and we hope to obtain definite results in the near future.

China must without doubt be considered a secondary center of cultivated wheat. The absence of a variety of cultivated forms of wheat indicates this. While in Transcaucasia we cultivate not less than a dozen different species of wheat, China grows mainly soft wheat alone. Two other kinds of wheat, club and shot wheat [*Triticum compactum* and *T. sphaerococcum*] are fairly rarely found, usually only occasionally. Within the Chinese secondary center, a special group of wheat has nevertheless developed as a result of a complicated set of conditions, which radically changes our ideas about the opportunities for improving this plant.

According to information from E. Talalayev, some Chinese wheat strains from the northern areas possess excellent frost-hardiness. Thus, there is in China a remarkable complex of important characteristics that are necessary for our own strains of wheat. This should be universally utilized as soon as possible.

The facts revealed are of yet another great importance. We have actually also found analogic appearances of barley. The similarities of the phenomena observed are amazing.

Southwestern Asia and northeastern Africa appear to be the centers of barley cultivation. The maximum diversity of intraspecific diversity is still found there. These areas can, with full reason, be called the native lands of barley cultivation. Closely allied wild relatives of cultivated barley can still be found there and these are definitely not found in China. At the same time, special forms of barley have been produced in China and Japan during thousands of years of cultivation. These are distinguished either by no awns or by short awns or even so-called furcated awns, which are not found in the main areas of type-formation of cultivated barley. These characteristics have, no doubt, been developed just within the countries mentioned. In spite of a thorough search, carried out by us for many years in the different countries of southwestern Asia and in Africa, we were never able to locate any short-awned or awnless barley there. Such forms appear exclusively in China and Japan, although these areas are undoubtedly only secondary centers for it. Why the barley has lost its awns in China and changed from awned to awnless and why this has not occurred in any other country outside east Asia is a riddle just like the development in China of definitely awnless or short-awned wheat. It is therefore interesting that, genetically, the lack of awns seems to be a dominant characteristic. It is also peculiar that, in the direction toward China, mainly naked-grained forms of barley have become isolated. Out of 1 100 000 hectares of barley cultivated in Japan, 600 000 hectares are occupied by naked-grained barley. This characteristic is definitely recessive and its selection is apparently linked to a definite preference for crops of naked-grained wheat.

It is interesting that we have observed similar phenomena in cultivated oats. Just in China, in its mountainous areas, a typical naked-grained group of large-grained oats has been isolated by the will of Man. It is unknown in Europe, although oats are, no doubt, a crop that has been introduced to China. This means that naked-grained oats, just like awnless and short-awned barley and the definitely awnless and short-awned wheat, appear to be endemic to China.

We also discovered a parallel in this direction in the case of rice. It is quite certain that the native land of rice is India, where we can still find links between wild and cultivated forms and where rice is represented by an amazing variety of forms. China and Japan adopted the cultivation of rice from India. However, during the course of centuries, a typical group of awnless cultivars has nevertheless developed in China and Japan. These are different from the ordinary awned types of rice in India and Inner Asia and are also characterized by fast ripening. In other words, within China not only has an important primary center of a rich cultivated flora been discovered, but we also find remarkable changes there of the plants adopted in a remote past which are of great importance for agriculture but which are still little appreciated or utilized. The nature of these changes is not well understood and requires thorough genetical investigation. From a practical point of view, the changes are apparently the result of effects exerted by the external environment but also by natural selection and the enormous role played by artificial selection carried out under the conditions of an extremely intensive cultivation, almost like that in the case of a vegetable garden. These facts concerning the major changes and the parallels with respect to these changes indicate a conscious direction of evolution and are of great importance and of exceptional practical importance. It makes it possible to find valuable characteristics necessary for us. Apparently, the changes occurred fairly rapidly, which can be judged by the fact that, e.g. in eastern Asia and even in our own Far East, kinds of gourds have been discovered that mature exceptionally early. These were unquestionably brought into eastern Asia from America only during the last century.

We have also established similar facts in the case of beans, imported from America, which are now distinguished by special waxy forms. Maize, introduced to China during the last century, has also developed interesting and peculiarly waxy forms, distinguished by the consistency of the kernels. Perhaps these facts bear witness to the great variability of the organisms due to the effects of the external environment. We still have no explanation for these facts. They may be the result of selection among the original populations or the initial heterozygotes and appear, in essence, to be the result both of selection, due to the effects of natural selection and of selection directed by Man himself. Perhaps there are direct effects here working in a particular direction. Far-ranging research is necessary for this and it will be done by us at the Institute of Genetics in connection with the practical tasks of plant breeding. Considering the importance of these facts, which pose new problems for geneticists and plant breeders, we have felt it necessary to bring this to the attention of the Physical–Mathematical Section of the Academy of the Sciences of the U.S.S.R.

A letter to S.M. Bukasov

Mexico City November 17, 1925

Dear Sergey Mikhaylovich,

Four months have passed from the time of your departure. Since that time a pair of letters [have arrived] with only a negligible content, just *as if you did not belong to the expedition*. At the same time, I have received half a dozen long letters from Yu. N. Voronov. I also had a letter from Vitmak [stating] that he did all he could for you. The moral sense is that it is necessary for you to write, in order that I will be able to take a personal interest in your journey. For me, Mexico is a country of great interest: the history of its agricultural crops, the composition of its cultivated plants, the complexes of maize, tobacco, solanaceous plants, beans and gourds are all new to me. What do they really represent? What do *you* find in the markets of the towns? Do you take photographs? We do not hear anything about all that. Yu. N. Voronov also complains that you write so little. Do you keep a diary? It is necessary that you turn into a writer and force us to take an interest in your excursions.

Your book has been set, finished and *printed* and should appear any day now. One colored picture was delayed; the color of the flowers was hard to reproduce. Three color plates were made, but it went badly. It became necessary to turn to chromolithography.

We have not received any material. Is there really nothing in the markets of Mexico that is of interest to us? Samples usually take three weeks to Mexico City. That period is sufficient in order to send something. Whatever you send, report it in a letter or on a postcard. *Voronov has sent many books.*

When you have received this letter, sit down and write a 10-page reply! We are interested in the endemic strains of America. Send photos, even postcards, of the plants species of Mexico. Are there really none? Send agricultural literature. Do not send anything via Borodin but send it directly; that is the best.

I am sending you all the numbers of *Trudy* that have been published.

Yours,

N. Vavilov.

Remarks by the Russian Editors:

The letter reproduced above demonstrates how important the creative energy of the learned must be in order to overcome the difficulties standing in their way when trying to master the wealth of the world's plants.

N.I. Vavilov always took a lively part in the planning and the programming of his colleague's expeditions. In far-away countries they were, in turn, delighted about the moral and material support on the part of N.I. Vavilov and to be informed by him about the life at the Institute. This led easily to tasks that were frequently almost beyond their abilities to carry out, since Nicolay Ivanovich was extremely demanding.

Index of Latin plant names

Synonymy according to Mansfield, R.: *Vorläufiges Verzeichnis landwirtschaftlich and gaertnerisch kultivierter Pflanzenarten* [*List of agriculturally or horticulturally cultivated plants*], *Die Kulturpflanze*, Berlin, 1959, or Terrell, E.E., Hill, S.R., Wiersma, J.H. and Rice, W.E.: *A Checklist of Names for 3000 Vascular Plants of Economic Importance*, U.S.D.A. Agriculture Handbook no. 505, rev. edn, 1986 and, in case of Triticeae, Á. Löve: Conspectus of the Triticeae, *Feddes Rep.* 95 (7–8), 425–521, 1984. All other names according to the original text. To locate name without page reference, look for synonym.

List of vernacular plant names

The vernacular name is followed by the equivalent Latin name and the synonym used by Vavilov in the text. For page-references, see Index of Latin plant names.

acacia, white, *Acacia farnesiana* Willd.
sweet, *Albizzia julibrissin* (Willd.) Durass.
achiote, anatto or lipstick tree, *Bixa orellana* L.
achira or edible canna, *Canna indica* L. (syn *C. edulis* Ker–Gawl)
achoca or pepino de comer, *Cyclanthera brachybotrys* (Poepp. & Endl.) Cogn.
C. pedata (L.) Schrad.
achotillo, almandra, *Caryocar amygdaliferum* Mutis (syn. *C.a.* Cav.)
agave, henequen, *Agave sisalana* Perr.
ixtlé, *A. ixtli* Karw.
magey, *A. atrovirens* Karw.
Tampico fiber, *A. lechugilla* Torr.
ailanthus or tree-of-heaven, *Ailanthus altissima* (Mill.) Swingle
'alfa', Egyptian, *Epilobium hirsutum* L.
alfalfa, white, *Medicago sativa* L. *s. lat.*
yellow, *M. sativa* L. *s. lat.*
almandra or achotillo, *Caryocar amygdaliferum* Mutis (syn. *C.a.* Cav)
almond, common, *Prunus dulcis* (Mill.) D.A. Webb (syn. *Amygdalis communis* L.)
Indian, *Terminalia catappa* L.
wild, *Amygdalis bucharia* Korsh.
A. fenzliana (Fritsch.) Lipsky
A. scoparia Spach
A. spinosissima Bunge
amaranth, *Amaranthus anardana* Wall.
livid, *A. lividus* L. (syn. *A. blitum* L. var. *oleraceum* Watt)
Indian, *A. tricolor* L. var. *gangeticus* (L.) Fiori (syn. *A. gangeticus* L.)
purple or huatli, *A. hybridus* L. ssp. *cruentus* (L.) Thell. (syn. *A. paniculatus* (L.) Thell. incl. var. *leucocarpus* Saff.)
pigweed or smooth, *A. hybridus* L. var. *frumentaceus* (Buch.–Ham.) Hand.–Mazz. (syn. *A. frumentaceus* Roxb.)
annato or lipstick tree, achiote, *Bixa orellana* L.
anise, *Pimpinella anisum* L.
anisette, *Pimpinella anisetum* Boiss.
annona or cherimoya, *Annona cherimola* Mill.
wild, *A. cinerea* Dunn.
A, purpurea Moq. & Sessé
anu, *Tropaeolum tuberosum* Ruiz & Pav.
apazote, *Chenopodium ambrosioides* L.
C. berlandieri (Saff.) Moq. & Sessé ssp. *nuttalliae* (Saff.) H.D. Wilson & Heiser (syn. *C. nuttaliae* Saff.)
apple, chinese, *Malus prunifolia* (Willd.) Borkh. var. *rinke* (Koidz.) Rehd. (syn. *M. asiatica* Nakai)
crab-, *M. pumila* Mill.
custard-, *Annona reticulata* L.
paradise, *Malus baccata* (L.) Borkh.
pond-, *Annona glabra* L.
apricot, *Prunus armeniaca* L.
Japanese, *P. mume* Sieb. & Zucc.
apricot-plum, *Prunus simonii* Carr.
araucaria, *Araucaria imbricata* Pav.
arracacha, *Arracacia xanthorriza* Bancr. (syn. *A. esculenta DC.*)
arrowhead, Chinese, *Sagittaria sagittifolia* L. var. *sinensis* Makino
arrow-root, *Maranta arundinacea* L. (syn. *M. edulis* Wedd.)
East-Indian, *Tacca leontopetaloides* (L.) Kuntze (syn. *T. pinnatifida* Forst. & Forst f.)
artichoke, common, *Cynara scolymus* L.
Jerusalem or sunchoke, *Helianthus tuberosus* L.
cardoon, or wild, *Cynara cardunculus* L.
Chinese, *Stachys sieboldii* Miq. (syn. *S. affinis* Bunge)
arum, giant white-spot, *Amorphophallus paeonifolius* (Dennst.) Nicolson (syn. *A. campanulatus* Roxb.)

ash, green, *Fraxinus pennsylvanica* Marshall
ashgon, *Trachyspermum ammi* (L.) Sprague (syn. *Ammi copticum* L.)
asparagus, club-shaped, *Asparagus lucidus* Lindl.
garden, *A. officinalis* L.
avocado, *Persea americana* L. (syn *P. gratissima* Gaertn. f.)
coyo-, *P. schiedeana* Nees

badan, *Bergenia crassifolia* (L.) Fritsch.
bael tree, Indian or Bengal quince, *Aegle marmelos* (L.) Corr.
bamboo, *Bambusa mitis* Poir.
hedge-, *B. multiplex* (Lour.) Räusch.
B. senanensis Fr. & Sav.
B. spinosa Roxb.
B. tulda Roxb.
B. vulgaris Schrad
Simon's, *Arundinaria simoni* (Carr.) A. & C. Riv.
Dendrocalamus asper (Schult.f.) Backer
Gigantochloa apus (Roem. & Schult.) Kurz.
G. ater, Kurz.
G. verticillata (Willd.) Munro
Japanese timber-, *Phyllostachya bambusoides* Sieb. & Zucc.
P. edulis, A. & C. Riv.
P. mitis A. & C. Riv.
P. nigra A. & C. Riv. var. *hononsis* Makino
P. puberula Munro
P. quiliot A. & C. Riv.
P. reticulata C. Koch
banana, Abyssinian, *Ensete ventricosum* (Welw.) Cheesman (syn. *Musa ensete* Gmel.)
common, *Musa acuminata* Colla (syn. *M. paradisiana* L., *M* × *sapientium* L.)
barberry, European, *Berberis vulgaris*, L.
barley, *Hordeum vulgare* L. s. lat.
awnless, *H. vulgare* L. (syn. *H. hexastichon* L.)
bulbous, *H. bulbosum* L.
distichous, *H. vulgare* L. (syn. *H. distichum* L.)
wild, *H. spontaneum* C. Koch
basil, *Ocimum basilicum* L.
camphor or hoary, *O. kilimandscharicum* Guerke
batat or sweet-potato, *Ipomaea batatas* (L.) Lam. (syn. *I.b.* Poir.)
bay-leaf or laurel, *Laurus nobilis* L.
beans, adzuki, *Vigna angularis* (Willd.) Ohwi & O. Ohashi (syn. *Phaseolus angularis* Wight)
V. sinensis Endl. ssp, *sesquipedalis* Piper
var. *catiang* (Walp.) Piper
var. *sinensis* (Stick.) Piper
Bertoni, *Vigna caracalla* (L.) Verdc. (syn. *Phaseolus caracalla* L.)
cluster, *Cyamopsis tetragonoloba* (L.) Taubert (syn. *C. psoraloides* DC.)
garden, *Phaseolus vulgaris* L.
hyacinth, *Lablab purpureus* (L.) Sweet (syn. *Dolichos lablab* L.)
jack-, *Canavalia ensiformis* (L.) DC., 316
kidney-, *Phaseolus vulgaris* L. var.
lima-, *Phaseolus lunatus* L.
gr. *macrospermus* N. Ivan
gr. *microspermus* N. Ivan
mat-, *Vigna aconitifolia* (Jacq.) Maréchal (syn. *Phaseolus aconitifolius* Jacq.)
mung, *Vigna radiata* (L.) Wilczek (syn. *Phaseolus aureus* (Roxb.) Piper)
rice-, *Vigna umbellata* (Thunb.) Ohwi & H. Ohashi (syn. *Phaseolus calcaratus* Roxb.)
scarlet runner-, *Phaseolus coccineus* L. (syn. *P. multiflorus* Willd.)
sword, *Canavalia gladiata* (Jacq.) DC.
tepary, *Phaseolus acutifolius* A. Gray, incl. v. *latifolius* Freeman
velvet, *Mucuna pruriens* (L.) DC, var. *utilis* (Wall.) Baker (syn. *M. utilis* Wall. & Wight)
winged, *Psophocarpus tetragonolobus* (L.) DC.
yam-, *Pachyrhizus erosus* (L.) Urban (syn. *P. angularis* Rich.)
yokohama-, *Mucuna hassjo* (Piper & Tracy) Mansfeld (syn. *Stizolobium hassjo* Piper & Tracy)
beet, incl. sugar beet, *Beta vulgaris* L.
wild, *B. vulgaris* ssp. *maritima* (L) Arcang. (syn. *B. maritima* L.)
berseem, *Trifolium alexandrinum* Juslen
bilberry, *Vaccinium uliginosum* L.
bilimbi, *Averrhoa bilimbi*, L.
blite, leafy, *Blitum virgatum* L. (syn. *Chenopodium foliosum* Asch.)
red, *B. rubrum* (L.) Rchb. (syn. *Chenopodium rubrum* L.)
strawberry, *B. capitatum* L. (syn. *Chenopodium capitatum* (L.) sch.)
blueberry, *Vaccinium myrtillus* L.
bomarea, *Bomarea acutifolia* Herb.
bromegrass, downy, *Bromus tectorum* L.
bouvardia, *Bouvardia tenuifolia* (Cav.) Schl.
Brazil nut, *Bertholletia excelsa* Humb. & Bonpl.
breadfruit tree, *Artocarpus altilis* (Parkins.) Forst. (syn. *A. communis* Forst.)
buckwheat, common, *Fagopyrum esculentum* Moench
tatar, *F. tataricum* (L.) Gaertn.
bulrush, Andean, *Scirpus riparius* J. & C.
bunchosia, *Bunchosia armeniaca* DC.
burdock, *Arcium lappa* L.

cabbage, balearican, *Brassica balearica* Pers.
Chinese, *B. pekinensis* (Lour.) Rupr.
Cretan, *B. cretica* Lam.
Japanese, *B. nipposinica* Bailey

cabbage (*cont.*)
sareptan or Indian mustard, *B. juncea* (L.) Czern. & Coss.
var. *sareptana* Sinsk.
salad, *B. juncea* (L.) Czern. & Coss, var. *japonica* (Thunb.) Bailey
wild, *B. oleracea* L. var. *oleracea*
cabellioda, *Eugenia tomentosa* Cambess.
cacao, *Theobroma cacao* L.
large flowered, *Th. grandiflora* K. Schum.
cacomite, *Trigridia pavonia* Ker-Gawl
cacti, *Cereus* Mill. *s. lat.*
Opuntia Mill.
Nopalea spp.
caigua, *Cyclanthera pedata* (L.) Schred
cainito or star-apple, *Chrysophyllum cainito* L.
calabash or bottle gourd, *Lagenaria siceraria* (Molina) Standley (syn. *L. vulgaris* Sér.)
camellia, common, *Camellia japonica* L.
sasanque, *C. sasanqua* Thunb.
camel's thorn, *Alhagi maurorum* Med. (syn. *A. camelorum* Fischer)
camphor tree, *Cinnamomum camphora* (L.) J.S. Presl (syn. *C. camphora* Nees & Eberm.)
cañahua or canihua, *Chenopodium canihua* O.F. Cook (syn. *C. pallidicaule* Aell.)
canary grass, *Phalaris canariensis* L.
candle-nut tree, *Aleuriters moluccana* (L.) Willd.
canicha, *Sesbania bispinosa* (Jacq.) W.F. Wight (syn. *S. aculeata* Pers.)
canna, edible or achira, *Canna indica* L. (syn. *C. edulis* Ker-Gawl)
Cape gooseberry, Peruvian gooseberry, *Physalis peruviana* L.
capulin or black cherry, *Prunus serotina* Ehrh. var. *salicifolia* (Kunth.) Koehne (syn. *P. capuli* Cav., or *P. capolin* DC.)
carambola or star-fruit, *Averrhoa carambola* L.
caraway, *Carum carvi* L.
cardamom, common, *Elettaria cardamomum* (L.) Maton & White.
Ceylon, *E. major* Smith
Java, *Amomum krevanh* Pierre
cardoon, *Cynara cardunculus* L.
carob or St. John's Bread tree, *Ceratonia siliqua* L.
carrot or (when wild) Queen Ann's Lace, *Daucus carota* L. ssp. *carota*
Peruvian, *Arracacia xanthorrhiza* Bancr. (syn. *A. esculenta* DC.)
casabanana, *Sicana odorifera* Naud.
cashew nut tree, *Anacardium occidentale* L.
castilloa rubber, *Castilla elastica* Sessé (syn. *C. elastica* Cerv.)
castor bean, *Ricinus communis* L.
Persian, *R. persicus* Pop (syn. *R. communis*)
casuarina or Australian pine, *Casuarina equisetifolia* L.
catalpa, northern, *Catalpa speciosa* Warde.
southern, *C. bignonioides* Walt
catiang, catjang or cow-pea, *Vigna unguiculata* (L.) Walp. (syn. *V. sinensis* Endl. and *V. catiang* Endl.)
catechu, *Acacia catechu* (L.f.) Willd.
cedar, Lebanon, *Cedrus libanotis* Loud.
cedrat or citron, *Citrus medica* L.
cedro, *Cedrela toona* Roxb.
celery, *Apium graveolens* L. (incl. var. *dulce* (Mill.) Pers.)
century plant, *Agave americana* L.
champedak or jack-fruit, *Artocarpus integer* (Thun.) Merr. (syn. *A. champeden* (Lour.) Spreng.)
chaulmoogra, *Hydnocarpus kurzii* (King) Warb. (syn. *Taraktogenos kurzii* King)
chayote, *Sechium edule* (Jacq.) Sw.
cherimoya, *Annona cherimola* Mill
wild, *A. purpurea* Moq. & Sessé
A. cinerea Dun.
cherry, Barbados or huesito, *Malphigia glabra* L.
bird, *Prunus padus* L.
black, *P. serotina* Ehrh. incl. var. *salicifolia* (Kunth) Koehne (syn. *P. capuli* Cav., *P. capolin* DC.)
Chinese, *P. tomentosa* Thunb.
choke-, *P. virginiana* L.
Peruvian (or Cape gooseberry), *Physalis peruviana* L.
sour, *Prunus cerasus* L.
Spanish, *Mimusops elengi* L.
sweet, *Prunus avium* L.
Surinam, *Eugenia uniflora* L.
cherry-plum, *Prunus cerasifera* Ehrh. var. *divaricata* (Ledeb.) Bailey (syn. *P. divaricata* Ledeb.)
chervil, *Anthriscus cerefolium* (L.) Hoffm.
chestnut tree, edible, *Castanea sativa* Mill. (syn. *C. vesca* (Gaertn.)
Chinese, *C. molissima* Bl.
Japanese, *C. crenata* Sieb. & Zucc.
chia, *Salvia potus* Epling (syn. *S. chia* Fern.)
chickling vetch or grass pea, *Lathyrus sativus* L. incl. var. *macrospermum* Zalk.
chick-pea, *Cicer arietinum* L.
incl. ssp. *pisiforme* Pop
chickory, *Cichorium intybus* L.
china-berry, *Melia azedarach* L.
china-root, *Smilax china* L.
chives, Chinese or rakkyo, *Allium odorum* L. (syn. *A. chinense* G. Don)
choke-cherry, *Prunus virginiana* L.
chrysanthemum, garden, *Chrysanthemum-* × *morifolium* Ram. var. *sinense* Makino
garland, *C. coronarium* L.
chufa or nutsedge, *Cyperus esculentus* L.
chupa-chupa, *Matisia cordata* Humb. & Bonpl.

cinnamon, Chinese, *Cinnamomum casia* L. (syn. *C. aromaticum* Nees & Eberm.)
ordinary, *C. verum* J.S. Presl (syn. *C. zeylanicum* Breyn.)
ciruela or mombin, red, *Spondias purpurea* L.
yellow, *S. mombin* L.
citron or cedrat, *Citrus medica* L.
citronella, *Cymbopogon nardus* (L.) Rendle
clover, berseem or Egyptian, *Trifolium alexandrinum* Juslen
crimson, *T. incarnatum* L.
Persian or shabdar, *T. resupinatum* L.
red, *T. pratense* L.
subterranean, *T. subterraneum* L.
white, *T. repens* L.
incl. var. *giganteum* Lag.-Foss.
clove tree, *Syzgium aromaticum* (L.) Merr. [syn. *Eugenia aromaticus* (L.) Merr.]
coca bush, *Erythroxylon coca* Lam., var. *coca*
cochineal cactus or nopal, *Nopalea cochinellifera* Salm-Dyck
cock's-head or sainfoin, *Onobrychis altissima* Grossh.
O. transcaucasia Grossh.
coffee, *Coffea arabica* L.
coix or Job's tears, *Coix lacryma-jobi* L.
coltsfoot, sweet, *Petasitus japonicus* Miq. (syn. *P.j.* (Sieb. & Zucc.) Maxim)
colza or turnip rape, *Brassica rapa* L. ssp. *rapifera* Metzq. var. *silvestris* (Lam.) Briggs (syn. *B. campestris* L. ssp. *oleifera* (Metzg.) Sinsk.)
comfrey, *Symphytum asperum* (syn. *S. asperrimum* Donn)
coriander or pesto, *Coriandrum sativum* L.
corn or maize, *Zea mays* L.
soft, *Z. mays* L. convar. *amylacea* (Sturtev.) Montg.
corn-spurry, *Spergula arvensis* L.
cosmos, *Cosmos bipinnatus* Cav.
C. caudatus H.B.K.
C. diversifolius Otto
C. sulfereus Cav.
cotton, bourbon, *Gossypium hirsutum* L. var. *punctatum* (Schum.) Hutch. (syn. *G. purpurascens* Poir.)
extra long staple or Egyptian, *C. barbadense* L. (syn. *G. peruvianum* Cav.)
levant, *G. barbadense* L. (syn. *G. peruvianum* Cav.)
sea island, *G. barbadense* L. (syn. *G. peruvianum* Cav.)
tree-, *G. arboreum* L.
upland, *G. hirsutum* L. (syn. *G. mexicanum* Tod.)
cow-pea or black-eyed pea, *Vigna unguiculata* (L.) Walpole (syn. *V. catiang* Endl., incl. var. *catiang* Endl. and var. *sinensis* (Stick.) Piper)
crab-apple, Siberian, *Malus bacccata* (L.) Borkh.
cress, garden or pepper grass, *Lepidium sativum* L.
croton, purging, *Croton tiglium* L.
cryptomeria, *Cryptomeria japonica* D. Don
cucumber, Anatolian, *Cucumis sativus* L. spp. *antasiaticus* Gabalev
Chinese, *C. chinensis* Pang.
Hardwick's, *C. sativus* L. (syn. *C. hardwickii* Royle)
ordinary, *C. sativus* L.
cumin, spicy, *Cuminum cyminum* L.
black, *Nigella sativa* L.
curry-leaf tree, *Murraya exotica* Koenig
M. koenigii (L.) Spreng.
custard apple, *Annona reticulata* L.

dahlias, *Dahlia coccinea* Cav.
D. excelsa Benth.
D. imperialis Reczb.
D. maximiliana Hort.
D. merckii Lehm.
D. pinnata Cav.
D. popenovii Saff.
date, Chinese or jujuba, *Ziziphys jujuba* Mill. (syn. *Z. sativus* Gaertn.)
(or Z. *vulgaris* Lam.)
date palm, cultivated, *Phoenix dactylifera* L.
wild, *P. sylvestris* (L.) Roxb.
date-plum, *Diospyros lotus* L.
devil's tongue, *Amorphophallus konjac* C. Kock
dill, *Anetum graveolens* L.
sowa, *A. graveolens* L. var. *submarginatum* Lej. & Const.
dogbane hemp, *Apocynum cannabinum* L.
dragoon, Spanish, *Lallemantia iberica* (Marsh. & Bieb.) Fisch (syn. *Dracocephalum ibericum* Marsh. & Bieb.)
duku or langsat, *Lansium domesticum* Corr.
dune grass, *Leymus racemosus* (Lam.) Tzvel, (syn. *Elymus giganteus* Vahl)
L. arenarius (L.) Hochst. (syn. *Elymus arenarius* L.)
durian, *Durio zibethinus* Murr.
durrah, *Pennisetum glaucum* (L.) R. Br. (syn. *P. typhoideum* L.C. Rich)
dyer's rocket, *Reseda luteola* L.
dyer's woad, *Isatis tinctoria* L.

elaeagnus, cherry-, *Elaeagnus multiflorus* Thunb. var. *hortensis* Maxim.
earth-almond or zulunut, *Cyperus esculentus* L.
eggfruit trees, *Pouteria campechiana* (Kunth) Baehni (syn. *Lucuma salicifolia* H.B.K.)
P. cainito R. & S. (syn. *Lucuma cainito* (R. & S.) Radik.
eggplants, *Solanum melongena* L.
S. muricatum Ait.

einkorn wheat, *Crithodium monococcum* (L.), Á. Löve (syn. *Triticum monococcum* L.)
  wild, *Crithodium monococcum* (L.) Á. Löve ssp. *aegilopoides* (Link.) Á Löve (syn. *Triticum thaoudar* Reut., T. *boëticum* Bioss. ssp. *thaoudar* (Reut.,) Schiemann)
elephant- or wood-apple, *Limonia acadissima* L. (syn. *Feronia elephantum* Corr.)
emblic or myrobalan, *Phyllanthes emblica* L.
emmer wheat, *Gigachilon polonicum* (L.) Seidl ssp. *dicoccon (Schrank)* Á. Löve (syn. *Triticum dicoccum* Schrank)
  wild, *Gigachilon polonicum* (L.) Seidl ssp. *dicocciodes* (Koern.) Á. Löve (syn. *Triticum dicoccoides* (Koern.) Aarons.), (and varieties)
endive, *Cichorium endivia* L.
  wild, *C. pumilum* Jacq.
erse, *Vicia ervilia* (L.) Willd., (syn. *Ervum ervilia* L.)
  single flowered, *V. articulata* Hornem. (syn. *Ervum monanthos* Desf.)
eucalyptus trees, *Eucalyptus* spp.

feijoa or pineapple guava, *Aca sellowiana* (Berg) Burret (syn. *Feijoa sellowiana* Berg)
fennel, *Foeniculum vulgare* Mill.
fenugreek, *Trigonella foenum-graecum* L.
fig, common, *Ficus carica* L.
  weeping, *F. benjamina* L.
fir, Douglas, *Pseudotsuga menziesii* (Mirkel) Franco
flax, fiber, *Linum usitatissimum* L. incl. var. *mediterraneum* Vav.
  false, *Camelina sativa* L.
    C. linicola Schimp. & Spenn.
  New Zealand, *Phormium tenax* Forst. & Forst. f.
  pale, *Linum angustifolium* Huds.
four-o'clock, *Mirabilis longiflora* L.

galanga, *Kaempferia galanga* L.
  roundish, *K. rotundata* L., [syn. Boesenbergia rotundata (L.) Mansf.]
galangal, *Alpinia galanga* (L.) Sav.
gama grass, *Tripsacum* L. spp.
gandaria, *Bouea gandaria* Bl. (syn. *B. macrophylla* Griff.)
garden cress, *Lepidium sativum*
garden rocket, *Eruca sativa* Mill.
garlic, *Allium sativum* L.
  elephant, *A. ampeloprasum* L.
geranium, *Pelargonium* spp.
ginger, mioga, *Z. mioga* Roscoe
  true, *Zingibar officinale* Roscoe
  zerumbet, *Z. zerumbet* (L.) Smith
ginseng, chinese, *Panax ginseng* C.A. Mey
goat grass, *Aegilops* L.
  barbed. *Ae. triuncialis* L. (syn. *Ae squarrosa* L.)
  coarse, *Ae. crassa* Boiss.
  jointed, *Ae. cylindrica* Host
gobo burdock, *Arctium lappa* L.
goldenrod, Edison's, *Solidago Edisoniana* Mack.
gooseberry, Cape or Peruvian, *Physalis peruviana* L.
  common, *Ribes uva-crispa* L. (syn. R. *grossularia* L.)
goosegrass, *Eleusine indica* (L.) Gaertn.
gorse, *Ulex europeus* L.
gourds, bitter, *Memordia charantia* L.
  bottle or calabash, *Lagenaria siceraria* (Mol.) Standley (syn. *L. vulgaris* Sér.)
  dishcloth, *Luffa acutangula* (L.) Roxb.
  figleaf, *Cucurbita ficifolia* Bouché
  pumpkin squash, *C. mixta* Pang.
    C. moschata Duch.
    C. pepo L.
  snake, *Trichosanthos anguina* L.
  sponge or luffa, *Luffa aegyptica* Mill. (syn. *L. cylindrica* Roem.)
    Cucurbita acutangula (L.) Roxb. (syn. *L. acutangula* L.)
  waxy, white or chinese, *Benincasa hispida* (Thunb.) Cogn.
gram, *Vigna mungo* (L.) Hopper (syn. *Phaseolus mungo* L.)
  horse-, *Vigna unguiculata* (L.) Walp. ssp. *cylindrica* (L.) Verdc. (syn. *Dolichus biflorus* L.)
granadilla or passion flower fruit, *Passiflora ligularis* Juss.
  giant, *P. quadrangularis* L.
grapefruit, *Citrus maxima* Merr. s. lat.
grape-tree, Brazilian, *Myricaris cauliflora* (Mart.) Berg
grapes, Amur, *Vitis amurensis* Rupr.
  wine, *V. vinifera* L.
  wild, *V. vinifera* L. var. *spontanea* M. Pop
    V. vinifera L. ssp. *sylvestris* (Gmel.) Berger (syn. *V. silvestris* Gmel.)
grass-pea or chickling vetch, *Lathyrus sativus* L.
groundnuts, peanuts, *Arachis hypogea* L.
grumichama, *Eugenia braziliensis* Lam. (syn. *E. dombei* (Spreng.) Skeels)
guava, *Psidium guajava* L.
  Costa Rican, *P. friedrichthalianum* (O. Berg) Niedenzu
  wild, *P. sartorium* (Berg) Niedenzu
guaymochil or Manila tamarind, *Pithecolobium* dulce (Roxb.), Benth (syn. *Pithecellobium d.*)
guayule rubber, *Parthenium argentatum* A. Gray
gutta-percha trees, *Eucommia ulmoides* Oliver
  Palaquium gutta (Hook. f.) Baillon

hausa potato, *Solenostemon rotundifolius* (Poir.)

J.K. Morton (syn. *Coleus tuberosus* (Bl.) Benth.)
hawthorn, Mexican or texocote, *Crataegus mexicana* Moq. & Sessé
Cr. pinnatifida Runge
Cr. stipulosa Steud.
hazelnut, filbert, *Corylus avellana* L.
giant filbert, *C. maxima* Mill.
Himalayan, *C. ferox* Wall.
Turkish, *C. colurna* L.
hemlock, mountain, *Tsuga mertensiana* (Bong.) Carr.
eastern, *T. canadensis* (L.) Carr.
western, *T. heterophylla* (Raf.) Berg
hemp, Bombay or sunhemp, *Crotolaria juncea* L.
Chinese, *Apocynum hendersonii* Hook.
Cuban, *Furcraea cubensis* Vent. (syn. *Fourcroya c.* Vent.)
dogbane, *Apocynum cannabinum* L.
Indian perennial, *Abroma angustata* (L.) L.f.
kenaf or mallow-, *Hibiscus cannabinus* L.
Manila, *Musa textilis* Née
marijuana, *Cannabis sativa* L., ssp. *sativa*
Cannabis sativa L., ssp. *indica* (Lam.) E. Small & Cronq., (syn. *C. indica* L.)
wild, *Cannabis sativa* L., ssp. *spontanea* Vav. (syn. C. *ruderalis* Janish)
henna, *Lawsonia inermis* L. (syn. *L. alba* Lam.)
hibiscus, *Hibiscus rosa-sinensis* L.
Hickory, black, *Carya texana* Bickl.
shagbark, *C. ovata* (Mill.) C. Koch
wild Chinese, *C. cathayensis* Sarg.
honey locust, *Gleditsia triacanthos* L.
hops, *Humulus lupulus* L.
horn nut, *Trapa bicornes* L.
horseradish, Japanese, *Wasabia japonica* Matsum. (syn. *Eutrema wasabi* (Sieb.) Maxim)
huatli or purple amaranth, *Amaranthus cruentus* L. (syn. *A. paniculatus* L. incl. var. *leucocarpus* Saff.)
huesito or Barbados cherry, *Malphigia glabra* L.
hyssop, *Hyssopus officinalis* L.

ilama, *Annona diversifolia* Saff.
Inca wheat, *Amaranthus caudatus* L.
incherto or green sapote, *Pouteria viridis* (Pitt.) Cronq. (syn. *Calocarpum viride* Pitt.)
indigo, Abyssinian, *Indigofera argentea* L., (According to V. Täckholm's *Student's Flora of Egypt*, 1974, in part *I. articulata* L., in part *I. caerulea* Roxb.)
Assam, *Strobilanthes cusia* (Nees), Imlay (syn. *S. flaccidifolium* Nees)
Chinese, *Polygonum tinctorium* Lour.
common, *Indigofera tintoria* L.
iris-root, *Iris pallida* Lam.
itsegek or barnyard millet, *Anabasis aphylla* L.
ixtlé agave, *Agave ixtli* Karw.

jaboticaba, *Myricaria jaboticaba* Berg
jack-bean, *Canavalia ensiformis* (L.), DC
jack-fruit or champedak, *Artocarpus integer* (Thunb.) Merr. (syn. *A. champeden* (Lour.) Spreng.)
Japan clover, *Lespedeza striata* (Thunb.) Hook. & Arn.
jasmine, Arabian royal, *Jasminum grandiflorum* L.
Java plum, *Syzygium cumini* (L.) Skeels (syn. *Eugenia cumini* (L.) Merr.)
Jerusalem artichoke, *Helianthus tuberosus* L.
jicama, *Pachyrhizus erosus* (L.) Urban (syn. *P. angulatus* Rich.)
Job's tears or coix, *Coix lacryma-jobi* L.
jojoba, *Simmondsia chinensis* (Link.) Schneider
jujuba, Chinese date, *Ziziphus jujuba* Mill. (syn. *Z. sativa* Gaertn., *Z. vulgaris* L.)
jute, Chinese, *Abutilon theophrasti* Med. (syn. *A. avicennae* (L.) Gaertn.)
Cuban, *Sida rhombifolia* L.
tussa, *Corchorus olitorius* Roxb.
white, *C. capsularis*

kale, Chinese, *Brassica alboglabra* L.H. Bailey
fodder-, *B. oleracea* L. var. *acephala* DC.
karanda, *Carissa carandas* L.
kapok or silktree, *Ceiba pentandra* (L.) Gaertn.
katuk, *Saurupus androgynus* (L.) Merr.
kenaf or mallow-hemp, *Hibiscus cannabinus* L.
khat, *Catha edulis* (Vahl) Forssk. (syn. Celastrus edulis Vahl)
kiwi, *Actinidia chinensis* Planchon
klaju, *Erioglossum edule* Blume (syn. *E. rubiginosum* (Roxb.) Blume)
kok-saghys, *Taraxacum bicorne* Dahlst. (syn. *T. kok-saghus* Rodin)
konjac or Japanese devil's tongue, *Amorphophallus konjac* C. Koch
kosso, *Hagenia abyssinica* Willd.
kumquat, Hong Kong, *F × crassifolia* Swingle (syn. *F. margarita × F. japonica)*
oval, *Fortunella margarita* (Lour.) Swingle.
round, *F. japonica*, (Thumb.) Swingle.
kunak or foxtail millet, *Setaria italica* (L.) PB. (syn. *Panicum italicum* L.)
kurrat, Egyptian, *Allium kurrat* Schweinf.

langsat or duku, *Lansium domesticum* Corr.
lanthana, *Lantana camara* L.
laquer tree, *Toxicodendron vernicifera* (Stokes) F. Barkley (syn. *Rhus verniciffuum* Stokes)
laurel, bayleaf, *Laurus nobilis* L.
cherry-, *Prunus laurocerasus* L. (syn. *Laurocerasus officinalis* Roem.)

laurel (*cont.*)
  Chinese, *Antidesma bunias* (L.) Spreng. (syn. *A. delicatulum* Hutch.)
lavender, *Lavandula angutifolia* Mill. (syn. *L. vera* DC.)
lechuguilla or fiber agave, *Agave lechugilla* Torr.
leek, *Allium fistulosum* L.
  Japanese, *A. porrum* L.
lemon, *Citrus limon* (L.) Burm. f.
lemon-grass, *Cymbopogon confertiflorum* (Standl.) Stapf.
lentils, *Lens culinaris* Med. (syn. *Ervum lens* L.)
  L. esculenta Moench, incl. var. *macrospermum* L.
  fodder- or bitter vetch, *Vicia ervilia* (L.) Willd. (syn. *Ervum ervilia* L.)
  wild, *Lens kotchyana* (Boiss.) Alef.
    L. lenticula (Schreb.), Alef
    L. nigricans (M.B.) Godr.
    L. orientale Boiss. (syn. *Ervum orientale* (Boiss.) Schmalh.)
lettuce, garden, *Lactuca sativa* L.
  Indian, *L. indica* L.
  prickly, *L. serrata* Torner, F. *integrifolia* Bogenh.
lily, tiger, *Lilium tigrinum* Ker.
  Maximowicz's, *L. maximowiczii* Regel
lime, Canton, *Citrus* × *limonia* Osb. (*C. limon* L. × *C. reticulata* Blanco)
  sour, *C. aurantifolium* (Christm.) Swingle
lingonberry, *Vaccinium vitis-idaea* L.
linseed or oil flax, *Linum usitatissimum* L.
  incl. gr. *macrospermum* Vav.
lipstick tree, archiota or annato, *Bixa orellana* L.
litchi nut, *Litchi chinensis* Sonn. (syn. *L. sinensis* Sonn.)
  Sansibar or rambutan, *Nephelium lappaceum* (L.) Wight (syn. *Litchi lappaceum* L.)
llacon or yacon, *Polymnia sonchifolia* Poepp. & Endl. (syn. *P. edulis* Wedd.)
loquat, *Eryobothrya japonica* (Thunb.) Lindl.
lotus, sacred, *Nelumbo nucifera* Gaertn.
lucrabão, *Hydnocarpus anthelminticus* Pierr.
luffa or sponge gourd, *Luffa aegyptica* Mill. (syn. *L. cylindrica* Roem.)
lulo, *Solanum quitoense* Lam.
lupine, blue, *Luplinous angustifolius* L.
  Bolivian, *L. mutabilis* Sweet
  Cunningham's, *L. cunninghamii* Cook
  white, *L. albus* (syn. *L. tenuis* Forssk.)
  yellow or egyptian, *L. luteus* L.

maca, *Lepidium meyenii* Walp.
madder, Indian, *Rubia cordifolia* L.
  true, *R. tinctoria* L.
madi or tarweed, *Madia sativa* Mol.
magnolia, southern, *Magnolia grandiflora* L.
  star, *M. stellata* (Sieb. & Zucc.) Maxim.
  sweet, *M. virginiana* L.
maguey or pulque agave, *Agave atrovirens* Karw.
maidenhair-tree, *Gingko biloba* L.
maize or corn, *Zea mays* L.
  sweet or flint, *Z. mays* L. convar. *vulgaris* Koern.
  soft, *Z. mays* L. var. *amylacea* (Sturtev.) Montg.
malanga, mangarito or yautia, *Xanthosoma sagittifolium* (L.) Schott. (syn. *X. edule* Meyer)
Malay-apple, *Syzigium malaccense* L. (syn. *Eugenia malaccensis* (L.) Merr. & Perr.)
mamey sapote, *Manilkara zapota* (L.) P. Royen (syn. *Achras zapota* L.)
mandarine orange, *Citrus nobilis* Lour. (syn. *C. reticulata* Blanco)
mangarito or malanga, *Xanthosoma sagittifolia* (L.) Schott (syn. *X. edule* Meyer)
mangistan, false, *Sandoricum koetjape* (Burn. f.) Merr.
mango, bachang, *M. foetida* Lour.
  binjai, *Mangifera caesia* Jack
  Indian, *M. indica*. L.
  kwini, *M. odorata* Griff.
mango chil, *Bromus mango* Desv.
mangosteen, *Garcinia mangostana* L.
manioc, *Manihot esculenta* Crantz (syn. *M. aipi* Pohl, *M. utilissima* Pohl)
maple, sugar-, *Acer saccharum* Marsh.
marigolds, wild, *Tagetes erecta* L.
  T. lucida Cav.
  T. minuta L.
  T. patula L.
  T. tenuifolia Cav. (syn. *T. signata* Bertl.)
marijuana, *Cannabis sativa* L. ssp. *indica* (Lam.) E. Small. & Cronq.)
mariola, *Parthenium incanum* Kunth.
maté, *Ilex paraguayiensis* A. St. Hil.
Mauritius raspberry, *Rubus rosifolius* Smith
medlar, *Mespilus germanica* L.
melons, Chinese, *Cucumis melo* L. ssp. *conomon* (Thunb.) Greb. (syn. *C. chinensis* Pang.)
  common, *C. melo* L.
  horned, *Cucurbita moschata* Duch. var. *toonasa* Mak. (syn. *C. moschata* Duch. var. *japonica* Zhit.)
melon pear or pepino, *Solanum muricatum* (L.), Ait.
melon tree or papaya, *Carica papaya* L.
  mountain, *C. pubescens*, (A. DC.) Solms-Laub.
menteng, *Bacaurea racemosa* (Bl.) Muell. (syn. *B. racemosa* (Reinw.) Muell.)
mesquite, *Prosopis juliflora* (Sw.) DC. (here more likely *P. glandulosa* Torr.)
  P. pubescens Benth.

Mexican hawthorn, *Crataegus mexicana* Moq. & Sessé
Mexican sunflower, *Tithania tubaeformis* Cass. (syn. *Helianthus tubaeformis* Ort.)
Mexican tea or apazote, *Chenopodium ambrosioides* L.
Ch. berlandieri Moq., ssp. *nutalliae* (Saff.) H.D. Wilson (syn. *Ch. nutalliae* Saff.)
mildew, *Erysiphe graminis* DC.
milk-tea, *Euphorbia candelabrum* Trem.
milkvetch, Chinese, *Astragalus sinicus* L.
milkweed, desert, *Asclepias subulata* Dcne.
millet, African or finger, *Eleusine caracana* (L.) Gaertn. (syn. *E. tocussa* Fres.)
barnyard, *Anabasis aphylla* L.
brown, *see* proso millet
foxtail, *Setaria italica* (L.) PB. (syn. *Panicum italicum* L.)
green, *S. viridis* (L.) PB.
yellow, *S. glauca* (L.) PB.
Japanese, *Echinochloa frumentacea* (Roxb.) Link. (syn. *Panicum frumentaceum* Roxb.)
kodo, *Paspalum serebiculatum* L.
pearl, *Pennisetum glaucum* (L.) Br. (syn. *P. spicatum* L.)
proso or brown, *Panicum miliaceum* L.
shama, *Echinochloa colona* (L.) Link
mint, *Mentha* × *piperita* L. (*M. arvensis* L. × *M. aquatica* L.)
mirabilis or four-o'clock, *Mirabilis longiflora* L.
miso, *Perilla frutescens* (L.) Britt. (syn. *P. arguta* Benth, *P. ocymoides* L.)
mispel, *Crataegus azarolus* L.
mock-orange, Mexican, *Philadelphus mexicanus* Schlecht.
mombin, Mexican plum or siruela, yellow, *Spondias mombin* L.
red, *S. purpurea* L.
monk's hood, Wilson's, *Aconitum carmichaelis* Debeaux (syn. *A. wilsonii* Hort.)
morning glory, *Ipomoea aquatica* L.
I. heterophylla Ort.
I. purga Wender (syn. *I. jalapa* Royle)
I. purpurea (L.) Lam. (syn. *I.p.* (L.) Roth.)
I. schiedeana Ham.
I. thyriantha Lindl.
mulberry tree, black, *Morus nigra* L.
Indian, *Morinda citrifolia* L.
paper, *Broussonetia papyrifera* (L.) Vent,
red, *Morus rubra* L.
silk, *Morus latifolia* Poir. (syn. *M. multicaulis* Perr.)
white, *Morus alba* L.
mu-oil tree, *Vernicia montana* Lour. (syn. *Aleurites montana* (Lour.) Wilson)
musk-mallow, *Abelmoschus manihot* (L.) Med.
mustard, Abyssinian, *Brassica carinata* A. Br.
black, *B. nigra* (L.) Koch incl. var. *pseudocampestris* Sinsk,
broadbeaked, *B. narinosa* Bailey
Chinese or Indian, *B. juncea* (L.) Czern. & Coss. incl. var. *sareptana* Sinsk., white, *Sinapis alba* L.
myrobalan, *Phyllanthes emblica* L.
Terminalia bellerica (Gaertn.) Roxb.
T. chebula (Gaertn.) Retz.
myrrh, *Commiphora habessinica* (Berg) Engler (syn. *C. abyssinica* (Berg) Engler)

naranjilla, *Solanum quitoense* Lam.
nectarine, *Prunus persica* (L.) Batsch. var. *nucipersica* (Suchow) G. Scheider (syn. *P. persica* L. var.)
nettle, dioecious, *Urtica dioica* L.
Niger seed or ramtil, *Guizotia abyssinica* (L. f.) Cass.
nightshade, black, *Solanum nigram* L.
nopal or cochineal cactus, *Nopalea cochinellifera* (L.) Salm-Dyck
nutmeg, *Myristica fragrans* Houtt.
nut-sedge, chufa or Zulu-nut, *Cyperus esculentus* L.

oak tree, cork, *Quercus suber* L.
red, *Q. rubra* L.
oats, Abyssinian, *Avena abyssinica* Hochst.
animated, *A. sterilis* L. ssp. *ludoviciana* (Dur.) Gillet & Magne (syn. *A. ludoviciana* Dur.)
mediterranean, *A. sativa* L. var. *byzantina* (C. Koch) (syn. *A. byzantina* C. Koch)
naked-grained, *Avena nuda* L., incl. var. *chinensis* Fisch. and var. *mongolica* Pissar.
oriental, *A. orientalis* Schreb. incl. var. *obtusata* Alef. and var. *tatarica* Ard.
sand, *A. strigosa* Schreb.
wild. *A. fatua* L.
oca, *Oxalis tuberosa* Molina (syn. *O. crenata* Jacq.)
oleaster or Russian olive, *Elaeagnus angustifolius* L.
olive, Chinese black, *Canarium pimela* Koenig
Chinese white, *C. album* (Lour.) Raeusch.
European, *Olea europaea* L.
Russian, *Eleagnus angustifolius* L.
ommu, *Trachyspermum ammi* (L.) Sprague (syn. *Carum copticum* Benth. & Hook.)
wild, *T. roxburghianum* (DC.) Wolff (syn. *Carum roxburghianum* Benth. & Hook.)
Onion, garden, *Allium cepa* L.
Peking, *A. sativum* L. var. *pekinense* (Prokh.) Maekawa (syn. *A. pekinense* Prokh.)
orach, Mexican, *Atriplex hortensis* L.
oranges, Chinese (endemic), *Citrus amblycarpa* (Hassk.) Ochse

oranges (*cont.*)
Chinese (endemic)
C. ponki Tan.
C. tarbiferos Tan.
king, *C. nobilis* Lour (*C. reticulata* Blanco × *C. sinensis* Osb.)
mandarine, *C. reticulata* Blanco (syn. *C. nobilis* Lour.)
papeda, *C. hystrix* DC.
pumelo, *C. maxima* (Burm.) Merr. (syn. *C. grandis* Osb.)
sweet, *C. sinensis* (L.) Osb.
Seville, *C. aurantiacum* L.
trifoliate, *Poncirus trifoliata* Raf.
orchids
oyster plant, *Tragopogon porrifolius* L.
Spanish, *Scolymus hispanicum* L.

pacay, *Inga feuillei* DC.
pak-choi, *Brassica chinensis* L.
palm trees, areca or betel nut, *Areca catechu* L.
cocos, *Cocos nucifera* L.
date, cultivated, *Phoenix dactylifera* L.
wild, *P. sylvestris* (L.) Roxb.
peach, *Bactris gasipaës* Kunth. (syn. *Guilielmia speciosa* Mart.)
sago, Japanese, *Cycas revoluta* Thunb.
starch, *Metroxylon sagu* Rottb.
salac, *Salacca edulis* Reinw.
sugar, *Arenga pinnata* (Wurm.) Merr. (syn. *A. saccharifera* Labill.)
windmill, *Trachycarpus fortunei* (Hook.) Wendl, (syn. *T. excelsus* Makino)
palmarosa, *Cymbopogon martini* (Roxb.) J.S. Wats. (syn. *C. martini* (Roxb.) Stapf)
pangi, *Pangium edule* Reinw. & Bl.
papaya or melon tree, *Carica papaya* L.
mountain, *C. pubescens* (A.DC.) Solms-Laub.
papeda, *Citrus hystrix* DC.
paradise apple, *Malus baccata* (L.) Borkh.
parsley, *Petroselinum crispum* (Mill.) Nyman (syn. *P. hortense* Hoff.)
horse-, *Smyrnium olusatrum* L.
parsnip, *Pastinaca sativa* L.
passion fruit, *Passiflora edulis* Sims.
patchouli, *Pogostemon cablin* (Blanco) Benth. (syn. *P. patchouli* Pellet)
peach, *Prunus persica* (L.) Batch. var. *persica*
wild, *P. davidiana* Franch.
peach palm, *Bactris gasipaës* Kunth (syn. *Guilielmia speciosa* Mart.)
peanuts or groundnuts, *Arachis hypogaea* L.
pears, Chinese, *Pyrus sinensis* Lindl.
common, *P. communis* L.
sand, *P. pyrifolia* (Burm. F.) Nakai (syn. *P. serotina* Rehd.)
snow, *P. nivalis* Jacq.
Syrian, *P. syriaca* Boiss.
Ussurian, *P. ussuriensis* Maxim.
peas, black-eyed or cow-pea, *Vigna unguiculata* (L.) Walp. ssp. *unguiculata* (syn. *V. sinensis* Endl.)
catiang, *Vigna unguiculata* (L.) Walp. ssp. *cylindrica* (L.) Verdc. (syn. *V. sinensis* Endl. var. *catiang* (Walp.) Piper, *Dolichus biflorus* L.)
chick-pea, *Cicer arietinum* L.
fodder pea, *Lathyrus cicera* L.
L. gorgonii Parl.
garden, *Pisum sativum* L. ssp. *sativum* (syn. *P. arvense* L.)
grass-pea or chickling vetch, *Lathyrus sativus* L. incl. var. *macrospermum* Zalk.
pidgeon, *Cajanus cajan* (L.) Huth (syn. *C. indicus* Spreng.)
sweet, *Lathyrus odoratus* L.
wild, *Pisum elatium* M.B.
P. fulvum Siebth. & Sm.
P. humile Boiss.
pecan nuts, *Carya illinoensis* (Wangenh.) Koch (syn. *Hircoria pecan* (Marsh.) Britt.)
peony, *Paeonia* spp.
pepino or melon pear, *Solanum muricatum* (L.) Ait.
pepino de comer or achoca, *Cyclanthera brachybotrys* (Poepp. & Endl.) Cogn.
C. pedata (L.) Schrad.
pepper, bell or pimento, *Capsicum annuum* L.
betel, *Piper betle* L.
black, *P. nigrum* L.
Bolivian, *Capsicum bolivianum* Hazenbusch
cayenne, *C. baccatum* L. v. *baccatum* (syn. *C. frutescens* L. var. *baccatum* L.)
chili, *C. pubescens* Ruiz & Pav.
Columbian, *C. columbianum* Hazenbusch
long Indian, *Piper longum* L.
Mexican, *Capsicum mexicanum* Hazenbusch
tabasco, *C. baccatum* L. (syn. *C. frutescens* L.)
pepper-grass or garden cress, *Lepidium sativum*
peppermint, *Mentha* × *piperita* L. (*M. aquatica* L. × *M. spicata* L.)
pepper tree, *Schinus terebinthifolius* Raddi
perilla, *Perilla frutescens* (L.) Britt. *s.lat.* (syn. *P. ocymoides* L.)
persimmon, Chinese, *D. sinensis* Bl.
Japanese, *D. kaki* L. F.
wild, *D. ebenaster* Retz.
pineapple, *Ananas comosus* (Stickm.) Merr. (Syn. *A. sativus* (Lindl.) Schult.)
pineapple guava, *Acca sellowiana* (Berg) Burret (syn. *Feijoa sellowiana* Berg)
pine nuts, *Pinus koraiensis* Sieb. & Zucc.
pine trees, Australian, *Casuarina equisetifolia* L.
jack- or Canadian, *Pinus banksiana* Lamb.
macedonian, *P. peuce* Griseb.

red, *P. resinosa* Ait.
pistacio nuts, *Pistacia vera* L.
pitanga or Surinam cherry, *Eugenia uniflora* L.
plane tree or sycamore, *Platanus occidentalis* L.
P. orientalis L.
plantain, *Musa acuminata* Colla (syn. *M. cavendishii* Lamb., *M. paradisiaca* L., *M. sapientium* L.)
plum trees, common, *Prunus domesticus* L.
Japanese, *P. salicina* Lindl.
P. triflora Roxb.
Javanese, *Syzigium cumini* (L.) Skeels (syn. *Eugenia jambolana* (Thunb.) Merr.)
Mexican, *Spondias mombin* L.
S. purpurea L.
poinsettia, *Euphorbia pulcherrima* Willd. (syn. *P. p.* (Willd.) Graham)
pomegranate, *Punica granatum* L.
pond-apple, *Annona glabra* L.
poplar, balsam, *Populus balsamifera* L.
black, *P. nigbra* L.
quaking, *P. tremuloides* Michx.
white, *P. alba*, L.
poppy, opium, *Papaver somniferum* L.
potatoes, *Solanum* spp.
potato mold, *Phytophtora infestans* (Mont.) DeBary
proso millet, *Panicum miliaceum* L.
pucoon, *Lithospermum officinale* L. ssp. *erythrorrhizon* (Sieb. & Zucc.) Hand.-Mazz. (syn. *L. erythrorrhizon* Sieb. & Zucc.)
pulasan, *Nephelium mutabile* Blume
pulque or maguey agave, *Agave atrovirens* Karw.
pumelo, *Citrus maxima* (Burm.) Merr. (syn. *C. grandis* (L.) Osb.)
pumpkins, *Cucumis melo*, L. ssp. *agrestis* (Naud.) Greb. (syn. *C. agrestis* Pang)
C. melo ssp. *microdudain* (L.) Greb. (syn. *C. microcarpus* (Alef.) Pang)
C. melo L. ssp. *flexuosus* (L.) Greb. (syn. *C. flexuosus* L.)
wild, *Cucumis trigonus* Roxb.
pumpkin squash, *Cucurbita moschata* Duchesne
puncha pat, *Pogostemon heyeaneus* Benth. (syn. *P. perilloides* (L.) Mansf.)
purslane, *Portulacca oleracea* L.
pyrethrum, camomille, *Anacyclus cotula* (L.) link.

Queen Anne's lace or wild carrot, *Daucus carota* L. ssp. *carota*
quince, *Cydonia oblonga* Mill.
Chinese, *Chaenomeles lagenaria* Koch
Indian or bael tree, *Aegle marmelos* (L.) Corr. & Serr.
Japanese, *Chaenomeles sinensis* (Dum.) Schneid.
quinine, *Cinchona calisaya* Wedd.
C. officinalis, L.
C. pubescens Vahl (syn. *C. succirubra* O. Kuntze)
quinoa, *Chenopodium quinoa* Willd

rabbit bush, *Chrysothamnus nauseous* (Pall. & Pursh) Britt.
radishes, garden, *Raphanus sativus* L. var. *sativus*
Indian, *R. indicus* Sinsk.
oil-, *R. sativus* L. var. *oleifera* Metzg.
rat-tail, *R. sativus* L., var. *mougri* Helm (syn. R. caudatus L.)
raisin tree, Japanese, *Hovenia dulcis* Thunb.
rakkyo or Chinese chives, *Allium odorum* L., syn. *A. chinensis* G. Don)
rambutan or Sansibar litchi, *Nephelium lappaceum* (L.) Wight
ramie, *Boehmeria nivea* (L.) Gaud.
incl. var. *tenasissima* (Gaud.) Miq. (syn. *B. tenasissima* Gaud.)
ramtil or Niger seed, *Guizotia abyssinica* (L. f.) Cass.
rape seed, *Brassica rapa* L. var. *rapa* (syn. *B. campestris* L. ssp. *oleifera* (Metzg.) Sinsk., *B. napus* L. ssp. *oleifera* Metzg. and ssp. *rapifera* Metzg, *B. rapa* L. var. *rapifera* Metzg.)
ratambi, *Garcinia indica* (Thouars) Choisy
redwood, *Sequoia sempervirens* (D. Don) Endl.
rhubarb, Chinese, *Rheum palmatum* L.
garden, *R. rhabarbareum* L.
Turkish, *R. officinale* Aill.
rice, *Oryza sativa* L.
Manschurian, *Zizania latifolia* (Griseb.) Turcz. (syn. *Z. caducifolia* (Turcz.) Hand.-Mazz.)
Philippine, *Coix lacryma-jobi* L.
wild, *Zizania aquatica* L.
rocket, dyer's, *Reseda luteola* L.
garden, *Eruca sativa* L., incl. var. *orientalis* Sinsk.
sea- or sea kale, *Cakile maritima* L.
rose apple, *Syzigium samarengense* (Blume) Merr. & Perr. (syn. *Eugenia javanica* Lour.)
Syzigium jambos (L.) Alston (syn. *Eugenia jambos* L.)
water rose-apple, *Syzigium aquaeum* (Burm. f.) Alston (syn. *Eugenia aquaea* Burm. f.)
rosemary, *Rosmarinus officinalis* L.
roses, *Rosa centifolia* L.
R. damascena Mill.
rubber plant, Indian, *Ficus elastica* Roxb.
rubber trees, *Hevea brasiliensis* (Willd.) Muell.
castilloa, *Castilla elastica* Sessé (syn. *C. e.* Cerv.)
rue, garden, *Ruta graveolens* L.

rukam, *Flaccourtia rukam* Zoll. & Mor.
rust, crown, *P. coronifera* Kleb.
rye, African, *Secale montanum* Guss. (syn. *S. africanum* Stapf)
cultivated, *S. cereale* L. ssp. *cereale*
mountain, *see* African
wild, *S. cereale* L., ssp. *ancestrale* Zhuk. (syn. *S. ancestrale* (Zhuk.) Zhuk.)
S. cereale L. ssp. *segetale* Zhuk. (syn. *S. segetale* (Zhuk.) Roshev., *S. vavilovii* Grossh.)
Dasypyrum villosum (L.) Candargv (syn. *Haynaldia villosa* (L.) Schur, *S. villosum* L.)

safflower, *Carthamus tinctorius* L.
saffron, *Crocus sativus* L.
sage, *Salvia officinalis* L.
saguey or pulque agave, *Agave atrovirens* Karw.
sainfoin, *Onobrychis viciifolia* Scop.
wild, *O. altissima* Grossh.
O. transcaucasia Grossh.
salsify, black, *Scorzonera hispanica* L.
rubber, *S. tau-saghus* Lipsch. & Bosse
wild, *S. deliciosa* Guss.
sandalwood, *Santalum album* L.
sansivieria, *Sansevieria hyacinthoides* (L.) Druce (syn. *S. zeylanica* Willd.)
sapodilla, *Manilkara zapota* (L.) Van Royen (syn. *S. sapotilla* (Jacq.) Cov. *Achras sapota* L.)
white, *Pouteria viridis* (Pitt.) Cronq. (syn. *Calocarpum viride* Pitt.)
sapote, *Manilkara zapota* (L.) Van Royen (syn. *Sapota sapotilla* (Jacq.) Cav.)
green, *Pouteria viridis* (Pitt.) Cronq. (syn. *Calocarpum viride* Pitt.)
white, *Casimiroa edulis* La Llave & Lex.
yellow, *Pouteria campechiana* (Kunth) Baehni (syn. *Lucuma obovata* H.B.K., *L. salicifolia* H.B.K.)
sappanwood, *Caesalpinia saapan* L.
sarsoon, yellow, *Brassica rapa* L., var. *trilocularis* (Roxb.) Kitam. (syn. *B. glauca* Wittm.)
savory, summer, *Satureja hortensis* L.
sea kale or sea rocket, *Cakile maritima* L.
senna, Indian, *Senna alexandrina* Mill. (syn. *Cassia angustifolia* Vahl)
sequoia, *Sequoiadendron giganteum* (Lindl.) J. Bucch.
serradella, *Ornithopus sativus* Brot.
sesame seed, *Sesamum indicum* L., incl. ssp. *bicarpellatum* Hillt.
shabdar or Persian clover, *Trifolium resupinatum* L.
shallot, *Allium ascalonium* L.
silkcotton tree or kapok, *Ceiba pentandra* (L.) Gaertn.
red, *Bombax ceiba* L. (syn. *B. malabaricum* DC.)
singhara nut, *Trapa natans* L. var. *bispinosa* (Roxb.) Makino (syn. *T. bispinosa* Roxb.)
siruela plum, purple, *Spondias purpurea* L.
yellow, *S. mombin* L.
sisal hemp, *Agave sisalana* Perr.
smut, *Ustilago avenae* P. Jens.
snapdragon, *Anthirrinum majus* L.
soapberry, Chinese, *Sapindus mukurossi* Gaertn.
sorghum, grain, *Sorghum bicolor* (L.) Moench
sorrel, common, *Acetosa pratensis* Mill. (syn. *Rumex acetosa* L.)
Jamaica, *Hibiscus sabdariffa* L.
sotol, *Dasylirion duranguense* Treb.
soursop, *Annona muricata* L.
soybeans, *Glycine max* (L.) Merr. (syn. *G. hispida* Maxim., *Soja hispida* Maxim.)
Ussurian, *G. soya* Sieb. & Zucc. (syn. *Soya ussuriensis* (Reg. & Maack) Kom. & Alis.)
spinach, garden, *Spinacia oleracea* L.
malabar, *Basella alba* L. (syn. *B. rubra* L.)
wild, *Spinacia oleracea* L. (syn. *S. tetrandra* Stev.)
spruce, Sitka, *Picea sitkensis* (Bong.) Carr.
squashes, figleaf, *Cucurbita ficifolia* Bouché (syn. *C. melanocarpa* A. Braun)
giant or Hubbard, *C. maxima* Duch. (syn. *C. andreana* Naud.)
pumpkin, *C. mixta* Pang.
C. moschata (Duch.) Dene
turban, *C. maxima* Duch. var, *turbaniformis* (Roem.) Alef. (syn. *C. turbaniformis* Roem.)
zucchini, *C. pepo* L.
star-anise, *Illicium verum* Hook. f. (syn. *I. anisatum* (L.) Gaertn.)
star-apple or cainito, *Chrysophyllum cainito* L.
star-fruit or carambola, *Averrhoa carambola* L.
St. John's bread tree or carob, *Ceratonia siliqua* L.
stock, *Matthiola* spp.
strawberry, *Fragaria chiloensis* Duch.
strychnine nut, *Strychnos nux vomica* L.
Sudan grass, *Sorghum* × *drummondii* (Steud.) Millsp. & Chase (syn. *S. sudanense* (Piper) Stapf)
sugar-apple or sweet-sop, *Annona squamosa* L.
sugar cane, *Saccharum officinarum* L.
Chinese, *S. sinense* Roxb.
sugar palm, *Arenga pinnata* (Wurmb.) Merr. (syn. *A. saccharifera* Labill.)
sulla or sweet vetch, *Hedysarum coronarium* L.
sumach, Sicilian, *Rhus coriaria* L.
tallow, *Toxicodendron succedaneum* (L.) Kuntze (syn. *Rhus succedana* L.)

sunchoke or Jerusalem artichoke, *Helianthus tuberosus* L.
sunflower, *Helianthus annus* L.
  Mexican, *Tithania tubaeformis* Cass. (syn. *Helianthus tubaeformis* Ort.)
sunhemp, *Crotolaria juncea* L.
Surinam cherry, *Eugenia uniflora* L.
sweet-clover, white, *Melilotus alba* Med.
  yellow, M. *officinalis* Lam.
sweet-potato, *Ipomaea batatas* (L.) Lam. (syn. *I. batatas* (L.) Poir.)
sweet-sop or sugar apple, *Annona squamosa* L.
Sweet vetch or sulla, *Hedysarum coronarium* L.
sycamore or plane tree, American, *Platanus occidentalis* L.
  Oriental, *P. orientalis* L.

tacacco, *Polakowskia tacacco* Pitt.
tallow sumach, *Toxicodendrum succedaneum* (L.) Kuntze (syn. *Rhus succedana* L.)
tallow tree, *Sapium sebiferum* (L.) Roxb.
tamarind, Indian *Tamarindus indica* L.
  Manila, *Pithesellobium dulce* (Roxb.) Benth. (syn. *Pithecolobium dulce* (Roxb.) Benth.)
  Sunda, *P. jiringa* (Jack) Prain (syn. *P. lobatum* Benth.)
taro, *Colocasia esculenta* (L.) Schott. (syn. *C. antiquorum* (L.) Schott.)
tarweed or madi, *Madia sativa* Moll.
tarwi or Bolivian lupine, *Lupinus mutabilis* Sweet
tau-saghys or rubber salsify, *Scorzonera tausaghus* Lipsch. & Bosse
tea, Chinese, *Camellia sinensis* (L.) O. Kuntze var. *sinensis*
  Mexican or worm-tea, *Chenopodium ambrosioides* L.
    C. berlandieri Moq. ssp. *nutalliae* (Saff.) H.D. Wilson (syn. *C. nutalliae* Saff.)
  Paraguayan or maté, *Ilex paraguayensis* A. St. Hil.
teasel, fuller's, *Dipsacus sativus* (L.) Scholler
  wild, *D. sylvester* Huds.
teff, *Eragrostis tef* (Zucc.) Trotter (syn. *E. abyssinica* Link)
temu mangga, *Curcuma mangga* Valet. & Zijp
teosinthe, *Zea mays* L. ssp. *mexicana* (Schrad.) Iltis (syn. *Euchlaena mexicana* Schrad.)
texocote or Mexican hawthorne, *Crataegus mexicana* Moq., & Sassé
  C. stipulosa Steud.
  C. pinnatifida Bunge
thistles, field, *Cirsium* spp.
  golden, *Scolymus hispanicus* L.
  Russian, *Salsola iberica* Sennen & Pau (syn. *S. kali* L.)
thorn-apple, *Brugmansia × candida* Pers. (syn. *Datura candida* (Pers.) Parquale)
Thuja, *Thuja occidentalis* L.
  T. plicata Donn.
thyme, *Thymus vulgaris* L.
tobacco, Aztec or peasant, *Nicotiana rustica* L.
  common, *N. tabacum* L., (an amphidiploid of *N. sylvestris* Speg. & Comes and *N. otophora* Griseb. or *N. rusbii* Brit.)
tomatillo, *Physalis ixocarpa* Brot. (syn. *P. aequata* Jacq.)
tomato, bush, *Lycopersicum lycopersicum* (L.) Karsten, var. *succenturiatum* Pasq.
  cherry, *L. cerasiforme* Dunal (syn. *L. esculentum* Mill. var. *cerasiforme* (Dunal) Alef.)
  common, *L. lycopersicum* (L.) Karsten (syn. *L. esculentum* Mill.)
  Peruvian, *L. peruvianum* (L.) Mill.
  tree-, *Cyphomandra betacea* (Cav.) Sendtner
tree-of-heaven, *Ailanthus altissima* (Mill.) Swingle
tuberose, *Polianthus tuberosa* L.
tuna cacti, *Opuntia* spp.
tung-oil tree, *Vernicia fordii* (Hemsl.) Airy-Shaw (syn. *Aleurites fordii* Hemsl.)
  Japanese, *V. cordata* (Thunb.) Airy-Shaw (syn. *Aleurites cordata* (Thunb.) R. Br.)
turban squash, *Cucurbita maxima* Dcne. var. *turbaniformis* (Roem.) Alef, (syn. *C. turbaniformis* Roem.)
turmeric, *Curcuma longa* L.
turnip, *Brassica rapa* L. var. *rapa* (syn. *B. campestris* L. var. *rapifera* (Metzg.) Sinsk.)

udo, *Aralia cordata* Thunb.
ulluco, *Ullucus tuberosus* Lozano
uvalha, *Eugenia uvalha* Cambess

vanilla, *Vanilla planifolia* Andrews, incl. var. *fragrans* (Salisb.) Ames
verbena, lemon-, *Aloysia triphylla* (l'Hér.) Britt. (syn. *Lippia triphylla* (l'Hér.) O. Kuntze)
vetch, bard-, *Vicia monanthos* (L.) Retz. (syn. *Ervum monanthos* L.)
  biennial, *Vicia picta* Fisch. & Mey
  bitter er erse, *Vicia ervilia* (L.) Willd. (syn. *Ervum ervilia* L.)
  chickling vetch, *Lathyrus sativus* L.
  common, *Vicia sativa* L. ssp. *sativa*
  French or Narbonne, *Vicia narbonnensis* L.
  hairy, *Vicia villosa* Roth. incl. var. *perennis* Tum.
  Hungarian, *Vicia pannonica* Crantz
  singleflowered, *Vicia articulata* Hook. (syn. *Ervum monanthos* L.)
  sweet or sulla, *Hedysarum coronarium* L.
veti-ver, *Vetiveria zizanioides* (L.) Nash (syn. *V. z.* Stapf.)

walnut trees, black, *Juglans nigra* L. var. *cordifolium* (Maxim.) Rehd.
Chinese, *J. sinensis* Dode
English, *J. regia* L.
Japanese, *J. ailanthifolia* Carr. (syn. *J. sieboldianua* Maxim.)
wampee, *Clausena lansium* (Lour.) Skeels
water-chestnuts, Chinese *Eleocharis dulcis* (Burm. f.) Trin. (syn. *E. tuberosa* (Roxb.) Schult.)
common, *Trapa natans* L. var. *bispinosa* (Roxb.) Makino (syn. *T. bispinosa* Roxb.)
watermelon, *Citrullus lanatus* (Thunb.) Matsum. & Nakai, var. *lanatus*
wild, *C. colocynthis* (L.) Schrad. (syn. *C. vulgaris* Schrad.)
wattle, *Acacia nilotica* (L.) Willd. (syn. *A. arabica* (L.) Del.)
silver, *A. dealbata* Link.
wax myrtle, *Myrica cerifera* L.
silver, *M. mexicana* Willd.
wax tree, *Toxicodendron succedaneum* (L.) Kuntze (syn. *Rhus succedana* L.)
wheat, bread or soft, *Triticum aestivum* L. (syn. *T. sativum* Lam., *T. vulgare* Vill.)
Chinese, *T. aestivum* L. var. *sinicum* (Vav.) (syn. *T. vulgare* L. var. *sinicum* Vav.)
club, *T. aestivum* L. ssp. *compactum* (Host) Thell. (syn. *T. compactum* Host)
durum or hard, *Gigachilon polonicum* (L.) Seidl, ssp. *durum* (L.) Á. Löve (syn. *Triticum durum* Desf.)
einkorn, *Crithodium monococcum* (L.) Á. Löve (syn. *T. monococcum* L.)
emmer, *Gigachilon polonicum* (L.) Seidl, ssp. *dicoccon* (Schrank) Á. Löve (syn. *T. dicoccon* Schrank, *T. dicoccum* (Schrank) Schuebl.)
Inca, *Amaranthus caudatus* L.
macha, *Triticum aestivum* L. ssp. *macha* (Dekapr. & Menabda) MacKey (syn. *T. macha* Dekapr. & Menabda)
oriental, *Gigachilon polonicum* (L.) Seidl ssp. *turgidum* (L.) Á Löve (syn. *T. orientale* Pers., *T. turgidum*, L.)
Persian, *Gigachilon polonicum* (L.) Seidl. ssp. *carthlicum* (Nevski) Á. Löve (syn. *T. carthlicum* Nevski, *T. persicum* Vav.)
Polish, *Gigachilon polonicum* (L.) Seidl. ssp. *polonicum* (syn. *T. polonicum* L.)
poulard, *Gigachilon polonicum* (L.) Seidl. ssp. *turgidum* (L.) Á. Löve (syn. *T. turgidum* L.)
shot, *Triticum aestivum* L. ssp. *sphaerococcum* (Perc.) MacKay (syn. *T. sphaerococcum* Perc.)
spelt, *Triticum aestivum* L. ssp. *spelta* (L.) Thell. (syn. *T. spelta* L.)
wild, *Gigachilon polonicum* (L.) Seidl ssp. *dicoccoides* (Koern.) Á. Löve (syn. *T. dicoccoides* Koern.)
wheatgrass, *Agropyrum tenerum* Vasey
crested, *A. desertorum* (Fisch.) Schult.
willows, *Salix cinerea* L.
S. triandra L.
woad, dyer's, *Isatis tinctoria* L.
wood or elephant apple, *Limonia acidissima* L. (syn. *Feronia elephantum* Corr.)
wormwood, *Artemisia ambrosioides* L.

yacon or llacon, *Polymnia sonchifolia* Poepp. & Endl. (syn. *P. edulis* Wedd.)
yam bean, *Pachyrhizus tuberosus* (Lam.) Spreng.
yama gobo, *Phytolacca acinosa* Roxb. (syn. *P. esculenta* Van Houtte)
yams, Chinese, *Dioscorea batatas* Decaisne
Goa, *D. esculenta* (Lour.) Burkill (syn. *D. aculeata* L.)
intoxicating, *D. hispida* Dennst.
Japanese, *D. japonica* Thunb.
potato, *D. bulbifera* L.
water or winged, *D. alata* L.
yang-mei, *Myrica rubra* Sieb. & Zucc.
yautia, *Xanthosoma sagittifolium* (L.) Schott (syn. *X. edule* Meyer)
ylang-ylang, *Cananga odorata* (Lam.) Hook. f. & Thomas.
ying-tao, *Prunus pseudocerasus* Lindl.
yucca, *Yucca aloifolia* L.
Y. elephantipes Regel

zeodary, *Curcuma zeodaria* (Christm.) Roscoe
zinnias, *Zinnia mexicana* Hart
Z. multiflora L.
Z. violacea Cav. (syn. *Z. elegans* Jacq.)
zulu-nut, nut-sedge or chufa, *Cyperus esculentus* L.

Index of animal names

Index of centers of origin